EAST AFRICAN
ECOSYSTEMS
AND THEIR
CONSERVATION

EAST AFRICAN
ECOSYSTEMS
AND THEIR
CONSERVATION

Edited by
T. R. McClanahan and T. P. Young

New York   Oxford
Oxford University Press
1996

Oxford University Press

Oxford   New York
Athens   Auckland   Bangkok   Bogota   Bombay   Buenos Aires
Calcutta   Cape Town   Dar es Salaam   Delhi   Florence   Hong Kong
Istanbul   Karachi   Kuala Lumpur   Madras   Madrid   Melbourne
Mexico City   Nairobi   Paris   Singapore   Taipei   Tokyo   Toronto

and associated companies in
Berlin   Ibadan

Library of Congress Cataloging-in-Publication Data
East African ecosystems and their conservation / edited by T.R. McClanahan and T.P. Young.
    p.   cm.
Includes bibliographical references and index.
ISBN 0-19-510817-5
1. Biotic communities—Africa, East.   2. Ecosystems management—Africa, East.
3. Nature conservation—Africa, East.   I. McClanahan, T.R.   II. Young, T.P.
QH195.A23E38   1996
333.95'09676—dc20   95-52333

9 8 7 6 5 4 3 2 1

Printed in the United States of America
on acid-free paper

Preface

Fifty years after the first national parks were established in East Africa we can reflect with some satisfaction on the role they play today. Our protected areas have become the *sin qua non* of conservation and no less important to the regional human economy. That at least is the common perception.

A closer look at the significance of protected areas quickly exposes their inadequacy. The reality is that our protected areas are wholly insufficient to guard against large-scale extinction, ecological disruption and biological impoverishment over the next fifty years. The threats to biological diversity and the productivity of our natural resources have as much to do with ecological ignorance as human numbers and over use.

The founders of our protected areas clearly had the recent extinction of the bison, bluebok and quagga in mind when they set aside parks and reserves to safeguard East Africa's great wildlife herds. We owe an enormous debt to their compassion and foresight. But, in all too many respects, the apparent safety of the great herds in Serengeti, Tsavo, Amboseli and other parks and reserves has beguiled us into thinking all was well, that conservation ends once we have set aside land for our most visible and vulnerable wildlife.

In reality, the protected areas are far from adequate. Most were set aside to protect the savanna's large mammal populations and largely ignore the many other habitats in East Africa, particularly forests and wetlands. Then again, few if any parks cover a complete ecosystem anyway, leaving entire communities or vital parts of the ecosystem outside of the protected-area boundaries.

For the most part, the chances of adequate biotic representation and complete ecosystem coverage by expanding protected areas is now remote. Land is too precious and the demands placed on it by a burgeoning population too great to depend on new areas for conservation.

New protected areas would, however, not necessarily solve the problems unless the areas were unimaginably large - an order of magnitude or two larger than present parks. The reason comes down to the importance of space. Species and ecological communities need space in order to escape local exigencies of climate, disease, predation and random disruptive events which cause localized extinction. Without space, and plenty of it, there is no escape from localized extinction, or source populations elsewhere to recolonize an area when conditions favor recovery. The requirement for large areas militates against the viability of parks as stand-alone conservation entities in the long run.

The maintenance of ecological processes is another constraint on conservation. Conservation is not only about protecting species, but also about maintaining the intricate biological and physical processes which support these species, and of course humans and our natural resources. We are now only beginning to appreciate how complex and intimately linked such processes are, and how vulnerable to disruption. Once disrupted, biological systems and the productivity of our natural resources - whether forestry, fisheries or wildlife - are at risk.

It is especially important to understand these processes in East Africa, for no place on Earth has such a long history of human occupation. Our present-day ecosystems have evolved over the last two million years, two million years in which humans have played an enormous and incompletely understood role in shaping these ecological communities. What we do know is that in many ecosystems human activity plays, or did play, a vital role in maintaining biological diversity. Remove the influence of fire, pastoralism, shifting settlement and hunting from protected areas and the community may fundamentally change. Many of our savanna parks are losing diversity not so much because of isolation, for few of them are yet entirely isolated, but because of the removal of people.

The other extreme, of course, is the adverse effect of too many people and over exploitation. Either way, people are the most important influences on biological diversity today. How then are we to conserve and use biological diversity in a sustainable manner in the face of these insurmountable problems within protected areas? Clearly the starting point is to understand the biology and ecology of all East Africa's natural ecosystems and the threats they face. The aim, ultimately, is to integrate land use and conservation such that we can sustainably use our natural resources and maintain biological diversity.

East African Ecosystems and their Conservation is the first attempt to do just this. Each major ecosystem is covered in sufficient detail to understand the basic structure and major processes at play. The role of people, both positive and negative, is underscored in virtually every chapter. Indeed, the volume begins with a description of the central role humanity has played in the evolution of East African ecosystems. It also touches on the main conservation problems and challenges facing each ecosystem. This volume, therefore, provides an invaluable reference for anyone interested in the ecology natural resource management and conservation of East African ecosystems.

D.C. Western

Introduction

It has become increasingly obvious over the past few decades that the biological diversity and ecosystems of the world are intricately interwoven with the fate of humans. The tremendous biological heritage that greeted humanity's arrival on planet earth at first provided a seemingly infinite bounty. We now know that this heritage includes species of plants and animals that number in the millions, and whose existence often makes our lives better every day. Ironically, this knowledge comes in the same generation that threatens to destroy much of that biological diversity, and greatly reduce the quality of human life. Nowhere is this conflict more evident than in East Africa. Kenya, Tanzania, Uganda, Ethiopia, Somalia, Rwanda and Burundi contain some of the best understood ecosystems in the tropics, and yet habitat destruction and loss of diversity in these countries is occurring at an alarming rate. There has, therefore, never been a better time for a synthesis of what is known about East African ecosystems and their prospects for conservation.

Students, teachers and researchers looking for an introduction to the scientific literature on the East African environment are likely to find it scattered and often difficult to access. Even the most persistent scholars will need to make a significant effort to accumulate the references and research papers that compose the scientific basis for understanding this environment for even some of the most well-studied ecosystems. Often interpretations of these data will differ when viewed with the hindsight of newer and improved findings and more recent theoretical developments. Students may need guidance with the interpretation of the various findings and how they fit into a coherent and contemporary understanding of the environment. Realizing this difficulty and the need for a current review of the East African environment, we asked a number of leading ecologists to tell the ecological story of their studied ecosystems as understood from the existing scientific literature. Authors were not asked to give an exhaustive review of the literature but rather to review the more notable findings in the last two decades and to summarize the relevance of these findings towards the further understanding and potential conservation of these ecosystems. Additionally, the authors were advised to make their summaries understandable to university-level students and the educated public.

The following chapters unveil the continuing story of humans and their environment in East Africa in the major ecosystems of East Africa. And, like all literature, the stories these ecologists tell are history-bound and, therefore, tell us as much about our ignorance as our knowledge. The reviews also reflect the way contemporary ecologists view their science - the important literature, concepts, contributors and reasons for doing ecological science. These summaries show the power of the scientific reasoning, methods, and instruments in illuminating the processes and history of nature, but they also remind us of the short time that science has been focused on the natural environment and the already irreversible consequences of the collective actions of humans on nature. The stories often tell us about plants and animals whose fascinating life histories are poorly known and yet on the edge of extinction. The reader will see that no ecosystem has been spared the influence of humans but, more hopefully, that humans have had their longest history in East Africa and yet, in some ways, his least destructive effect on nature. Humans and nature have coexisted here in a uneasy relationship for the whole of human history. The great animal herds that roamed the earth more extensively in the last geologic era have their last refuge in East Africa. This has been one of the great attractions that has given East Africa both a faunal and scientific heritage that matches any tropical region.

Although East Africa has the greatest diversity and concentrations of large mammals in the world and a tremendous diversity of other plants and animals, it also has the world's highest human population growth rates, which threaten to squander its otherwise impressive gains in evolutionary development. In these chapters, you will learn of the varied and powerful threats to East Africa's tremendous biological heritage, and the prospects for conservation. In all likelihood, the coming generation of East African citizens and scientists will determine, for a very long time, the fate of these remarkable ecosystems and biota.

This book is written for students who seek a contemporary summary of the East African environment. More specifically, it was written for undergraduate and first-year graduate courses in tropical ecology, environmental study issues of tropical or developing countries, and as a source book on East Africa for any courses on general ecology, geography, environment, resource issues, and people and the environment. Professionals and teachers will also find the book useful for summarizing the literature in their own fields and in ecosystems outside their speciality. And finally, more serious tourists (such as ecotourists) or naturalists will find the book a useful reference for understanding the environments that they visit while in East Africa. These visitors may find the ecological reviews in this book more challenging and, hopefully, more stimulating than the cursory descriptions of the environment found in the common field guides. An effort has been made to keep ecological jargon to a minimum and a glossary is provided for defining some of the less familiar terms. Some of the key ecological and conservation concepts are briefly described in accompanying boxes to briefly review ecological concepts that are important to fully appreciate the authors' summaries. A basic knowledge of ecological science will, however, make the relevance of the authors summaries more coherent. Consequently, teachers using this book should either brief their students in ecological theory or use the book for students already prepared in this science.

The way one reads the book depends on the reader's ecological knowledge. Professionals and college-level teachers can read chapters in the order that interests them. Chapters are organized into sections that often correspond to courses in ecological science and teachers may wish to use these sections as part of these specific courses. Some sections are organized such that the first and second chapter in the section contains the most summary information to help the reader with subsequent chapters. Additionally, the first chapter in the Marine Ecosystems section briefly summarizes some of the concepts of seasonality and resource harvesting. This summary is written within the context of fisheries but similar or slightly modified concepts apply to harvesting in terrestrial ecosystems. The first two chapters will also help readers put the following chapters within an historical and evolutionary context which is so important for understanding contemporary ecosystems. We hope that this book will be read by a large audience and that it will help forge the next critical generation of questions, naturalists, ecologists, and conservation biologists in East Africa.

T.R. McClanahan & T.P. Young
Mombasa and Nanyuki, Kenya

Acknowledgments

Production of this book was assisted by S. Bashir, C. Billington, J. Fanshawe, H. Glaescel, J.B. Hall, D. Harrison, K. Jensen, L. Lockwood, N.A. Muthiga, E. Onyango, P. Reyniers, J. Romero, and S. Wiseman. Most chapters were reviewed by the authors and editors but assistance for reading and commenting on various chapters we wish to thank L. Pigott Burney, G. Howard, L. Isbell, Q. Luke, P. Martin, S. Njuguna, M. Risk and I. Tattersall. Authors received financial support for their research and writing from a number of organizations. These include the East African Wildlife Society, the Food, Conservation and Health Foundation, the Phillip M. Hampton International Research Scholarship Fund, the Human Dimensions of Global Change Program, National Oceanic and Atmospheric Administration, U.S. Dept. of Commerce, Miami University, the National Geographic Society, the United States National Science Foundation, the New England Aquarium, The Pew Charitable Trust, the Smithsonian Institution, the University of Florida, USAID Program in Science and Technology, and The Wildlife Conservation Society. The inspiration of L.D. Harris, H.T. Odum and A.P. Smith also deserve thanks.

Permission to reprint figures was granted by Springer Verlag (Fig. 7.2), Blackwell Scientific Publications (Fig. 7.3) and Academic Press (10.1 - Box).

List of Contributors

M. Atsedu, Natural Resource Ecology Laboratory, Colorado State University, Fort Collins, CO., 80523, USA

N. Burgess, Centre for Tropical Biodiversity, Zoological Museum, University of Copenhagen, Universitetsparken 15, DK 2100, Copenhagen, Denmark

D.A. Burney, Department of Biological Sciences, Fordham University, Bronx, NY, 10458,USA

C.A. Chapman, Department of Zoology, University of Florida, Gainesville, Fl., 32611, USA

L.J. Chapman, Department of Zoology, University of Florida, Gainesville, Fl., 32611, USA

P. Clarke, Frontier - Tanzania Coastal Forest Research Programme, c/o Society for Environmental Exploration, 77 Leonard Street, London EC2A 4QS, England

S.D. Cooper, Department of Biological Sciences, University of California at Santa Barbara, Santa Barbara, Ca. 93106-9610, USA

M.B. Coughenour, Natural Resource Ecology Laboratory, Colorado State University, Fort Collins, CO., 80523 USA

C.D. Fitzgibbon, Large Animal Research Group, Department of Zoology, University of Cambridge, Downing Street, Cambridge, CB2 3EJ, England

C. Gakahu, c/o African Conservation Centre/WCI, P.O. Box 62844, Nairobi, Kenya

H. Gichohi, African Conservation Centre/ WCI, P.O. Box 62844, Nairobi, Kenya

I. Gordon, The Kipepeo Project, P.O. Box 57, Kilifi, Kenya

D.M. Harper, Ecology Unit, Zoology, University of Leicester, University Rd., Leicester LE1 7RH, England

D.J. Herlocker, Range Management Handbook Project, GTZ/Ministry of Agriculture, Livestock Development and Marketing, P.O. Box 47051, Nairobi, Kenya

J.C. Hillman, Resources & Environment Division, Ministry of Marine Resources, P.O. Box 18, Massawa, Eritrea

F.M.R. Hughes, Department of Geography, University of Cambridge, Cambridge, CB2 3EN, England

L.A. Isbell, Department of Anthropology, University of California at Davis, Davis, Ca., USA

L. Kaufman, New England Aquarium, Edgerton Research Laboratory, Central Wharf, Boston, Massachusetts, 02110-3309, USA

D.A. Livingstone, Zoology Department, 243 BioSci, Duke University, Durham, N.C., 27708-0325, USA

J.C. Lovett, Department of Environmental Economics and Environmental Management, University of York, Heslington, York, England, Y01 5DD

K.M. Mavuti, Department of Zoology, University of Nairobi, P.O. Box 30197, Nairobi, Kenya

T.R. McClanahan, The Wildlife Conservation Society, Coral Reef Conservation Project, P.O. Box 99470, Mombasa, Kenya

K.E. Medley, Department of Geography, Miami University, Oxford, Ohio, 45056, USA

J.M. Melack, Department of Biological Sciences, University of California at Santa Barbara, Santa Barbara, Ca., 93106-9610, USA

E. Mwangi, Jomo Kenyatta University, College of Agriculture and Technology Nairobi, Kenya

D.O. Obura, Indian Ocean Conservation Program, New England Aquarium, Central Wharf, Boston, Massachusetts, 02110- 3399, USA

W.A. Rodgers, FAO-Biodiversity Program, P.O. Box 2, Dar es Salaam, Tanzania

R.K. Ruwa, Kenya Marine & Fisheries Research Institute, P.O. Box 81651, Mombasa, Kenya

D.M. Swift, Natural Resource Ecology Laboratory, Colorado State University, Fort Collins, CO., 80523, USA

A. Vedder, The Wildlife Conservation Society, 185th Street & Southern Blvd, Bronx, New York, 10460, USA

J.M. Waithaka, African Conservation Centre/WCI, P.O. Box 62844, Nairobi, Kenya

W. Weber, The Wildlife Conservation Society, 185th Street & Southern Blvd, Bronx, New York, 10460, USA

D.C. Western, Kenya Wildlife Service, P.O. Box 40241, Nairobi, Kenya

S.W. Wiseman, Department of Biological Sciences, University of California at Santa Barbara, Santa Barbara, Ca., 93106-9610

T.P. Young, Louis Calder Center, Fordham University, Drawer K, Armonk, New York, 10504, USA and Mpala Research Centre, P.O. Box 555, Nanyuki, Kenya

Contents in Brief

Section IV: Grass, Shrub, and Woodland Ecosystems

Section V: Forest Ecosystems

Contents

Chapter 4. *Coral Reefs and Nearshore Fisheries* 67
T.R. McClanahan & D.O. Obura

Chapter 5. *Intertidal Wetlands* 101
R.K. Ruwa

Section III: Inland-Water Ecosystems

Chapter 6. *Rivers and Streams*

S.D. Cooper

Chapter 11. *Savanna Ecosystems* 273
H. Gichohi, E. Mwangi & C. Gakahu

Chapter 12. *The Miombo Woodlands* 299
W.A. Rodgers

Section V: Forest Ecosystems

Chapter 13. *Coastal Forests*
N. Burgess, C.D. Fitzgibbon & P. Clarke

Section I:

Environmental and Human History

Chapter 1

Historical Ecology

D.A. Livingstone

It is often assumed that the natural environment does not change. Temperature and rainfall, for example, are expected to vary around some climatic mean. Actually, climate has changed, is changing, and will change in the future. It would be safer to assume constant change than climatic constancy, and the geologic record reveals many climatic changes great enough to affect human affairs. Periods of very large and rapid change are not the norm, but one scans the geologic record in vain to find long periods of climatic constancy. On this restless earth there seems to be no qualitative boundary between the short-term small-amplitude changes we perceive as weather and longer-term phenomena such as ice ages.

Human alteration of the atmosphere is imposing additional changes, possibly very rapid and large ones, on this naturally variable background. The consequences to climate of increasing atmospheric concentrations of carbon dioxide, methane, and other greenhouse gases are not predictable in detail on a global scale; exact prediction on the regional scale of government policy is even more difficult. Most students of global change, however, believe that world climate will soon be several °C warmer (23). The biological consequences of such warming look very serious indeed.

Modelling

Attempts to estimate the climatic consequences of present and future changes in the composition of the atmosphere depend on accurate computer models of the global atmospheric circulation. Relatively cheap and fast computers make it possible to construct models of global circulation based on first principles such as the laws of motion and thermodynamics, but computer speed and cost still limit our ability to model future climate. Models do not provide, so far, spatial resolution nearly so fine as one would like: a typical model divides the earth into areas of about 500 km^2 at the equator, and the atmosphere into nine levels along a vertical axis (23). Further, models are limited to dealing with equilibrium states, so that it is possible to model the temperature to be expected after a doubling of carbon dioxide, but not possible to model the dynamic transition from present conditions to warmer ones, which is unlikely to be simple.

Despite their limitations, the performance of the models is very impressive. They make no use of curve-fitted constants, and incorporate no fine tuning by trial and error. Rather, they are based solely on first principles such as the known distribution of solar radiation, land, water and mountains and the known thermodynamic behavior of the atmosphere. The pattern of modern climate the models generate is very similar to our climate based on meteorological observations.

A more rigorous test of the model's predictions would compare model reconstructions of past climate with the geologic record. For example, the full-glacial temperature on land can be reconstructed, using the atmospheric composition and distribution of sea surface temperature indicated by geologic evidence, and compared with the known geologic record of terrestrial temperature (27). The models survive such tests reasonably well for much of the world. They agree, for example, with geologic evidence that during ice-age times, the glacial temperature of temperate lands was 5 to 15° C cooler, that the ice-age jet stream split into two branches around the continental ice-sheet of North America, and that the monsoons were much weaker. However, as we will see below, there is some disagreement between models and geologic evidence regarding ice-age temperatures over tropical lands.

These models deal with global conditions, and their spatial resolution is too coarse to permit predictions on the regional scale that affects much state or national policy. For example, the models are spatially detailed enough to include the Himalayas and the American Cordillera but not the smaller mountainous areas of Africa. Mapping the topographic grain of Africa, and incorporating it in global models of future climate, is beyond the reach of present computers.

Past climate was controlled by CO_2, other greenhouse gases and dustiness, all of which we are changing today, but also by factors we do not alter. Because past climates were controlled partly by these other factors, such as changes in seasonality and the distribution of solar energy over the world, no past time is a close analogue of likely future conditions. Present model predictions of global warming will not receive a really rigorous test until global warming is upon us - or until many decades have passed without it.

Biological consequences of climatic change are even less certain. It is predicted that the greenhouse effect is likely to raise global temperature in a very short time by something around 3° C, and one would like to see how the biota of the world adjusted to natural changes of such a speed and magnitude in the past. The warmest temperatures of the last 100,000 years, however, were only about 1 degree above that of today. To find temperatures 3 degrees warmer we must go back several million years to the Pliocene, when modern species of birds and mammals had not yet evolved. There is no satisfactory way to use the fossil record to predict the response of modern species, especially tropical ones, to a climatic change that they have never experienced.

The natural geography of the present world differs in one important respect from that of any previous time - human activities have made it much harder for organisms to adjust their ranges in the face of climatic change. The ranges of many animals have been drastically reduced, with the main populations surviving in national parks and preserves separated by extensive tracts of urban or agricultural land, industrial development and highways. Faced with the rapid climatic changes at the end of the

Box 1.1. Geologic Time Table

When scientists speak or write about events long ago, they often use a special terminology to describe particular time periods. For non-specialists, this special terminology can be frustrating, because it requires special knowledge of the temporal sequence of the Periods and Epochs. Below is a brief summary of the geological time table that may help in understanding this terminology. There are a couple of reasons that this terminology is used, instead of simply giving dates. First, each period or epoch was characterized by distinct sets of organisms that often form a distinct layer in the studied rocks or sediments. These sets of organisms often underwent sudden changes, such as the extinction of most dinosaurs at the end of the Cretaceous, and a earlier extinction of most marine animals at the end of the Permian. Second, these names are easier to remember (once you learn them!), and not dependent on the current estimates of their dates. Even today, there are occasional (minor) changes in the dates associated with these periods and epochs.

T.P. Young

Start of epoch (millions of years ago)	Period	Epoch	Characteristics
0.01 (10,000 years)	Quaternary	Holocene	Agriculture, industrialization
1.6		Pleistocene	Ice Ages
5.3	Teriary	Pliocene	Age of mammals, origins of humans
23.7		Miocene	
36.6		Oligocene	
57.8		Eocene	
66.4		Paleocene	
144	Cretaceous		Dinosaur extinctions, rise of flowering plants
208	Jurassic		Age of dinosaurs, first mammals, birds
245	Triassic		Rise of dinosaurs
286	Permian		Rise of reptiles
352	Carboniferous		Rise of land plants, insects, amphibians
408	Devonian		Age of fishes
438	Silurian		First land animals
505	Ordovician		Seaweeds, first vertebrates
570	Cambrian		Algae and invertebrates
Earlier	Precambrian		Unicellular life

last ice age these animals, and the plants that support them, adjusted their ranges by migration. With the many modern barriers to migration they could not adjust today.

The Global Record of Past Climatic Change

Instrumental records of rainfall and temperature from meteorological stations are much too short for assessing the size and timing of natural changes in climate. Africa is fortunate in having long runs of data for the level of water in the Nile, but this by itself is insufficient to provide a good assessment of the amount, periodicity and net direction of climatic change in our region. For that we must turn to natural sources of information, proxy weather records that can be interpreted in terms of climate. Not all of these sources are available in East Africa, so it will be useful to consider the global picture before turning to the African record of climatic change.

Tree-rings produced by the annual pulse in growing conditions can provide an excellent chronology and useful climatic information (19). The thickness of a ring depends on the weather of a few years before and during its growth so the tree-ring record can be interpreted in terms of climate, especially rainfall in regions where rain is scarce. Year-to-year changes in climate produce a sequence of ring widths that is specific for a specific time, permitting correlation from one woody stem to another, - from a living tree, for example, to a cathedral beam hewn from a tree whose lifetime overlapped that of the living one. Under the best circumstances one can build such a record that covers close to 10,000 years, but in most places the run of proxy tree-ring data is much shorter. In East Africa a comparable tree-ring record hardly exists. The equable climate, at least of forest reserves and plantations, does not commonly generate visible tree-rings. Possibly other methods, such as isotopic (20) and x-ray examination, might reveal growth rings where none are visible to the naked eye. A vigorous research effort, starting with material that does show visible rings, such as *Acacia* trees around Machakos (T. McClanahan, personal communication), might generate a very useful climatic record. The initial steps in such an investigation would require no tools beyond pencil and paper, a sharp panga and a sharp eye.

A good chronology of environmental change has been extracted from Greenland and Antarctic glaciers. Annual dust bands formed during the season of lowest snow accumulation provide the chronology, and the contents of the ice, such as dust, acid, and bubbles of trapped air, reveal the past composition of the atmosphere. It is much more expensive to collect and preserve ice cores than wood, but the ice core record is much longer. Glacier ice gives a particularly good record of polar temperatures, because the ratio of ^{16}O to ^{18}O depends on the local temperature at the time of snow formation (21).

The isotopic ratios of polar glaciers reveal very large and rapid changes in temperature. It has been apparent for some time that about fifty percent of the temperature change between full glacial and full interglacial conditions occurred during a period of less than fifty years. Now there is reason to believe that much of this change occurred within five years (1)! Such rapid changes, which may be due to a radiation-driven shift from one stable state of oceanic circulation to another, remove the end and the onset of ice ages from a comfortable time-scale of millennia to one of immediate human concern.

Concentration of atmospheric carbon dioxide has been estimated from ice cores over a complete glacial-interglacial oscillation, and it appears that this gas changed enough to account for most of the apparent glacial-interglacial temperature difference (4). High-amplitude but shorter-term oscillations of carbon dioxide also occurred every few thousand years during the last ice age. If carbon dioxide oscillated with the intrinsic oscillation in the size of the continental ice mass then these dramatic oscillations are unlikely to recur today, when ice sheets are restricted in extent. If, however, the past carbon dioxide and temperature changes are due to oscillations between stable states in the circulation of the sea they might recur at any time. There was apparently such a false alarm during the last interglacial period when glaciers were briefly even smaller than they are today (14,21). This finding temporarily compounded the continuing fear that earth's climate is poised so delicately between two stable states that we could, with our current manipulation of the atmosphere, inadvertently tip it from one state to the other. These findings have all come from polar glaciers. The same methods yield less insight when applied to the smaller, thinner and younger accumulations of equatorial high-mountain ice (47).

Oceanic sediment provides records that are less accurately dated, but that permit study of climatic change over very long spans of time. Most of the paleoclimatic information in marine sediments comes from calcareous microfossils and two principal methods have been used to extract it, isotopic analysis and characterizing microfossil species assemblages.

Oxygen in the ocean includes a mixture of ^{16}O and ^{18}O isotopes. Calcareous organisms do not use these isotopes in their shells in the proportions in which they are found in the surrounding sea water, and the extent to which they preferentially select isotopes depends on water temperature. Stratigraphic changes in the isotopic ratio in fossils once suggested that Caribbean surface temperature had fallen by 6° C during each ice-age (11). Unfortunately there is no independent way of knowing the past isotopic ratio of sea water, although it was certainly richer in the heavier isotope during each ice-age when much of the more easily evaporated lighter isotope was tied up on land in glacier ice. Most of the glacial-interglacial difference in oxygen isotopes is now attributed to waxing and waning of ice sheets rather than change in ambient temperature (40).

The species composition of assemblages of surface-dwelling foraminifera has been used more often than isotopes to estimate past temperature. These assemblages differ in cold and warm oceans, so statistical analysis of them can be combined with modern temperature records to develop an equation relating species assemblages to temperature. This equation is then used to estimate the sea surface temperature of past times from microfossil assemblages, and maps of past sea surface temperature are drawn by contouring the microfossil results (20).

This method, or some less objective variant of it, is standard in paleoecology, but it is not without its logical problems. The statistical analysis generates a hypothesis about the relation between microfossil assemblage and temperature, but nothing in the procedure tests that hypothesis. It is quite possible that environmental variables other than temperature also affect the assemblages of foraminiferal species, and that the relation of these other things to temperature, though now constant, has not been constant through time. Because of the difficulty of rearing foraminifera in the laboratory it is not possible to test the temperature hypothesis by ecological experiment. An additional problem is that the "modern" assemblages are taken from core-top samples of sediments (11). If the coring operation fails to collect currently-accumulating sediment the core-top assemblage will not be modern, and comparisons between it and instrumental temperatures may be misleading.

Despite such possible problems, faunal analysis has been used widely to plot the past distribution of sea-surface temperature, and the resulting maps display dramatic changes of temperature in the high-latitude oceans. North Atlantic faunal analyses suggest that during each ice age the southern boundary of the pack ice descended from Iceland to Portugal (38). The equatorial ocean, however, does not seem to have cooled nearly so much - by only one or two °C in any part, and in some places not even that (9).

For purposes of geological correlation it does not matter whether most of the change in isotopic composition depends on temperature or on the mass of glacial ice. In either case a high ratio of ^{18}O to ^{16}O indicates ice-age conditions. The core data provide a reliable index of the extent of glaciation and can be used to count the number of ice ages, to determine their periodicity, and to plot changes in the severity of ice ages through the past several million years.

Much of the variation in extent of glaciation can be attributed to three irregularities in the motion of the sun, with periods close to 20,000, 40,000 and 100,000 years. Correlation is so good that few paleoclimatologists now doubt that these astronomical irregularities account for glacial changes. Irregularities are caused mainly by the changing gravitational pull of Jupiter, Neptune and Uranus and drive the atmosphere from glacial to interglacial states. Exactly how they do so is not yet clear. The astronomical irregularities change the distribution of solar energy over the earth rather than the total amount, and some changes of solar radiation should be out of phase in the two hemispheres. Actually, the advances and retreats of continental glaciers around the north and south poles seem to be in phase. It is very likely that radiation-driven changes in oceanic circulation, especially of deep-water mixing at high latitudes, controls the availability of nutrients and so the productivity of phytoplankton in critical areas such as the Greenland Sea. This change in the productivity of oceans would alter the partial pressure of CO_2 in sea water enough to control the CO_2 content of the atmosphere and so modulate the world-wide greenhouse effect (5).

Many more ice ages - about 22 during the 1.8 million years of the Pleistocene alone - are recorded in the marine record than by glacial deposits on land. Later glaciations tend to destroy the evidence of earlier ones, so the glacial record of climatic change on land is very incomplete. Little evidence from land records contradicts the assumption that temperature trends indicated by the marine record were world-wide. The last 10,000 years have been much warmer than most of the last million years, and throughout the evolutionary history of *Homo sapiens* the usual condition of the earth has been much more like an ice age than like the present interglacial.

Changes in East Africa

The best East African indicators of past climatic conditions are glaciers and lakes, but modern ranges of organisms sometimes give revealing hints about past conditions at some unknown time. We will now consider these three sources of historical evidence one by one.

Glacial Evidence

Osmaston, in a very thorough study of the glacial history of the Ruwenzori Mountains, developed an ingenious method for determining past temperature from geomorphological indications of former more extensive glaciation. Instead of using the sometimes misleading lower limit of each glacier as a paleoclimatic indicator, Osmaston (34) computed statistically the line of equilibrium between melting and accumulation for a group of glaciers and demonstrated very substantial changes of climate. On the assumption that glacial advances were due entirely to cooling, with no change in precipitation, he concluded that temperature had gone through the following changes since the height of the last ice age:

Ruwenzori	$+4°C$
Mt. Kenya	$+5°C$
Kilimanjaro	$+6°C$

Changes in lake level show that precipitation has not been constant, but was very much lower during the last glaciation. Allowance for this change in precipitation could increase these estimates of temperature change by 50%.

Evidence from Lakes

A lake may respond to climatic change by changes in water level, in the composition and concentration of dissolved salts, or in its biological community, all of which leave a geologic record. As basins of continuous sedimentation, lakes also accumulate debris blown and washed off the surrounding countryside, particularly pollen grains of flowering plants and charred leaves of grasses. These provide a record of external conditions that can sometimes be interpreted in terms of climate.

Changes in Lake Level. Raised strand lines - either beach deposits or wave-cut cliffs (49) - around East African lakes were one of the first recognized indicators of tropical climatic change. They were once grouped into four age classes, correlated with the four glacial periods that were then known from Europe, and used as the basis for a climatic stratigraphy of tropical Africa. This procedure did not follow standard geological practice, and marine geologists have since shown that there have been many more than four ice ages.

The youngest set of extensive raised strand lines was once believed to date from the last ice-age. Radiocarbon dating of fossil shells from those strand lines and a wealth of chemical and microfossil evidence from lake sediments showed that the strandlines were actually formed during the first half of postglacial time (22,37). The high water levels of the wet years 1961 to 1964 show how fast the lakes of East Africa can rise and fall, but the geological strand lines show that many lakes were much deeper at times in the past than they have ever been in the recent historic period.

Lakes without a surface outlet, like Lake Turkana, Lake Manyara and the lakes of the Nakuru-Naivasha depression, are most sensitive to climatic change, and are surrounded by several sets of strand lines of known geologic age. Radiocarbon (6) and uranium-thorium (7) dating has shown that high water levels correspond generally to times 6000 to 10,000, 23,000 to 27,000 and 90,000 years ago, when world glaciers were relatively small. Many older strand lines are still undated, although it should be possible to date at least some of them with argon-argon analysis, thermoluminescence, or electron spin resonance techniques.

Strand lines give a reasonable record of times when the water level was high, but times of low water level are registered only in the sediments, often as unconformities, gaps in deposition due to periods of non-deposition, erosion or weathering. Many lakes, including Lakes Tanganyika, Kivu, Albert, Victoria, Nakuru, Naivasha, and Elmenteita were much lower during the last ice age (15,37). The three deepest lakes of the Western Rift, Malawi, Tanganyika and Kivu, held water during that dry time. However, the deepest parts of Lakes Victoria, Albert and Edward have not been cored, so it is uncertain whether these lakes dried. If they did, the entire flock of endemic Victoria fishes must have evolved very rapidly during the last 14,000 years.

Even though some of the large lakes held water during the dry time at the end of the last ice age, Lakes Malawi and Tanganyika may not have provided suitable fish habitat for very much longer (Chapter 8). Deep-penetrating echo sounding reveals a number of times when water level fell so far that erosion exceeded deposition over all but the deepest parts of both lakes. Since the last of these erosional periods less than 100 m of sediment has been deposited (39). This suggests a time of the order of 100,000 years since the last major water-level reduction. Extensive erosion associated with this hiatus indicates that the water level remained low for a very long time. No dry period in East Africa or elsewhere seems to correspond to this low-stand of Lake Malawi and Lake Tanganyika.

The question of how the fish fauna survived through such lake-level fluctuation is of interest as in the absence of refuges the rich fauna must have evolved over a relatively short time. Although some water may have remained, evaporation during a long-lasting low-stand should have made the water much more saline, perhaps too saline for the endemic fish. Students of East African ichthyology are less hostile than most biologists to the idea of very rapid speciation, but even they might find it unlikely that the rich endemic fauna of Tanganyika, which includes endemic genera as well as endemic species, could have evolved during so short a time as 100,000 years. Perhaps the unconformity represents a fluctuating water level rather than a continuously low one, with the water being freshened occasionally during episodes of overflow that left no clear sedimentary record.

New analytical methods should provide a very detailed record of changing levels in East African lakes. Some of the lakes, such as Turkana, Albert, Tanganyika and Malawi seem to be among the oldest lakes in the world, and have held water at least intermittently during 5,000,000 or more years (10). Their sedimentary record is resolvable to the nearest year for at least part of that time.

The results in hand reveal a very clear pattern for East African lakes, with the latter part of the last ice age, before 12,500 years ago, being drier and early post-glacial time, 6000 to 10,000 years ago, being wetter than the historic period over an area from Lake Tanganyika and Lake Rukwa north to Ethiopia (43). This is accordant with similar paleoclimatic information across the continent to West

Africa, but different patterns seem to prevail from Lake Malawi and Madagascar south to the Cape.

Conditions Within Lakes. The broad-brush lake history provided by changes in water level can be refined by identifying and counting the microfossils in lake sediments and deducing from them the physical and chemical characteristics of the lakes. Microfossils can be very abundant - 10,000,000 per cm^3 - and very well preserved in the dark, wet, anoxic environment of lake sediment. All East African lakes that have been investigated contain a good fossil record of diatoms, microscopic algae with a siliceous skeleton that permits very precise identification. Some of the physical and chemical conditions of growth of these algae are very well understood, and they can be used to reconstruct the alkalinity of water and the amount of silica dissolved in it (18,24,25,48). The ratio of free-floating and bottom-dwelling diatoms reflects changing water level. These biological insights can be extended by a variety of chemical means (45).

East African lakes are very diverse (see Chapters 7 and 8; 31,46). With respect to concentration of dissolved salts, they range from fresher than rain to ten times as concentrated as the sea. The range in chemical composition, transparency, mixing, productivity, depth and biotic richness is also great, and this wealth of modern analogues can be used to interpret the sedimentary record. The risk of accepting a fortuitous coincidence as evidence of causality is reduced somewhat by the present diversity of lakes (13). Further, diatoms can be cultured in the laboratory and their generation time is short, so it is possible to test hypotheses that emerge from correlation analysis by physiological and ecological experiment (25). Not all of the stratigraphic changes in the diatom assemblage are interpretable: problems arise with fossil assemblages that have no known modern analogue, such as those dominated by *Cyclotella ocelata*, a diatom that is nowhere so abundant today. Despite this, diatoms are the most useful fossil indicators of conditions within lakes. After diatoms, the commonest aquatic organisms in the East African fossil record are a group of green algae, including *Pediastrum*, *Botryococcus*, and *Coelastrum*, with cell walls made of sporopollenin, a carotenoid polymer also found in pollen and spores. Animal fossils are scarce by the standards of temperate and arctic lakes, but ostracodes are abundant and useful as indicators of salinity, and potentially of temperature, in shallow-water sediments.

Conditions Around Lakes. The commonest microfossil indicators of vegetation on the surrounding countryside are pollen grains of flowering plants and the spores of ferns and their allies (16). These are well preserved in the dark, moist and slightly acidic conditions of most lake sediment, but they are less precisely identifiable than diatoms. In general, they can be identified only to genus, although in a few groups species can be distinguished and in some others, such as the grasses, even the genera are not separable.

Most pollen being deposited in the center of a small lake originates within one kilometer, but a substantial part of the pollen comes from ten to a hundred times as far in a large lake. Plants differ in their pollen productivity, with wind-pollinated ones such as grasses, *Podocarpus*, and some Moraceae (the fig family which also includes the *mvule* tree) being profligate, and the ones pollinated by animals, such as *Acacia*, coffee and bananas, much more parsimonious. Much of East African vegetation is rich in leguminous trees, which are light pollen producers, and in grasses, which are very heavy pollen producers. Fossil assemblages tend to be dominated by grass pollen even if they come from woodland or wooded grassland.

These complexities are met in temperate parts of the world by sampling the surface sediments of many lakes and using them to develop an equation relating pollen assemblages to present-day vegetation, or even directly to climate (50). In East Africa this is not easy to do, because lakes are clumped, being particularly abundant in grassland and wooded grassland, and scarce in lowland forest. This can be remedied by investigation of the pollen rain from lake-rich forests. A more serious problem, however, is the topographic grain of the East African countryside. The pollen assemblage entering a lake is affected very much by local altitude and may bear little resemblance to the vegetation at a national or continental scale. Pollen assemblages should be calibrated against a net of vegetational and meteorological measurements much finer than are presently available.

Despite these difficulties, and the rich diversity of the flora, considerable progress has been made in working out the history of East African vegetation by pollen analysis (16). The result is most successful where the surrounding vegetation has included a substantial component of wind-pollinated trees, which are common in evergreen and semi-deciduous forest (22). In woodland and wooded grassland, trees are swamped by grass and sedge pollen. The signal of vegetational and climatic change is borne by only the few percent of the total pollen that comes from trees, so large numbers of pollen must be counted to yield acceptable estimates of these significant traces.

One of the best pollen studies of vegetational change was carried out by Kendall on sediment cores taken from the northern waters of Lake Victoria (22). His data were very well controlled by radiocarbon dating, and he was able to show that late Pleistocene pollen assemblages older than 12,500 years were dominated by grasses with a component of *Acacia* pollen, indicating wooded grassland. Much of the Holocene part of the core was dominated by forest taxa, such as *Celtis*, olive and *mvule* trees, until about 3000 years ago when grasses began to rise once again. This forest regression could be due to climatic causes, fire or agriculture. If pollen-producing forest trees were displaced by crop plants, most of which are not heavy pollen producers, grasses would become relatively more abundant. With so many carbon dates Kendall could compute rates of sedimentation of the different pollen types, and show that both climate and agriculture contributed to the modern prominence of grass pollen.

At all present-day East African localities in or near forest used for pollen analysis, forest pollen is very scarce at levels dating from the last ice-age (28). Evidently there was a considerable contraction of forest vegetation at that time. This does not necessarily mean that there was an equal contraction in the ranges of the constituent species of forest, some of which may have persisted widely in favorable micro-habitats as populations too small to register in the pollen record. It is not clear that this contraction of forest was severe enough to reduce the main Central and West African blocks of forest to small refuges (29).

Grass pollen, which is so abundant in many modern and fossil East African pollen assemblages, is not commonly identifiable much below the level of family. This makes determination of past conditions difficult. Only the post-Colombian introduced cereal grass *Zea mays* is identifiable to genus by its pollen. However, this deficiency can be repaired by using the charred cuticles of grass leaves, which are sometimes preserved well enough for identification to genus (35,36). Accelerator mass spectrometry of pure concentrates of these cuticles might be used to estimate past temperature, because the ratio of C_3 to C_4 grasses in a flora is highly temperature-dependent, and the isotopic composition of the photosynthate depends on which of these two photosynthetic pathways is used by a plant. Research techniques different from those used in temperate lakes may be necessary to distinguish past plant assemblages and climatic patterns.

Biogeographic Evidence

Mount Cameroon and the mountains of East Africa have many species of plants and animals in common-although those common species are presently absent from intervening lowlands (8,33). This pattern could result if there had been a past time when the climate was cool enough to permit a continuous range of montane organisms across Africa. Because African mountains rise so sharply only a very substantial fall in temperature, between 5 and 10° C, would explain the discontinuous range this way (30).

Three kinds of evidence, glacial geology, fossils, and geographic ranges, all agree in suggesting that the temperature of the last ice age was between five and ten °C cooler than the present. The marine evidence suggests that the tropical ocean was not more than two degrees cooler. Global circulation models are unable to reconcile these very different estimates, so there must be some fault in either the models, the marine conclusion or the terrestrial conclusion. If 5 to 10 °C of cooling occurred, the present flora of tropical Africa must be composed exclusively of plants that could cope with cooler temperatures by some combination of rapid genetic selection, physiological plasticity, or by refuge in especially favorable microhabitats.

Conclusions

The climate and ecology of a region such as East Africa changes, sometimes very rapidly, in response to global climatic change. It is not wise to base policies of resource use, including conservation, on an assumption of climatic and ecological constancy. For example, expensive lake port facilities should be designed to cope with considerable and rapid changes in lake level, and park managers should be prepared to make extensive transfers of animals as climatic change alters their present rangelands. There is every reason to expect future climatic changes, some due to natural causes, some due to human activities. Prediction of the nature, timing and extent of these changes is not yet possible, but a substantial rise in temperature seems likely.

Most paleoecological work in East Africa has been aimed at long-term questions, changes over millennia, but the same methods can be applied to much shorter spans of time (2,41). In lakes with a well-developed bottom fauna, or lakes shallow enough to be stirred to the bottom by wind, post-depositional reworking of the sediment destroys fine stratigraphic detail. The sediments are so mixed that it is not easy to identify past events separated by less than ten or a hundred years. In lakes that are stratified (no mixing), however, it is sometimes possible to separate the events of individual years. Under the best circumstances, where the climate is sufficiently seasonal, one can recognize annual bands, count them to provide a chronology, and work out the history, year by year, of the lake and the country around it. Such annual bands were formed in Crescent Island Crater of Lake Naivasha during the early Holocene although they are not forming there today. Annual bands are to be expected in any small deep lake, such as many of the volcanic crater lakes of East Africa. These bands could provide a proxy record of climate very much longer than the meteorological one against which to detect the first signs of artificial climatic change, and a very detailed record of the introduction and spread of some crop plants, especially maize.

Such short-term paleoecology has just begun in East Africa. Lake Victoria has recently undergone a catastrophic reorganization, with introduced Nile Perch and Nile Tilapia replacing native fish, a weaker circulation and a completely different phytoplankton from the one it had only thirty years ago. It may seem reasonable to blame these changes on introduced fishes as their proliferation could change the rest of the ecosystem. Hecky showed, however, by close interval analysis of a core from the deep water of Lake Victoria, that the changes in plankton were already underway before the exotic introductions (17).

Summary

(1) The climate of East Africa, like the climate of the rest of the world, is unstable. That instability has produced great changes in vegetation, in the level of lakes, and in the organisms inhabiting them.

(2) Artificial changes in atmospheric greenhouse gases such as carbon dioxide and methane are likely to impose large and sudden changes on future climate and cause ecological instability.

(3) In the absence of long weather records the best source of background information about such changes in East Africa is the sediment of stratified lakes. Their use for this purpose has barely begun.

(4) Present strategies of conservation depend very heavily on preserves of limited area, separated by long distances where the conserved biota is scarce or lacking. If climatic change makes preserves unsuitable for part of their present biota, the affected species must be moved to suitable new habitat if they are to survive.

References

1. Alley, R.B., Meese, D.A., Shuman, C.A., Gow, A.J., Taylor, K.C., Grootes, P.M.,White, J.W.C., Ram, M., Waddington, E.D., Mayewski P.A., Zielinski, G.A. 1993. Abrupt increase in Greenland snow accumulation at the end of the Younger Dryas event. *Nature* 362:527

2. Battarbee, R.W. 1990. The causes of lake acidification, with special reference to the role of acid deposition. *Philosophical Transactions of the Royal Society of London,* Series B 327:339-347

3. Bonnefille, R., Riollet, G. 1988. The Kashiru pollen sequence (Burundi): paleoclimatic implications for the last 40,000 yr BP in tropical Africa. *Quaternary Research* 30:19-35

4.   Broccoli, A.J., Manabe, S. 1987. The influence of continental ice, atmospheric CO_2, and land albedo on the climate of the last glacial maximum. *Climate Dynamics* 1:87-99

5.   Broecker, W.S., Denton, G.H. 1990. What drives glacial cycles? *Scientific American* Jan: 48-56

6.   Butzer, K.W., Isaac, G.L., Richardson, J.L., Washbourn-Kamau, C. 1966. Radiocarbon dating of East African lake levels. *Science* 175: 1069-1076

7.   Casanova, J., Hillaire-Marcel, C. 1992. Chronology and paleohydrology of Late Quaternary high lake levels in the Manyara Basin (Tanzania) from isotopic data (^{18}O, ^{13}C, ^{14}C, Th/U) on fossil stromatolites. *Quaternary Research* 38:205-226

8.   Clayton, W.D. 1976. The chorology of African montane grasses. Kew Bulletin 31:273-288

9.   CLIMAP Project Members. 1981. Seasonal reconstructions of the Earth's surface at the lst glacial maximum. *Geological Society of America Map and Chart Series* MC-36

10.   Cohen, A.S., Soreghan, M.J., Scholz, C.A. 1993. Estimating the age of formation of lakes: An example from Lake Tanganyika, East African Rift system. *Geology* 21:511-514

11.   Emiliani, C. 1955. Pleistocene temperatures. *Journal of Geology* 63:538-578

12.   Emiliani, C. 1992. Pleistocene Paleotemperatures. *Science* 25: 1462

13.   Gasse, F., Talling, J.F., Kilham, P. 1983. Diatom assemblages in East Africa: classification, distribution and ecology. *Revue Hydrobiologie Tropicale* 16: 3-34

14.   Grootes, P.M., Stuiver, M. White, J.W.C. Johnsen, S., Jouzel, J. 1993. Comparison of oxygen isotope records from the GISP2 and GRIP Greenland ice cores. *Nature* 366:552-554

15.   Haberyan, K.A., Hecky, R.E. 1987. The late Pleistocene and Holocene stratigraphy and paleolimnology of Lakes Kivu and Tanganyika. *Palaeogeography, Palaeoclimatology, and Palaeoecology* 61: 169-197

16.   Hamilton, A.C., 1982. *Environmental History of East Africa.* London: Academic Press

17.   Hecky, R.E. 1993. The eutrophication of L. Victoria. Kilham Memorial Lecture, *Proceedings of the International Association for Theoretical and Applied Limnology* 25, 39-48

18.   Hecky, R.E., Kilham, P. 1973. Diatoms in alkaline saline lakes: ecology and geochemical implications. *Limnology and Oceanography* 18:53-71

19.   Hughes, M.K., Kelly, P.M., Pilcher, J.R., Lamarche, V.C. Jr., 1982. *Climate From Tree Rings.* Cambridge: Cambridge University Press

20.   Imbrie, J., Kipp, N.G. 1971. A new micropaleontological method for quantitative paleoclimatology: application to a Late Pleistocene Caribbean core. In *The Late Cenozoic Glacial Ages.*, ed. Turekian, K.K., p.71-181. New Haven and London: Yale Universiity Press

21.   Johnsen, S.J., Clausen, H.B., Dansgaard, W., Fuhrer, K., Gundestrup, N., Hammer, C.U., Iversen, P., Jouzel, J., Stauffer, B., Steffensen, J.P. 1992. Irregular glacial interstadials recorded in a new Greenland ice core. *Nature* 359:311

22.   Kendall, R.L. 1969. An ecological history of the Lake Victoria Basin. *Ecological Monographs* 39: 121-176

23.   Kerr, R.A. 1986. Greenhouse warming still coming. *Science* 232:573-574

24.   Kilham, P., Kilham, S.S., Hecky, R.E. 1986. Hypothesized resource relationships among African planktonic diatoms. *Limnology and Oceanography* 31: 1169-1181

25.   Kilham, S.S. 1984. Silicon and phosphorus growth kinetics and competitive interactions between *Stephanodiscus minutus* and *Synedra* sp. *Proceedings of the International Association of Theoretical and Applied Limnology* 22:435-439

26. Krishnamurthy, R.V., Epstein, S. 1985. Tree ring D/H ratio from Kenya, East Africa and its palaeoclimatic significance. *Nature* 317:160-162

27. Kutzbach, J.E., Wright, H.E. Jr. 1985. Simulation of the climate of 18,000 yr BP: results for the North American/European sector and comparison with the geologic record of North America. *Quaternary Science Reviews* 4:147-187

28. Livingstone, D.A., 1975. Late Quaternary climatic change in Africa. *Annual Review of Ecology and Systematics* 6:249-280

29. Livingstone, D.A., 1982. Quaternary geography of Africa and the refuge theory. In *Biological Diversification in the Tropics*, ed. Prance, G.T., p. 523-536. New York: Columbia University Press

30. Livingstone, D.A., 1993. Evolution of the African climate. In *Biological Relationships Between Africa and South America*, ed. Goldblatt, P., p.455-472. London and New Haven: Yale University Press

31. Livingstone, D.A., Melack, J.M. 1984. Some lakes of Subsaharan Africa. In *Lakes and Reservoirs*, ed. Taub, F.B., p. 467-497. Amsterdam:.Elsevier

32. Manabe, S., Broccoli, A.J. 1985. A comparison of climate model sensitivity with data from the last glacial maximum. *Journal of Atmospheric Science* 42:2643-2651

33. Moreau, R.E., 1963. Vicissitudes of the African biomes in the late Pleistocene. *Proceedings of the Zoological Society of London* 141:395-421

34. Osmaston, H., 1989. Glaciers, glaciations and equilibrium line altitudes on the Ruwenzori. In *Quaternary and Environmental Research on East African Mountains*, ed Mahaney, W.C., p. 31-104. Rotterdam: A.A. Balkema

35. Palmer, P.G. 1976. Grass cuticles: a new paleoecological tool for East African lake sediments. *Canadian Journal of Botany* 54: 1725-1734

36. Palmer, P.G., Gerbeth-Jones, S. 1989. A scanning electron microscope survey of the epidermis of East African grasses, V and West African supplement. *Smithsonian Contributions to Botany* 67: iv + 157

37. Richardson, J.L., Harvey. T.J., Holdship, S.A. 1978. Diatoms in the history of shallow East African lakes. *Polish Archives of Hydrobiology* 25: 341-353

38. Ruddiman, W., MacIntyre, A. 1980. The North Atlantic Ocean during the last deglaciation. *Palaeogeography, Palaeoclimatology, and Palaeoecology.* 35:145-214

39. Scholz, C.A., Rosendahl, B.R. 1988. Low lake stands in Lakes Malawi and Tanganyika, East Africa, delineated with multifold seismic data. *Science* 240:1645-48

40. Shackleton, N.J. 1967. Oxygen isotope analyses and Pleistocene temperatures re-assessed. *Nature* 215:15-17

41. Smol, J.P., Brown, S.R., McNeely, R.N. 1983. Cultural disturbances and trophic history of a small meromictic lake from central Canada. *Hydrobiologia* 103:125-130

42. Stager, J.C., Reinthal, P.N., Livingstone, D.A. 1986. A 25,000-year history for Lake Victoria, East Africa and some comments on its significance for the evolution of cichlid fishes. *Freshwater Biology* 16:15-19

43. Street-Perrott, F.A., Marchand, D.S., Roberts, N., Harrison, S.P. 1989. *Global Lake-level Variations from 18,000 to 0 Years Ago: a Palaeoclimatic Analysis.* U.S. Department of Energy, Office of Energy Research DOE/ER/60304-H1

44. Talbot, M.R., Livingstone, D.A. 1989. Hydrogen index and carbon isotopes of lacustrine organic matter as lake level indicators. *Palaeogeography, Palaeoclimatology and Palaeoecology* 70:121-137

45. Talbot, M.R., Johannessen, T. 1992. A high resolution palaeoclimatic record for the last 27,500 years in tropical West Africa from the carbon and nitrogen isotopic composition of lacustrine organic matter. *Earth and Planetary Science Letters* 100:23-37

46. Talling, J.F. 1992. Environmental regulation in African shallow lakes and wetlands. *Revue Hydrobiologie Tropicale* 25:87-144

47. Thompson, L.G. 1979. Ice core studies from Mt. Kenya, Africa, and their relationship to other tropical ice core studies. In *Sea Level, Ice, and Climatic Change,* ed. Allison, A., pp. 55-62. International Association of Hydrologic Sciences Publication no. 131

48. Tilman, D., Kilham, S.S., Kilham, P. 1982. Phytoplankton community ecology: the role of limiting nutrients. *Annual Review of Ecology and Systematics* 13:349-372

49. Washbourn-Kamau, C.K., 1975. Late Quaternary shorelines of Lake Naivasha, Kenya. *Azania* 10:77-92

50. Webb, T. 1988. Eastern North America. In *Vegetation History*, eds. Huntley, B., Webb, T., p. 385-414. Dordrecht: Kluwer Academic Publisher

Chapter 2

Paleoecology of Humans and their Ancestors

David A. Burney

Humans and their ancestors have interacted with East African environments longer than with any other ecosystems in the world. This is perhaps the supreme irony of human ecology: the megafauna-dominated "living Pleistocene" of East Africa is in fact the environment that the Hominidae have affected for the longest period of time. For those students of the past who are interested in analyzing the connections between humans, species extinction, and environmental change, East Africa is thus a special and enigmatic case.

The long and complex prehistoric roots of human activity in the region are a prime stimulus for paleoecological investigations of the human-environment relationship, but present a virtually insurmountable methodological challenge. The fundamental problem in this region, as elsewhere, is to separate in the fossil and historical record the purely background-level environmental change from that which is caused by humans. The kind of "before and after human impact" scenarios that make it possible to delineate some past human effects with relative clarity on the other continents and oceanic islands are absolutely out of the question in Africa (13,42). This ambiguity of the evidence has been a factor in the inconclusive debates that have held center stage in the East African environmental community for many years concerning three important issues: (1) causes of desertification and other changes in vegetation; (2) the roles of fire and large-mammal herbivory in savanna ecosystems; and (3) the causes, timing and consequences of faunal decline and extinction. These are not topics that scientists and managers can afford to dismiss simply because they are on scales too large to address experimentally. Instead, they are so important that we must do the best we can to extract whatever is possible from past records that might be helpful in making management decisions.

Many human-environment interactions are so ancient in East Africa that separating human influences and responses from background changes seems virtually impossible, but much can be learned by comparing this venerable case to parts of the world in which many of the same or similar processes have operated, but in which the human time-depth is much shallower. Seeing East Africa from this global perspective may allow us to view human ecology as a series of experiments in adaptation, with this region as the longest-running "treatment." Until relatively recent times, other tropical areas, including remote oceanic islands and the

Box 2.1.   East Africa: The Cradle of Humanity

One of the reasons that the ecology of East Africa is of particular interest to humans is because eastern and southern Africa is the place where humankind first evolved. Somewhere around 5 to 8 million years ago, the human line diverged from our closest living relatives, the chimpanzees and gorillas. Associated with this divergence arose a number of adaptations unique to humans. Chief among these were bipedalism and large brain size, but they also include large body size, a decrease in body hair, and the development of complex language and technology. There has been considerable speculation about the evolution of all of these traits.

The earliest hallmark of this divergence was bipedalism, appearing at least 3.7 million years ago. Among the proposed selective pressures favoring its evolution: 1) carrying things, including dependent offspring or food, 2) provisioning of group members, 3) changing feeding postures, 4) decreased thermal loading, and 5) increased locomotory efficiency (1,2). Because there was a pronounced drying trend in East Africa at this time (see main text) most scenarios invoke environmental change as the underlying cause of the evolution of bipedalism, and some require a shift in habitat use from forest or woodland to savanna. Recent evidence suggests, however, that the earliest hominids lived in wooded habitats (3), and any appraisal of these theories must take this into account (2).

Another hallmark of humans, large brain size, shows more gradual evolution. All the australopithecines had relatively small brains. At about 400 cm^3, their brains were similar in size to that of the chimpanzee. Even *Homo habilis*, the first member of our genus, had a small brain. Nonetheless, throughout most of hominid evolution, brain size increased, until anatomically modern humans appeared by 35,000 years ago, with a brain size of about 1400 cm^3. What selective pressure(s) favored the evolution of large brain size (and its correlate, greater intelligence)? Again, a number of hypotheses have been suggested, including tool use, cooperative hunting, and complex social systems. The evidence for any of these is weak. Hominids apparently began modifying objects into tools as early as 2.5 million years ago, long before the hominid brain increased in size, suggesting that large brains were not required for this skill. Chimpanzees, with their relatively small brains, have been observed both using tools and hunting cooperatively. Many primates have apparently complex social systems.

Perhaps greater intelligence has always been advantageous to hominids, but constraints had limited its evolution until relatively recently. Large human brains 1) have great energetic demands and 2) may even today be constrained by birthing difficulties. Is it possible that either the increased ability to have more animal matter in the diet, or a bipedal posture, may have released humans from either of these constraints on brain size? Definitive answers are premature, but it is likely that something in the ecology of East Africa caused our ancestors to evolve into modern humans.

L.A. Isbell

References

1.   Foley, R. 1989. *Another Unique Primate*. London: Blackwell

2. Isbell, L.A., Young, T.P. 1995. The evolution of bipedalism in humans and reduced group size in chimpanzees: alternative responses to decreasing resource availability. *Journal of Human Evolution*, in press

3. WoldeGabriel, G., White, T.D., Suwa, G., Renne, P., de Heinzelon, J., Hart, W.K., Helken, G. 1994. Ecological and temporal placement of early Pliocene hominids at Aramis, Ethiopia. *Nature* 371: 330-333

neotropics, served as "controls" for certain key variables by virtue of their historical isolation from humans.

The history of human physical and cultural evolution in East Africa, to a greater extent than any other part of the planet, could be said to encompass most of the key events in human development. Many investigators point to Africa as the most likely place of origin for the human family, genus and species (Box 2.1; 53), as well as a seat of many early adaptations in human social organization and technology. Yet, despite all this human activity, a great deal of nature remains on this continent. In many respects, it seems on the surface, at least, that this cradle of humanity has withstood the human environmental onslaught better than many parts of the world in which man is a relative newcomer. How can this be?

This chapter will examine this all-pervasive question, not with the hope of any definitive answer, but as a tool for exploring the paleoecological backdrop for many important conservation issues. Let us begin with a brief inventory of what we know or think we know about the effects of human ancestors and *Homo sapiens* on East African environments and vice versa, a story that begins millions of years ago.

The Paleoenvironmental Setting of Hominid Evolution

During the first half of the Cenozoic era, which began 65 million years ago (=mya), warm, moist conditions are thought to have prevailed over much of Africa, with the result that lowland-type rain forests were probably more widespread than they have been at any time since (1). Human ancestors at this time were tree-adapted early anthropoids, exemplified by fossils of *Aegyptopithecus zeuxis* from the Fayum Desert of Egypt, an austere wasteland today that was a lush forest in the Oligocene (8).

By around 18 mya, however, relatively dry, open woodlands had interrupted the more continuous African forest of earlier times (6). Both human ancestors and their environments were changing considerably, with the appearance of ape-like hominoids. African climates, especially after mid-Miocene time (around 12 mya), were becoming dryer and more seasonal in character, as on other continents. Changes in African climate and hominoid evolution paralleled changes in the landscape itself, as rifting and uplift produced the highlands of East Africa, perhaps cutting off some of the monsoonal moisture that reached the interior in earlier times. The result, in terms of vegetation types inferred from the fossil record, is an increased diversity of open and ecotonal vegetation, with some East African paleofloras indicating predominantly grassland environments (see 6) and others indicating that lowland closed forests were still persistent (see 28). Carbon isotope evidence for the past distribution of grasslands and woodlands, from soil carbonates in the Tugen Hills of Kenya, suggests that mosaic-type environments have been a feature in the locality for the last 15.5 million years or more (33).

By the early Pliocene (around 5 mya), the trend to cooler, drier environments was apparently well established, and open grasslands were perhaps becoming a more prevalent feature of the vegetation mosaic. This notion is supported by the great evolutionary radiation of grazing bovids occurring about this time (56). For the earliest well-dated hominids, found recently in Ethiopia, arboreality was probably still very important. Fossils of *Australopithecus ramidus* (59), that date to 4.4 mya (60), were associated with paleobotanical evidence suggesting that it frequented wooded habitats.

The more complete anatomical evidence from the next hominid ancestor, *Australopithecus afarensis*, shows that it was bipedal, able to move swiftly through open terrain and forage on the ground, but with extremities still well-adapted to climbing trees (30). The now-familiar "Lucy" lived from roughly 3 to 4 mya.

During the late Pliocene and earliest Pleistocene, (especially around 2.4 and 1.8 mya) further cooling and drying of the climate resulted in a major expansion of grassland and desert environments (7,55), enhancing the opportunity for our ancestors to evolve feeding strategies adapted to open country. During this time of climatic upheaval, hominids faced new environmental challenges with new adaptations.

Some of the species that evolved during this time of maximum hominid diversity (between around 2.5 to 1.5 mya), perhaps adapted to a coarse diet consisting primarily of roots and other vegetable matter (the robust australopithecines, *Paranthropus* spp.), with increasingly heavy jaws and large molars. Others (the gracile australopithecine, *A. africanus*, and early *Homo* species such as *H. habilis*) apparently took the familiar "human" route toward dietary diversity, larger brain size, and material culture (37).

These Plio-Pleistocene hominids probably occupied terrestrial omnivore niches and developed strategies that may or may not have been akin to those of modern hunter-gatherers. Exactly what role carnivory played in the diets of various hominid species is still a matter of contention among paleoanthropologists, but evidence for cut-marks on bones of large mammals, apparently the result of butchery with stone tools, date back to 1.8 mya at Olduvai Gorge in Tanzania and elsewhere (9,50). Although these and other authors have cautiously suggested that some hunting skills may have been employed by these meat-eating hominids, it is generally acknowledged that scavenging of kills of other carnivores could have been as important, if not more so (see 2,5). A few researchers (3) maintain that scavenging was the primary, if not exclusive, mode of meat-procurement by Plio-Pleistocene hominids, on the basis of the lack of direct evidence for actual killing of animals with the crude stone tools available.

Whatever consensus eventually emerges regarding hominid ecology and behavior, it is probably safe to say that, in the early stages of hominid evolution, our ancestors were evolving strategies to cope with increasingly savanna-dominated environments in Africa. The evolutionary strategy eventually shifted, at least on the Pleistocene *Homo* line, toward developing adaptations that allowed our ancestors to do what humans have been doing ever since: responding to environmental challenges with expanding technologies that permit increasing control over the external environment.

The Stone Age: Tools for Environmental Change

With the ascendancy of the highly successful (widespread and long-persisting) species *Homo erectus* (and related forms such as *H. ergaster)* in the Lower Pleistocene (1.8 to 1.0 mya), we begin to see the first controversial hints of "environmental impacts" by a human ancestor (Table 2.1). Before that time, we can imagine that australopithecines and early *Homo* may have competed for resources among themselves and with less intelligent quadrupedal herbivores and omnivores, and perhaps stolen kills from carnivores or even competed with them for prey. Some time after the evolution of *H. erectus,* the other hominid species gradually disappear from the fossil record. The apparent direct ancestor of our own species eventually becomes the only surviving member of the hominid family, and becomes established virtually throughout the Old World. Pleistocene fossils of this species occur all the way from South Africa to Java, and northward into western Europe. We may never know for certain whether *H. erectus'* success, and the other hominids' demise, reflects the outcome of interspecific resource competition among hominids, predatory interactions, or other evolutionary events. The intriguing possibility exists that *H. erectus* may have been the first human ancestor to do what subsequent humans do so well - affect the populations of other species and even drive them to extinction through direct predation or modification of the environment. With a simple tool-kit of choppers and handaxes, and possibly the use of fire (see below) *H. erectus* made its mark on the world.

Martin (41) generated some controversy (see 36) with his assertion that *H. erectus* may have created a wave or waves of megafaunal extinctions in the Middle Pleistocene. Although the list of large-mammal extinctions Martin compiled to make his case for Pleistocene overkill in Africa showed a much less pronounced extinction catastrophe than those documented for the late Pleistocene of Australia and the Americas, or the Holocene of many oceanic islands (42), some notable disappearances did occur. Martin (41) estimates that only about 60% of the megafaunal genera present in Africa during *H. erectus'* heyday are still there today (subsequent taxonomic revisions would probably modify this percentage, but not very much). A modern naturalist would encounter many familiar beasts on the early Pleistocene African savanna, but also quite a few that are gone forever, such as sabertooths *(Dinofelis* and *Homotherium*), giant buffalo (*Pelorovis*), antlered giraffe (*Sivatherium*) and a distant cousin of the elephants with down-pointing tusks (*Deinotherium*). Large primates were particularly hard-hit, including the australopithecines and the giant baboons (*Theropithecus*; 35). Small sample sizes and a lack of dates on these extinctions make it difficult to say more about the possible role of hominid hunters in this faunal decline. Bones of some of the extinct creatures are fairly numerous in sites where our ancestors are believed to have butchered their quarry, but such associations or presumed associations span a million years of the fossil record (35), and could, therefore, be taken as evidence of long coexistence. This differs markedly from the "swift and devastating" kind of event that eradicated most of the megafauna of the Americas around 12,000 years ago (42).

The issue of hominid control of fire is as controversial as that of hunting impacts in the African Lower and Middle Pleistocene. Early evidence of fire use is typified by the 1.4 mya hominid site near Kenya's Lake Baringo at Chesowanja (24). This and other early sites with fire-hardened clay, charcoal deposits, or other indirect

Table 2.1. Inferred stages of anthropogenic activity in East Africa (mya = millions of years ago; kya = thousands of years ago; B.C. = before Christian era; A.D. = since birth of Christ)

Stage	Time Interval	Taxa	Lifeways - effects on environment
Early hominid evolution to Early Stone Age	Pliocene to Lower Pleistocene (~5mya to 700 kya B.P.)	*Australopithecus* spp. *Paranthropus* spp. early *Homo* spp.	Foragers, scavengers, hunters (?) - interspecific food competition, predation.
Early Stone Age to Middle Stone Age	Middle Pleistocene (~700 to 125 kya)	*Homo erectus* early *H. sapiens*	Hunter-gatherers - predation on megafauna, use of fire.
Middle Stone Age to Late Stone Age	Upper Pleistocene (125 to 11 kya)	*H. sapiens*	Advanced hunter-gatherers - may have caused some megafaunal extinctions, increased use and control of fire.
Neolithic	Early to mid-Holocene (11 to 4kya)	"	Mixed subsistence -increased population density and sedentarianism; some forest clearance and overgrazing; introduction of livestock.
Iron Age Asian contact (coastal area)	Late Holocene (2000 B.C. to 1500 A.D.)	"	Pastoralists, agriculturalists, hunter - gatherers-increased population pressure and human migration; deforestation; introduction of exotic crops; soil erosion; ivory trade begins.
European domination	Colonial period (1500 A.D. to mid-20th century)	"	Plantation agriculture, extractive industries, hunting with firearms - increased deforestation; introduction of rinderpest and other exotic organisms; loss of wildlife habitat and overhunting of some species; establishment of reserves.
Independence	Late 20th century	"	Manufacturing, expansion of traditional agriculture and agribusiness, tourism - industrial chemical and fossil-fuel pollution; deforestation and land-conversion escalates with population explosion; park and reserve system expands; poaching drives rhinos and elephants to local extinction in many areas.

evidence of fire use have been disputed by many anthropologists, since it is not generally feasible in these cases to rule out the possibility that the fires were of natural origin, rather than being evidence of fire deliberately brought to the site by human ancestors. James (29) questions the interpretation of these early African sites and also the classic Asian site of Zoukoudian (4), often cited as proof that *H. erectus* was using fire extensively because of charcoal deposits in apparent association.

Although this debate is unresolved, it is generally acknowledged that fire use probably developed in two stages: 1) *opportunistic use*, in which early hominids may have picked up firebrands from natural fires and used them to start other fires; and 2) *fire control*, in which *H. erectus* (probably) and early *H. sapiens* (certainly) kept fire to cook food and heat shelters, as well as to light fires in their habitats. Evidence for the control of fire is accepted by virtually all anthropologists from sites dating back about 250,000 years (=250 kya), and actual hearths turn up in early *H. sapiens* sites (29).

By about 40 to 30 kya, evidence for humans with an advanced tool kit (streamlined projectile points, awls, scrapers, and other tools) appears at many eastern and southern African sites. These people were probably anatomically modern humans (*H. sapiens*). That they were more effective hunters than those in the Middle Stone Age is inferred from studies of the species composition and age-distribution of their kills (reviewed in 35). This line of evidence suggests that, while earlier hunters preyed upon old and juvenile age-classes, and particularly more docile animals such as the eland, the later hunters took more prime adults of a wide array of species, including those that are potentially dangerous, such as the buffalo.

Whereas perhaps 20% of the Pleistocene megafauna disappeared over the more than a million years of *H. erectus* and early *H. sapiens* activity in Africa, an approximately equal number of taxa disappeared in just 20 or 30 kya at the end of the Pleistocene. This is still far less than on any of the other continents in this period (13,42). This was also a time of maximum aridity and rapid climate change in East Africa (25,38), yet human populations must have been thriving, based on the relatively large number of known archaeological sites. Organized hunting of big game and periodic intentional firing of grasslands and bushlands were frequent practices, by analogy to modern hunter-gatherers in the savanna. The pace of technological innovation also picks up near the close of the Pleistocene, as inferred from the appearance of new, specialized types of stone projectiles (probably including points for arrows), bone implements, and jewelry (35). The probable increase in human population and lethal hunting technology at this time set the stage, only a few millennia later, for major human migrations and the spread of new technological ideas.

Holocene Humans and Their Influence on the Environment

In East Africa, as elsewhere, the last ten millennia have been a time of unparalleled technological progress. It has also been a time in which the various major phenotypes and cultures of *H. sapiens* have moved about and interacted a great deal. For instance, 10,000 years ago, the indigenous East Africans were probably members of two human groups no longer prevalent in the region, with pygmies in the western, heavily-forested part, and bushmen in the savannas and deserts. At the beginning of the Holocene, the other groups we think of as East Africans, such as the Bantu, Nilotic, and Cushitic peoples, were in West Africa, the Nile Valley, and along the southern portion of the Red Sea, respectively (44).

Pygmies and bushmen, as well as a few tribes from other groups, have remained primarily hunter-gatherers up to contemporary times. Many other human groups, in the broad range of mesic ecosystems between the humid forests and the most arid deserts, begin to show evidence of new lifestyles in the early and mid-Holocene. The appearance of pottery in archaeological sites, ranging from 8 to 2 kya at various sites in eastern and southern Africa (48), probably signals increased sedentarianism, reflecting reliance on more concentrated food sources and long-term food storage. The history of early agriculture is not as well-known for East Africa as for the Middle East, where evidence of cultivation of cereals and domestication of sheep and goats appears about 10 to 12 kya. But, by 7 kya, people in the eastern Sahara were supplementing their hunting and gathering with husbandry of sheep and goats (probably imported from the Middle East) and perhaps some cultivation (57). As the climate turned drier in mid-Holocene times, these people migrated southward, reaching East Africa by 5 kya or soon after, bringing their goats and sheep. Cattle appear in East African sites by 4 kya (40). This period in East Africa, often labeled with a term borrowed from European archaeology as the "Neolithic" (35), was a time of diversified lifestyles based on the use of highly-developed stone and bone implements and domesticated plants and animals, both exotic and indigenous. It is likely that human population density was greater than ever before in East Africa, as elsewhere, and that deforestation began in some areas. Hamilton and coworkers (27) present fossil pollen and stratigraphic charcoal data from a site in Uganda as evidence that agriculturists may have contributed to a decline in forest near the site as early as 4.8 kya.

Beginning around 500 B.C., additional changes came to East Africa, with the influx of Bantu migrants from the west. It was about this time that Iron Age technology and evidence for increased dependence on agricultural crops (especially sorghum and millet) and livestock appears in some archaeological sites (48). It is also about this time that many pollen diagrams show evidence for forest decline (see 32). Lake-level evidence shows that East African rainfall was declining about the same time (52). It is reasonable to assume that, as the populations of Iron Age farmers and pastoralists increased during this time of climatic desiccation, such now-familiar ecological maladies as overgrazing, soil erosion and the decline of wildlife habitat increased in the region. Despite the inferred local decline of some large mammals, however, no species are known to have gone extinct in East Africa during the Iron Age (35).

Asian seafarers probably began visiting the East African coast more than two millennia ago (19). A new kind of impact - elephant hunting for the ivory trade - was fully underway by the ninth century A.D., with the arrival of Arab traders. Asian contact probably had effects on more than elephant populations. New crops were introduced, such as the banana (51). The establishment of trading routes deep into the interior, for the extraction of ivory and slaves, probably brought changes to human settlement patterns and lifestyles. By the time Vasco da Gama arrived with Portuguese ships in 1498, Malindi already had been a major port for centuries. A Swahili network stretched from Mozambique to Somalia and included such urban centers as Mombasa and Lamu on the Kenya coast. These population centers maintained large plantations on the coastal mainland, and developed extensive trade networks in the interior (19,20).

Colonial Resource Extraction

There is no need to summarize here the written history of East Africa, but some of the major effects of human activity since 1500 A.D. bear mentioning. The Portuguese, German, English, French, Italian and other European colonizers of East Africa wrought many changes on this landscape in the space of only a few centuries. These impacts can conveniently be grouped into four categories: introduction of exotic organisms, accelerated resource extraction, changes in indigenous human population structure, and, somewhat belatedly, the initiation of conservation efforts.

One of the first ways that Europeans influenced East Africa was through the introduction of crops, mostly native to the Americas (such as maize, beans, groundnuts, sweet potatoes and cassava) that transformed traditional agriculture (51). Another early influence was the spread of firearms for big-game hunting. Hunting for ivory and other game products, that had been well underway for centuries in the coastal hinterland, became more efficient and had more impact inland as Europeans, Asians, and Africans used the lethal technology to turn hunting into a cash-earning industry. For instance, Johnson (31) noted that in the 1920s, ivory was fetching the equivalent of $5.00 (U.S.) per pound in Nairobi. "It can be understood," the author remarked dryly, "why elephant hunting is so popular."

Early in the twentieth century, with the help of railroads into the interior, timber extraction and mining became important in parts of East Africa. About the same time, plantation agriculture, previously restricted to the coastal region, was introduced to the highland regions of the interior as white settlers began to stake out large farms in tribal lands. Many of these highland areas appeared to arriving Europeans to be very sparsely populated. One reason, in many cases, was that the indigenous people had abandoned their pastures to avoid starvation, because their livestock herds had died a few years before in the epidemics of rinderpest, a devastating disease of artiodactyls that is related to measles (49). The first known cases in Africa occurred in cattle imported to Somalia by the Italian army in 1889, and, in only a decade, it had spread all the way to the Cape, leaving in its wake more than 5 million dead cattle and decimating herds of buffalo, wildebeest and other wild ungulates. The effect of this exotic disease, in terms of starvation and displacement of indigenous people and disruption of the ecology of savanna regions and wildlife populations, far outstripped the effects of other devastating livestock and human diseases, including sleeping sickness, smallpox and malaria, that also plagued many regions of Africa around the same time.

By the 1950s, however, these and many other diseases had come under at least partial control through the effects of inoculation, improved sanitation, and insect control. Insect control, especially in the form of insecticide use, wildlife eradication, and bush clearance to control tsetse fly, exacted its own toll on the environment. But human population growth, through lowered infant mortality, had begun an exponential climb that, although a sign of progress at the time, was to generate a new set of problems in the late twentieth century.

The quickening pace of deforestation, wildlife loss, and other impacts in the colonial period was to some extent countered by the establishment of reserves of various types. Within reserve areas, resources such as water, forests, and game were managed in larger land units than might have been possible otherwise. Some of the more scenic of these evolved into the largest and most pristine of East Africa's national parks. Nevertheless, many resources were squandered during the period of European rule. For instance, Langdale-Brown (cited in 26) estimated that the area of Uganda covered with forest and moist thicket declined between 1900 and 1950 from 12.7% to 4.6%.

The seeds of preservation had been sown in East Africa, however, by such dedicated early conservationists as David and Daphne Sheldrick, Joy and George Adamson, and a host of others. Fortunately, the independence movements that swept East Africa in the mid-twentieth century also permitted and ultimately encouraged this work to continue. A new generation of indigenous conservationists, spawned by the efforts of the East African Wildlife Society, the African Wildlife Foundation, the National Museums, regional universities, Wildlife Clubs, and other organizations dedicated to the concept of Africanization of the conservation movement, emerged to carry on the mission of conservation. In recent years, the greatest challenge has been to encourage wise development in tandem with conservation. The last three decades have seen the continued explosive growth of the human population in East Africa, while adjustment to the new realities of industrial pollution, urbanization, agribusiness, and tourism has summoned heroic efforts from a dedicated cadre of local, regional, and international conservationists.

East African Environmental History from a Global Perspective

Despite the uncertainties outlined above regarding the timing and extent of various human interactions with the East African environment, the general pattern is clear. As in the rest of the world, human effects have accelerated as the population density has increased, and as expanding technological capabilities have increased each individual's demand for and control over resources. Anthropogenic effects began hundreds of thousands of years ago, and have increased steadily ever since, with new types of environmental stress emerging with each new stage of technology (see last column of Table 2.1).

While it is obvious that humans have played a key role in many environmental changes in East Africa, the fossil record also shows clearly that many biotic changes occurred before humans could have played any role at all. In the recent human period, the rapidity and magnitude of some of the observed changes (such as deforestation, local climate change, and soil erosion) have defied easy explanation. Consequently, resource managers have not always agreed on what has gone wrong and particularly on what should be done about the problems. Perhaps an examination of the East African historical evidence from a global perspective would provide some useful backdrop to these areas of uncertainty, that might be conveniently treated in three categories: 1) the role of humans versus climate change in vegetation changes; 2) the search for optimum strategies in the application of fire and grazing in savanna ecosystems and the preservation of forest ecosystems; and 3) the threat and consequences of species loss, especially declining numbers of large animals.

Humans Versus Climate in Recent Environmental Changes

As noted above, some East African sites show evidence for human-caused vegetation changes several thousand years ago, the earliest being a site in Uganda providing a pollen and charcoal record of possible deforestation beginning 4.8 kya (27). Some authors have argued that the decline in African forests and the expansion of deserts are a consequence of increasing human population and expanding technology - that, in effect, Africa's environmental troubles begin with the advent of agriculture (see 46). On the other hand, investigators who have sought to relate human cultural developments to climatic constraints have sometimes argued cause and effect the other way around: that plant cultivation and pastoralism, whose roots in Africa stretch back to the early Holocene or before, became a more compelling option in the late Holocene as a result of the declining natural productivity of desiccating environments (see 57). It is likely, of course, that both human and non-human factors have been at work simultaneously. Understanding past ecological changes is more than an academic exercise. Coping with current and future environmental crises will require knowledge of long-term trends and ecological feedbacks between nature and humans that we do not presently have.

As noted in the introduction, separating human and "natural" trends is especially difficult in East Africa. Perhaps recent information from other regions, that are similar in most respects except human history, might be useful. The island of Madagascar, which lies off the east coast of Africa in the latitudes 12° to 26° S, is broadly similar to Africa in terms of climate and soils, and these similarities are expressed by the vegetation communities. Despite the peculiar endemic character of this long-isolated biota, it could be said that, in many respects, Madagascar is a miniature version of sub-Saharan Africa, particularly southern Africa (12). It is likely that the long-term vicissitudes of climate on the two land masses have followed similar basic patterns. For instance, East Africa showed a wet early Holocene, with forest expansion (38), while southern Africa was generally dry in this time period (16). This dry Southern Hemisphere pattern has been observed as far north as Lake Malawi (22). Virtually all tropical African paleoecological sites sensitive to changes in precipitation seem to show increased aridity after about 3 kya (25). Madagascar tracks these trends well, showing the same Holocene pattern as comparable latitudes in Africa (10). There is just one big difference: until around 2 kya, there were apparently no people on Madagascar (39) to confound the issue as to a human role in Holocene changes. For instance, a site in Madagascar's southwestern desert region shows pollen evidence for forest decline at about 3 kya, a full millennium before the first evidence for humans on the island (14).

This type of evidence, using remote islands at similar latitudes to "control" for human impacts on Holocene vegetation, shows that such places underwent changes that nicely parallel those of Holocene Africa, yet could only be accounted for by natural climate change. For the reasons outlined above, Madagascar is the most compelling case, but other remote tropical landmasses only recently colonized by humans, such as Hawaii (18) and various South Pacific islands (23,34) also show major vegetation changes in the Holocene in the absence of humans. But all these sites show the mark of human arrival quite clearly in the sediments of the late Holocene, in the form of microfossil evidence for an increase in fires, a decline of forests, and the advent of plants introduced by humans (reviewed in 15).

The lesson contained in these comparisons for East Africa is that, even in the complete absence of humans, rapid changes in vegetation have occurred in tropical environments during the Holocene. The general trend toward drier climates in the last three millennia has probably contributed to the surprisingly rapid and thorough human modification of some arid and semiarid environments in Africa, making ecological problems much worse than they would have been otherwise. It is likely that this "climatic feedback" accounts for the apparent ease with which pastoralists, even at surprisingly low population densities, have appeared to enhance the spread of desert conditions in the Sahel in recent decades.

Fire and Grazing in Savanna Ecosystems

A related controversial topic among East African scientists and managers is that of fire and grazing ecology. It has been widely stated in the literature of Africa, with little or no evidence, that such pyrogenic communities as the miombo woodlands of southern Africa and the mesic derived grasslands of East Africa are not "natural" communities, but rather artifacts of our species' widespread, persistent and long-standing practice of burning any vegetation that can be burned (58). It is obvious that overly-frequent burning of vegetation at some sites in recent decades has resulted in soil erosion and a decline in the productivity and diversity of the vegetation. But is all burning "unnatural" and undesirable in reserves that are managed for a diversity of plant species and wild herbivores? What is the connection between burning and large-mammal herbivory, and what is the proper balance between the two? These have not been easy questions to answer, owing to the long temporal and large spatial scales involved. To begin with, the history of fire ecology in East Africa, using fossil charcoal particles in sediments, has been more difficult to determine than in regions with a shorter human history, since there is no certain way to distinguish the fossil evidence of anthropogenic wildfires from natural fires.

Again, remote islands in which humans are a relatively recent arrival provide some interesting parallels and contrasts. The charcoal stratigraphies of an array of lakes and bogs throughout Madagascar show that, in the absence of humans, wildfires nevertheless occur when the necessary conditions of low fuel moisture and a natural ignition source (for example volcanoes and lightning) are met (11). The seasonally dry highland regions of the island were particularly prone to burning. Sediment analyses often yield as much or more charcoal in the prehuman late Pleistocene and early Holocene strata as after human arrival. Some wetter areas of the northwest, with a shorter dry season, and arid southwestern regions lacking enough litter to burn easily, show a much less frequent occurrence of prehuman fire and a marked increase in fire after human colonization (see 14). This is also the same trend found in a wet site with low seasonality (equable precipitation) in Puerto Rico (17). Here the sudden appearance of charcoal in the sediments apparently signals the arrival of the first indigenous peoples of the West Indies about 5 kya. On the other hand, an alpine grassland site near Haleakala Crater, Maui, in the Hawaiian Islands, shows that the nearby active volcano (or perhaps lightning) set fires long before the arrival of the Hawaiians (18), who apparently did not set fires at this remote montane site. These and other studies from tropical islands (reviewed in 15) tell us something important about East Africa: areas with an intermediate amount of rainfall (about 500 to 1200 mm/y) and a long dry season, probably burned periodically even before humans and their ancestors began setting fires. Once our ancestors began firing the landscape, it is likely that fire frequency increased, and areas too wet or too sparsely vegetated to burn frequently without a human source were also affected by fire. A similar trend has been postulated for the Upper Pleistocene of Australia, also based on stratigraphic charcoal evidence (21).

The species compositions of savanna communities are very different depending on the fire return intervals (see Chapters 11 and 12). Current research indicates that fire benefits many savanna ecosystems by returning the nutrients in dead or slow-growing plant biomass to the soil for stimulation of new growth (45). Managers must pay careful attention to the seasonal and interannual timing of these fires, however, as the choice of fire regimes can have drastic effects on plant structure and species composition.

Ecologists are also turning their attention to the interactions between fire and grazing ecology. Some experimental evidence (reviewed in Chapters 10 and 11) suggests that there are optimal levels of rangeland herbivory for maximizing productivity, biodiversity, and resource sustainability. Some progress has been made in characterizing the relation between Africa's large grazers and browsers and fire ecology. Owen-Smith (47), based on analogies between the fossil record and modern large-herbivore systems, has proposed the "Keystone Herbivore Hypothesis". Very briefly, he makes a reasonable case for the idea that in African gamelands (and in the Pleistocene of other continents), the megaherbivores (such as elephants and rhinos) foster the diversity and productivity of rangelands by promoting the turnover of coarse plant biomass and the maintenance of a fine-scale heterogeneity in the environment. This mosaic creates opportunities for many grazing and browsing specialists. In the absence of this top-down-influenced megaherbivore system, Owen-Smith speculates, flammable plant biomass accumulates and fire becomes a major influence on savanna structure, perhaps resulting in less heterogeneous vegetation. Indirect support for this idea comes from the observational and experimental work of McNaughton (see 45) in the Serengeti ecosystem, showing that fire and grazing represent alternate pathways of nutrient recycling. Careful measurements are needed, however, on longer temporal scales and over wider geographic regions, with and without megaherbivores. Much work remains to be done in this fertile area of inquiry, and a complete discussion is beyond the scope of this paper. One thing is clear, however: the decline in recent decades of megaherbivores and diversified grazer/browser communities in many parts of East Africa has almost certainly affected plant community structure and may have increased the prevalence and ecological importance of fire in these savannas.

Megafaunal Decline and Extinction

An often-cited consequence of human activities in Africa is the decline of large-mammal communities and the potential acceleration of extinction rates. Poaching and land conversion have certainly resulted in many local extinctions, and only a few African nations have retained even beleaguered examples of the once-widespread megafaunal communities of the continent. Yet, as noted previously, Africans can take some small comfort in the knowledge that their continent has experienced far fewer large animal extinctions since the late Pleistocene than any other habitable continent and even some oceanic islands (41,42). Unfortunately, the awesome weaponry and increased international trade introduced in recent centuries has initiated a modern crisis of species survival in Africa. It might therefore be useful to ask why Africa's extinction history is so different, and whether this provides any clues as to appropriate contemporary strategies for preservation of Africa's "living Pleistocene." Once again, patterns observed elsewhere may shed some light on the seeming paradox of megafaunal survival in Africa - where animals have been hunted and environments have been modified by humans and their ancestors for by far the longest period of time.

Forty thousand years ago, every continent except Antarctica supported a large mammal fauna with a diversity comparable to that of the game parks of present-day East Africa (13). First Australia and New Guinea, then North and South America, lost most of their large mammals near the end of the Pleistocene. Europe and Asia experienced major losses as well, but more gradually. Most of the larger animals that inhabited the Caribbean and Mediterranean islands became extinct several thousand years ago, and the giant flightless birds of New Zealand disappeared in the present millennium. Madagascar was probably the last place on earth in which a diverse large animal fauna (and some smaller species) became extinct in prehistoric times, including at least two dozen mammals, birds and reptiles (13). Most if not all of the apparently rapid late Quaternary extinction events seem to track closely the archaeologically estimated time of human arrival in each of these places. Many scientists point the accusing finger at *Homo sapiens*, although the issue has not been completely laid to rest anywhere, with various authors invoking overhunting, habitat modification, ecosystem and evolutionary dynamics, climate change, or some combination of factors (see articles in 43).

Only Africa appears to have survived these late Pleistocene extinction events relatively intact. Thus people flock to East Africa from around the world to view wildlife spectacles similar to those that their own ancestors probably helped eliminate on some other land mass thousands of years ago.

How could Africa, which has probably suffered human impacts longer than any other continent, escape this prehistoric catastrophe? Perhaps we will never know for sure, but work on the extinctions in other parts of the world has provided some indirect clues. Based on limited evidence, the late Quaternary patterns appear to range between two extremes: relatively sudden catastrophic extinction events, such as in the Americas, and more gradual or attritional extinctions, as in Eurasia. The difference may lie in the state of human technology at the time of first contact between humans and these now-extinct faunas. Most archaeologists agree that the first native Americans were well-armed and organized big-game hunters. In contrast, the history of our genus in Eurasia goes back to our stone-wielding ancestor, *H. erectus*, who may have been there as early as 1.8 mya (54).

Martin (42) and other advocates of the "Blitzkrieg Hypothesis" (*blitzkrieg* is German for "lightning war," in reference to the military technique of attacking along a fast-moving front) conclude that the faunas in greatest danger were those that made their first contact with humans when the latter had already attained the stage of advanced hunters. It follows that those faunas most likely to survive were the ones that were less naive, having adopted appropriate flight responses to human ancestors while humans were still in the process of developing the skills that eventually culminated in the lethal technologies of big-game hunting. In a sense, Africans and

African wildlife have co-evolved to an uneasy balance. Many of the surviving species have had a much longer time than animals in other parts of the world to gradually adjust to the increasingly destructive capacity of the genus *Homo*.

Conclusions

The record of Africa's past exhibits, with greater time-depth than any other part of the world, the entire slowly unfolding saga of human ecology. The story begins with the evolutionary adaptations of hominoid primates and early hominids to changing Cenozoic climates and increasingly open vegetation. Within the genus *Homo*, the evolution of intelligence and tool use over the last few million years represented the beginning of what humans have been doing ever since: modifying the environment to suit the needs of the organism, rather than passively adapting to the environment. By the Upper Pleistocene, if not well before, humans were employing hunting skills and fire in ways that probably influenced many other species and perhaps fundamentally altered biotic communities. The last 10 kyr has seen a rapid progression of emerging technologies associated with pastoralism and cultivation that have sometimes been destructive to wildlife. The advent of iron tools, coupled with human population growth, migrations, and contact with peoples and domestic species from other continents, has made the last two millennia a time of unprecedented influence on the environment. This has culminated in the environmental crisis of the twentieth century and the belated emergence of a conservation movement aimed at redressing some of the deleterious effects of our progressively expanding demands on the environment.

Perhaps there are some valuable lessons for conservationists in this drawn-out experiment in human ecology. First, climate and vegetation change have been a major factor in East Africa, as elsewhere, for far longer than human ancestors have been exerting any control over these processes. This observation emphasizes the need to consider not only human but also natural factors in dealing with ecological changes in Africa. In the past, as is likely in the future, humans precipitated environmental catastrophes when they failed to account for the naturally large ecological variation that may be amplified by humans. Inputs to the system that seem relatively innocuous when viewed in static isolation have a different effect when human and natural influences interact synergistically.

Second, natural fires and pyrogenic plant communities play an important role in many tropical environments even in the complete absence of humans. The effect of the long human use of fire in East Africa has probably been to expand fire's role through increasing fire frequency and encouraging fires in wetter and drier environments than those that regularly burned in prehuman times. Studies in the Serengeti ecosystem (45) have demonstrated that fire benefits megafaunal herbivory through reduction of the unpalatable standing crop in grasslands and the cycling of sequestered nutrients back to the soil. Little is known, however, about the long-term interactions between fire and herbivory, although it has often been noted that fire seems to increase in importance when large herbivores are decimated by poaching or disease (see 45). In particular, we need to know more about possible top-down influences on savanna ecosystems by megaherbivores, and past interactions between fire, megafauna, human activities, and disease.

Third, an uneasy coexistence between humans and wildlife appears to be a long-standing feature of African life. The overall situation for wildlife preservation in

Africa appears bleak, but there may be some hope in this historical difference between Africa and most other places, if technological change is not too rapid and destructive. If the hominid line originated in Africa, then the usual distinction between "human" and "natural" ecology is in the evolutionary sense artificial here. This is in contrast to most other indigenous ecosystems around the globe, in which humans are clearly an invasive exotic organism. This point, aside from its philosophical appeal, may also be relevant in terms of evolutionary biology, ecosystem dynamics, and conservation strategies. We should not lose site of the fact, however, that the influx of modern technology for habitat modification and destruction of wildlife has seriously threatened this already uneasy coexistence.

The widespread environmental degradation and catastrophic faunal losses of the earth's other land masses in recent millennia suggest that human and natural ecological trends may interact in ways more devastating and irreversible than anyone at the time could have foreseen. The coalescence of the deleterious effects of human population growth, lethal technologies, and exotic organisms has already overwhelmed most of the biota of the rest of the planet. It remains to be seen whether, as if coming full circle, our species will also extinguish in the near future the more intact biotic communities of East Africa, where it all began.

References

1. Axelrod, D.I., Raven, P.H. 1978. Late Cretaceous and Tertiary vegetation history of Africa. In *Biogeography and Ecology of Southern Africa*, ed. Werger, M.J.A., pp. 77-130. The Hague: Junk
2. Behrensmeyer, A.K. 1987. Taphonomy and hunting. In *The Evolution of Human Hunting*, eds. Nitecki, M.H., Nitecki, D.V., pp. 423-450. New York: Plenum
3. Binford, L.R. 1985. Human ancestors: Changing views of their behavior. *Journal of Anthropological Archaeology* 4: 292-327
4. Binford, L.R., Ho, C.K. 1985. Taphonomy at a distance: Zhoukoudian, "The cave home of Beijing man"? *Current Anthropology* 26: 413-429
5. Blumenschine, R.J. 1987. Characteristics of an early hominid scavenging niche. *Current Anthropology* 28: 383-407
6. Bonnefille, R. 1985. Evolution of the continental vegetation: The palaeobotanical record from East Africa. *South African Journal of Science* 81: 267-270
7. Bonnefille, R., Lobreau, D., Riollet, G. 1982. Fossil pollen of *Ximenia* (Olacaceae) in the Lower Pleistocene of Olduvai, Tanzania: Palaeoecological implications. *Journal of Biogeography* 9: 469-486
8. Bown, T.M., Kraus, M.J., Wing, S.L., Fleagle, J.G., Tiffney, B.H., Simons, E.L., Vondra, C.F. 1982. The Fayum primate forest revisited. *Journal of Human Evolution* 11: 603-632
9. Bunn, H.T., Kroll, E.M. 1986. Systematic butchery by Plio/Pleistocene hominids at Olduvai Gorge, Tanzania. *Current Anthropology* 27: 431-452
10. Burney, D.A. 1987. Pre-settlement vegetation changes at Lake Tritrivakely, Madagascar. *Palaeoecology of Africa* 18: 357-381
11. Burney, D.A. 1987. Late Quaternary stratigraphic charcoal records from Madagascar. *Quaternary Research* 28: 274-280
12. Burney, D.A. 1989. The piece of East Africa that slipped away. *Swara* 12: 8-13
13. Burney, D.A. 1993. Recent animal extinctions: Recipes for disaster. *American Scientist* 81: 530-541
14. Burney, D.A. 1993. Late Holocene environmental changes in arid southwestern Madagascar. *Quaternary Research* 40: 98-106

15. Burney, D.A. 1995. Tropical islands as paleoecological laboratories: Gauging the consequences of human arrival. *Human Ecology*, in press
16. Burney, D.A., Brook, G.A., Cowart, J.B. 1994. A Holocene pollen record for the Kalahari Desert of Botswana from a U-series dated speleothem. *The Holocene* 4: 225-232
17. Burney, D.A., Burney, L.P., MacPhee, R.D.E. 1994. Holocene charcoal stratigraphy from Laguna Tortuguero, Puerto Rico, and the timing of human arrival on the island. *Journal of Archaeological Science* 21: 273-281
18. Burney, D.A., DeCandido, R.V., Burney, L.P., Kostel-Hughes, F.N., Stafford, T.W., Jr., James, H.F. 1995. A Holocene record of climate change, fire ecology and human activity from montane Flat Top Bog, Maui. *Journal of Paleolimnology* Vol.13 (in press)
19. Chittick, H.N. 1967. Discoveries in the Lamu Archipelago. *Azania* 2: 37-68
20. Chittick, H.N. 1974. *Kilwa: An Islamic Trading City on the East African Coast.* Nairobi: British Institute in Eastern Africa, memoir 5
21. Clark, R.L. 1983. Pollen and charcoal evidence for the effects of Aboriginal burning on the vegetation of Australia. *Archaeology of Oceania* 18: 32-37
22. Finney, B.P., Johnson, T.C. 1991. Sedimentation in Lake Malawi (East Africa) during the past 10,000 years: a continuous paleoclimatic record from the southern tropics. *Palaeogeography, Palaeoclimatology, Palaeoecology* 85: 351-366
23. Flenley, J.R., King, A.S.M., Teller, J.T., Prentice, M.E., Jackson, J., Chew, C. 1991. The Late Quaternary vegetational and climatic history of Easter Island. *Journal of Quaternary Science* 6: 85-115
24. Gowlett, J.A.J., Harris, J.W.K., Walton, D., Wood, B.A. 1981. Early archaeological sites, hominid remains and traces of fire from Chesowanja, Kenya. *Nature* 294: 125-129
25. Hamilton, A.C. 1982. *Environmental History of East Africa: A Study of the Quaternary.* New York: Academic
26. Hamilton, A.C. 1984. *Deforestation in Uganda.* Nairobi: Oxford University Press
27. Hamilton, A.C., Taylor, D., Vogel, J.C. 1986. Early forest clearance and environmental degradation in south-west Uganda. *Nature* 320: 166-167
28. Jacobs, B.F., Kabuye, C.H.S. 1987. A middle Miocene (12.2 m.y. old) forest in the East African Rift Valley, Kenya. *Journal of Human Evolution* 16: 147-155
29. James, S.R. 1989. Hominid use of fire in the Lower and Middle Pleistocene: A review of the evidence. *Current Anthropology* 30: 1-26
30. Johanson, D.C., Edey, M.A. 1981. *Lucy: The Beginnings of Humankind.* New York: Simon and Schuster
31. Johnson, M. 1928. *Safari: A Saga of the African Blue.* New York: Putnam
32. Kendall, R.L. 1969. An ecological history of the Lake Victoria basin. *Ecological Monographs* 39: 121-176
33. Kingston, J.D., Marino, B.D, Hill, A. 1994. Isotopic evidence for Neogene hominid paleoenvironments in the Kenya Rift Valley. *Science* 264: 955-959
34. Kirch, P.V., Flenley, J.R., Steadman, D.W., Lamont, F., Dawson, S. 1992. Ancient environmental degradation. *National Geographic Research and Exploration* 8: 166-179
35. Klein, R.G. 1984. Mammalian extinctions and Stone Age people in Africa. In *Quaternary Extinctions: A Prehistoric Revolution*, eds. Martin, P.S., Klein, R.G., pp. 553-573. Tucson: University of Arizona Press
36. Leakey, L.S.B. 1966. Reply to Martin, 1966. *Nature* 212: 1615
37. Leakey, R.E. 1981. *The Making of Mankind.* London: Michael Joseph
38. Livingstone, D.A. 1975. Late Quaternary climatic change in Africa. *Annual Review of Ecology and Systematics* 6: 249-280
39. MacPhee, R.D.E., Burney, D.A. 1991. Dating of modified femora of extinct dwarf *Hippopotamus* from southern Madagascar: implications for constraining human

colonization and vertebrate extinction events. *Journal of Archaeological Science* 18: 695-706

40. Marshall, F. 1994. Archaeological perspectives on East African pastoralism. In *African Pastoralist Systems: An Integrated Approach*, eds. Fratkin, E. and coworkers. Lynne Reinner: Boulder

41. Martin, P.S. 1966. Africa and Pleistocene overkill. *Nature* 212: 339-342

42. Martin, P.S. 1984. Prehistoric overkill: The global model. In *Quaternary Extinction: A Prehistoric Revolution*, eds. Martin, P.S., Klein, R.G., pp. 354-403. Tucson: University of Arizona Press

43. Martin, P.S., Klein, R.G., eds. 1984. *Quaternary Extinctions: A Prehistoric Revolution*. Tucson: University of Arizona Press

44. McEvedy, C. 1980. *The Penguin Atlas of African History*. London: Penguin

45. McNaughton, S.J. 1985. Ecology of a grazing ecosystem: the Serengeti. *Ecological Monographs* 55(3): 259-294

46. Morrison, M.E.S. 1968. Vegetation and climate in the uplands of southwestern Uganda during the later Pleistocene Period. 2. Forest clearance and other vegetational changes in the Rukigu Highlands during the past 8000 years. *Journal of Ecology* 62: 1-31

47. Owen-Smith, R.N. 1988. *Megaherbivores: The Influence of Very Large Body Size on Ecology*. Cambridge: Cambridge University Press

48. Phillipson, D.W. 1977. *The Later Prehistory of Eastern and Southern Africa*. London: Heinnemann

49. Plowright, W. 1982. The effects of rinderpest and rinderpest control in Africa. In *Animal Disease in Relation to Animal Conservation*, eds. Edwards, M.A., McDonnell, U., Symposia of the Zoological Society of London 50: 1-28

50. Potts, R., Shipman, P. 1981. Cutmarks made by stone tools on bones from Olduvai Gorge, Tanzania. *Nature* 291: 577-580

51. Purseglove, J.W. 1976. The origins and migrations of crops in Tropical Africa. In *Origins of African Plant Domestication*, ed. Harlan, J.R., pp. 291-309. The Hague: Mouton

52. Street, F., Grove, AT. 1976. Environmental and climatic implications of late Quaternary lake-level fluctuations in Africa. *Nature* 261: 385-390

53. Stringer,. 1992. Replacement, continuity and the origin of *Homo sapiens*. In *Continuity or Replacement: Controversies in* Homo sapiens *Evolution*, eds. Brauer , G., Smith, F.H., pp. 9-24. Brookfield, CT: Balkema

54. Swisher, C.C., III, Curtis, G.H., Jacob, T., Getty, A.G., Widiasmoro, A.S. 1994. Age of the earliest known hominids in Java, Indonesia. *Science* 263: 1118-1121

55. Van Zinderen Bakker, E.M, Mercer, J.H. 1986. Major late Cenozoic climatic events and palaeoenvironmental changes in Africa viewed in a world wide context. *Palaeogeography, Palaeoclimatology, Palaeoecology* 56: 217-235

56. Vrba, E.S. 1985. African Bovidae: Evolutionary events since the Miocene. *South African Journal of Science* 81: 263-266

57. Wendorf, F., Schild, R. 1984. Conclusions. In *Cattle-keepers of the Eastern Sahara: The Neolithic of Bir Kiseiba*, eds. Wendorf, F. Schild, R., Close, A., Dallas: Southern Methodist University

58. West, O. 1971. Fire, man, and wildlife as interacting factors limiting the development of climax vegetation in Rhodesia. *Tall Timbers Fire Ecology Conference* 11: 121-145

59. White, T.D., Suwa, G., Asfaw, B. 1994. *Australopithecus ramidus*, a new species of early hominid from Aramis, Ethiopia. *Nature* 371: 306-312

60. WoldeGabriel, G., White, T.D., Suwa, G., Renne, P., de Heinzelin, J., Hart, W.K., Heiken, G. 1994. Ecological and temporal placement of early Pliocene hominids at Aramis, Ethiopia. *Nature* 371: 330-333

Section II:

Marine Ecosystems

Chapter 3

Oceanic Ecosystems and Pelagic Fisheries

T. R. McClanahan

Most of us view the ocean as strangers who can only visit for a short time before returning to our more solid and airy homes. But ours is largely a wet and salty planet where life has had its greatest success, in terms of phyletic biological diversity and its persistence over geologic time, in the ocean. Many of the plant and animal forms we presently find in the ocean failed to produce the more recent land-invading relatives - which may explain the "extra-terrestrial" feeling divers who visit coral reef or seagrass ecosystems experience. But, for many, such as fishermen, the ocean is their work place, and understanding the various intricacies of life on the seafloor or the patterns of waves, currents and tides may be critical to their successful resource use and their personal survival. East Africa and Africans have had a long relationship with the Indian Ocean not only as a means of transport to and from the web of commercial sea-faring routes that have long connected Indian Ocean and Red Sea inhabitants, but also as an important resource for food and other commercial products. These natural and commercial resources, their distribution, the ecological basis of their production, and their use by humans are themes of the following discussion.

Geography

The equatorial position and eastern continental edge (oceanographers prefer the term western boundary as they have an ocean-centered view of the earth) assures East Africa a number of its characteristic climatic and oceanographic conditions. These include prevailing winds from the east which produces a warm equatorial current converging on the continent in Tanzania. This current leaves the continent in Somalia (Fig. 3.1) and northern Kenya and draws up deep, cool and nutrient-laden seawater during part of the year. In contrast, the current reaching the Tanzanian coastline has traversed the Indian Ocean and the majority of nutrients in the seawater have sunk to the ocean floor. When this current encounters the coastline currents are driven south and north along the African coast and some water is also forced down (called downwelling). The warm and clear waters in Tanzania and Kenya produced areas suitable for coral growth and most of the coast and coastal towns lie on ancient

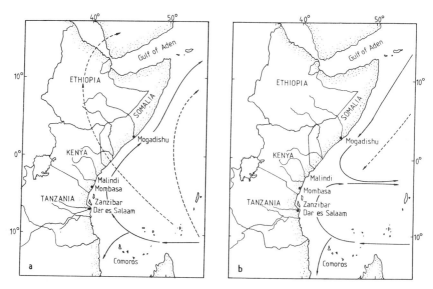

Figure 3.1. Map of the East African region and the patterns of monsoon seasonality showing ocean currents and wind patterns during (a) the southeast monsoon and (b) the northeast monsoon.

reefs formed around 130,000 year ago during interglacial periods when sea levels were higher than today (7). Conversely, coral growth and reef development (see Chapter 4), which requires warm temperatures, has been poorer in the Somali region and, therefore, Somalia is a depositional area for sands eroded from weathered rocks of the continent - transported by wind, rivers and ocean currents. Geologically, the coastline is largely distinguished by ancient reefs in southern Kenya and Tanzania while northern Kenya and Somalia are largely formed by recent eroded sand and sediments interspersed with ancient and actively growing coral reefs. These coral reef areas are usually patchy and this same patchy distribution can be seen in the ancient reefs which formed the Lamu archipelago of northern Kenya and southern Somalia over 100,000 years ago. These differences in oceanography and geology create essentially two coastal regions with the Malindi area as the boundary. The northern and southern regions are geologically, meteorologically, oceanographically and ecologically distinct as will become evident from the following discussion and the following chapters.

Oceanography

Oceanography is the study of the ocean and has been divided, by academic discipline, into physical (water motion and light), chemical, and biological oceanography. Each discipline, however, recognizes that all oceanographic factors interact to create the properties and processes of the earth's oceans. Consequently, a comprehensive understanding of the ocean must include a basic knowledge of each discipline. Similarly, to understand the East African coastal and oceanographic region, like any region, one must understand the interaction of atmosphere and ocean. Radiation from the sun causes differential heating and cooling of the land and ocean surface;

this in turn drives wind which in turn drives the earth's major ocean currents. Changes in the earth's heating and cooling are determined by the geographic position of continents and oceans, their heat absorption and reflection capacities and the annual change in the earth's inclination towards the sun. East Africa and the Indian Ocean are unique in that most of the Indian Ocean is in the southern hemisphere and does not extend above the 15th parallel due to the Asian Continent. Most oceans in the northern hemisphere have currents which turn clockwise most of the year due to consistent Easterlies at the equator and Westerlies in temperate latitudes. The northern Indian Ocean current, however, is not as affected by temperate-latitude Westerlies and current patterns follow local tropical monsoon winds that migrate among tropical latitudes (around 15° north and south of the equator).

The Monsoon Annual Cycle

As the earth rotates around the sun the earth's inclination towards the sun changes and causes differential heating of the earth's surface. The Inter-Tropical Convergence Zone (ITCZ) is the area created by the convergence of winds coming from the north and south replacing the rising air (heating = air expansion and rising) formed by this heating. As air rises it cools, water condenses, clouds are formed, and rainfall is usually highest directly under this ITCZ. Most East Africans are well aware of the effect of these meteorological phenomena on terrestrial ecosystems (see previous chapters), but they may be less aware that this pattern also has a profound influence on the western Indian Ocean and its ecology (2,15). At a particular point along the coastline the climate will reflect the direction of the wind by the position of the ITCZ. The latitudinal position of the ITCZ creates the north and south directions of the wind and the rotation of the earth gives winds its easterly direction. When the ITCZ is to the north, wind is from the southeast (SE), when the ITCZ is to the south, winds are from the northeast (NE), and the short period when the ITCZ is directly overhead is called the inter-monsoon - of which there are two periods per year. For the sake of simplicity, we will largely discuss the two extremes of the northeast (from October to March) and southeast monsoons (from March to October).

During the SE monsoon, winds travel a long distance across the Indian Ocean and become saturated with water. High cloud cover and rainfall follow soon after the monsoon's onset, and, although rainfall drops after June, cloud cover and low sunlight persist until September (Fig. 3.2). High winds during this season have a profound influence on physical, chemical, and biological oceanographic parameters. High winds create large waves and a thorough stirring of the seawater which brings phosphorus closer to the surface. Low sunlight along with this stirring of deeper and cooler water reduces surface-water temperatures. In general, with the exception of the Somali-current upwelling system that draws deeper and more nutrient-laden water, these oceanographic conditions of high water mixing, low water temperatures and sunlight make conditions for planktonic growth difficult because of a lack of nitrogen and sunlight. Throughout most of the western Indian Ocean this season is characterized by clear, cool water and low biological productivity in the water column.

The Somali upwelling system, in the northern Indian Ocean, is an exception to lower primary productivity in the south. During the SE monsoon, currents travel north along the East African coastline and leave the coastline in northern Somalia.

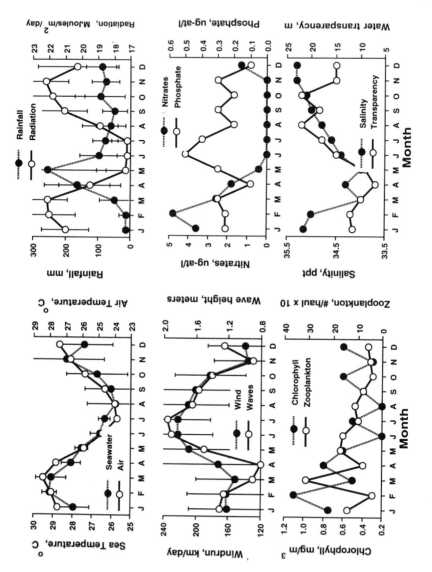

Figure 3.2. Seasonal cycle of meteorological and oceanographic processes along the East African coast (data from 2,15).

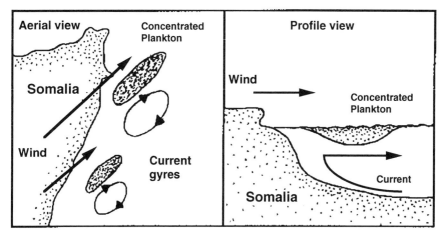

Figure 3.3. Diagram showing mixing and associated processes in the water column during the Somalia south-east monsoon.

Winds traveling over Somalia typically curve back eastward towards the low air pressure created by the heating of the Asian continent and, therefore, travel from the southwest off of Somalia (Fig. 3.1). When winds blow offshore and currents leave a coastline at speeds up to 3.5 meters/second, water from great depths is drawn up to replace the water moving offshore (Fig. 3.3). Water moving offshore also creates clock-wise rotating circulation gyres that also partially replace water moving offshore (28). This deep water is typically rich in nutrients important for plant growth, such as phosphorus and nitrogen, which previously settled to the seafloor. This chemical enrichment results in seasonal blooms of microscopic algae that form the basis for a planktonic food web leading to high fish production. When the monsoon winds change direction this upwelling system breaks down and planktonic productivity is reduced to around half of its upwelling level.

Winds during the NE monsoon travel greater distances over the cool upwelled water in the northern Indian Ocean and the dry land of Somalia and therefore do not have the same strength or water vapor as during the SE monsoons. Consequently, many of the conditions of the SE monsoon are reversed and the NE monsoon is characterized by high sunlight, water temperatures, and lower wave energy and water-column mixing (Fig. 3.2). These conditions are more conducive to planktonic growth except for somewhat lower nutrient levels in the seawater. Nonetheless, nitrogen is often one of the more limiting nutrients and some blue-green algae are able to use nitrogen gases dissolved in seawater and convert this nitrogen into forms utilizable by plants. These algae need to stay close to the water surface where sunlight and dissolved gases are abundant, and species typically have adaptations for floating. Calmer conditions during the NE monsoon favor blue-green algae and their increased abundance forms the basis for overall increased productivity and the establishment of a planktonic food web. Zooplankton, which feeds on phytoplankton, appears to peak late in the northeast monsoon season probably lagging behind the phytoplankton peak (16,23). Blue-green algae may eventually use up other necessary nutrients, such as phosphorus, resulting in productivity declines. Eventually the onset of the SE monsoon, and associated water-column stirring and reduced sunlight, causes a decline in blue-green algal abundance, lowered

Box 3.1. Production and Resources

The complexity of life is exemplified in the way that organisms are able to convert material and energy resources into living organisms. It is not sufficient for organisms to simply undertake and maintain this production process but they must do so in a way that allows them to compete effectively with other organisms or species undertaking a similar process (1). Consequently, a resource is any consumable factor that allows an organism to maintain itself and/or reproduce. In the case of plants, there may be no more than 30 resources required for life, but over 300,000 species of terrestrial plants. Some of these resources are essential, others are substitutable, and some are antagonistic in that the combination of two or more resources negates the effect of one resource or perhaps causes mortality when the use of only one resource would not. Essential resources include elements such as carbon, phosphorus, nitrogen, oxygen and others elements, but could also include non-elemental items such as light and a safe place to live from predators or competitors. Where species are competing for a single resource, and do not directly interact, then the species able to maintain itself and/or reproduce at the lowest resource level will eventually win the competition. Species sharing resources can coexist because a competitive equilibrium is never reached due to disturbances such as predation, or because organisms use multiple resources such that each species uses the resource, that limits itself most, faster than its competitors.

Early studies of resource use suggested that an organism's production was often limited by a single resource and that only the addition of this limiting resource would increase the organism's production. In more recent years, however, it has become evident that organisms have greater flexibility in their use of resources and can often substitute resources within limits. In fact, some ecologists believe that some organisms can store resources and adjust their allocation of resources such that all resources, rather than a single resource, are equally limiting. Adjustments include short-term physiological acclimation or long-term genetic adaptations. Storage of resources will allow species to acquire resources (that might otherwise be acquired by competitors) that can be used later when the resource is in short supply. The complexity and interaction between different biochemical pathways within organisms make it possible to create the needed resources from other chemical building blocks.

T.R. McClanahan

References

1.   Tilman, D. 1982. *Resource Competition and Community Structure*. Princeton: Princeton University Press

productivity, and may result in migration of larger animals such as jacks, sailfish, and tuna at the top of the food chain.

Mozambique, at the southern end of the tropical monsoon influence, is affected by both monsoonal and the temperate south African climate. The interaction of these winds on this boundary creates wind gyres that stir up seawater and create small upwelling areas. The combination of this upwelling and major discharges from the Zambezi river creates a productive area for fisheries in the Sofala, Maputo, and Beira Bays.

Many aspects of the monsoonal cycle are not well understood and further research is certain to produce some surprises and changes in our understanding. Further these patterns may be consistent within years, but the magnitude of the variation may change greatly between years and is predicted to increase with global warming. How between-year variation changes with global weather patterns such as the El-Nino oscillation and sunspot cycles requires more attention (3). Of particular interest is how these seasonal cycles affect fish production and migrations. How do fish cope with periods of reduced food abundance? Do large-bodied species follow migration paths (like the wildebeest) determined by the changing productivity of the ocean? If so, how do they know where to go and what guides their movements? How variable are these patterns between years compared to within-year variation? If humans are to use the fisheries of this region in an intelligent way, we must understand the seasonal patterns and their annual variability.

Production Ecology

Productivity in oceanic areas is the result of interaction between physical factors such as water movement and sunlight, nutrient availability, and biological factors such as the types and abundance of various species (Fig. 3.4). The combination and interaction of these factors creates the ecological food web that is the basis for fisheries production. In oceanic plankton-based ecosystems phytoplankton (microscopic algae) need to overcome problems of sinking of themselves and their nutrients. Consequently, biological production can often be explained by the source and concentration of nutrients. In coastal waters nutrients come from 1) upwelling or mixing of the water column, 2) fixation from the air (such as nitrogen fixation and carbon dioxide acquisition), and 3) terrestrial runoff. Many chemicals are necessary for biological production, but some chemicals are more limiting than others and nitrogen appears to be one of the chemicals frequently most limiting in coastal environments (11,25). In oceanic areas off East Africa nitrogen-fixation is likely to be the most important source of nitrogen (see above) except in upwelling areas and in coastal areas where nitrogen can come from terrestrial sources such as river discharge. Because nitrogen comes from different sources there can be productivity peaks in different areas at different times.

Terrestrial runoff and associated nutrients will experience peaks during the beginning of the SE monsoon when rainfall and river discharge reach their maximum. A study of zooplankton in Kenyan creeks found zooplankton peaks 2 to 4 weeks after rainfall peaks but generally greater abundance during the NE monsoon (14,23). Greater plankton abundance in the NE monsoon may result from water in creeks mixing with oceanic water. Consequently, patterns in creeks may reflect the interaction of mixing between oceanic and creek waters and their nutrients and productivity. River discharge areas such as the Tana and Sabaki Rivers have long been known as areas of high productivity and fisheries yield (18,19) and nutrient inputs from land must contribute to this productivity. Nonetheless, it should be appreciated that, relative to the volume of ocean currents, these river discharge volumes are small, and may not be as important as oceanic processes such as nitrogen fixation and mixing in deeper water and on the larger scale of the

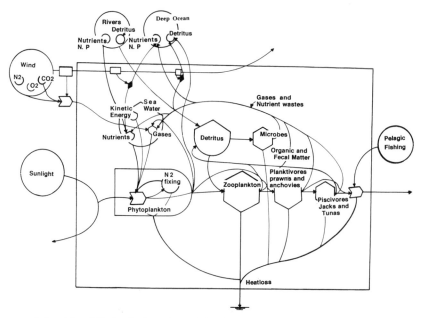

Figure 3.4. Simplified plankton-based food web (see Box 4.1 for explanation of symbols).

continental shelf. There is some positive association between Kenyan fish landings and river discharges 3 to 4 years previously that may suggest an increase in the reproductive success or growth of fishes with increasing land-derived nutrients (15). The mechanisms of this relationship should, however, be subjected to further scientific scrutiny.

Few studies have been undertaken to determine the environmental factors that affect fish production in coastal waters (21). Fisheries data from Kenya, do not include the numbers of fishermen or boats used in the catch. Consequently, it is difficult to determine the various factors of fishing effort, fish production, and various environmental parameters in determining productivity and subsequent catch. It is clear from data collected within years that, in Kenya, fish catch is seasonal as is fish reproduction (Fig. 3.5a,b). Both catch and reproduction are highest during the sunny and warm NE monsoon. Factors that can effect fish catch on a seasonal basis include 1) changing effort of the fishermen who may not fish as often or as extensively in the SE monsoon, 2) fish migrating to more productive areas (such as Somalia) during the SE monsoon, and 3) decreased fish population density due to a deeper thermocline, and 4) lowered fish movement, feeding activity and increased starvation-related mortality during the cooler SE monsoon. All of these factors may play a role in observed catch patterns. Surveys have shown that catch drops during the SE monsoon even when effort was kept constant (13,21) or the catch per unit effort (CPUE) was calculated, suggesting environmental factors may more important than fishermen effort.

In Kenya and Tanzania fish reproductive activity is also most common during the NE monsoon. Some species, however, breed throughout the year, while a few species like the speckled parrotfish (*Leptoscarus vaigensis*), which inhabits seagrass beds, appear to reproduce during the SE monsoon (22,24,29). The reason for high

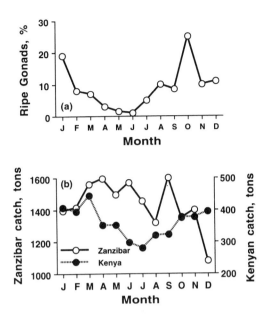

Figure 3.5. (a) seasonal pattern of fish reproduction and (b) fish catch (22).

reproduction during the NE monsoon may be warm water temperatures during the NE monsoons, but it should be appreciated that the higher abundance of phytoplankton and zooplankton during this time may also improve the chances of juvenile fish survival. At present, most reproduction studies have been descriptive and further experimental work is required to determine the importance of various factors such as food availability, for both adults and juveniles, and other environmental cues (such as light, temperature, and plankton concentrations) on reproductive patterns.

Fisheries Production and Management

Fisheries management is difficult throughout the world due to the difficulties in predicting fish abundance and the difficulty of coordinating fishermen, but is further complicated by the diversity of the catch in tropical areas and the lack of alternative employment opportunities. Further, the basic assumptions of fisheries management are very tenuous due to complicated environmental interactions and the cultural beliefs of fishermen. Nonetheless, an understanding of the basic theory of fisheries management can lend insights into the objectives of "rational management", its shortcomings, and alternative methods for management. The objective of most fisheries management plan is usually to maximize either the gross yield of a desired fish (Maximum Sustained Yield = MSY) or the "profits" (Maximum Economic Yield = MEY; 5,17) of this yield once the "costs" of the catch have been calculated (Fig. 3.6). Fisheries biologists typically focus on the MSY concept while fisheries economists find MEY the most useful concept as it gives an indication of the human effort and costs required to access fish. Maximizing employment is also a major consideration in many fisheries because of the political advantage of this position (5,17). Unfortunately, social and political objectives of free-market economics and concepts of personal freedom only rarely result in catches that improve the livelihood of individual fishermen (10,17).

The economic cost of fishing is important in determining the feasibility and profitability of catching fish. Costs are what economists call opportunity costs or choosing among the activities with the highest profits among competing options. This can be illustrated with the conceptual model in Figure 3.6 which presents 3 different lines for the cost of fishing - high, medium, and low costs. In the high cost situation it is not economically profitable to fish at all and fishing will only be a recreational activity. This might occur when the equipment required to fish the resource is expensive and/or the biological productivity of the fisheries is low. Additionally, if the labor force can make more money or acquire more resources from other employment opportunities than there will be insufficient labor to undertake fishing due to competition with better employment opportunities. This situation is likely to occur in developed countries but less likely in poor countries without other lucrative employment opportunities. Most of the people reading this book have high opportunity costs (better paid options because of their education) relative to fishermen and will therefore not consider fishing as an employment opportunity - although they might enjoy fishing or studying fish as a form of recreation. In the case of medium opportunity costs, the MEY is close to the MSY and the fisheries should be exploited close to its maximum level of biological productivity (MSY). In the case of low costs the fisheries are often exploited beyond its biological capacity. This is common in nearshore fisheries where poor people can fish with little investment in boats and gear, and may fish whenever there are no better options such as planting and harvesting crops, or working in the tourist trade.

In many cases competition among fishermen in an open-access fisheries may drive the fisheries to a bioeconomic equilibrium where costs are equal to gains (Fig. 3.6). High cost often stops the fisheries from being overexploited as only fishing with a high return per unit of effort will maintain a profit. In the case of virgin fisheries, where a new species or area has been discovered, a small investment in effort will result in high yields. When a fishery is opened, fish are abundant and profits are high, this usually results in further investment in fishing effort and the capacity (such as boats and labor) for harvesting increases until the reduced catch per effort (such as reduced return on the investment) eventually reduces profits to the level where profits equal costs. Due to the hump-backed shape of the theoretical fish-production curve this bioeconomic equilibrium occurs at twice the effort of the maximum economic yield. At this high level of effort it is likely that the fisheries will collapse due to the high mortality of the fish stocks and changes in species interactions within the ecological food web that are detrimental to the preferred species (9,16). A recent theoretical model suggests that due to the complexity of species interactions most fisheries will collapse before the MSY is reached but it may take many years for this collapse to occur (16). Consequently, in areas where a fishery is well developed most management focuses on ways to reduce the effort or effectiveness of fishermen.

When a major investment in fishing capacity has been made it may be cheaper to continue harvesting even at a monetary loss in operating expenses than to allow the fishing capacity to remain idle. An idle ship can be a greater loss to the fishing industry than a ship that is harvesting fish but not making a profit above operating expenses. This problem is called overcapitalization and is a major fisheries

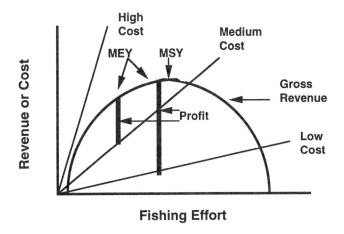

Figure 3.6. Theoretical model of the revenue, costs, and profits of fishing based on three estimates of the cost of fishing.

problem in many parts of the world but less so in the western Indian Ocean. In nearshore fisheries the major cost is the persons' time or labor as boats are often optional or small and without engines. Where the major investment in fishing is labor, fishing will be considered a useful expenditure of time when other alternatives (such as farming, business, and school) are not profitable or available to fishermen for reasons of culture or social status, and the state of the local economy. If there are people with few options for maintaining a living, fishermen will frequently overfish their resources if left unchecked (see Chapter 4). In the case of open-ocean pelagic fisheries the major investment may be fuel and boat maintenance and operating expenses, and labor and time are lower costs.

Shrimp Trawling

An example of shrimp trawling in Ungwana Bay in northern Kenya in waters less than 20 meters deep indicates that the high costs of trawling (such as fuel and capital investment) make fishing below the MSY a more profitable strategy for individual trawlers (Fig. 3.7; 28). This profit is also only possible if shrimp are sold on the international market ($7/kilogram in this analysis) where the price is well above the local price. At the local price it would not be economical to trawl for shrimp. The theoretical maximum sustained yield for this fishery is 350 tons per year found at around 1400 days of trawling while the MEY is somewhere between 600 and 900 days of trawling per year depending on the cost estimate. One trawler can fish somewhat more than 200 days per year making the optimal number of trawlers for Ungwana Bay around 6 for obtaining MSY but around 3 for MEY. In reality, catch and catch per unit effort (CPUE) rates are highly variable because of changing environmental conditions (3.7d). MSY predictions will change with changing environmental conditions such as the quantity of freshwater and nutrient discharges from estuaries.

Actual catch and effort statistics indicate the difficulties of predicting both the MSY and the level of fishing that will be reached by the fishing industry (Fig. 3.7c). The actual number of trawlers has varied from 5 to 20 between 1981 and 1990 with the maximum catch per trawler in 1986 when only 6 trawlers were working and around 400 tons of shrimp were caught. The increase in shrimp trawlers in recent years indicates overcapitalization and excessive effort in this fishery. In such a situation catches are likely to be highly variable between years depending on environmental conditions and recruitment of shrimp populations during each year. Excess trawlers may begin to utilize other less productive areas due to competition among trawlers - thereby increasing the total catch for the coastline which has been around 500 tons between 1987 and 1991. This yield is however becoming unprofitable and the use of petrol, labor, and capital investment may be better invested in other fisheries or economic activities in the region. Therefore, the MSY and MEY for the whole Kenyan and East African coastline needs to be determined, but environmental conditions for shrimp populations are frequently limited to areas with rivers discharging into the sea (such as Tana, Zambezi and Rufiji Rivers). Ungwana Bay is the largest single area along the Kenyan coast with appropriate conditions and other areas are likely to be less productive. This analysis indicates just a few of the difficulties in managing and harvesting the bountiful seas.

Regional Fisheries

Fisheries production is not uniform among or within regions. Therefore, to understand differences in actual and potential catches, the fishery is frequently divided into sectors that reflect the ecosystems and the way fish are harvested. The main categories of fisheries in East Africa are artisanal, pelagic, and demersal fisheries. Artisanal fisheries are the nearshore fisheries that are harvested by fishermen frequently using fishing gear and boat designs that have been in use for some time. In East Africa, many fishermen can travel by foot to fishing grounds, but boats are common where reefs are further offshore and include small canoes (hori and ngalawa) and sailboats (mashua or jahazi). These artisanal boats lack refrigeration facilities and, therefore, typically stay within a few kilometers of coastline where they can return their catch to market before it spoils. Exceptions are fish that can be salted or dried such as mullet and shark. Artisanal fishermen typically use and make basket-type traps made of palm fronds (madema or malema traps) baited with algae or sea urchin carcasses, various types of nets that were introduced into East Africa in the 1950's (gill and seining nets), and hook and line. These nearshore areas are frequently the most productive fisheries grounds, require the least investment, and, therefore, produce the highest yields per effort. Nonetheless, because of the narrowness or steepness of the continental shelf, in many areas, the total area of these fishing grounds may be small relative to the open ocean.

The more expansive open ocean, but less productive (per unit area), fisheries are divided into demersal and pelagic fisheries. Demersal ones are associated with the deeper seafloor while pelagic fisheries exploit the water column and are driven by planktonic productivity. Demersal and pelagic fisheries have lower fish biomass and production because of low light on the seafloor and low nutrients in the water column, which are constantly sinking to the seafloor, and high water movement that

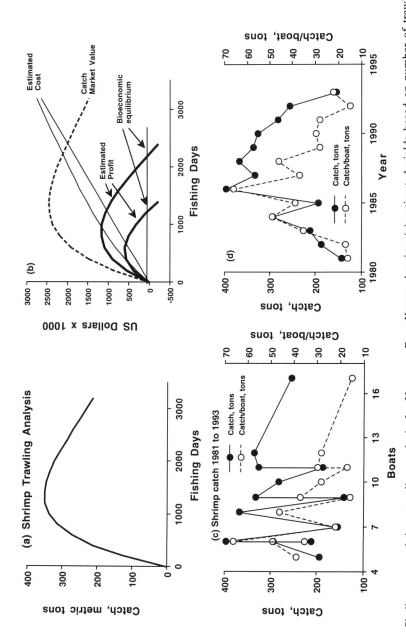

Figure 3.7. Shallow-water shrimp trawling analysis for Ungwana Bay, Kenya showing (a) estimated yields based on number of trawling days, (b) the revenue, costs, and profits of trawling (c) catch per unit effort and actual shrimp catch as a function of the number of registered boats between 1981 and 1993, and (d) catch per boat over the entire study period.

frequently causes plankton to sink below the optimal light levels for photosynthesis. These areas can be extensive and harvesting profitability often depends on the efficiency of harvesting sparsely populated fish or crustaceans, or knowing about migrations and areas of high seasonal or spatial concentrations of fish and crustaceans.

An analysis of the fisheries catch and potential for the western Indian Ocean shows that artisanal fisheries have both the top catch and potential for catch followed by small pelagics, such as sardines and anchovies, demersal fish, shallow-water shrimp (as described above), and lastly deep-water crustaceans (Fig. 3.8; 29) This review also suggests that most East African countries, with the exception of Somalia, Madagascar, and Mozambique, are near the upper limit of fisheries productivity based on Maximum Sustained Yield calculations (Fig. 3.8). Most nearshore areas are well fished and near their biological limits of production while areas offshore are more variable. Consequently, the success of offshore fishing depends on the economic viability of harvesting sparsely distributed resources and the willingness of fishermen to participate in mechanized or deep-water fishing. The estimated potential catch in Figure 3.8 is based on MSY and MEY may be below these MSY potentials due to high costs associated with deep-water fishing. Consequently, the potential for increased fish catch may be deceptive unless low cost and efficient harvesting methods agreeable to fishermen are developed.

Country Profiles

Somalia. Somalia has a long coastline (32,000 kilometers) and presently underutilized fisheries due to the pastoral traditions of most Somali people. The projected potential maximum catch for Somalia is 150,000 tons per year of which only 25,000 tons were being harvested prior to the civil war of 1991 (1). Large-scale industrial fishing, done by foreigners, was the main form of catch until development projects in the 1980s increased the efforts of small-scale artisanal fishermen - often by resettling migratory people. The artisanal catch had equaled or exceeded industrial offshore fishing prior to the civil war (Fig. 3.9). A large part of the industrial catch of foreign vessels is probably landed in other countries and may not be reported in the Somali statistics. Therefore, the actual additional potential catch of 150,000 for Somalia's economic use may be an overestimate unless all catch is landed in Somalia. Measurements of catch per unit effort (4 to 21 tons per boat per day) are high suggesting a high potential for increased fisheries yields regardless of present levels of effort and export (1). Efforts to involve nomadic pastoralists in fishing have increased the artisanal catch but few pastoralists are willing to leave their pastoralist roots entirely. Only about 20% of the landed fish is eaten in Somalia - most is exported to Europe.

Post-war fish catches are often exceptionally high due to the respite that fish populations experienced during the war. In a country suffering from famine, fish remains a potential and available source of nourishment. The same situation also applies in Eritrea, on the Red Sea, where 30 years of civil war caused a slow decay of investment in fisheries, and yields decayed to a fraction of prior ones. Since Independence in 1993, however, peace has prevailed and the catch is increasing rapidly such that conservation measures will soon be required. In these two cases

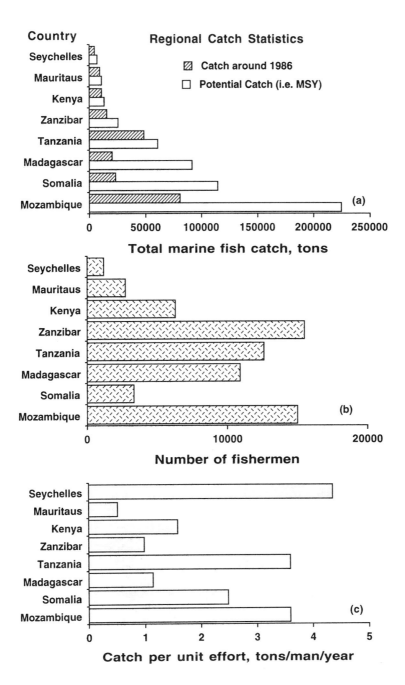

Figure 3.8. Regional statistics (around 1986) of (a) actual catch and potential catch (based on Maximum Sustained Yield estimates), (b) the estimated number of fishermen, and (c) the annual catch per fisher.

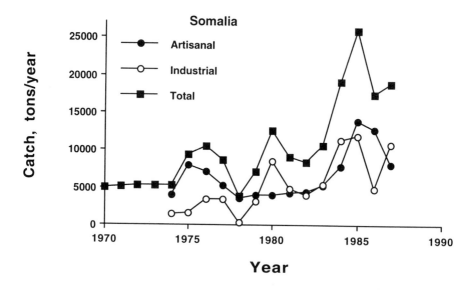

Figure 3.9. Total, industrial (such as trawlers), and artisanal marine fish catches from Somalia between 1971 and 1991. Data from Food and Agricultural Organization and (1).

and Mozambique, marine resources are abundant and fishing and tourism remain ways to improve human living standards. In areas without a history of war, fishery resource are scarce, and the need for managing fishing effort remains the central concern to high fish yields.

Kenya. Kenyan marine fisheries (reported around ~10,000 tons per year) have been overlooked compared to its freshwater fisheries (~180,000 tons) due to inaccurate reports of actual catch at the coast. A study intended to improve fish catch data collection methods found that actual catches are around twice those reported (4). The number of fishermen is also probably underestimated as estimates are based on the number of boats. Kenya's fringing reef is close to shore and many fishermen do not rely on boats to reach fishing grounds. Despite these inaccuracies most indicators suggest that Kenya's marine fisheries yield is at, near, or beyond MSY depending on the fishery. A recent evaluation suggested a total potential MSY catch of 37,000 tons of which 10,000 tons per year were already reported as harvested (20). Artisanal fishing of nearshore reefs is probably beyond the MSY (see Chapter 4), high-priced shrimp, lobsters, and crab are at their MSY and beyond their MEY (see above), while the less profitable offshore pelagic and demersal fish catches are probably below their MSY but, perhaps, not far from the MEY. Pelagic fish catch has vacillated around 600 to 1200 tons per year which is around 10% of the artisanal catch and composed largely of the planktivorous sardines (Clupeidae) and mullets (Mugilidae), and a variety of piscivores such as Jacks and Kingfish (Carangidae), and Tuna (Scombridae). Much of the pelagic catch is by-catch (fish caught while trawling for another species) from shrimp and lobster trawlers. Because of the questionable economic benefits of harvesting these pelagic species it would be unwise to project future Kenyan catches on an estimated MSY rather than the MEY of these species. Major fisheries trends over the past 25 years (Fig. 3.10) indicate that catches have vacillated, but steadily increased in the last decade. The dramatic drop in catch in 1972 is curious, and the importance of fishing effort, fisheries data collection, and environmental factors are not known. It does suggest, however, that unexpected changes can occur, and that catch predictions based on yields in the prior year can be unreliable. During this same 25-year period the coastal human population has more than doubled. Although catch has increased in the last decade to accommodate the demand it is clear that prior to 1972 the ratio of fish to humans was much higher than in recent years. Kenya has now reached the stage where management and control of its existing fisheries may be the only way to maintain present catches or to increase catches by a small fraction.

Tanzania. Like Kenya's, Tanzania's and Zanzibar's artisanal marine fisheries have approached their maximum levels of exploitation (12,20,27). Tanzania's economy is also highly dependent on fish catch which is a major food item, and the most important protein source for most coastal Tanzanians (25 to 40 kilograms/person/year). Ninety-five percent of the estimated 55,000 tons of catch comes from the estimated 16,000 artisanal fishermen and, with exception of shrimp, most seafood is consumed in the country. Traditional boat building is highly developed, as most reefs are not accessible by foot, and an estimated 7300 boats are used by the above fishermen in contrast to only 1800 boats used in Kenya.

An analysis of fisheries data from mainland Tanzania completed in 1986 suggested that the MSY of the coastline will be achieved at around 10,000 to 12,000 fishermen (Fig. 3.11b; 20). Nonetheless, despite an increase in fishermen since 1986 the catch has continued to rise (Fig. 3.11b). This may be due to the use of new fishing areas or changes in ecological production of the marine environment. Some potential for increased yields of demersal and pelagic fishes is feasible, particularly in the Mafia channel. The feasibility of obtaining these yields, however, depends on where the MEY lies in relationship to the MSY. Additionally, it may be that the ecological productivity of a region changes or oscillates such that a MSY for one time period may be different for another. Perhaps ecological production has increased in Tanzanian waters since 1986. Nonetheless, production is likely to drop again and fishermen will find it very difficult to sustain their livelihood and numbers when it does. Future "development" efforts need to focus on optimizing effort and conserving the resource for its long-term sustainability and to reduce effort when drops in production occur. Data on boats and fishermen suggest fluctuations, but that fishermen populations have doubled between 1975 and 1992 while boats have fluctuated or dropped from a high of 4500 vessels in 1978 to 3500 vessels in 1992. (Fig. 3.11b). This trend is an indication of lower investment or cost per fishermen, and may be a natural method of restricting effort when the fishery has already reached its maximum yield and profits are reduced.

Madagascar and Mozambique. In contrast to the above two countries, Mozambique and Madagascar remain areas with greater potential for increased marine fish catches (Fig. 3.8). Both areas have the potential for highly productive fisheries since actual catches in 1986 were less than half of estimated potential yields. Since the early 1980s catches in both countries have increased and

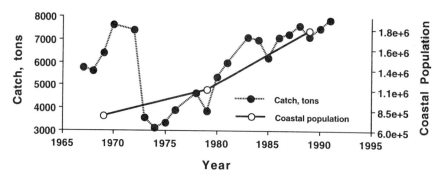

Figure 3.10. Total Kenyan marine catch and the coastal population (Lamu, Kilifi, Mombasa, and Kwale Districts) between 1971 and 1991. Data from Kenyan Fisheries Department and Kenyan Population Census.

it may not be long before these two countries have also reached their maximum levels of fish catch. In the case of Mozambique, the potential of the artisanal fisheries is estimated at 55,000 tons of which greater than 95% is already being caught. Mozambique's pelagic MSY is estimated at 130,000 tons and its demersal catch at 27,000 tons of which less than 10% was being caught in the mid 1980s. Madagascar, in contrast, has a potential artisanal yield of 65,000 tons. Between 1970 and 1983 only 12,000 tons were being caught annually, but since 1983 the catch has more than quadrupled to around 50,000 tons in 1988 - where it has remained constant until 1992. Again, Madagascar may have potential for increased yields in demersal and pelagic fishes, but the economic feasibility of using these fish resources requires further study.

   In general, it can be seen that in stable countries such as Kenya and Tanzania, fish, which are easily caught with low capital and technological investments, are already being exploited near their maximum sustainability by "simple" fishing technology. Therefore, technological improvements aimed at increasing catch of individual fisheries will not increase the total catch of the region unless they allow exploitation of a previously unfished stock. Presently, most unfished stocks require greater costs and risks that may not be acceptable or profitable for artisanal fishermen. In historically war-effected or politically unstable countries such as Eritrea, Somalia, and Mozambique the fisheries are under-utilized, but the response to political stability may be a dramatic increase in catches over a short period of time - largely due to increased fishing effort. Future increased catches in stable countries may require harvesting pelagic and demersal fish of lower biomass and productivity, although associated with greater costs. The potential yield of this region and particularly offshore fisheries should not be judged on MSY estimates (which is the common Food and Agricultural Organization practice), but rather MEY estimates. MEY estimates will produce more sobering potential yield estimates and will highlight the need for conservation in maintaining the sustainability of existing yields. Future technological innovation must focus on methods for reducing the costs of exploitation (such as increased fuel efficiency) of these pelagic and demersal resources.

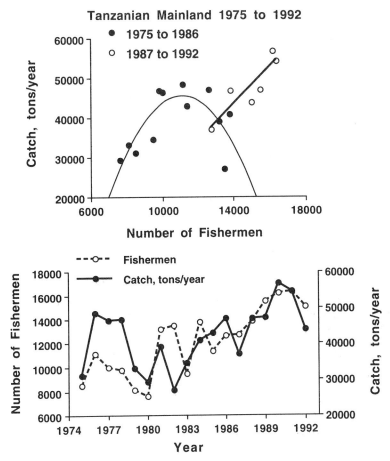

Figure 3.11. (a) mainland Tanzania marine fishermen and catch between 1975 and 1992 and (b) catch as a function of the total number of fishermen. Graph suggests a Maximum Sustained Yield curve with a maximum catch at around 11,000 fishermen for the period 1975 to 1986 but since 1986 catch has increased despite fishermen increasing to 16,000. Data from (21) and the Tanzanian Fisheries Department.

Managing Fisheries

As suggested in the above regional analysis future fisheries effort may require a greater focus on conservation than on exploitation to maintain or increase existing fisheries production. This will require a reversal in thought for many fishermen and fisheries organizations that have historically looked towards increased exploitation as the major focus of management. When a resource reaches its maximum level of production, the rules of the management game need to change in order to avoid afisheries collapse and resulting social problems (5,10). Conservation of marine fisheries often requires reducing the effort or effectiveness of fishermen or restricting access to the fishery. Methods for reducing effort include a variety of indigenous and government regulations.

Many of these regulations can be self-imposed by the fishermen themselves due to decreased profits while others may require government intervention. In most cases, restraint may require the realization that the fisheries are being over-exploited and that conservation is required to insure that catches remain high and stable to maintain an acceptable living standard. The point where conservation supersedes increased exploitation as a management priority may depend on individual fishermen and the opportunities that their society and environment provide. As catch per unit effort declines, fishermen can decide to pursue alternative activities that produce a greater return on their investment of time, money, and energy. Consequently, job alternatives (such as economic pluralism) are an important form of fisheries regulation. If there are few additional opportunities, or these opportunities are declining at the same rate or faster than fisheries yields, then fishermen will continue to harvest despite declining profits. When people are landless (or lack sufficient land), uneducated, and part of a poor rural economy, alternatives are often limited. Further, as human population increases, natural resources are reduced and opportunities for resource use are simultaneously reduced. Beyond the economic rationale, fishermen may simply continue to fish for cultural or traditional reasons even in the face of great economic adversity.

The above discussion suggests that the bioeconomic equilibrium varies with fishermen because individuals have different opportunity costs as well as fishing skills that will make profits and costs dependent on individual fishermen. Consequently, the theoretical bioeconomic equilibrium must be seen as an average derived from the population of fishermen and their various profits and costs. Nonetheless, despite this individual variation and the self-reliance of fishermen, fishermen are part of larger society which can make decisions and establish regulations that can greatly effect the costs and benefits of operation to individuals.

Most common efforts to control fishing effort or access include (17):

(1) Closing areas
(2) Closing seasons
(3) Restricting gear and technology
(4) Establishing quotas on the total catch
(5) Influencing effort through subsidies or taxes
(6) Limiting the number and type of fishermen who can enter the fishery
(7) Instituting various forms of property rights

All of these management techniques have been used by indigenous as well as cosmopolitan and industrialized societies. Closing areas permanently or seasonally includes marine parks and reserves but also includes temporary closure of breeding grounds or heavily exploited areas that require recovery periods. These techniques are useful if they somehow increase the total fisheries production of the region, but they may otherwise concentrate fishermen into smaller areas, causing increased overexploitation in unprotected areas. They may also cause discontinuous market supply that can disrupt an established market.

Gear restrictions include regulations on fish nets or trap mesh sizes, and the types of gear used. Fishing reserves in Kenya allow only "traditional" methods that includes madema traps and hook and line fishing. These restrictions are often useful in conservation, but they may also cause some fish groups to be underutilized due to a lack of appropriate fishing gear. Also, abrupt enforcement of gear restriction policies can cause a temporary reduction in yields when policies first go into effect. Gear restriction policies can also create additional expenses for fishermen by retiring old but still useful gear, and forcing the purchase of new gear.

Quota establishment is often a fairly complicated endeavor that requires a good knowledge of the resource and its variability, and depends on the manager's ability to enforce quota restrictions. It is therefore more likely to succeed in countries with a high level of fisheries expertise and enforcement ability for monitoring the resource and surveillance of the catch. Nonetheless, even with this expertise, catches have a certain amount of unpredictability which, at times, makes quotas difficult to justify. Quotas are most frequently used in the high-seas or single-species fisheries of industrialized nations.

Fishing can be encouraged or discouraged by a subsidy or a tax, as appropriate. When fish are under exploited, subsidy is appropriate, while taxation is appropriate for overexploited fisheries. Most developing countries find it difficult to resist the temptation to seek outside assistance in order to increase exploitation. However, if the fishery is already near or at the MSY or MEY then subsidies can have adverse consequences by causing overcapitalization, further resource depletion, and subsequent social problems. Subsidies and foreign aid must be used only where lack of indigenous capital prevents exploitation to maximum capacity. Taxes have the effect of increasing the cost of fishing and therefore may be used in low-cost fisheries where there is potential for overexploitation. In principle, tax money can be used to augment fish production techniques such as habitat improvement, the development of fish hatcheries, or other applied fisheries research. Taxation is often hard to legislate and enforce and can be an unrealistic burden for poor fishermen who have fished their resource to the bioeconomic equilibrium. Therefore, taxation should be aimed at 1) fish consumers, in the hope that this taxation will reduce the demand for fish products, or 2) at equipment, such as boats and fishing gear, to prevent overcapitalization.

Limited entry and property rights are perhaps the most frequent restrictions used by indigenous people, but have become less common due to changing cultural and political ideologies associated with westernization and foreign religions. Western views of individual freedom have assumed that free access to common property would result in "the tragedy of the commons" (10) or overexploitation of resources to a bioeconomic equilibrium in the absence of individual property rights. In contrast, most indigenous cultures had a variety of social mechanisms for excluding outsiders as well as ways to restrict members from overfishing. Although these mechanisms may work in small homogeneous ethnic groups, the inequality that arises from these "political decisions" can be a source of conflict particularly in diverse multi-ethnic nations. Additionally, restrictions to a certain number of vessels or people does not necessarily reduce effort if vessels become larger and people become more effective through improved technology. Consequently, entry and property rights may also require simultaneous gear and effort restrictions. The recent initiation of the Exclusive Economic Zone (EEZ), or the 200 mile exclusive right to marine resources, is a form of property right that nations may agree to (in stark contrast to the "Freedom of the Seas" doctrine), but enforcement by poorer nations remains problematic.

Adaptive Fisheries Management

Previously presented data on shrimp trawling and Tanzanian marine fisheries indicate that the MSY fisheries model has both some predictive ability and some troubles in predicting fish catches. Further, it is clear from the above discussion that no single method of fisheries regulation is without problems. Because of the human-environment interactive nature of the fishing problem, successful management cannot simply focus on biological arguments of productivity, but must also consider human culture and organization, and find management strategies that are acceptable to people using the resource. Successful management is likely to employ a mixture of strategies and should involve fishermen in decision-making and regulation by giving them the information that will result in appropriate conservation, management and high yields.

In order to cope with the uncertainty of fisheries some fisheries biologists suggest that an experimental management approach is likely to have the greatest resilience to changing conditions. Experimental management requires establishing areas where fisheries activities are regulated and sufficient data collected on characteristics of the fisheries yields in order to distinguish changes caused by fisheries regulations. By comparing different management methods and adjusting them over time the fisheries can accumulate knowledge that will allow fishers to adapt to changing conditions. Ecological models and field studies that monitor ecological parameters and populations of species can further understanding of the resource and suggest ways to alleviate unsustainable use. It may be most beneficial to have marine areas zoned under different resource uses in order to get the maximum sustained use of marine resources and to avoid resource user conflict (see Chapter 5 for a further discussion of zoning). Clearly, this experimental approach requires expenditures on information gathering, training and human organization.

Pollution and Habitat Changes

The preceding analysis focuses on humans as predators, but our interaction with the marine environment is not restricted to predation. In many cases we can change significantly the structure and chemical composition of marine ecosystems with both positive and negative effects on the productivity, aesthetics, and fisheries production of these ecosystems. Chemical composition of the water can be effected by human waste from a number of activities (6,8,25). Wastes include 1) biologically degradable wastes such as sewage, fertilizers, oil, and food processing wastes, 2) dissipating wastes such as heat, acids and alkalis, and cyanide, 3) particulates such as clay, silt, and ash, and 4) conservative wastes, that are neither dissipated nor degraded by bacteria, such as heavy metals (mercury and lead), halogenated hydrocarbons (DDT and PCBs (polychlorinated biphenyls)), and radioactivity. These wastes can enter the sea from 1) direct outfalls from coastal towns and industries, 2) river discharges, 3) ships, and 4) atmospheric inputs such as lead and hydrocarbons from vehicle exhausts. These inputs are frequently given the term "pollution" which can be defined as the introduction, by man, of substances or energy that results in deleterious effects such as hazards to human health, the aesthetic basis of tourism, or harvesting of resources such as fish. Many of these wastes would naturally enter the ocean without human influence and many have positive effect on production. Therefore, the issue is usually one of concentration of waste and its resultant effect on man and the marine ecosystems rather its simple existence in oceanic waters. The response of productivity to pollutants is either 1) a direct negative response from some initial stressful amount or more frequently 2) a hump-shaped response with productivity increasing for small amounts of pollution, but eventually being reduced to a low level at high concentrations.

Fortunately, the low level of industrialization in East Africa makes many forms of coastal pollution that damage industrialized countries such as the burning of coal and petrol, and release radioactive wastes, of low environmental concern at present. Nonetheless, some forms of pollution are of concern at least, for their potential harm in the future. These include 1) the release of organic wastes from coastal towns and tourist resorts, 2) sediment, fertilizers, pesticides, and organic matter from rivers, surface flow and groundwater seepage, and 3) oil pollution. Fortunately, with the exception of the poorly studied sediments and pesticides the above wastes are susceptible to biological degradation and can be treated with proper planning.

Eutrophication

Much organic matter and fertilizer waste is oxygen-demanding because the bacteria that degrade require large quantities of dissolved oxygen. Oxygen is required by marine plants, animals, and aerobic (oxygen-demanding) microorganisms for respiration. Seawater has a lower concentration of oxygen than air and a high density (1.026 grams/cm^3). Consequently, marine organisms require more energy to pump water over gills ("breathe") which can make oxygen a limiting element for animal growth and survival. When oxygen is depleted by abundant oxygen-consuming organisms it can reduce the number of species that require high oxygen levels. Many marine organisms, such as fish, require high levels of oxygen and expend a significant proportion of their body energy pumping water over their gills to extract oxygen. When nutrients such as phosphorus, nitrogen and organic matter become abundant, bacteria and phytoplankton increase so that dissolved oxygen is reduced to low levels - particularly near dawn (Fig. 3.12). Mortality of oxygen-demanding organisms may lead to polluted areas devoid of higher animal forms. This can be particularly severe in warm tropical waters as the quantity of oxygen dissolved in seawater is reduced with increasing water temperature while oxygen requirements of most organisms increase. Consequently, tropical aquatic ecosystems may be less tolerant to these types of biodegradable pollutants than temperate ones. If organic and chemical wastes are sufficiently diluted there is less reason for concern. Dilution will naturally occur due to tidal and current mixing. This makes pollution a less likely occurrence in the open ocean, but a more likely hazard in poorly circulated bays, creeks, lagoons, and estuaries adjacent to areas of high human population, farming, or industrial activity. Fortunately, East Africa has a large tidal range (4 meters in Kenya) and strong oceanic currents that flush and dilute pollutants in lagoons and creeks. This, however, is not sufficient rationale to dispose of untreated sewage and other human wastes.

In most coastal areas, sewage is primarily treated by septic tanks where much of the organic matter and nutrients settle and are periodically removed and dumped. Remaining "dirty water" seeps into the ancient reef, sand or groundwater. This water is partially cleaned by bacterial degradation before finding its way into rivers or the ocean, but groundwater in coral rock typically has high nitrogen concentrations that can stimulate productivity of phytoplankton and benthic algae once released into

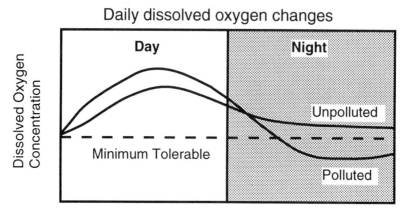

Figure 3.12. Graphs of hypothetical diurnal oxygen curves and the effect of oxygen-demanding pollution on these curves.

rivers, estuaries, and the ocean. The effect of high groundwater nutrient concentrations and its effect on coastal ecosystems in East African has been poorly studied. Dumping solid waste also causes problems when it is poorly regulated. In many third-world countries solid waste is used as fertilizer for crops, but it is sometimes dumped directly into nearby creeks and causes severe pollution. Spreading solid waste on pasture is perhaps its safest and most productive use.

Sewage and solid wastes contain bacteria, pathogens and viruses (such as those causing cholera and typhoid), and eggs of intestinal parasites that can infect humans if inadvertently swallowed with seawater, while bathing, or more commonly through eating contaminated seafood such as oysters and some types of fish. Additionally, pollution in coastal waters seems to be associated with an increased frequency of "red tides". Red tides are phytoplankton blooms that have the detrimental effect of 1) reducing seawater oxygen levels or 2) stimulating the growth of dinoflagellates toxic to fish and seabirds. Sometimes dinoflagellate neurotoxins are concentrated in important fisheries species sufficiently to kill people who eat them. Fortunately, red tides have not been reported in the East African region.

Oil Pollution

Negligible amounts of oil are produced in East African coastal waters, but the Red Sea and western Indian Ocean are the world's largest shipping lanes for Middle Eastern oil. (Fig. 3.13). Oil spills have occurred from refineries in the harbors of Mombasa, Dar es Salaam and Matola, and could occur at Mogadishu. Most oil discharged into the sea is from seawater ballast tankers discharge on their return journey to the Middle East. An average tanker would have approximately 8 tons of oil leftover or stuck to its hull after it discharged oil at a port. This oil would eventually be dumped in the ocean before refilling. An estimate of the oil discharge for the region in 1981 suggested that 33,000 metric tons were released from the 5200 large and medium size tankers traveling the Indian Ocean per year. This was around 5% of the total oil discharged into the sea by man that year. Since 1981 oil exports from the Middle East have increased by nearly a factor of ten. Despite this increased production and associated transport the discharge of oil ballast has been

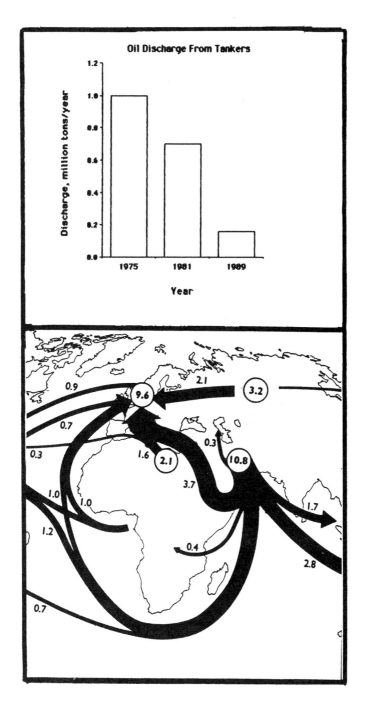

Figure 3.13. Major oil routes at sea (in millions of barrels per day) in 1987 and the total estimated amount of oil discharge into the sea by tankers over the past 15 years (6,8).

reduced nearly to 10% of mid 1970 levels. This is due to different tanker designs and cleaning methods which retrieve the left-over ballast oil. It should also be appreciated that phytoplankton produces dispersed and low molecular weight hydrocarbons that, although perhaps not as toxic as oil products such as aromatic hydrocarbons, are exceed by 10,000 times the tonnage of human-induced oil spills. Even atmospheric fall-out of petroleum hydrocarbons from natural and anthropogenic sources is estimated at 50 to 2000 times the rate due to oil spills. Nonetheless, it is often the concentration of oil spills and unsightly "tarballs" found on beaches that most upset tourists. Tarball abundance should be considerably lower in the last decade because of the lower discharge of oil.

Despite the above statistics, the effect of oil spills close to shore can be dramatic on the short term. Most studies indicate that, with the exception of mangroves, that have air-breathing roots that can be smothered with oil, many marine plants and algae are not greatly effected by oil. Ecological and species composition changes in primary producers is often subtle, and, in some cases, growth increases because of nutrient inputs from biological degradation of oil. Herbivorous copepods and snails, and marine birds and mammals are most affected, but populations can recover a few years after the spill. Unfortunately, attempts to clean up oil spills are often more damaging than the spills themselves and, therefore, containment, nontoxic dispersants, and "oil-eating" bacteria are usually best for oil spills. East African harbors and refineries should keep and maintain stocks of equipment to control future harbor spills.

Habitat Changes

The previous analysis suggests that a number of environmental challenges remain for research and management, but that management of fisheries is among the most pressing concerns in view of the rising population, starvation, and reduced warfare in the region. Fish have a natural capacity to recover from high mortality as long as there is appropriate habitat and refugia, and that extinctions of important species (such as keystone species) do not occur. Consequently, over the long term it remains important to protect fish habitat, to maintain refugia or unexploited populations and prevent species extinctions. Both habitat and species losses have occurred over geologic and recent time due to changing climatic conditions and destructive overfishing (see Chapter 4). Ecosystems can change abruptly in species composition and productivity with changes in fish stock detrimental to fisheries yields. Many theoretical and field studies support this idea (9,16). In view of the variability and unpredictability of fish stocks, the reduced human population to fish catch ratio in Kenya and Tanzania over the last 2 decades, and the potential role of habitat and climate in affecting fish yields it is desirable to protect fish habitat and create refugia. This can be achieved by creating marine parks and reserves, eliminating destructive fishing methods, and undertaking further studies on the influences of climate, ecosystem dynamics, and fisheries yields on populations of marine organisms.

References

1. Ahmed, O.H. 1988. The fisheries resources of Somalia. In *Ecology and Bioproductivity of the Marine Coastal Waters of East Africa*, ed. Mainoya, J. R., pp. 102-106. Dar es Salaam: University of Dar-es-Salaam Press

2. Bryceson, I. 1982. Seasonality of oceanographic conditions and phytoplankton in Dar es Salaam waters. *University Science Journal* (Dar University). 8: 66-76

3. Cadet, D.L. 1985. The Southern Oscillation over the Indian Ocean. *Journal Climatology* 5: 189-212

4. Carrara, G., Coppola, R.S. 1985. *Results of the First Year of Implementation of the Kenya Catch Assessment Survey.* FAO/UNDP (TA), (3006)

5. Clark, C.W. 1985. *Bioeconomic Modelling and Fisheries Management.* New York: A Wiley - Interscience Publication

6. Clark, R.B. 1992. *Marine Pollution.* Clarendon Press, Oxford

7. Crame, J.A. 1981. Ecological stratification in the Pleistocene coral reefs of the Kenya Coast. *Palaeontology.* 24: 609 - 646

8. Ferrari, J. 1983. Oil in troubled waters. *Ambio* 12: 354-357

9. Gulland, J.A. 1988. *Fish Population Dynamics.* John Wiley & Sons Chichester

10. Hardin, G. 1968. The tragedy of the commons. *Science* 162: 1243 - 1248

11. Hecky, R.E., Kilham, P. 1988. Nutrient limitations of phytoplankton in freshwater and marine environments: a review of recent evidence on the effects of enrichment. *Limnology and Oceanography.* 33: 796-822

12. Jiddawi, N.S., Pandu, V. M. 1988. Summary of fisheries resources information for Zanzibar. In *Proceedings of the Workshop on the Assessment of the Fishery Resources in the Southwest Indian Ocean.* eds. Sanders, M.J., Venema S.C. FAO/UNDP: RAF/79/065/WP/41/88/E

13. Kamanyi, J.R. 1975. Biological observations of Indian mackerel *Rastelliger kanagurta* (Cuvier) 1816 (Pisces: Scombridae) from East African waters. *African Journal of Tropical Hydrobiology and Fisheries.* 4: 61-78

14. Kimaro, M.M. 1986. The composition and abundance of near-surface zooplankton in Tudor Creek, Mombasa. M.Sc Thesis, University of Nairobi

15. McClanahan, T.R. 1988. Seasonality in East Africa's coastal waters. *Marine Ecology Progress Series* 44: 191-199

16. McClanahan, T.R. 1992. Resource utilization, competition and predation: a model and example from coral reef grazers. *Ecological Modelling.* 61: 195-215

17. McGoodwin, J.R. 1990. *Crisis in the World's Fisheries: People, Problems, and Policies.* California: Stanford University Press, Stanford

18. Morgans, J.F.C. 1959. The North Kenya banks. *Nature* 184: 259-260

19. Morgans, J.F.C. 1962. Ecological aspects of demersal tropical fishes off East Africa. *Nature* 193: 86-87

20. Ngoile, M.A.K., Bwathondi, P.O.J., Makwaia, E.S. 1988. Trends in the exploitation of marine fisheries resources in Tanzania. In *Ecology and Bioproductivity of the Marine and Coastal Waters of East Africa*, ed. Mainoya, J.R., pp. 93-100. Dar es Salaam: University of Dar es Salaam

21. Nhwani, L.B. 1980. A gillnetting survey of an inshore fish stock off Mombasa coast. *University Science Journal of Dar es Salaam* 6: 29-45

22. Nzioka, R.M. 1979. Observations on the spawning seasons of East African reef fishes. *Journal of Fish Biology* 14: 329-342

23. Okera, W. 1974. The zooplankton of the inshore waters of Dar es Salaam (Tanzania, S.E. Africa) with observation on reactions to artificial light. *Marine Biology.* 26: 13-25

24. Rubindamayugi, M.S.T. 1983. The biology of the blue-speckled parrotfish *Leptoscarus vaigiensis* (Family: Scaridae) occurring along the coast of Tanzania. M.Sc. Thesis, University of Dar es Salaam, Dar es Salaam

25. Ryther, J.H., Dustan, W.M. 1971. Nitrogen, phosphorus, and eutrophication in the coastal marine environment. *Science* 209: 597-600

26. Sanders, M.J., Gichere, S.G., Nzioka, R.M. 1991. *Report of Kenya Marine Fisheries Sub-sector Study.* FAO RAF/87/008/DR/ 65/E

27. Sanders, M.J., Sparre, P., Venema, S.C. 1988. *Proceedings of the Workshop on the Assessment of the Fishery Resources in the Southwest Indian Ocean.* FAO/UNDP: RAF/79/065/WP/41/88/E

28. Smith, S.L., Codispoti, L.A. 1980. Southwest monsoon of 1979: chemical and biological response of Somali coastal waters. *Science* 209: 597-600

29. Talbot, F.H. 1960. Notes on the biology of the Lutjanidae (Pisces) of the East African coast, with special reference to L. bohar (Forskal). *Annals of the South African Museum* 45: 549-573

30. Walters, C. 1986. *Adaptive Management of Renewable Resources.* New York: MacMillan

Chapter 4

Coral Reefs and Nearshore Fisheries

T. R. McClanahan & D.O. Obura

Due to high numbers of species, productivity, and complexity of species interactions some ecologists refer to coral reefs as the "ultimate ecosystem". This complexity and diversity fascinates naturalists and scientists, but this same complexity also makes a comprehensive understanding and subsequent management of coral reefs difficult. Nonetheless, recent efforts since the development of SCUBA diving have made the coral reef more accessible to scientists, naturalists, and tourists, and our interest and understanding has increased greatly.

The ecological view of coral reefs has changed greatly since the first observations and measurements of naturalists and scientists. Early naturalists argued that, due to the low levels of nutrients in seawater passing over coral reefs, that reefs would exhibit low productivity and be the equivalent of underwater deserts. But, the first measurements to test this idea in 1949 found the exact opposite (38). This early study and subsequent measurements have confirmed that coral reefs are among the earth's most productive ecosystems and near the theoretical limits of primary productivity.

Coral reefs are also one of the earth's oldest ecosystems, as determined from the geologic record (30) and this led early observers to suggest that the high diversity of reefs resulted from the reef's great age and long stability. Stability and reef age were hypothesized to result in a fine partitioning of coral reef species resources (see Box 4.2). Subsequent work has shown however that a number of large and small-scale species extinctions have occurred over the earth's geologic record and in various regions. Similarly, a great deal of variability in location and time is found for some populations of coral-reef species, and in their use of resources. The coral reef has, therefore, been an enigmatic ecosystem, full of surprises, and challenges in scientific understanding and challenging for management. The following discussion will review some of our current knowledge of this ecosystem and the progress and problems associated with its management focusing on the coral reefs of East Africa.

Geography

Although hard corals (Scleractinia) are found throughout the earth's oceans, calcium carbonate reefs, formed by corals and algae, are generally restricted to tropical latitudes (<25° N and S of the equator). Appropriate conditions of temperature, light, water, and nutrients are most frequently found in the tropics. Yet it should be appreciated that the distribution and diversity of coral reefs throughout the tropics is by no means uniform (see Fig. 4.1). Where conditions of warm and clear water are not found, coral reefs are absent or poorly developed. These suboptimal areas include areas close to river discharges and upwelling areas (Chapter 3) which, in East Africa, are most common in northern Kenya and Somalia and sporadically along the coast of Tanzania and Mozambique. Consequently, the most developed reefs in East Africa are found in southern Kenya and Tanzania. These areas are influenced by the Indian Ocean's equatorial current and its low nutrients and warm water. As this current is diverted north and south along the coastline it receives pulses of riverine inputs (Chapter 3, Fig. 4.1) which, if large enough, will reduce light penetration to the seabottom and reduce the growth of benthic algae and corals.

The East African coastline was formed largely by the deposition of calcium carbonate by corals. Towns such as Mombasa and Dar es Salaam are built on ancient reefs (~130,000 years old) formed during a period when polar ice caps were smaller and sea levels higher than today (Fig. 4.2(a); 5,6,7,9). Polar ice caps in the northern hemisphere change their size with changes in northern-hemisphere summer solar insolation influenced by small changes in the earth's orbit around the sun. Ice cap changes appear to follow predictable cycles driven by cyclical variations in the eccentricity of earth's orbit and the tilt and precession of the earth's axis (7). Sea level has changed constantly and often abruptly over the earth's geological history and the result is a tiered coastline (9). Three tiers have been formed during the last 200,000 years (9). The top, and presently exposed tier (~5 meters above present sea level, see Chapter 5), was formed during the last interglacier 130,000 year ago. The present-day reef is the second tier, and a third tier lies 11 to 18 meters beneath the present sea level and was formed during the peak of glacial expansion around 80,000 to 117,000 years ago. Each of these tiers experienced times of growth but today only the present reef is actively growing while the others are eroding. The bottom tiers have largely eroded to the point where their distribution along the coastline is patchy. The breakdown and erosion of these reefs is the source of calcium carbonate sand that forms beaches and reef lagoons - a major source of attraction for East Africa's numerous tourists. The upper-most exposed tier is the source of coral stones, limestone and cement used for building.

Viewed from the air, the coastline is characterized by rocky promontories or cliffs often bisected by deep creeks and stretches of coral sand beaches. Beaches represent areas where ancient reefs have largely eroded (or were not well formed) and coral sand is moved, deposited and stored on beaches by waves, currents, and tides. These beaches change their shape and the amount of stored sand based, largely, on the amount of wave energy impacting the beach. This is affected by tidal (or lunar), seasonal, and interannual changes in wave and current energy (see Chapter 3). Rocky cliffs are often undercut by present-day erosion - also forced by waves and water movement. Creeks which bisect this ancient rocky reef (such as Kilifi Creek) were formed by ancient rivers which discharged greater quantities of water in times past - when coastal rainfall was more abundant. Today these creeks may also discharge local rainfall from coastal hills, but the quantities of discharge are small compared to the Rufiji, Tanga, Sabaki and Tana Rivers. The outlets of these larger rivers are typically surrounded by depositional sands of terrestrial origin which, when deposited on land, form large dunes shaped by monsoonal winds.

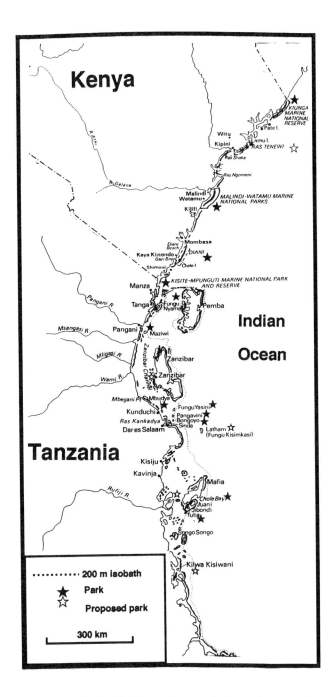

Figure 4.1. Coral reef distribution and coastal rivers from southern Somalia to southern Tanzania. Marine Parks and Reserves are shown. Reef locations show only the general locations and not actual reefs. Redrawn from (40).

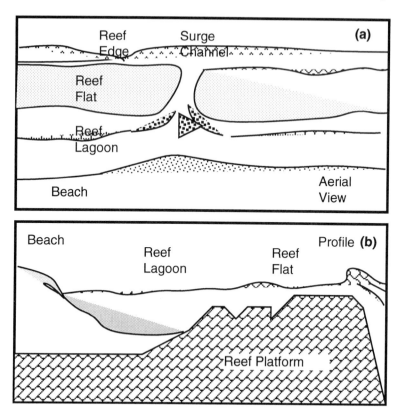

Figure 4.2. (a) aerial and (b) profile view of a hypothetical fringing reef along Kenya's southern coastline.

Coral reefs are uncommon or absent at the outlet of large freshwater discharges areas due to the lack of rocky substrate for corals to attach themselves to and grow, and reduced light on the seafloor. Corals are often abundant, however, near the outlets of smaller creeks suggesting that smaller discharge quantities may not be detrimental to coral survival, and perhaps some nutrient increases are beneficial to corals. Soft corals (Alcyonacea), that require less light and more detritus, often thrive on hard substrate areas within creeks. There is probably a replacement of hard corals by soft corals as one moves closer to creek and river outlets. Expansive areas of living coral are, however, generally found below low-tide levels on seaward and leeward edges of the offshore fringing and patch reefs (Fig. 4.2(b)) common to East Africa.

Fringing and patch reefs are the two most common reef forms found in East Africa. The southern portion of Kenya's coastline from Msambweni, in the south, to Malindi, in the north, is the earth's largest continuous fringing reef. This fringing reef borders the coastline at a distance of a few meters to 2 kilometers from shore. Reefs are frequently divided into 3 zones: the reef crest (or edge), the reef flat, and the reef lagoon (Fig. 4.2(b)). The fore reef and reef crest are the most seaward sections and are exposed to the full force of waves and currents. During low tides the top of this fringing reef, named the reef flat, is exposed to air and has very little live coral cover. Algae, sponges and invertebrates such as snails and sea urchins are abundant on reef flats. Reef lagoons are less exposed to physical forces, but they may experience high water temperatures, salinities or nutrients due to reduced water circulation. Beneath low-tide water level corals and associated fish are more abundant.

The Kenyan fringing reef is only occasionally bisected - usually near creek outlets. The fringing reef probably formed in southern Kenya due to the low level of freshwater discharge. Coastal rainfall increases to the south, towards Tanzania, where patch reefs further offshore are most common. Frequent and larger discharge areas may cause reefs to form further from shore and may stop reef patches from joining into a continuous fringing reef. Nonetheless, the reef geology of Tanzania is more complicated than Kenya and has resulted in a more abundant and elaborate indigenous boat-building culture to access these fertile coral-reef fishing grounds. Reefs north of Malindi up to Somalia are also composed of scattered patch reefs that usually have lower coral cover and diversity than southern reefs.

Biogeography

Biogeography is the study of the global distribution of species and the factors that affect species distribution and diversity patterns on the regional or global scale. Species composition of the flora and fauna at a specific site is a mixture of historical factors that cause speciation over evolutionary time, chance events such as drifting of the earth's plates, and environmental and biological factors that maintain species over ecological time. Existing marine biogeographic theory is in a state of change with biogeographers trying to distinguish between various factors that create and maintain species diversity at the scale of oceans and ocean basins. Existing explanations for observed patterns include (2,36,39,41).

(1)   The **Origin-dispersal Hypothesis** states that areas of high species diversity represent areas of species origination and that species diversity decreases away from these centers.

(2)   The **Vicariance Hypothesis** stresses the importance of barriers to dispersal and isolation in creating new species.

(3)   The **Stress Hypothesis** maintains that environmental factors (such as cold temperatures) that increase mortality and species extinction will control the abundance of species over both geologic and ecological time.

(4)   The **Productivity Hypothesis** contends that high benthic or planktonic primary productivity will support large numbers of species and prevent them from going extinct

The above hypotheses may, in many cases, interact to produce the observed patterns of faunal diversity rather than being alternative or competing explanations.

Biogeographers have long known that the highest species diversity in the marine environment occurs in tropical latitudes. Within the tropics the central and western Pacific (Indonesia and Philippines) contains the highest diversity with diversity decreasing irregularly away from this center (41). The Origin-dispersal Hypotheses was the dominant explanation for this pattern until recently when the importance of isolation and therefore vicariance factors in creating new species was appreciated. Diversity in the Indian Ocean does not show a constant decline in species away from the central Pacific but rather sites of high diversity around southern India (Sri Lanka and Maldives), the northern Red Sea and East Africa including Madagascar (2,36,39).

There have been few comprehensive biogeographic studies in the western Indian Ocean and East Africa. A compilation of coral genera distribution suggests that East Africa, the northern Red Sea, and islands of southern India (Chagos, Maldives, Seychelles and Sri Lanka) have the highest coral diversity (Fig. 4.3(a); 2, 39). Originally, this diversity pattern was suggested to result from either 1) a center of species origin for the Indian Ocean or 2) as a dispersal corridor from the major center of origin in the central Pacific. Both explanations are variations on the Origin-dispersal hypothesis but no evidence was presented to suggest that species origination occurred in this region.

A study of damselfish (Pomacentridae) distribution and diversity patterns showed that East Africa has a few more species of damselfish than areas off of southern India (Fig. 4.3(b);1). Further, species in this high diversity belt from southern India to East Africa were composed of species with widespread distribution patterns. In contrast, peripheral areas such as Mauritius/Madagascar, the Red Sea, and the Arabian Gulf had the highest numbers of endemic species. This suggests that vicariance or isolation factors are important in species origination, but data conflict with the classic Origin-dispersal explanation as few endemic species were found in areas with the highest species diversity. Further research is needed to distinguish the importance of the various speciation and species maintenance mechanisms and their interaction in creating observed Indian Ocean diversity patterns.

On the smaller biogeographic scale of the African coastline faunal studies are scant. Some patterns however may be deduced from existing studies. Coral reef-associated fish and snails (Gastropoda) are less diverse in South Africa than Tanzania and southern Kenya. Species richness may however decrease towards the equator in southern Somalia. Studies of fish and marine snails indicate that northern Kenya has fewer species than southern Kenya and Tanzania (Fig. 4.3(c); 17,31,43). Northern Kenya and Somalia are influenced by major river discharges and upwelling areas which, although beneficial to planktonic ecosystems, may be an environmental stress (such as reduced light or cooler water temperatures) for bottom living tropical fish and snails. Upwelling areas in northern Somalia (between 8 to 12° N latitude) have surface water temperatures as low as 17° C that will restrict many coral reef species from inhabiting this region (see Chapter 3). Consequently, along the continental African coastline the peak of species diversity is probably found on the islands of Tanzania but the lack of extensive faunal studies from this area makes this statement tentative.

Historically, patterns of marine species diversity in East Africa have changed over geologic time influenced by changing sea levels and the associated environment (6,17). Studies of fossil corals and snails have been undertaken on ancient interglacial reefs formed ~130,000 years ago when sea levels were higher than present. Coral studies indicate that common and dominant genera during the last interglacial were similar to those seen today. The coral genus *Porites* dominated in shallow disturbed environments and formed the calcium carbonate structure that was colonized by other encrusting and branching coral species. This general pattern persists until today but is now influenced by human activities (28). Comparative

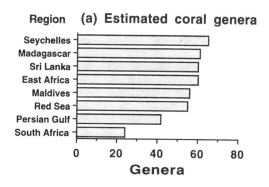

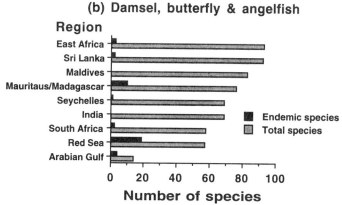

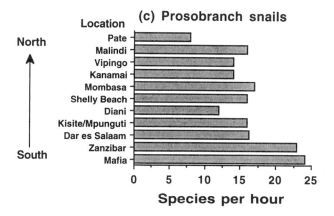

Figure 4.3. Patterns of species diversity in the western Indian Ocean (a and b) and along the East African coastline. (a) predicted numbers of coral genera and (b) abundance of common coral-reef inhabiting fishes in various regions. (c) numbers of species of snails (prosobranch gastropods) found at different sites along the coast from north on the left to south on the right.

Box 4.1.   Ecological Food Webs and Modelling Nature

The science of ecology has its historical roots in the field studies of American and British natural historians and evolutionary biologists who focused their research on species behavior and interactions between species (1). Nonetheless, many important contributors to ecological theory and the associated mathematics were schooled in other academic disciplines such as geology, chemistry, and physics. The research of these scholars made it clear that ecosystems are composed of both living and nonliving matter that are interrelated through the feeding, excretion and respiration of living organisms. Interactions between living and nonliving matter create food webs based on the feeding relationships of species in the ecosystem. The primary source of energy for most food webs is solar energy that is captured by plants through photosynthesis and stored as chemical bonds. The energy stored in plants is then captured by herbivores that are, in turn, fed on by other predators and parasites. Food webs typically form a hierarchy of energy in which the largest quantity of energy is the incoming physical energies (such as sunlight and water and air movement) followed by energy stored in plant matter. The total stored energy in the ecosystem decreases along the hierarchy from sunlight to plants to herbivores to carnivores. Ecosystems, however, are seldom linear chains of energy flow but are more web-like because predators choose their prey and can switch their choices dependent on prey availability and the energetic or nutrient benefits of the prey relative to the cost of capture. The predator's ability to choose prey gives the predator's population greater stability. Each predator has multiple prey species and each prey species has multiple predators.

   Natural historians like Charles Elton (his ideas originating from around 1920), intrigued by animals' use of resources and predator-prey relationships, diagrammed ecological food chains and webs. In Elton's diagrams, boxes represented species with related feeding habits and arrows showed the direction that prey or energy were passed in the predatory interaction. The concept of ecosystems as energy storage compartments and flows became recognized and formalized by the more physical science educated group of G.E. Hutchinson (around 1944) and his graduate students such as R. Lindeman and H.T. Odum. Odum modified the box and arrow symbols to distinguish between consumers (hexagons) and producers (bullets) and to further use the language to describe many different types of interacting systems (2). Odum argued that most groups of interactions (systems) consist of storages (tanks in his language), flows (lines) and their interactions (arrows in his language). Self-replicating systems (organisms) invest some energy or power (energy/time) to gain a larger amount of energy as prey to use for future power needs. Organisms, therefore, have the potential for exponential growth, but environmental or resource limitations can cause populations to fluctuate or stabilize around an equilibrium determined by their resources and their predators. To account for the heat losses of organisms and the total energy flow, Odum added a heat sink to his diagrams so that all the energy entering an ecosystem can be accounted for as heat generated by organisms through their metabolic activities. Examples of the different diagramming languages are used in this book because the different symbols reflect ecologists' preferences and education. The box and arrow language is frequently associated with population or net-production models where the units of interest are the number of individuals or the animals' gained production (3). The energy-circuit language is frequently used when the energy content or weights of organisms or ecosystem components are the focus of study and is often used by researchers interested in harvesting organisms.

T.R. McClanahan

References

1. Hagen, J.B. 1992. *An Entangled Bank: The Origins of Ecosystem Ecology.* New Brunswick: Rutgers University Press
2. Odum, H.T. 1983. *Systems Ecology: An Introduction.* New York: John Wiley & Sons
3. Christensen, V., Pauly, D. eds. 1993. *Trophic Models of Aquatic Ecosystems.* Manila: ICLARM

(a)

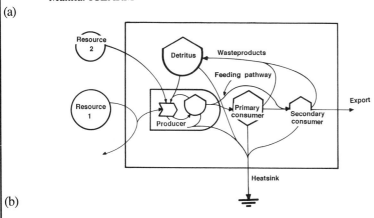

(b)

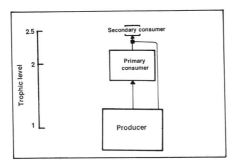

Figure 4.1.1. Two commonly used diagramming systems for depicting the feeding pathways in ecosystems. (a) the energy-circuit language organizes the energy or feeding hierarchy horizontally from left (most abundant) to right (least abundant), differentiates producers (bullet symbols) from consumers (hexagons) and symbol sizes are proportional to the total weight of the organisms on an area basis. Lines exiting the bottom represent heat losses or the heat sink of the ecosystem. (b) the box model system organizes feeding groups along a vertical feeding level axis and the size of the boxes are proportional to the logarithm of the total weights of the organisms. The total energy flow is rarely accounted for in this language and associated mathematics and, therefore, no heat loss lines are included.

coral diversity patterns between the Pleistocene and now are difficult due to incomplete sampling, but thorough studies of present-day and Pleistocene marine snails indicates that there were 45% more marine snail species during the last interglacial period than today (17). Perhaps higher sea levels, stronger currents, and warmer conditions enhanced the diversity of coral reef-associated species in East Africa.

The Food Web and its Diversity

Algae, corals, fish, snails, starfish, sea urchins and other invertebrates are among the most common animal groups on coral reefs and may number from just a few to 100s of species per group (Fig. 4.3 and 4.4). This high species diversity usually forces scientists to focus on particular taxa of which algae, corals, snails, sea urchins, starfish, and fish are among the most studied groups. Despite this diversity there are some species that share similar ecological functions. For example, grouping species into functional groups such as primary producers and grazers allows one to conceptualize the ecological state and processes of the reef in a simple manner. One can aggregate one's view of the reef at many different levels depending on the questions of interest and the desired level of understanding. For example, grazers of algae may include a diversity of groups such as damselfishes, parrotfishes, sea urchins, crustaceans and snails with different evolutionary histories, levels of abundance, and energetic characteristics (such as metabolic or respiration rates). Each of these groups may also affect ecological processes differently. Both corals and algae convert calcium ions and carbon dioxide gases dissolved in seawater into solid calcium carbonate at different rates (mg $CaCO_3$/hour/gram wet or dry weight of organism). Therefore, the abundance of the different groups will have different effects on the coral reef's calcium carbonate balance. This will, in turn, affect the structure and physical complexity of the reef and the subsequent fauna that occupy the reef.

A diagram showing some of the main functional groups and their interactions is presented in Figure 4.5. Water motion, sunlight and nutrients are the most important physical and chemical sources for primary production. The main primary producers are benthic algae, coral, and to a lesser extent, phytoplankton. Many species of algae fix nitrogen from nitrogen gas dissolved in seawater (see Chapter 3), and nitrogen is frequently exported from coral reefs (41). Secondary consumers include fish and invertebrate grazers. Sea urchins are the dominant invertebrate and parrotfish (Scaridae), rabbitfish (Siganidae), and surgeonfish (Acanthuridae) are the dominant fish grazers (Fig. 4.4). Zooplankton, and fish that feed on zooplankton, such as many damselfish, can be abundant in some reefs, particularly on reef edges. Plankton productivity is, however, often lower than one tenth the productivity of benthic algae which is among the most productive primary producer groups on earth. Coral reefs typically house a high diversity of fish that feed on other fish (piscivores), and in deeper open-water reefs, large groupers (Serranidae), sharks (Chondrichthyes), jacks (Carangidae), barracuda (Syphyraenidae) and other large piscivores can form the top of the food web. Many species of fish such as sweetlips (Gaterinidae), scavengers or emperors (Lethrinidae), wrasses (Labridae), and triggerfish (Balistidae) feed on invertebrates such as sea urchins, crabs, and shrimp. Energy and nutrient cycling in the reef is tight due to warm water temperatures and high levels of predation. Nonetheless, even unconsumed detritus is an important

Figure 4.4. Photograph of a coral reef (Malindi MNP) showing a variety of corals on the bottom and the reef's major grazers, parrotfish, at the top. (Credit: T.R. McClanahan).

food pathway for invertebrates such as sea cucumbers (Holothuroidea) and sand-dwelling worms.

With the exception of marine parks and some remote reefs, humans are at the top of the reef's food web and humans can feed at many different trophic levels, but generally harvest the high trophic level fish and invertebrates. Zanzibarians, however, have also begun to culture algae (*Euchemea*) that is used in food and for laboratory culturing purposes (such as agar).

Each of the components in the food web diagram may contain tens to hundreds of species aggregated into one feeding group. How these species coexist, in view of the competitive exclusion principle, has fascinated many coral-reef biologists (Box 4.2). The early suggestions of naturalists and scientists was that over the long geologic history of coral reefs species had developed a number of morphological, behavioral and feeding differences that allowed the reef's resources to be finely partitioned by each species. This view placed interspecific competition as the primary force controlling the diversity of species. Subsequent experimental studies failed to find strong interspecific competitive interactions between many species as well as less partitioning of food resources than previously thought. Species

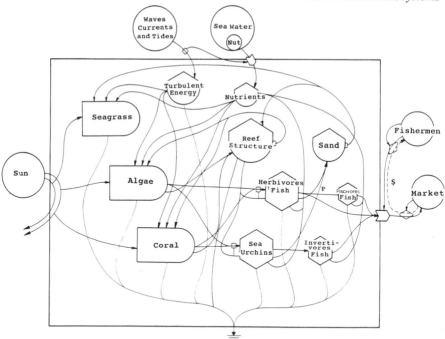

Figure 4.5. An aggregated energy-circuit diagram of the coral reef ecosystem showing the major energy and chemical inputs (circles) and storages (tank shapes), primary producers (bullet shapes), and consumers (hexagons).

populations were also found to be highly variable in time and space. Subsequent work has shown that factors such as predation, periodic diseases, and physical disturbances, such as waves and currents, can cause high mortality and keep populations below the level at which interspecific competition for food resources becomes a source of mortality. Consequently, coral-reef biologists have recently focused their efforts into studying factors such as predation, disturbances and the species reproductive and recruitment patterns in order to determine factors that control species populations.

A study of coexistence in sea urchins was undertaken in Kenya's reef lagoons to help determine how three closely related species managed to coexist while living close to each other and sharing similar food resources (16,26). The three species, *Diadema setosum*, *D. savignyi*, and *Echinometra mathaei*, are the most common sea urchins found in Kenya and many tropical areas in the Indo-Pacific. Analysis of their distributions showed no clear zonation patterns across reef lagoons and that their stomach contents were largely indistinguishable from each other - each species feeding on a mixture of algae, seagrass, and detritus. On closer examination, however, it was found that each species had different microspatial preferences (crevice sizes), that corresponded with differences in the species body morphologies and behavior (Fig. 4.6(a)). Further, the species differed in their susceptibility to predation and their body shapes and behavior could be best explained by this differential susceptibility to predation. The smallest species, *E. mathaei*, is solitary and defends small self-created burrows from intruders. It is also the species most susceptible to predation when outside its burrows. The two larger species of *Diadema* have more subtle body-shape differences with the largest species, *D.*

Box 4.2. Resource Partitioning

Organisms require energy and nutrient resources to support metabolic costs and to reproduce. Chemical requirements for life are fairly similar for many complex organisms but evolutionary processes have produced numerous species that use these resources. The ecological principle of competitive exclusion states that if species are using the same resource(s) then one species will be locally extirpated given sufficient time. How then do individuals and species coexist when they share similar resource requirements? This problem has intrigued ecologists and has been the focus of a great deal of field and theoretical research (1).

Studies during the 1950s and 1960s suggested that each species has a niche or a set of resources which the species is able to utilize more effectively than any other species. Species coexist, therefore, because they have slightly different resource requirements and take advantage of the natural variability of resources in time and space. If resources were uniform, then those species able to tolerate the lowest resource levels and still grow or reproduce would eventually eliminate competitors. If resources are not uniform or equilibrium is never reached then species can coexist (2). Superior competitors, however, often have the ability to reduce resources to a low and uniform level if they are given sufficient time without disturbances. Low colonization rates, natural physical disturbances (such as waves and extreme temperatures) and predators, however, have the ability to stop or slow superior competitors from dominating a resource. Consequently, a species' niche may be more a window of opportunity created after disturbances than an adaptation to stable resources. This concept of the niche is referred to as the regeneration niche. Because resources are limited, species frequently have morphological or physiological attributes that suggest trade-offs between traits that give species superior competitive ability versus faster colonization rates or tolerance to disturbances. These organismic attributes or trade-offs partially determine the niche possibilities for a species within the changing ecosystem or landscape.

T.R. McClanahan

References

1. Kingsland, S.E. 1985. *Modeling Nature: Episodes in the History of Population Ecology.* Chicago: University of Chicago Press
2. Tilman, D. 1988. *Plant Strategies and the Dynamics and Structure of Plant Communities.* Princeton: Princeton University Press

setosum, living most frequently in the open and in groups of individuals with intertwined spines. The intermediate-sized *D. savignyi* was sometimes found in social groups but more frequently occupied larger naturally occurring crevices and overhangs.

Experiments on predation and competition showed that there was an inverse relationship between competitive dominance and predator susceptibility with *E. mathaei* being the top competitor but most susceptible to predators. It was therefore concluded that these three species are able to coexist due to differences in spatial preferences. Competitive exclusion of weaker competitors, however, is unlikely due to greater predator susceptibility of the competitive dominant. The competitive

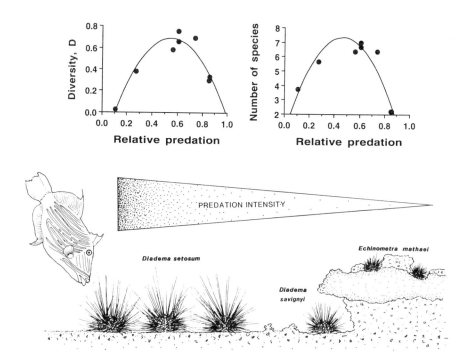

Figure 4.6. Numbers of species and species diversity index (0 being lowest and 1 being highest) of coral-reef inhabiting sea urchins as a function of predation intensity on sea urchins. Bottom picture shows the distribution patterns of 3 of the most common sea urchin species as a function of predation intensity from sheltered to open environments.

dominant, *E. mathaei*, is forced to stay within burrows or face the high chance of being eaten. The larger *Diadema* species are more tolerant of predation due to a combination of their large body sizes, long spines and the formation of social groups that makes these species difficult prey.

This study suggests that *E. mathaei* is capable of expanding its distribution and extirpating competitors under conditions of low predation. Predation, however, is a major force promoting sea urchin coexistence. To test this idea, the population numbers and diversity of sea urchin were studied along a predation gradient. The experiment (Fig. 4.6) produced a hump-shaped curve between the intensity of predation and the number and diversity of species (26). At the lowest levels of predation, reefs were dominated by *E. mathaei* and other species were rare. At intermediate levels of predation most of the species appeared to be present in intermediate numbers while at the highest levels of predation, only a few cryptic-species were found. This study implicates an interaction between the factors of resource use, competition, and predation in creating observed species diversity. A study of Kenyan seagrasses also supports this contention (27).

Coral reefs are often associated with relatively stable environmental conditions that allows the higher trophic level carnivores to develop, persist, and control many of the species populations at lower trophic levels. Consequently, coexistence may depend less on partitioning food resources and more on partitioning "predator-free space" or refuge from these carnivores. Further studies will help uncover other important interactions and population processes that control the abundance and diversity of coral reef organisms.

Production and Ecological Processes

As mentioned previously coral reefs are among the earth's most productive ecosystems, but coral-reef species also consume most of this productivity. Consequently, very little energy is lost from a coral reef before it is consumed and much of the energy and nutrients are recycled in the reef in an effective manner. Equally surprising is that to maintain this productivity little or no nutrients are consumed from seawater currents passing over the reef. This has been shown by measuring the quantity of nutrients such as phosphates and nitrates in seawater on the upstream and downstream sides of the reef. Studies have shown that there is no net difference in the phosphorus concentrations on either side of a current passing the reef and that nitrates are higher on the down than the upstream side of the reef (33,42). Although nutrients in seawater may be consumed by reef organisms, organisms are releasing nutrients at or greater than the rate of nutrient consumption. In the case of nitrogen, lawns or turfs of nitrogen-fixing benthic algae are converting nitrogen gases dissolved in seawater into the forms required by organisms. This nitrogen is finally lost to the passing seawater. Consequently, productivity and nutrient studies, to date, present a picture of the coral reef as a highly productive, nutrient self-sufficient ecosystem.

What is the basis of this production and what eventually limits it? Unlike many terrestrial ecosystems that require freshwater the coral reef is more limited by light and the rate of uptake of necessary elements such as nitrogen. Light is the largest single energy source at 7.12×10^9 Joules/m^2/year but the energy of water motion created by waves, currents, and tides combined is also significant at 1.76×10^9 Joules/m^2/year, or 25% of the sunlight's energy (Fig. 4.7). Water motion is necessary in the renewal and loss of necessary elements such as oxygen, carbon dioxide, and nitrogen and plays an important role in production (3,12). Nonetheless, the species composition and abundance of a site may play an equally important role in productivity by changing the balance between gross and net productivity. Figure 4.7(b) shows a hypothetical relationships between production and the biomass of a primary producer such as benthic algae. Net production, or production available to the next trophic level, is highest at low to intermediate biomass and frequently reflects the abundance of herbivores such as parrotfish or sea urchins. Consequently, production in coral reefs is a result of the interaction between the physical and biological environment at each reef. Studies on algal turf productivity to date, however, often find the highest net production with the lowest algal biomass (14). The point at which reduced biomass results in lowered production has not been found because some algae grow inside the pores of dead coral (named endolithic algae). Consequently, herbivory may limit primary production more than nutrients or light.

A second ecological process of interest is calcium carbonate ($CaCO_3$) or skeletal deposition formed by corals and other shelled organisms. Calcium carbonate forms the physical structure of the reef that is essential for organisms that require hard substrate to attach to and crevice space to avoid predators. As described above the rates of calcium carbonate deposition by algae and corals depends on light.

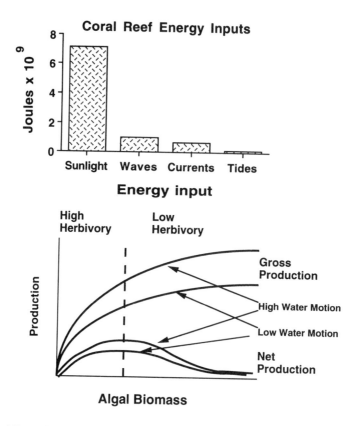

Figure 4.7. Major energy inputs into the coral reef ecosystem (18) and a hypothetical example of the effects of water motion on algal gross and net production as a function of algal biomass. Algal biomass is often a function of the abundance of herbivores.

Therefore, total calcium carbonate deposition is likely to reflect the total weight of these organisms and their respective rates of deposition - average algal deposition rates being approximately 5% of coral on a wet-weight basis. Contributions from coral and algal plus minor contributions from mollusc, crustacean and echinoderm shells make up the gross deposition or production of calcium carbonate in reefs.

Losses of calcium carbonate are caused by the burrowing, boring, and feeding activities of a number of fish and invertebrates and the balance or net calcium carbonate deposition is achieved by the balance of depositors and eroders. The major contributors to reef erosion are sea urchins and parrotfish that excavate dead coral as part of their feeding activities. Physical erosion and damage of coral rock is caused by wave and current energy. Physical abrasion also occurs but is highly variable between reef locations - being most important on the reef crest. The major portion of the calcium carbonate budget, however, is a result of reef organisms and their response to environmental factors such as light and water motion. Consequently, the ecological processes of productivity and the calcium carbonate balance are open to influence by human activities both in and far away from coral reefs. The species composition of reefs can be affected by human fishing activities and pollution (see Chapter 3). Sediment erosion can affect water clarity thereby influencing light penetration and its subsequent availability for photosynthesis and calcification on the seafloor.

Human Use of Resources

The most direct way that humans affect coral reefs is by removing organisms and their skeletons - which are composed of calcium carbonate. This is done through fishing (removal of bony and cartilaginous fish), and the collection of living and dead corals, starfish, and snails. As described above, predation is naturally high in pristine coral reefs and little energy is lost from the reef. Therefore, people are simply another predator in the reef - albeit a very effective one. The effect of humans can be very dramatic particularly when the number of fishermen is high and their use of gear unrestrained. Management of coral reefs is hampered by many of the problems that make managing any fishery difficult (see Chapter 3), but is further complicated by the high diversity of harvested organisms and the complexity of ecological interactions. This makes management decisions based on traditionally accepted techniques (such as mesh-size regulations, restricted seasons, and size restrictions) difficult to justify unless a good understanding of the ecological interactions and the subsequent effects of management decisions are understood. Progress has been made in some areas of management-related research and this understanding combined with an experimental management approach may eventually result in sustainable use of coral reefs.

Shell and Coral Collecting

Human predators are a relative newcomer to coral reefs over evolutionary time and many species may have few adaptations to avoid human predation and particularly our recent technological advancements. Coral reefs harbor an abundance of organisms with ornate and colorful skeletons. It has been hypothesized that many of the elaborate body shapes and spines of coral reef species serve as defenses against predators (41). Additionally, color may play a role in communication by warning predators of toxic substances (aposomatism) or as a means for recognizing potential mates or competitors. While this ornateness may deter coral-reef predators it may have the effect of increasing attractiveness to humans - which results in collection and sale. Consequently, collection of some ornamental species could cause significant population reductions and result in an ecological imbalance in the reef ecosystem. This condition, however, has rarely been convincingly documented.

The three most collected ornamental groups include marine snails, corals, and aquarium fish. Each of these groups has different evolutionary histories and ecological attributes that must be considered in determining the effect that their removal will have on the reef. Corals produce elaborate but sometimes delicate skeletons that are used as ornaments when dried and bleached. It is this same complex structure that provides homes and refuge for many of the coral-reef fish and invertebrates. The loss of this structural complexity will undoubtedly reduce the abundance and species richness of coral reef fishes and therefore coral collection is often regarded as a destructive harvesting method. Nonetheless, corals, like all living organisms, have the ability to tolerate mortality and replace individuals lost from the population. Therefore, it is theoretically possible to harvest corals in a sustainable way. This possibility, however, has not received much attention and most shallow-water coral harvesting is either unregulated or banned. In the Mediterranean and the western Pacific precious coral fisheries have occurred since Paleolithic times, but their abundance is limited in East Africa (10). These largely unregulated fisheries have been typified by a boom and bust behavior as precious corals have slow growth rates and fishermen usually remove all accessible corals.

Studies of coral diversity and abundance in Kenyan reefs have shown differences in the abundance of some taxa that may result from collection and other human-induced disturbances (28). Marine parks have slightly higher coral cover, reef physical complexity, and generally more genera and species of coral than unprotected reefs (Fig. 4.8). Coral genera most abundant in marine parks include *Galaxea*, *Montipora,* and *Acropora.* Differences in the abundance and species composition of corals may be due to trampling or net entanglement by fishermen, a high abundance of sea urchins in unprotected reefs, coral collection, or perhaps environmental differences between reefs. Branching *Acropora* are among the most prized corals of shallow-water reefs and their low abundance in unprotected reefs may be due to coral collection, trampling and net entanglement. Further studies are required to distinguish between the various environmental and human factors that produced these findings. In Tanzania, where dynamite fishing has been extensive, no rigorous studies are available, but dynamiting has probably resulted in the loss of coral cover and associated reef complexity. The long-term consequences of this destructive fishing method are not known but are likely to have a negative effect on the already heavily exploited fishery. Developing fisheries management and enforcement that reduces coral destruction is a high conservation priority for the region.

The collection of marine snails (molluscs) for their ornamental shells is also a conservation concern as overcollection could potentially result in species extinctions or an ecological imbalance in the reef. Existing East African studies, do not however, support either of these contentions (17,31). Studies of mollusc abundance and diversity in Kenya's marine parks and unprotected reefs found an overall higher mollusc abundance in the unprotected reefs where collection is occurring. The exception was three large ornamental species that typically live exposed on the reef's surface - *Lambis truncata, Lambis chiragra* and *Ovula ovum* (17). The reasons for the limited human influence was unclear until studies of predators and predation rates were undertaken. Studies showed that unprotected reefs had fewer predators of molluscs such as triggerfish and that many molluscs were benefiting from the removal of their predators by fishermen. Additional studies in Tanzania found reduced numbers of a few ornamental species in areas with and without heavy collection (31). Again, it is difficult to distinguish natural environmental and predatory factors from human effects.

The Tanzanian and Kenya species that exhibited population reductions due to collection have distribution patterns extending throughout the Indian and Pacific Oceans. They are, therefore, unlikely to suffer from extinction due to overcollection. In the central Pacific, however, a few highly prized species of edible giant clam (*Tridacna gigas* and *Tridacna derasa*), with restricted distribution patterns, are facing the possibility of extinction due to overcollection (29). In contrast, the money cowrie (*Cypraea moneta*) used as payment for slaves in west Africa between the 16th and early 19th century, was heavily collected from the Maldive Islands for hundreds of years and has persisted over this time. It is estimated that ~ 12,000

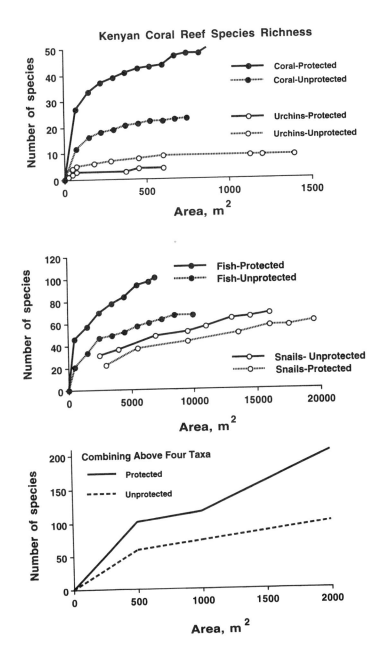

Figure 4.8. Numbers of species of sea urchins, coral, fish and snails in protected (marine parks) and unprotected reefs (heavy fishing) as a function of the area sampled data from (18,20,28). Corals and fish show high species in protected reefs while sea urchins and snails have slightly lower numbers of species in protected reefs. Combining all taxa (bottom) indicates more species in protected coral reefs.

metric tons or 10 billion (1×10^{10}) individual cowries were imported into West Africa by British and Dutch exporters between 1700 and 1790 during the peak of the slave trade (11). Consequently, conservation of molluscs may need a species-specific approach because each species may respond differently to collection and other environmental factors.

The importance of molluscs in the balance of the coral reef ecosystem has also probably been overstated. Some researchers hypothesized that the removal of molluscs that prey on pestilent echinoderms, such as the crown-of-thorns starfish (*Acanthaster planci*) and the rock-boring sea urchin (*Echinometra mathaei*), would result in more severe and frequent outbreaks of these pests. Subsequent research, however, has largely shown that predatory molluscs such as the giant triton (*Charonia tritonis*) and the bullmouth helmet shell (*Cypraecassis rufa*) are typically found at low population densities and have low rates of feeding on these echinoderm pests relative to finfish predators such as the red-lined triggerfish (23,25,26). Evolutionary studies suggest that molluscs are a low-metabolism ancestral lineage (dating back over 500 million years ago). Many mollusc species have been forced to find refuge from competition and predation from the more recent evolutionary proliferation and diversification of predatory bony fish (Subdivision = Teleostei) that occurred since the last mass extinction at the Mesozoic-Cenozoic boundary around 65 mya. (41). Therefore, it remains more likely that these high-metabolism and more recently evolved bony fish are controlling mollusc and other invertebrate populations rather than vice versa.

The above discussion suggests that the collection of molluscs as a group is not a major conservation concern but that some species may require conservation efforts that should be specific to certain species or regions. Species that are heavily collected and have restricted geographic distributions are likely to have the greatest chances of extinction. Overexploitation of populations beyond a sustainable basis is likely for species that 1) take a long time to sexually mature, 2) live exposed on the surface, 3) lack deep-water populations, and 4) lack pelagic larvae or the ability to move among regions or reefs (4). Species that display indications of overexploitation in East Africa have most of these characteristics. Management of marine molluscs may require a combination of existing fisheries management techniques such as size restrictions, harvesting quotas or limited entry restrictions, stock enhancement, or aquaculture (see Chapter 3). In the Pacific, aquaculture of the giant clam (*Tridacna*) has proved to be an effective way to produce both shells and meat of highly prized and easy-to-grow animals that face local and global extinction due to overharvesting (29).

Fishing

Fishing has traditionally been a source of food but more recently aquarium fishing has become an export business of some economic value. The world's marine aquarium trade was valued at U.S. $ 40 million dollars in 1990 and the value of ornamental fish is considerably higher than food fish (U.S. $ 0.20 to 50.00 per fish compared to less than $1 per kilogram for food fish). Consequently, the aquarium trade remains a potential profitable and sustainable use of coral reefs if collection does not use destructive techniques such as breaking corals or the use of poisons (34). Care must be taken to protect rare and endemic species through marine protected areas or harvesting restrictions.

Fishing in East Africa is most frequently focused at shallow-water coral and seagrass beds. Fishing areas are within the first kilometer of shore in Kenya, but often further offshore in Tanzania. Fishermen mostly rely on a combination of hook and line, hand-woven madema traps, and nets of various purposes and mesh sizes. The major demersal catch includes carnivorous scavengers (Lethrinidae), herbivorous rabbitfish (Siganidae), and a high diversity of reef-associated fishes ("mixed demersals") such as herbivorous parrot and surgeonfish (Scaridae and Acanthuridae), invertebrate-eating wrasses (Labridae), piscivorous snappers, groupers and goatfish (Lutjanidae, Serranidae, Mullidae), and detritivorous mullets (Mugilidae), and catfish (Siluriformes; Fig. 4.9). The diversity of the catch is very high and although fishermen will focus on the preferred fish of their customers (such as Siganidae and Lethrinidae) almost all fish species are caught. Even some seemingly inedible fishes such as porcupine (Didondontidae) and triggerfish (Balistidae) are eaten once their tough skins are removed.

Fishermen using simple technology generally behave like optimal foragers - optimizing their net gain by reducing expenditures such as travel distance and time, and maximizing their gross catch. They do, however, need to consider market times and customer fish preferences into their fishing patterns. Consequently, fishermen will frequently exploit nearshore areas to their maximum capacity before traveling further offshore to less fished and more risky areas. Nonetheless, as nearshore areas begin to produce lower yields fishermen may need to travel greater distances. This entails greater hazards and the potential for losing their catch before it becomes inedible. This often leads to an equilibrium where costs and gains are nearly equal and to poverty that is so pervasive among fishing societies throughout the world. Technologies such as engines and frigeration will extend the distance that can be traveled. Travel and technologies will, however, also increase the cost of fishing and will result in a new equilibrium often with a higher total cost and yield but not necessarily with a greater profit. A study in Tanzania showed that fishermen using small outboard engines had lower profits than those relying on sails (13).

Many of the nearshore areas in Kenya and Tanzania are heavily utilized and overfished (catches beyond their Maximum Sustained Yield) while some areas further offshore have potential for increased yields. What is often not appreciated by fishermen and their managers is that ecosystem state and processes, which the fish are an important part of, is often controlled by the feeding rates and prey choices of the fish. Consequently, their removal or reduced abundance in the ecosystem can effectively alter the state of the coral reef and seagrass ecosystem. This indirect effect of fishing on ecosystem state and processes has been studied in some detail in coral reefs of Kenya's fringing reef lagoons (16,17,18,20,22,23,25,26,27,28).

The example of sea urchin species coexistence described above reflects an even larger problem in coexistence among all reef herbivores including edible herbivorous fishes, such as parrotfish, surgeonfish and taffy (Siganidae) as well as unharvested sea urchins. As presently understood, herbivorous fishes and sea urchins have different feeding and metabolism rates that give them advantages under different conditions of predation and competition. Herbivorous fishes have high consumption and metabolism that gives these species high dispersal and growth rates and greater tolerance to predation. Sea urchins, on the other hand, have lower consumption and metabolisms that allows them to persist better under conditions of low food abundance but make them less tolerant of predators. Consequently, each of these groups may have different minimum resource levels of predator-free space and food that can be tolerated (19,21). The two functional groups will have competitive

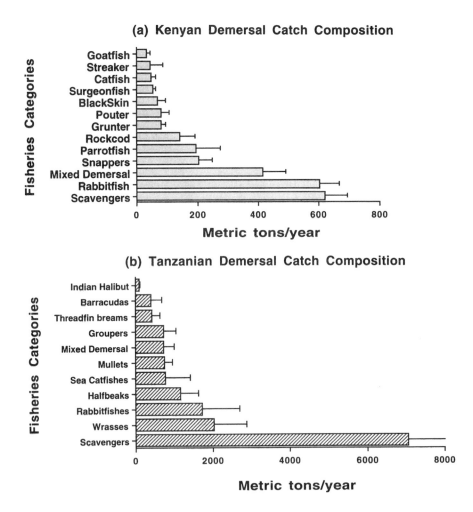

Figure 4.9. Tanzanian and Kenyan average demersal fish catch analyzed by fish categories.

advantages under different conditions - sea urchins being superior competitors when food is limiting and herbivorous fish being superior when predator-free space is limiting.

In pristine or lightly fished reefs the high abundance of predators insures that herbivorous fishes have a competitive advantage over sea urchins. As predators are reduced with increased fishing, sea urchins have the competitive advantage for algal resources. At some level of fishing pressure (estimated at >7 fishermen/km^2) the reef may become dominated by sea urchins through the direct effect of fishing and the indirect consequence of resource competition among reef herbivores. Most of Kenya's southern fringing reef lagoon is dominated by sea urchins, as are large reef areas of Zanzibar and Tanzania. This has been hypothesized to result from the above described fishing and competitive interactions (16,21). Some coral reefs, however, have neither a high abundance of sea urchins or herbivorous fishes and are typified by high brown algae abundance.

A simple simulation model (difference equations describing species interactions solved by computer programs) of the East African coral reef ecosystem has been created to help understand potential interactions of fishing on the coral reef ecosystem and the effect that fishing intensity and catch selection have on the coral reef ecosystem (19,20). The general structure of the model is a further simplification of the energetic pathways shown in Figure 4.5 where the dominant primary producers and depositors of calcium carbonate are coral and algae, dominant herbivores and eroders of the calcium carbonate substrate are herbivorous fishes and sea urchins, and the top carnivores are piscivores and "invertivores" (species that feed on invertebrates such as sea urchins). The model allows one to change fishing pressure (number of fishermen) and the types of catch that fishermen take (piscivores, invertivores or herbivorous fishes) and to see what effect this has on the resultant community and ecological processes.

The model's output for various fishing experiments suggests three major ecological states resulting from fishing. These states are described by a conceptual diagram in Figure 4.10. At low fishing intensity or fishing only piscivores the ecosystem remains in a state of high coral and fish abundance. When all groups are fished intensely the model predicts a switch to sea urchin dominance when sea urchin predators are no longer able to maintain sea urchin prey at low levels. The third ecological state of algal dominance is reached if fishermen leave invertivores unfished and focus on piscivores and herbivores. This last fishing method also produces the highest and most predictable yields of fish and may be the least objectionable fishing method.

Outputs from these types of simulation models are useful in thinking about possible management scenarios and in planning management experiments (a management experiment is anything that people do to an ecosystem, see Chapter 3). Species and ecological communities often exhibit unpredictable behavior that cannot be predicted by simple simulation models. For instance, many fished Caribbean reefs had a high dominance of the sea urchin *Diadema antillarum* prior to 1984 (believed by some marine scientists to result from heavy fishing) when a sea urchin disease suddenly killed 99.9% of all individuals throughout most of the Caribbean in a 3-month period (15). Many reefs became dominated by brown algae that colonized reefs in the absence of sea urchins and herbivorous fish grazers. Models can be augmented and modified to include additional interactions or periodic events such as diseases as we learn about the ecosystems through ecological modeling, experiments, and monitoring ecological variables. Models are one of the experimental scientific tools for understanding ecosystems rather than the final summary of years of monitoring and field research.

The above model does augment existing field studies suggesting the important role of predation on sea urchins in controlling sea urchin abundance in many Kenyan coral reef lagoons (24,26). Field studies indicate that a number of coral-reef species feed on sea urchins including important fisheries species of wrasse and scavengers. Observations in marine parks suggest, however, that some triggerfish (mostly the red-lined triggerfish) are the main predators of sea urchins (23,26). Triggerfish act as keystone species by keeping sea urchins below the level that they outcompete herbivorous fishes. They also maintain coexistence between herbivorous sea urchins (see above). Since triggerfish are not a preferred prey of fishermen in this region it may be best to leave them unfished in order to increase yields of herbivorous fishes.

CORAL REEF METAMODEL

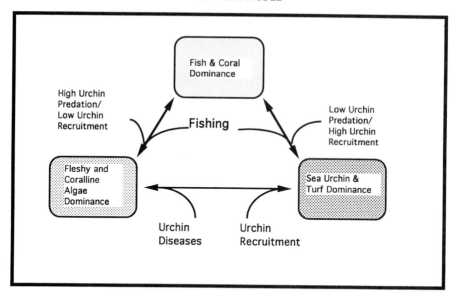

Figure 4.10. A conceptual or metamodel of the different ecological states that a coral reef will move between as a function of fishing and or sea urchin recruitment and diseases.

Patterns of Species Richness

Associated with the changes in fishing and herbivory described above are changes in species abundance and composition. Numbers of species in protected and unprotected reefs differ, but the difference is not a simple reduction of species numbers in unprotected reefs. Studies have been completed on fish, corals, snails, and sea urchins (16,17,20,26,28). Results of these surveys suggest that fish and coral diversity are considerably lower but sea urchin and snail species richness are similar in unprotected reef lagoons (Fig. 4.8). In deep water and less intensely fished reef edge sites, only fish have been surveyed and few differences in populations were detected among reefs based on their management status apart from the groupers (37). Additionally, snails on shallow reef flats and edges seem to exhibit little relationship with reef management (17).

The most-likely explanation for reduced number of fish species in reef lagoons is the direct effect of fishing itself combined with reduced bottom topographic complexity and coral cover and high sea-urchin abundance in unprotected reefs (20). Reduced bottom complexity may be due to a combination of factors including tides and water depth, reduced coral cover due to competition with algae, increased reef erosion rates - partially influenced by abundant sea urchin populations and increased trampling and breakage of corals and reef substrate by fishermen and tourists. Reduced coral cover in unprotected reefs is due to competition with algae, grazing by sea urchins, and breakage and mortality by human trampling. The overall effect is lowered coral and fish species richness in unprotected reefs and a reduced aesthetic appeal to tourists.

The Value of Marine Parks

Although marine parks may, in principle, be able to tolerate a low level of fishing and resource harvesting the above discussion suggests that commercial fishing and marine parks as local and international tourist attractions are largely incompatible uses of reef space. Given the pressing problem of food abundance and security in most African and undeveloped countries how can marine parks be justified, and, if justified, how much area should be placed under this form of management? The justification for marine parks, like terrestrial parks, is that they 1) preserve a genetic, biological and ecological heritage of a country that developed over millions of years of evolution, 2) attract tourists and therefore income and foreign exchange, 3) they are areas where scientists can study organisms in a relatively undisturbed environment, and 4) they provide refuge for a mature breeding stock that may enhance fisheries recruitment in adjacent unprotected reefs.

The economic arguments of increased tourism and fish catches are perhaps the most compelling for nonscientists, politicians and resource planners. Marine parks can create greater income from tourism than the income from fishing when tourist visitation is high. In Kenya, where around 115,000 visitors entered the 4 marine parks in 1991, the gross income to boat owners and Kenya Wildlife Service, on a constant area basis, was around 2.5 times the income from fishing (Fig. 4.11). This income does not include indirect economic effects that the existence of marine parks will have on enhancing the tourism-based economy or the net incomes. Marine parks only occupy 54 km² of Kenya's ocean or around 5% of the potential fishing grounds. Given the high income of parks compared to fishing it is a wise national economic policy to increase the area in marine parks. This analysis, however, is highly dependent on the number of visitors. But, at Kenya's current 115,000 visitors/year, the area in marine park could be increased and still be more profitable than fishing. Furthermore, marine parks located near tourist sites will increase park income even further. This was found during the Mombasa MNPs creation in 1990. This park became the top income earner among Kenya's marine parks in less than 2 years after establishment - largely due to its nearness to a major tourist beach. The difficulty with starting marine parks is compensation or creation of additional economic opportunities for fishermen who lose access to fishing grounds.

River Discharges and Sedimentation

In addition to the direct effects of harvesting on coral reefs, human activities at a great distance from reefs can potentially influence their ecology. Rivers carry human waste and soil lost due to erosion, affected by changing land uses and associated changes in vegetation cover. Changes in human population and land use can have downstream effects (Chapter 3) that will eventually influence marine organisms and communities in river discharge areas. The distribution of coral reefs on the East African coast reflects the long term (geological time scale) effects of river discharge and fresh water influence on coastal ecosystems (Fig. 4.1). The greatest influence of freshwater is in the Tana and Sabaki river discharge areas that lack developed reefs. Intermediate levels of freshwater results in interspersed reefs in Tanzania, Mozambique, and the Lamu archipelago of northern Kenya. The lowest

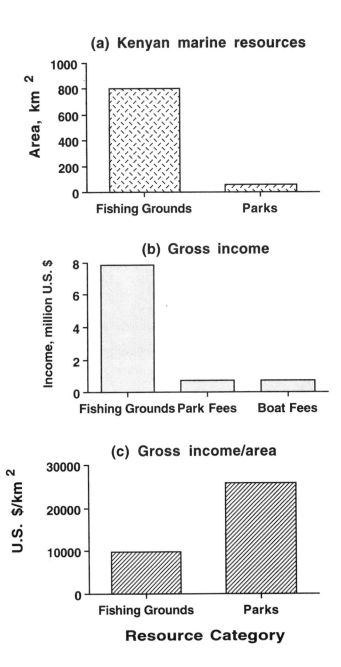

Figure 4.11. Analysis of the area and income from Kenya's nearshore fisheries and marine parks. Based on data from Kenya's Fisheries Department and Kenya Wildlife Service for the year 1991. The analysis suggests that marine parks gross income is around 2.5 times that of fisheries grounds on a per area basis.

concentration of freshwater discharge has resulted in the long fringing reef of southern Kenya.

Freshwater discharge also varies seasonally, caused by the migration of the ITCZ and it's effects on rainfall and coastal currents (see Chapter 3). Soil erosion is caused by rainfall and subsequent run-off, but the amount depends on several factors - land relief, vegetation cover, and soil type being the main influences. In the tropics, however, land use is the major factor influencing soil erosion (8). Soil erosion rates decrease from open rangeland or grassland to agricultural land, to forested land (Fig. 4.12). The conversion of much of Kenya's land to agriculture and grazing, often without consistent soil conservation practices, has led to faster water runoff and soil erosion resulting in increased river discharge and sediment loads at the coast (see Chapter 3).

River discharge is greatest during the long and short rains, showing two annual peaks (Fig. 4.13). These peaks occur during the two different monsoon seasons - the long rains (April to July) fall during the southeast monsoon, while the short rains (October to December) fall during the northeast monsoon. The northward-flowing currents during the southeast monsoon pushes river discharge produced by the long rains northwards along the coast. South-flowing currents of the northeast monsoon, in Somalia and northern Kenya, push river discharge, produced by the short rains southwards (Fig. 4.13) towards the Malindi reefs. Due to these seasonally reversing currents, the coastline between the Tana and Sabaki Rivers (about 100 km) experiences river discharge during both seasons. As a result, there are no continuous fringing reefs, and the coastline is dominated by sand dunes and beaches formed from river sediments and mangroves. South of the Sabaki River, Malindi experiences river discharge for only a few months (December to March) of the year, resulting in a transition from the sand-dominated shoreline, of the north, to the fringing coral reefs of the southern coast.

The seasonal predictability of the monsoons hides a high degree of uncertainty and variation in annual cycles of rainfall and river discharge. This unpredictability, or stochasticity, has important implications for the reef ecosystem. High rains in 1961 and 1962, after several years of low rainfall, coupled with increasing population pressure and land clearing for agriculture, resulted in the first recorded massive increase in sediment discharge from the Sabaki River. Since 1961 the Malindi beach has grown seaward from sediment delivery, and Malindi Bay waters have become increasingly brown and turbid.

Rivers also carry high concentrations of nutrients from plant litter, soil humus, and human waste. Where rivers meet the sea, ecosystems of high net production (plankton, mangroves, estuaries) occur and river mouths are frequently the site of productive fisheries. Increased nutrient and sediment loads can thus increase biological production in surrounding seas, and evidence has been found for increased fisheries yields in the years following heavy river discharge. Excessive nutrients, however, can cause eutrophication and a shift in the species assemblage from one ecosystem state to another - with high-diversity benthic ecosystems such as coral reefs often being the first to suffer.

Coral Reefs and River Discharge

Coral - reef organisms have adaptations to low nutrients, high light, and high predation. Corals and some other marine animals (such as some sponges,

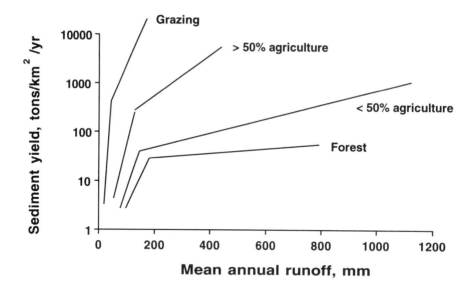

Figure 4.12. Effect of land-use category on river sediment yields for forested, agricultural and grazing lands. Adapted from (8).

ascidians, and giant clams) contain symbiotic algae within their tissues, that use sunlight for photosynthesis. This symbiosis enables a tight cycling of nutrients and energy between host and symbiont without losses to the environment. Calcium carbonate deposition appears related to photosynthesis in these organisms and the calcium carbonate skeletons provide protection from the many predators inhabiting coral reefs. Under low nutrient, high light, and high predation conditions, these symbiotic organisms are competitively superior to algae and other benthic animals such as sponges and soft corals, and are a major portion of the reef's substrate cover (~10 to 50%).

Sediment and nutrients in the water column increase productivity and biomass of the water column's planktonic community. High plankton abundance along with suspended sediments, reduces light penetration to the bottom, resulting in less light energy available to benthic photosynthesizing organisms. Sediments settle onto the bottom and smother animals and plants resulting in higher metabolic costs to remove sediments, reduced efficiency of gas exchange with seawater, or reduced food capture. Higher nutrient plankton concentrations give a competitive edge to fast growing heterotrophic filter and suspension feeders (sponges and soft corals), that are able to increase their growth rates to a greater extent than corals under turbid conditions. If the predators of filter feeders and algae are not abundant some corals may be outcompeted for space. The result is a shift in the benthic community such that hard corals are no longer as dominant. In the long term, this may lead to erosion of the reef structure itself as most reef-building is done by hard corals.

The above ecological scenario suggests three main effects of river discharge on coral reefs: 1) that corals will grow more slowly and may suffer higher mortality, 2) that other organisms (such as sponges and soft corals) with less dependence on sunlight will replace corals as the dominant organisms, and 3) that the presence of herbivores may be a critical factor controlling the switch in dominance from coral to

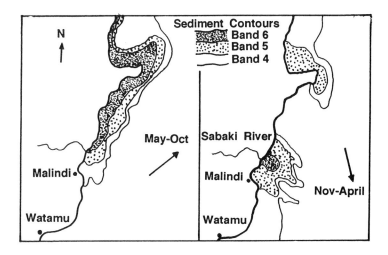

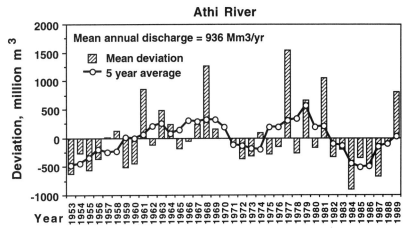

Figure 4.13. Athi/Sabaki river discharge and visibility of reef waters in Malindi. Top) transport of sediment plume from the Sabaki River in June (left) and December (right). Only sediment discharge during the northeast monsoon (November to April) is transported to the Malindi reefs. Black, stippled and clear contours show water with decreasing concentrations of sediment. Bottom) annual variation in discharge in the Athi River plotted as deviations from the overall average of 936 million cubic meters.

non-coral dominated reefs.

Scientific literature on the effects of sediments on coral reefs is large and diverse (36). Effects ranging from none to total reef death have been recorded over a wide variation of sediment and river discharge levels. The only pattern that emerges from these studies is that the outcome of sediment stress cannot be predicted without a thorough knowledge of particular circumstances. Currents, waves and tides, seasonal oceanographic patterns, sediment quantities and duration, fishing intensity and other environmental stresses all play a role in determining the outcome of increased sediment stress.

Table 4.1. Comparison of community structure, coral growth rates and coral stress indicators for coral communities in Watamu and Malindi.

Variable	Watamu (no sediment)	Malindi (sediment) Site 1	Malindi (sediment) Site 2
Hard coral cover, %	25	40	
Soft coral cover, %	1.4	10.7	
Number of coral genera			
per transect	24	23	
Searching	36	36	
Diversity indices			
H - Shannon-Weiner index	2.27	2.06	
D - Simpson's index	0.83	0.82	
Coral growth, grams/day			
Mean	.15	.14	.034
Standard deviation	.021	.097	.038
t-test of significance	Watamu = Malindi site 1	Malindi site 1 > Malindi site 2	Malindi site 2< Watamu and Malindi 1
Stress indicators per 120 coral individuals			
Mucus sheaths	1	7	
Partial mortality	8	16	
Mortality	8	11	

Malindi Marine Park as a Case Study.

Research in the Malindi and Watamu Marine Parks has focused on coral growth rates to determine if river sediments are reducing growth rates or changing species composition (32). Corals in Watamu act as controls (no sediment influence) in comparison to Malindi (no information exists on sediment conditions in Malindi before the initial concern began in 1961). Visibility and sedimentation rates in Watamu were used as a baseline for comparison with Malindi conditions. Low water visibility in Malindi compared to Watamu (Fig. 4.13) during the northeast monsoons indicates the effects of sediment on water clarity. Sedimentation rates are highest in November and March, during the calmer weather between monsoon seasons.

Results of a growth study illustrate the difficulties of determining sediment influence from growth studies alone (Table 4.1). Though 'average' growth rates are widely known and reported (about 10 centimeters per year in *Acropora* staghorn corals, 0.5 to 2 centimeters per year for *Porites* boulder coral, equivalent to 0.2 millimeter per day of surface diameter growth in Table 4.1) the degree of variation found is high, including 'negative' growth - a result of breakage, competition with other animals and plants, predation, and other factors. Three studied coral species indicate that growth rates during the northeast monsoon were highly variable and not different from studied Watamu corals at one site but much lower at a second Malindi reef. Further, coral cover was higher in the sedimented Malindi reefs than the Watamu sites. Total and partial mortality of experimental corals, however, was higher in Malindi than Watamu (Table 4.1). Corals in the genus *Porites* characteristically coat themselves in mucus sheaths when stressed by sediments - this behavior was 7 times more prevalent in Malindi than in Watamu.

Reduced biological diversity is one of the clearest and most easily recognized indicators of a stressed ecosystem. Increased stress levels due to sediments would be expected to decrease diversity by driving sediment-intolerant species to extinction.

The shallow reefs (less than 3 meters depth) of Malindi, however, were no less diverse than those of Watamu in terms of numbers of coral genera and of community dominance (as exemplified by the diversity index, Table 4.1). As one goes deeper (> 3 meters), coral diversity decreases on the Malindi study reef, while it increases in Watamu. Highest coral diversity is usually at intermediate depths (10 meters) on coral reefs studied in different areas around the world. The diversity decrease in Malindi probably reflects the reduced availability of light due to water turbidity at greater depths.

Though the 'magnitude' of diversity is not different between Malindi and Watamu reefs, the coral species on the reefs differ to some degree. Such patterns in species distributions, if they can be related through the ecology of the species to environmental conditions, may give some information on the effects of sediments. The Malindi reef had higher abundance of certain coral genera (*Galaxea, Hydnophora, Millepora, Turbinaria*) that may be more tolerant of sediments. Malindi also has a higher cover of soft corals. An expected increase in algal cover in Malindi has not occurred, and may be due to high grazing pressure. Fishing on Malindi's reefs would probably exacerbate problems of sedimentation by removing the controlling effects of herbivory on algal abundance.

Research, so far, suggests that corals on shallow coral reefs in the Malindi park are not heavily stressed by river discharge and sediments but conditions could worsen. Though growth rates are lower in some areas and evidence of stress and mortality are higher during sediment events, no conspicuous change in the coral community has occurred and Malindi still maintains high coral cover. Deeper reefs, however, receive less light and the diversity and cover of corals often declines with depth.

Broader questions in coral reef ecology - whether sediment-tolerant organisms are beginning to take over, how deep-water corals fair, and how river discharge controls processes such as herbivory, are the subject of on-going research. A further aspect to note is that though the ecology of Malindi reefs has not yet been drastically altered by Sabaki River discharge, the importance of the Marine Park, as a tourist attraction, is reduced as water conditions are unsuitable for tourism and snorkeling during peak sediment seasons (about 1 to 3 months per year). Perception problems and economic losses due to Park closures and a bad reputation may be more important than ecological effects in determining future management and conservation options.

Conclusions

East African coral reefs are presently heavily used by fishermen and tourists and experience other potentially detrimental forces such as river discharge and pollution. Research has been useful in determining the relative importance of human influence including sediment in rivers and, coral, shell and fish collecting. Tourism appears to be the most benign use of the coral reef as it earns the highest per area income and causes the least ecological damage and species losses. Increasing the area designated to marine protected areas is a good national economic policy and ways to compensate excluded fishermen should be a priority for future conservation efforts. Most East African nearshore reefs are being fished at or beyond their Maximum Sustainable Yields and some conservation efforts such as mesh size regulation, the protection of keystone predators (such as triggerfish), and the elimination of dynamite fishing need to be implemented to maintain harvests. Stricter regulations on commercial food fish can be economically compensated by encouraging the aquarium trade fishery, park service employment, tourist-related work, and aquaculture programs.

References

1.   Allen, G.R. 1991. *Damselfishes of the World*. Hans A. Baensch, Mille
2.   Belasky, P. 1992. Assessment of sampling bias in biogeography by means of a probabilistic estimate of taxonomic diversity: application to modern Indo-Pacific reef corals. *Palaeogeography, Palaeoclimatology, Palaeoecology* 99: 243-270
3.   Carpenter, R.C., Hackney, J.M., Adey, W.H. 1991. Measurements of primary productivity and nitrogenous activity of coral reef algae in a chamber incorporating oscillatory flow. *Limnology* and *Oceanography* 36: 40-49
4.   Catterall, C.P., Poiner, I.R. 1987. The potential impact of human gathering on shellfish populations, with reference to some NE Australian intertidal flats. *Oikos* 50: 114-122
5.   Crame, J.A. 1981. Ecological stratification in the Pleistocene coral reefs of the Kenya Coast. *Palaeontology* 24: 609-646
6.   Crame, J.A. 1986. Late Pleistocene molluscan assemblages from the coral reefs of the Kenya coast. *Coral Reefs* 4: 183-196
7   Crowley, T.J., Kwang-Yul, K. 1994. Milankovitch forcing of the last interglacial sea level. *Science* 265:1566-1568
8.   Dunne, T. 1979. Sediment yield and land use in tropical catchments. *Journal of Hydrology* 42: 281-300
9   Gallup, C.D., Edwards, R.L., Johnson, R.G. 1994. The timing of high sea levels over the past 200,000 years. *Science* 263: 796-800
10. Grigg, R.W. 1984. Resource management of precious corals: a review and application to shallow water reef building corals. *Marine Ecology* 5: 57-74
11. Hogendorn, J., Johnson, M. 1986. *The Shell Money of the Slave Trade*. London: Cambridge University Press
12. Jokiel, P.L. 1978. Effects of water movement on reef corals. *Journal of Experimental Marine Biology and Ecology* 35: 87-97
13. Kamukuru, A.T. 1992. *Costs and Earnings of Basket Trap and Handline Fishery in the Dar-es-salaam Region of Tanzania*. University of Kuopio, Finland. 1-43
14. Klumpp, D.W., McKinnon, A.D. 1992. Community structure, biomass and productivity of epilithic algal communities on the great barrier reef: dynamics at different spatial scales. *Marine Ecology Progress Series* 86: 77-89
15. Lessios, H.A., Robertson, D.R., Cubit, J.D. 1984. Spread of *Diadema* mass mortality through the Caribbean. *Science* 226: 335-337
16. McClanahan, T.R. 1988. Coexistence in a sea urchin guild and its implications to coral reef diversity and degradation. *Oecologia* 77: 210-218
17. McClanahan, T.R. 1990. Kenyan coral reef-associated gastropod assemblages: distribution and diversity patterns. *Coral Reefs* 9: 63-74
18. McClanahan, T.R. 1990. *Hierarchical Control of Coral Reef Ecosystems*. PhD Dissertation, University of Florida, Gainesville
19. McClanahan, T.R. 1992. Resource utilization, competition and predation: a model and example from coral reef grazers. *Ecological Modelling* 61: 195-215
20. McClanahan, T.R. 1994. Kenyan coral reef lagoon fish: Associations with reef management, substrate complexity, and sea urchins. *Coral Reefs* 13: 231-241
21. McClanahan, T.R. 1995. Harvesting in an uncertain world impact of resource competition on harvesting dynamics. E*cological Modelling* 80: 1-19

22. McClanahan, T.R. 1995. A coral reef ecosystem-fisheries model: impacts of fishing intensity and catch selection on reef structure and processes. *Ecological Modelling* 80: 21-26

23. McClanahan, T.R. 1995. Fish predators and scavengers of the sea urchin *Echinometra mathaei* in Kenyan coral-reef marine parks. *Environmental Biology of Fishes* 43: 187-193

24. McClanahan, T.R., Muthiga, N.A. 1988. Changes in Kenyan coral reef community structure due to exploitation. *Hydrobiologia* 166: 269-276

25. McClanahan, T.R., Muthiga, N.A. 1989. Patterns of predation on a sea urchin, *Echinometra mathaei* (de Blainville), on Kenyan coral reefs. *Journal of Experimental Marine Biology and Ecology* 126: 77-94

26. McClanahan, T.R., Shafir, S.H. 1990. Causes and consequences of sea urchin abundance and diversity in Kenyan coral reef lagoons. *Oecologia* 83: 362-370

27. McClanahan, T.R., Nugues, M., Mwachireya 1994. Fish and sea urchin herbivory and competition in Kenyan coral reef lagoons. *Journal of Experimental Marine Biology and Ecology* 184: 237-254

28. McClanahan, T.R., Mutere, J.C. 1994. Coral and sea urchin assemblage structure and interrelationships in Kenyan reef lagoons. *Hydrobiologia* 286: 109-124

29. Munro, J.L. 1989. Fisheries for giant clams (Tridacnidae: Bivalvia) and prospects for stock enhancement. In *Marine Invertebrate Fisheries: Their Assessment and Management*, ed. Caddy, J. F., pp. 541-558. New York: Wiley

30. Newell, N.D. 1972. The evolution of reefs. *Scientific American* 226: 54-65

31. Newton, L.C., Parkes, E.V.H., Thompson, R.C. 1993. The effects of shell collecting on the abundance of gastropods on Tanzanian shores. *Biological Conservation* 63: 241-245

32. Obura, D.O. 1995. *Environmental Stress and Life History Strategies, a Case Study of Corals and River Sediment from Malindi, Kenya*. PhD Dissertation, University of Miami, Miami

33. Pilson, M.E.Q., Betzer, S.B. 1973. Phosphorus flux across a coral reef. *Ecology* 54: 581-584

34. Randall, J.E. 1987. Collecting reef fishes for aquaria. In *Human Impacts on Coral Reefs: Facts and Recommendations*, ed. Salvat, B., pp.29-39. French Polynesia: Antenne Museum E.P.H.E.

35. Rogers, C.S. 1990. Responses of coral reefs and reef organisms to sedimentation. *Marine Ecology Progress Series* 62: 185-202

36. Rosen, R. 1971. The distribution of reef coral genera in the Indian Ocean. *Symposium of the Zoological Society of London* 28: 263-299

37. Samoilys, M.A. 1988. Abundance and species richness of coral reef fish on the Kenyan coast: The effects of protective management and fishing. *Proceedings of the 6th International Coral Reef Symposium* 2: 261-266

38. Sargent, M.C., Austin, T.S. 1949. Organic productivity of an atoll. *American Geophysics Union Transactions* 30: 245-249

39. Sheppard, C., Price, A., Roberts, C. 1992. *Marine Ecology of the Arabian Region*. London: Academic Press

40. UNEP/IUCN. 1988. *Coral Reefs of the World* Vol. 2. Gland Switzerland: IUCN/UNEP

41. Vermeij, G.J. 1978. *Biogeography and Adaptation: Patterns of Marine Life*. Cambridge, MA: Harvard University Press

42. Wilkinson, C.R., Willams, D.M., Sammarco, P.W., Hogg, R.W., Trott, L.A. 1984. Rates of nitrogen fixation on coral reefs across the continental shelf of the central Great Barrier Reef. *Marine Biology* 80: 255-262

43. Yaninek, J.S. 1978. A comparative survey of reef-associated gastropods at Maziwa Island, Tanzania. *Journal of the East Africa Natural History Society and National Museum* 31: 1-16

Chapter 5

Intertidal Wetlands

R.K. Ruwa

Intertidal wetlands (Fig. 5.1) are an ecotone or transitional zone between marine, estuarine and terrestrial ecosystems. The transitional nature of the environment from land to saltwater creates gradients or zones in which the species composition and ecological processes change (15,25,26). Ecotones often allow numerous species to coexist along resource or environmental stress gradients. Consequently, the intertidal zones of tropical regions are well endowed with numerous forms of marine life of different evolutionary origins. This rich biological diversity increases from sandy shores to rocky shores to mangrove and seagrass ecosystems. Various biotic and abiotic factors have been used to explain the presence, absence or abundance of species in the intertidal zone (25). The following attributes have significance to the uniqueness of the intertidal ecology in the East African region.

Shoreline Geomorphology

The geology and shape of the shoreline in the region is variable. Shorelines along the coasts of the continent and the Madagascar subcontinent are characterized by deep indented creeks or ancient river valleys that have been drowned by the sea (45). Found along these coasts are sandy shores, dunes, rocky shores and mangrove forests or vegetation. Most shorelines of both continents and islands are sheltered from heavy wave action by fringing coral reefs that run almost continuously parallel to the coasts - only being broken at river mouths (see Chapter 3).

Rocky shores in the East African region are predominantly fossil Pleistocene coral rocks while sandy shores are a mixture of land-derived soils and sand created from the erosion of coral skeletons (45,50). Limestone cliffs of rocky shores are heavily eroded giving them a characteristic cavernous undercut shape. The rocky coasts of some oceanic islands are, however, granitic rocks (such as Mahe Seychelles) or basaltic rock (such as Mauritius and Comoros Islands). Some coral islands are atolls formed from corals (such as Aldabra Atoll Seychelles) and have limestone cliffs of interglacial origin.

Box 5.1. Environmental Stress

Stress is a broad term used by ecologists to describe a number of environmental processes and changes that reduce growth rates or induce mortality of organisms inhabiting the ecosystem (3,4). Stress is, therefore, often relative to the organisms or assemblage of species and their specific environmental adaptations, although some extreme environmental conditions are a stress to most complex organisms (such as hurricane winds and high ultraviolet radiation). Stress is often a rare event that produces environmental conditions at or beyond the limits of the species' range of tolerance. Environments that experience rare environmental events more often are considered more stressful. Species' evolutionary adaptations to stress depend on the particular environmental stress but often include some combination of traits. If the organism has a high metabolism it should also have a small body size or short life and produce a large numbers of small propagules; however, if the organism has a low metabolism it may have a long life and tolerance to low resource levels. These species may often be evolutionary old or ancestral forms that lack energy-expensive adaptations of more complex organisms adapted to a predator-prey arms race. Stressed ecosystems may have short and simple food chains or webs or an abundance of species feeding low on the food chain. They may also have low predation intensity, reduced numbers of species and reduced diversity because of the high dominance of the few species adapted to the environmental stress. Stressful environments may also produce positive associations among species because species may be able to ameliorate the habitat such that one or more species can facilitate the existence of other species (1,2).

T.R. McClanahan

References

1. Bertness, M.D. 1994. Positive interactions in communities. *Trends in Ecology and Evolution* 9: 191-193.
2. Bertness, M.D., Hacker, S.D. 1994. Physical stress and positive associations among marsh plants. *American Naturalist* 144: 363-372.
3. Grime, J.P. 1979. *Plant Strategies and Vegetation Processes.* Chichester: John Wiley & Sons
4. Vermeij, G.J. 1978. *Biogeography and Adaptation: Patterns of Marine Life.* Cambridge: Harvard University Press

---

Tides

Tides in the region are mainly semidiurnal (there are two high and low tides per day; 50). Tidal ranges during spring or extreme low and high tides vary between 3 and 4 meters in Kenya, Tanzania and Madagascar. Most of the oceanic islands of the Seychelles, and Mauritius experience lower tidal ranges, ranging from 1.5 to 2.0 meters during spring tides.

The implication for the large tidal range is that a large volume of seawater periodically drifts back and forth replacing nearly all the water in shallow lagoons

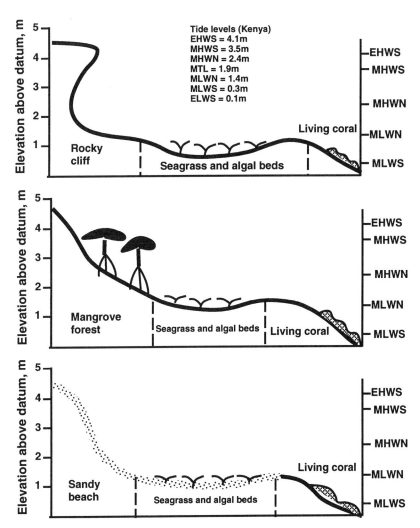

Figure 5.1. Three typical intertidal wetland shore profiles and habitat types. Graphs indicates tidal levels for East African shores on the right vertical axis where EHWS = Extreme high water spring tide, MHWS = Mean high water spring, MHWN = Mean high water neap, MLWN = Mean low water neap, MLWS = Mean low water spring. The high of the shore above datum (an arbitrary point around spring low tide) is shown of the left vertical axis.

and rock pools along the shore. At extreme high tides, large waves can pass over the coral reef and wave energy can dislodge littoral organisms, break cliffs, erode and transport sediments. These impacts are most pronounced during the SE monsoon when high winds create large waves (38). Consequently, the greatest amount of physical damage is likely to occur when the highest spring tides coincide with the SE monsoons. In the macro-tidal shores the lowest tides also occur during the hottest part of the day (2). Low tides occurring during the hot and dry NE monsoon can create extremes of temperature and desiccation for shore-dwelling organisms. Finally, the vertical distribution zones of littoral organisms are wider on shores with larger tidal ranges than shores with small tidal ranges.

Cyclones

The cyclone zones are between the 10° S and 30° S latitude in the southern hemisphere and 10° N to 30° N in the northern hemisphere (50). Cyclones cause physical destruction of habitats, increase water turbulence and increase the sediment in seawater. Cyclones inhibit development of mangrove forests that require a stable and building sandy substrate. Limitations set by cyclones are only important in the southern section of this region (such as Madagascar and Mozambique) while the northern section (< 10° N) of this region does not experience cyclones.

Hydrography

Whereas tidal currents and longshore currents play the most significant and immediate role in transport and dispersal of seeds and eggs, food, nutrients, plankton and gases, the major oceanic circulation patterns are responsible for the biogeographic distribution and biological diversity in the region (53). For example, the temperate geographic areas of southern Africa around Durban support a tropical faunal component owing to the warm Mozambique Current branching from the South Equatorial Current (Fig. 3.1).

Based on the major oceanic circulation patterns (see Fig. 3.1, Chapter 3) there are three hydrographic regimes. These are:

(1)   Somali marine water regime - this is influenced by the annual reversing of the Somali current, cold upwelling water, and high salinity originating from the Red Sea and Indian subcontinent through the North Equatorial Current. This current, however, only reaches around 2° S before being diverted seaward and does not affect the southern section of East Africa.

(2)   East African marine water regime - this is delineated in the north by the Equatorial Counter Current and the south by the South Equatorial Current.

(3)   East Southern Africa marine water regime - the northern boundary is determined by the geographical position of the South Equatorial Current that flows southwards west of Madagascar in the Mozambique channel. The second current branch flows east of Madagascar.

Salinity

Seawater salinity can play an important influence on the flora and fauna of the intertidal zone (16). Significant sources of freshwater are from rain, rivers and underground aquifer discharges. Major river drainages include the Tana and Sabaki rivers in Kenya, the Rufiji and Ruvuma rivers in Tanzania and the Zambezi and Limpopo in Mozambique. In Madagascar, rivers with major drainage basins include Mangoky, Tsiribihina and Betsiboka rivers which all flow to the Western Coast.

The wetter coasts include Tanzania, Mozambique and eastern Madagascar. The northern coasts of Kenya and Somalia are dry and freshwater discharges are largely restricted to river mouths and seepage out from underground aquifers. In various areas throughout the region freshwater seepage or submarine groundwater discharges have been observed.

Ecological Aspects

Intertidal zones are often referred to by the name of the species of flora or fauna that characteristically dominate the substrate in terms of their numbers or percent cover. Thus, vegetated intertidal zones may be referred to as mangrove shores, seagrass or algal beds but, in some cases, sparsely vegetated or unvegetated intertidal zones may be referred to by faunal organisms (such as oysterbeds). Otherwise, the shores are often named by the geologic substrate (such as rocky shore or cobble bed) or sediment deposits (such as sand and mud). This chapter compares the distribution and ecology of the flora and fauna of these littoral biotopes.

Regional studies of intertidal ecology have mostly focused on the distribution and behavior of a few organisms. In most cases, these are short-term studies conducted during expeditions by scientists from developed countries. The short-term nature of previous studies often causes scientists to focus on the taxonomy or nomenclature of the studied organisms as this is often the starting point for most ecological studies. The taxonomy of many organisms is still not well known and name changes are a frequent occurrence as the evolutionary relationships between species become better understood. Significant initial attempts at the nomenclature of intertidal algae (14), seagrasses and mangroves (13), molluscs (9,29) and crabs (1,7,8,28,56) have been completed. The South African fauna has been described in more detail (9) but these studies are usually incomplete for the more equatorial parts of East Africa. Consequently, the fledgling marine biologist will often have to consult a number of taxonomic studies compiled for other parts of the world before learning the names of the less conspicuous organisms. Older taxonomic works should generally be avoided if newer publications exist because of recent changes in the nomenclature.

Rocky Shores

Rocky shores are high water energy environments with high geologic erosion that do not allow the deposition and build up of sediments and coral sand. Ecological studies on rocky shores have been carried out at Mahe and Aldabra in the Seychelles (51,52,53), Dar es Salaam, Tanzania (12), Mombasa, Kenya (27,31) and Mogadishu in Somalia (6). Early studies by Lewis (15) described behavioral and morphological adaptations to living on rocky shores. He also developed a universal terminology for describing rocky shore zonation (Fig. 5.2).

In his scheme the shore is divided into the following zones:

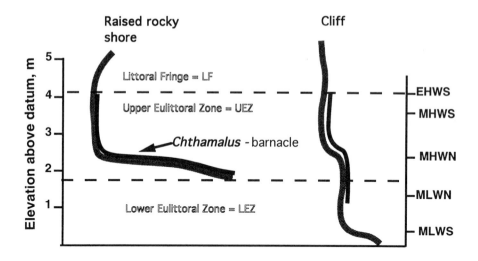

Figure 5.2. Description of shore levels using the terminology proposed by Lewis (15). An example of a raised rocky shore and a rocky cliff that extends to the lower shore before joining the platform.

(1)   Littoral fringe whose upper limit is delineated by the upper limit of the littorinid gastropods and the lower limit by the upper limit of the barnacles (mostly *Chthamalus* or *Euraphia* spp.).

(2)   Eulittoral zone whose upper limit is delineated by the upper limit of the barnacles and its lower limit by the upper limit of the fleshy algae *Sargassum or Turbinaria*, the seagrass, *Syringodium* or soft corals. In some cases the eulittoral zone can be split at the mean tide level to form an upper eulittoral zone and lower eulittoral zone.

(3)   The sublittoral zone which begins at the lower limit of the eulittoral zone and extends into shallow waters at low tide.

Behavioral, physiological and morphological adaptations often result in species being confined to specific zones on the rocky shore. These zones, however, may be static for sessile or inactive organisms but dynamic for the more active fauna. Therefore, the ecology of intertidal organisms is often predictable based on the activity level of the specific organism. I briefly discuss the ecology of rocky-shore organisms based on three levels of activity - inactive, moderately active, and active fauna.

The Faunal Zones

Sessile or Inactive Fauna. Sessile fauna occur abundantly on rocky shores and throughout this region (Table 5.1). Slow-moving snails of the genera *Littorina* and *Nodolittorina* occur abundantly in the littoral fringe in East Africa as in the rest of the world (15). Abundant organisms in the eulittoral zone include the oysters (such as *Crassostrea cucullata*), the bivalve mollusc *Isognomon* *dentifer* and the ubiquitous barnacle - *Chthamalus* spp. Although these species are generally eulittoral organisms their zonation patterns may vary geographically. Studies on the settlement patterns of barnacles and oysters indicate that the early settlers have broader distribution patterns than adults but that biological processes of competition and predation result in more narrow adult distribution patterns (41). For example thaid and muricacean snails are important predators of barnacles in the eulittoral zone of East Africa (54) and their abundance may restrict barnacles and oysters to zones where conditions for predator survival are difficult. Consequently, many organisms may be able to survive the physical conditions in different zones but species interactions often reduce species abundance outside of zones of greatest abundance.

The earliest studies of rocky shores focused on abiotic factors such as temperature and desiccation in controlling species ecology. More recent studies, however, suggest that biotic interactions of predation and competition are very important but are mediated by the abiotic environment and the rock-dwellers reproductive success (such as recruitment; 21,22).

The eulittoral zone supports the most numbers of species but these species are frequently found at low population densities suggesting low reproductive success or high mortality of recruiting juveniles. Records of settlement patterns of oysters and barnacles indicate periodically profuse settlement from the plankton during the SE monsoons but populations are rapidly diminished through competition and predation (41). Predatory muricacean and thaid snails that inhabit the higher eulittoral zone prevent barnacles and oysters from colonizing the higher tidal levels while at lower levels the greatest sources of mortality may be competition with other fouling organisms. Consequently, only a narrow zone or band of barnacles and oysters is formed at the eulittoral zone. These zones can vary with region and are often absent depending on abiotic and biotic variations that are not well understood. Many questions about species interactions and its interaction with abiotic factors await to be unraveled by future field studies. Algae are also sessile organisms but will be discussed in a separate section on flora.

Moderately Active Fauna. A cursory study of rocky shores might suggest that many of the common mollusc species are inactive or slow moving. However, if one marks the spots where individuals are resting and then follows their movements over time one will begin to recognize that many species undertake migrations of different lengths. For example, studies of neritid snails (Fig. 5.3a) showed significant upward and downward migrations following day and night periods, the spring and neap tidal cycle and the even on the annual monsoonal cycle (37,38,39,58,62). During the SE monsoon, when wave energy is high, neritids rest at a higher tidal level than during the calmer NE monsoons. Depending on the time, nature of tides and prevailing season and winds the resting position of the of three closely related species of herbivorous snail, *Nerita plicata*, *N. undata* and *N. textilis* are found at the upper eulittoral zone and littoral fringe levels. These species of neritids have interspecific zonation patterns with a downward slope gradient as follows: *Nerita plicata*, *N. undata*, *N. debilis*, *N. textilis* and *N. albicilla* (see Table 5.2). Some of these species may often be found in the same zones but often inhabit different sized holes in the rock substrate. Consequently, these species display distribution patterns that suggest some partitioning of resources but the mechanisms

Table 5.1. Distribution of sessile and inactive fauna in relation to the zones in which they occur abundantly at various geographical areas in the Western Indian Ocean (30). The acronyms LF, UEZ, LEZ, EZ and SLZ stand for littoral fringe, upper eulittoral zone, lower eulittoral zone, eulittoral zone and sublittoral zone.

Species	Seychelles	Aldabra	Tanzania	Kenya	Somalia
Littorina kraussi	-	-	LF	LF	LF
Nodolittorina natalensis	-	-	LF	LF	LF
Littorina scabra	LF	LF	-	LF	-
Isognomon dentifer	UEZ	LF	UEZ	UEZ	LF
Acmaea profunda	-	EZ	UEZ	UEZ	UEZ
Crassostrea cucullata	LEZ	EZ	UEZ	UEZ	UEZ
Acanthopleura spinigera	-	-	UEZ	UEZ	UEZ
Tetraclita squamosa	LEZ	EZ	LEZ	UEZ	-
Pyrene ocellata	-	-	-	UEZ	-
Strigatella paupercula	-	EZ	-	UEZ	-
Turbo coronatus	-	-	-	UEZ	-
Vasum turbinellus	-	-	EZ		UEZ
Planaxis sulcatus	-	EZ	-	UEZ	-
Cerithium caeruleum	-	-	-	LEZ	-
Conus ebraeus	-	-	-	LEZ	-
Engina mendicaria	-	-	-	LEZ	-
Oestrea hyotis	-	-	-	LEZ	-
Echinometra mathaei	-	SLZ	LEZ	LEZ	-

that cause this resource partitioning are not well known at present.

Thaid snails are the second most abundant group of snails on rocky shores. Unlike neritid snails they do not make regular cyclic migrations but wander about the eulittoral zone in search of their molluscan and barnacle prey. The thaids *Drupa ricina*, *Drupa morum*, and *Mancinella alouina* are confined to the lower eulittoral zone of rocky shores and are also common in shallow coral-reef areas. The species *Thais savignyi*, *Morula fenestra*, *M. granulata*, *M. anexeres*, *Thais aculeata* and *Purpura panama* move about in the lower eulittoral zone and the upper eulittoral zone and do not appear to have any interspecific zonation patterns (31). How these species partition resources remains to be studied.

Active Fauna. Crabs, skinks and the occasional predatory birds are the most common active fauna feeding on rocky shores. Skinks are a reptile of terrestrial origin that occupy the highest shoreline levels and are commonly seen feeding on small invertebrates (such as amphipods) just below the terrestrial vegetation line. Crabs are conspicuous members of the rocky shores. The crab *Grapsus maculatus* undergoes daily migrations following the rise and fall of the tides (Fig. 5.3b; 30). This species often forms groups and grazes on algae recently exposed to the air on the edge of the breaking waves. The clamoring of crab legs against limestone as these species climb up and down the shore with the breaking waves is a common site. Wading birds are occasionally seen feeding along rocky shores but their ecology and influence on permanent rocky shore fauna has been poorly studied in East Africa.

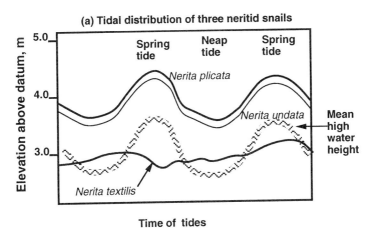

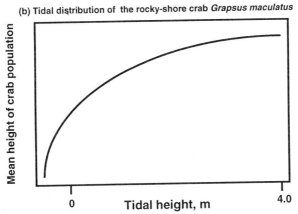

Figure 5.3. (a) mean resting positions during the daylight of the gastropod *Nerita plicata*, *N. undata* and *N. textiles* during spring and neap tidal cycles. (b) relationship between the mean height above datum of the abundance of the herbivorous crab *Grapsus maculatus* near Mombasa, Kenya.

Influence of Topography on Faunal Zones.
Zonation patterns can be influenced by the shape of the rocky shore. Overhangs and platforms are common shoreline features on coasts with Pleistocene cliffs. Where the rocky cliff is cut by waves the extent of the vertical and horizontal surfaces will influence the extent of ecological zones and species abundance and distribution patterns. A sloping cliff (Fig. 5.4a) has a larger surface exposed to the tidal fluctuations than a vertical cliff (Fig. 5.4b) and ecological zones, may, therefore, be broader.

Reef platforms may broaden species distribution patterns but species must tolerate the extreme heat and desiccation common to these platforms (15). Some species appear to avoid reef platforms and are nearly always found on vertical

Table 5.2. Distribution of moderately active fauna in relation to the zones in which they occur abundantly at the various geographical areas in the Western Indian Ocean region (30). The acronyms LF, UEZ, LEZ, EZ and SLZ are as explained in Table 5.1.

Species	Seychelles	Aldabra	Tanzania	Kenya	Somalia
Nerita plicata	UEZ	LF	LF	LF	LF
Nerita undata	-	EZ	LF	LF	-
Nerita debilis	LEZ	-	-	UEZ	LF
Nerita textilis	-	EZ	UEZ	UEZ	UEZ
Nerita albicilla	LEZ	EZ	UEZ	UEZ	UEZ
Thais savignyi	-	UEZ	-	UEZ	-
Morula fenestra	-	-	-	UEZ	-
Thais aculeata	-	-	-	UEZ	-
Morula granulata	SLZ	LEZ	LEZ	UEZ	-
Morula anexeres	-	UEZ	-	UEZ	LEZ
Purpura rudolphi	-	UEZ	-	LEZ	UEZ
Drupa ricina	SLZ	LEZ	-	LEZ	-
Drupa morum	-	LEZ	-	LEZ	-
Onchidium verruculatum	-	-	-	LEZ	-
Mancinella alouina	-	-	-	LEZ	-

surfaces (12,31). Both Hartnoll and Ruwa found that neritids such as *Nerita plicata*, *Nerita textilis* and *Nerita undata* were largely restricted to vertical surfaces in Tanzania and Kenya. In contrast, the bivalves *Crassostrea cucullata* and *Isognomon dentifer* and the barnacle *Chthamalus* are most abundant on reef platforms. Bivalves and barnacles can close their shell apertures firmly during low tides to mitigate against desiccation. Neritids can also avoid desiccation by storing water in a reservoir in their shells (43) and reduce temperatures through evaporative cooling from the surfaces of the exposed foot (60,61). They can also tightly seal their shells with their very tight-fitting operculi. Their distribution patterns suggest, however, that these adaptations may not be adequate for occupying horizontal surfaces during the day. Thus, at night these species make feeding migrations to lower level platforms to graze but return to vertical surfaces at flood tides and during daylight (39,58). Additionally, vertical surfaces often have greater topographic complexity or more holes than horizontal surfaces. This allows these non-attached species to find refuge from predators and desiccation. Attached bivalves and barnacles may have less need of these hide outs. Figure 5.5a summarizes the distribution patterns of fauna on rocky shores emphasizing the critical factors of habitat discussed earlier.

Floral Zones

Surfaces of most rocky shores are covered by both microscopic and macroscopic algae that are the major prey for many of the grazing crabs and snails. Pleistocene cliffs support a conspicuous, almost maritime zone of blue-green algae or Cyanobacteria (*Entophysalis* spp.) followed by a littoral fringe zone of red algae (Rhodophyta such as *Bostrichia*) which is sometimes covered by a green algae *Chaetomorpha*. The upper eulittoral zone is dominated by mixed species of red algae particularly on vertical or overhang surfaces. Red algae appear to be most common in the low sunlight conditions of overhangs and vertical surfaces. Sometimes the

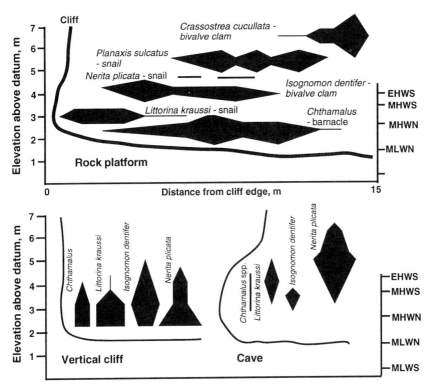

Figure 5.4. Distribution of organisms on (top) a horizontal rocky shore platform and (bottom) vertical rocky cliffs.

lower eulittoral zones and horizontal platforms are dominated by green algae (27). The common species of algae and their various zones are depicted in Figure 5.5b.

Influence of Topography on Floral Zones.

Vertical and horizontal zonation affected by tidal oscillations appears to be a better predictor of rocky shore fauna than algae distribution. Algae are directly affected by the orientation of rocky shore surfaces to the sun and which effects both the primary production and desiccation stress (55). Thus, algal distributions are better predicted by the shore profile and surface complexities. Cliffs with overhangs have richer algal cover than cliffs without overhangs that are exposed to high solar insolation and desiccation during low tides. Exposed platforms are frequently dominated by green and blue-green algae but these can also be common in shaded areas. Fully exposed cliffs are nearly bare of macroscopic algae save for pits and crevices that are wetted and cooled by seepage. Unfortunately, the role that rocky-shore herbivores play in algal distribution has been poorly studied in this region but future research will help to understand their potentially important role. Studies in other regions suggest that both herbivorous snails and fishes can have significant roles in affecting the abundance and diversity of rocky-shore algae (21,22).

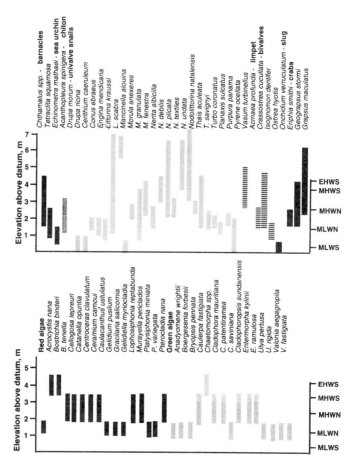

Figure 5.5. Summary of the distribution of dominant (top) fauna and (bottom) flora on rocky shores near Mombasa.

Sandy Shores

East African beaches are formed largely from the breakdown of the shells of corals, calcareous algae, and other shelled organisms. Beaches are formed between rocky-shore headlands and are depositional areas for shell fragments and decaying plants and algae dislodged from rocky shores, coral reefs and seagrass beds. Beaches are formed by waves and tidal fluctuations but, because of their sandy substrate, the fauna associated with them is very different from rocky shores. Crabs have been an evolutionarily successful macrofauna because they can burrow into the sand and therefore avoid predators, damage from waves and desiccation of their gills. Beach ecosystems produce very little of their own energy so detritus from other ecosystems forms the basis for the food chain. Consequently, detritivorous crabs, sand hoppers, various worms, sand dollars and snails and their predators are the major faunal groups found on beaches.

Zonation. Ecological studies on sandy beach crabs have been undertaken in Somalia, (57) Kenya, (32) Aldabra Atoll, (52) Mozambique (19) and South Africa with a tropical species component (3,10,23). Based on these studies a consistent pattern of crab zonation has been described from the upper to lower shore as 1) *Ocypode* Zone, 2) *Dotilla* Zone, and 3) *Uca-Macrophthalmus* Zone. Crab zones (Fig. 5.6) are related to tide levels, the position of the water table and submarine groundwater discharge or seepage.

Ocypode crabs feed on algae and seagrass drift and detritus that accumulates and decays on sandy shores. The large body size of *Ocypode* crabs allows them to build very large tunnels to the groundwater where they can wet their gills during low tides. Consequently, they are found from just above extreme high water to the mean high water of neap tides. *Dotilla* which generally feeds on finer detrital particles is, therefore, found at intermediate tidal levels. *Dotilla* has a smaller body size and less burrowing ability than *Ocypode* and may require being closer to the water table. Studies suggest that *Dotilla* is rarely further than 10 to 20 cm from the water table at low tide (32). The lowest zone occurs between the mean tidal level and spring tides and the sand or soils near the surface is always wet through seepage. Faunal distributions shown in Table 5.3 indicate that many sandy beach macrofaunal species are encountered in many reported areas within this region.

Ocypode crabs show an interesting distribution pattern in that whenever more than one species are present on a beach one species far outnumbers the others. The two common *Ocypode* crabs *O. ryderi* and *O. ceratophthalmus* were found together at Inhaca, Mozambique, Durban and Richards Bay, South Africa (10,19,23) but on each beach one species far outnumbered the other. Similarly at a beach of Sar Uanle, Somalia where *O. cordimanus* and *O. ryderi* were found it was observed that in a mixed catch only 7.4% were *O. cordimanus* (57). In mixed catches of *O. ceratophthalmus* and *O. cordimanus* at Mahe, Seychelles showed the former far outnumbered the latter (52). The role of biogeography and local competitive interactions on these abundance patterns has yet to be studied in sufficient detail to explain the above curious pattern.

Both the upper *Ocypode* and the intermediately placed *Dotilla* zone contain only one or two species. The most species rich zone is the lowest and wettest areas or the *Uca-Macrophthalmus* zone (Table 5.3). This zone can shift with the changing tides and water table levels. Therefore, some aspect of water availability and the frequency of wave and water table changes or disturbances may create the abiotic conditions for more species to coexist.

Unfortunately, studies on other sandy-shore faunal groups in this region have been poorly studied from an ecosystem viewpoint. Sandy-shore species play an important role in the breakdown and decay of plant matter which is fed on by a variety of microbes and invertebrates which, in turn, becomes food for fishes and shore birds. Many view the build up of seagrass and algae on the beach as unsightly but the decay and subsequent utilization of these organisms by the intertidal fauna is one of the energetic pathways of intertidal ecosystems that sustains its bountiful productivity.

Mangroves

Beaches typically develop in wave-exposed environments that make the establishment and growth of plants difficult. When sand or silt substrate is protected

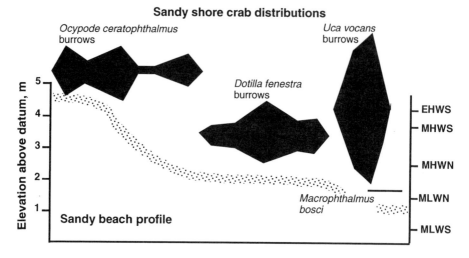

Figure 5.6. Distribution of dominant crab fauna on East African sandy shores.

from strong waves, however, mangrove trees usually flourish. Consequently, many of the creeks, river mouths or sheltered shores create the appropriate environment of sandy substrate, shelter and lowered salinities that are favored by mangrove species. Mangroves are therefore largely estuarine plants that have evolved from terrestrial environments and have numerous adaptations to saturated soils of high salinity relative to the terrestrial environments of their evolutionary origin.

Different taxonomic families and genera of mangroves have separately evolved morphological and physiological adaptations that enable them to occupy intertidal saltwater wetlands (Fig. 5.7). Consequently, mangrove is an ecological term to describe vegetation adapted to brackish-water environments rather than an evolution-based taxonomic classification. Mangroves grow and survive well under freshwater conditions but are poor competitors with other terrestrial plant species. In fact, mangrove growth rates are highest under low salinity conditions but can survive salinities well above seawater (18). High salinities are common in estuaries where evaporation concentrates salts and are a severe abiotic stress that mangroves must and are able to contend with. Consequently, it is often competition between mangroves and other terrestrial vegetation that causes mangroves to occupy brackish water rather than the inability to survive the abiotic conditions of the terrestrial environment. Mangroves are superior competitors in non freshwater environments and have been successful in colonizing and flourishing in wave-sheltered saltwater wetlands throughout the tropics. Temperate saltwater wetlands are dominated by marsh rushes and reeds rather than mangrove trees that are killed by freezing temperatures.

In brief the following are significant adaptations of mangroves. They have elaborate rooting systems which firmly anchor them in the unstable substrate. Their spongy root structures may grow above the low-oxygen ground to allow gaseous exchange. The roots that grow above the ground are called breathing roots or pneumatophores (Fig. 5.7). To facilitate their dispersal by water, they have spongy seeds some of which are even viviparous propagules. All mangroves are salt excluders and some can even secrete salt through the leaves such as *Avicennia marina*. This is an adaptation to salt tolerance. Mangroves have maintained most of

Table 5.3. Geographical distribution in the Western Indian Ocean region of sandy beach macrofauna (31). The letter (P) indicates where the organism was seen and recorded: The Countries are as follows: S.Afr South Africa), Moz (Mozambique), Tan (Tanzania), Ken (Kenya), Som (Somalia), Mad (Madagascar), Sey (Seychelles), Ald (Aldabra Atoll).

Species	S. Afr	Moz	Tan	Ken	Som	Mad	Sey	Ald
Ocypode zone								
O. ceratophthalmus	P	P	P	P	P	P	P	P
O. cordimanus	P	P	P	P	P	P	P	P
O. ryderi	P	P	P	P	P	P	-	-
Dottila zone								
D. fenestrata	P	P	P	P	-	P	-	-
Hippa adactyla	P	-	P	P	-	P	-	-
Emerita austroafricana	P	P	P	P	-	-	-	-
Uca-Macrophthalmus zone								
Uca lactea	P	P	P	P	P	P	P	P
U. vocans	P	-	P	P	-	P	-	-
U. tetragonon	-	-	P	P	-	-	P	P
Macrophthalmus bosci	P	P	P	P	-	P	-	-
M. grandidieri	P	P	P	P	-	P	-	-
M. parvimanus	-	-	P	P	-	P	-	-
M. milloti	-	-	P	P	-	P	-	-
Thalamita crenata	P	P	P	P	P	P	P	P
T. gatavakensis	-	-	-	P	-	P	-	-
Scylla serrata	P	P	P	P	P	P	-	-
Natica gualtierana	-	-	P	P	P	-	-	-
Polinices mammila	P	-	P	P	P	-	P	-
Nassarius arcularius	P	P	P	P	P	-	P	-
N. coronatus	P	P	P	P	P	-	P	-
N. margaritifera	-	-	P	P	-	-	-	-
Strombus mutabilis	-	P	P	P	-	-	-	-
S. gibberulus	P	-	P	P	P	-	P	-
Nerita polita	P	P	P	P	-	-	P	-
Echinodiscus bisperforatus	P	-	-	P	-	-	P	-

their terrestrial characteristics. For example, their flowers are pollinated by insects and bats, hence the flowers are sweet smelling and colorful. To photosynthesize effectively their canopy is positioned just above high tide.

The center of mangrove diversity is in the central Pacific, Indonesian region. Some species from this region have expanded their distributions into the Western Indian Ocean and East Africa contains eight species of mangrove (Table 5.4) while only two species, *Avicennia marina* and *Rhizophora mucronata*, are found in the Red Sea (4). Mangroves in this region extend as far south as Durban in South Africa due to the warm-water currents traveling from the tropics. Most sites in East Africa, however, will harbor fewer than the eight species found in the region.

Mangroves are most extensive where big rivers discharge into wave-sheltered estuaries, such as the Rufiji delta in Tanzania (Fig. 5.8). There are also extensive mangroves in wave-sheltered bays or the leeward side of islands far away from estuaries or deltas. Mida Creek and the Lamu Archipelago in northern Kenya are examples of extensive mangroves living in brackish water conditions created by seepage. The northernmost extent of mangroves along the Indian Ocean coast is the

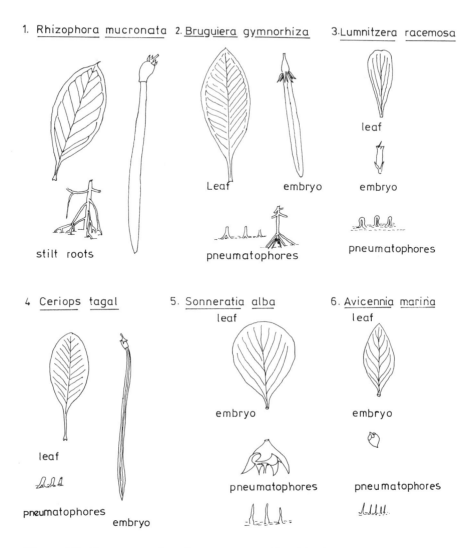

1. Rhizophora mucronata   2. Bruguiera gymnorhiza   3. Lumnitzera racemosa

stilt roots   Leaf   embryo   leaf / embryo

pneumatophores   pneumatophores

4. Ceriops tagal   5. Sonneratia alba   6. Avicennia marina

leaf   leaf   leaf

leaf   embryo   embryo

pneumatophores   pneumatophores   pneumatophores

pneumatophores embryo

Figure 5.7. Common species of mangrove trees found in the mangrove ecosystem showing the embryo, leaf and root (pneumatophores) morphology.

Juba/Shebele estuary in southern Somalia. It is significant that all species of mangrove except for *Heritiera littoralis* in this region are encountered on both the coasts of the African mainland and offshore islands. In the Red Sea, however, where little freshwater inflow occurs only two species are found. The roles of dispersal and abiotic stress on species richness deserves more research attention.

Zonation. The two major types of mangrove formations in this region include creek and fringe formations that are dependent on the geomorphology of the coastline (18). Fringe mangroves are solitary or small clusters of trees composed of one or two species growing in the front or at the bases of rocky cliffs or high-energy shores

Table 5.4. Distribution of mangrove species and cover in the East African region: Seychelles (Sey), Mauritius (Mau), Reunion (Reu), South Africa (S.Afr Durban), Madagascar (Mad), Mozambique (Moz), Tanzania (Tan), Kenya (Ken) and Somalia (Som) including crab and prawn fisheries production. The acronym: MSY = maximum sustainable yield as summarized by Ruwa (34).

Species	Sey	Mau	Reu	S.Afr	Mad	Moz	Tan	Ken	Som
1. *Sonneratia alba*	+	-	-	-	+	+	+	+	-
2. *Rhizophora mucronata*	+	+	–	–	+	+	+	+	+
3. *Bruguiera gymorhiza*	+	+	–	+	+	+	+	+	–
4. *Ceriops tagal*	+	–	–	+	+	+	+	+	–
5. *Xylocarpus granatum*	+	–	–	–	+	+	+	+	–
6. *Avicennia marina*	+	-	-	+	+	+	+	+	+
7. *Heritiera littoralis*	-	-	-	-	+	+	+	+	-
8. *Lumnitzera racemosa*	+	–	–	–	+	+	+	+	–
Total number of species	7	2	–	3	8	8	8	8	2
Total mangrove area (km²)	–	–	–	–	3207	850	960	530	
MSY (tons/y) for crabs	–	–	–	–	5760	1530	1728	1057	–
Landing of prawns (tons/y)	–	–	–	–	1700	1000	979	350	–

exposed to the open sea. High wave energy, lack of sandy substrate and high salinity probably interact to reduce the extent of these small mangrove patches. Creek mangroves are found in sheltered bays with shallow topographic gradients that can form extensive forests exhibiting species zonation patterns from low to high water. In arid environments, such the Red Sea, few species are present to create zonation patterns. Many areas sheltered from waves are sparsely vegetated due to the salty soils that develop during hot times of the year.

Zonation patterns of common creek mangroves in this region (18,34) indicate a species gradient from low tide to high tide as follows: *Sonneratia alba*, *Rhizophora mucronata*, *Ceriops tagal* and *Avicennia marina* (Fig. 5.8a). Because *Sonneratia alba* is uncommon on many shores it follows that *Rhizophora mucronata* is frequently the species in the lowest shore level. The lower leaves and limbs of *Rhizophora* often mark the high spring tide water mark. Crabs are the most conspicuous macrofauna in the mangroves and crab species may also exhibit some zonation patterns (Fig. 5.8b). Crabs and potamidid gastropods feed on the fallen mangrove leaves and propagules such that few seeds survive on the mangrove floor for more than a tidal cycle. Recent studies indicate that high predation on mangrove propagules by crabs may play an important role in the survival and establishment of mangroves and subsequent zonation patterns (46). Therefore, the distribution and abundance of crabs is very important for mangrove survival, perhaps more than the abiotic environment. The way that abiotic and biotic factors influence crab distribution is not well studied. It has also been suggested that the burrowing activity of crabs increases the dissolved oxygen reaching the mangrove roots in

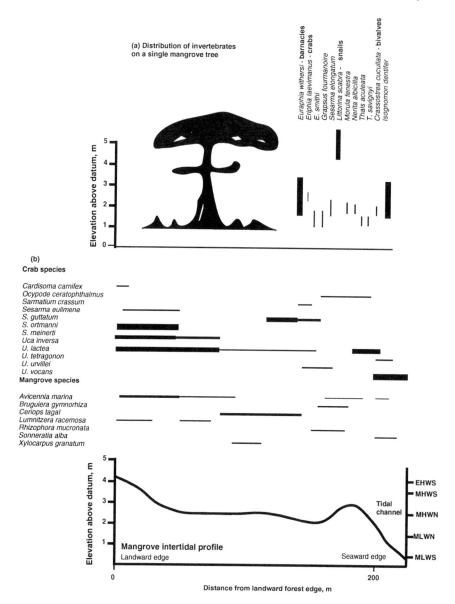

Figure 5.8. Distribution of mangrove fauna (a) on a mangrove tree and (b) across a mangrove creek in Gazi, Kenya.

these low-oxygen sediments. Thus increasing the productivity of mangroves. These recent species-interaction studies suggest the important role that consumers play in affecting the distribution, diversity and productivity of mangrove ecosystems.

Ecological Functions

One of the notable attributes of mangroves is their high net primary productivity compared to other aquatic ecosystems (Table 5.5). This production results in high leaf and litter fall that forms the basis for the detritus food web of microbes, crabs, snails, and fish. Large amounts of detritus are consumed in the estuary but some is exported to other nearshore habitats and fisheries. Mangrove fisheries are usually more productive than other coastal fisheries because of their high net production and the relatively low abundance of piscivorous fish. Consequently, the protection of mangroves is important for the maintenance of nearshore fisheries as well as mangrove forestry.

Besides enhancing biological diversity through supporting the food chain, binding between roots and sediments creates a stable habitat for burrowing organisms (Fig. 5.8a) and their trunks are often sanctuaries to various arboreal organisms (Fig. 5.8b). The forests also support terrestrial fauna that includes birds, reptiles, mammals (such as bush pigs and monkeys) and insects (such as wasps, spiders, mosquitoes, and ants). The terrestrial flora is less diverse and is mainly composed of fungi, lichens and mistletoes. By acting as nutrient traps, through absorbing nutrients from effluents and rivers, the growth of epiphytes on seagrasses is discouraged. Thus, the luxuriant growth of seagrasses would otherwise be diminished by competition for light with the seagrass epiphytes. If high nutrient levels also reach corals, similar deleterious effects may occur - particularly when herbivores have been reduced through excessive fishing.

Mangrove vegetation structure provides important landmarks that give clues for coastal and marine birds where to find watering points. Seepage of freshwater from underground aquifers is common around mangrove vegetation area as evidenced from measurements from seepage outlets around such biotopes at low tide (40). A summary of faunal distribution in mangrove ecosystems is diagramatically presented in Figure 5.9. It is important to note that whereas some species are encountered in mangrove vegetation biotopes only others extend their distribution to brackish water or estuarine biotopes away from mangrove vegetation. This raises the question as to whether there is typical mangrove fauna. Additional studies need to focus on this question.

Seagrass Beds

Seagrasses (Fig. 5.10a) are flowering plants that have adapted to life in the sea on hard bottoms with sandy substrates. They are easily distinguished from algae by their roots or rhizomes that connect the growing blades and is the main form of plant expansion. Flowers and seeds are rarely observed except by the avid and observant snorkeller. Most seagrass beds are dominated by a few species but there are 12 species in East Africa (13). Seagrasses are distributed throughout the intertidal zone but also grow luxuriantly subtidally. Sandy intertidal platforms do not significantly slant and therefore the influence of tidal levels on their distribution and species coexistence is less significant compared to previously discussed

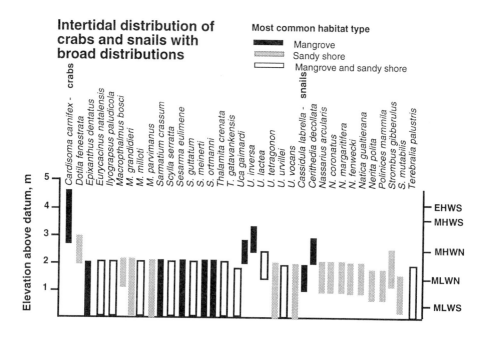

Figure 5.9. Distribution of fauna in brackish water. Combines fauna in mangrove forests, sandy shores and those species that occupy both habitats.

ecosystems. Factors such as disturbance of the sand, by waves and currents, and the existence of standing water during low tide (such as pools or seepage outlets) can be more important factors in maintaining coexistence patterns among these species. Different species prefer different microhabitats or successional seres (the time since disturbance) and have different susceptibilities to herbivorous fish and sea urchins (20). These factors interact to create the patchy distribution of seagrasses.

From a birds-eye view seagrass species distribution in the intertidal zone is patchy giving no distinct zones. In the less wave-disturbed subtidal areas, however, either one of the two species *Thalassodendron ciliatum* or *Thalassia hemprichii* is dominant in sandy substrate. The tallest-growing species *Enhalus acoroides* grows luxuriously in muddy substrate, may also be encountered in deep sandy pools but rarely forms the extensive beds of the other two species. *Thalassia hemprichii* is preferred by parrotfish and may be restricted from inhabiting deeper lightly fished areas by high herbivory.

The intertidal zone is rarely dominated by a single species (44) and frequent wave and sediment disturbances may stop any species from outcompeting others. Ecological studies of species composition changes after a disturbance suggest that there are colonizing seagrass species such as *Syringodium isoetifolium* that are replaced through competition with more robust and taller species such as *Thalassia hemprichii* and *Thalassodendron ciliatum*. The three smallest or early successional genera *Syringodium*, *Halodule* and *Halophila* are frequently found on recently-disturbed sand in shallow water. *Halodule* appears to prefer fine sand whereas *Halophila* is often surrounded by recently disturbed muddy substrates. These early

Table 5.5. Primary productivity in various tropical marine ecosystems (as grams of Carbon).

Ecosystem	Productivity, gC/m^2/y	Reference
Mangrove forest	5112	17
Mangrove leaf litter	366 to 1464	45
Phytoplankton		
Open ocean	50	40
Continental shelf	100 to 150	40
Upwelling areas	300 to 500	40
Seagrass/algal beds	900 to 4650	63
Sandy beach	5	47

successional species are less abundant in subtidal sandy areas and may not be able to tolerate the more intense herbivory and competition, with the more herbivore-resistant species, at greater depths. Less abundant species such as *Cymodocea serrulata* and *C. rotundata* are mostly found in these disturbed intertidal pools or divots formed by waves.

Seagrass beds offer a habitat to a variety of invertebrates (such as molluscs (Fig. 5.11a), echinoderms (Fig. 5.11b), crustaceans (Fig. 5.11c) and fishes (Fig. 5.10b). A few species are permanent residents which often have coloration or behavioral patterns for hiding in seagrass. Other species make foraging excursions into seagrass beds and return to resting or hiding spots in the more topographically complex coral reef. Grazers include those species that eat seagrass directly and those that feed on algal epiphytes growing on the seagrass leaves. Seagrass is tough and difficult to digest for many species and therefore a large amount of seagrass is not eaten and forms detritus that supports a diverse detrital food web. Grazers included both vertebrates and invertebrates. Grazing invertebrates include the sea urchins *Tripneustes gratilla* and *Echinothrix diadema* while detritivores include sea cucumbers (Holothurians), snails and a wide variety of worms. Herbivorous fishes include the speckled parrotfish (*Leptoscarus vaigensis*) and rabbitfishes (*Siganus*). Carnivores are mostly species of fish from the Mullidae, Lethrinidae, Lutjanidae, Balistidae and Diodontidae families. These species migrate into the intertidal zone during high tides and feed on invertebrates. Invertebrates find a partial refuge from fish predators during low tide periods when these vertebrate predators are restricted to subtidal areas. The diversity of fish species in seagrass beds is not nearly as high as coral reefs but seagrass beds provide foraging grounds for many of the coral-reef associated species, and, therefore, indirectly support coral reef species diversity (20).

Human Influences on the Intertidal Zone

The major common environmental problems arising from human influences on the intertidal zones are as follows:-

(1) Overexploitation of intertidal resources. This includes overcutting of mangroves which leads to excessive soil erosion and subsequent loss of habitat. Also, destruction of coastal forests and sand dune vegetation has lead to an increase in soil and beach erosion especially on coasts bordered by sand dunes.

a) Common seagrass-associated fishes

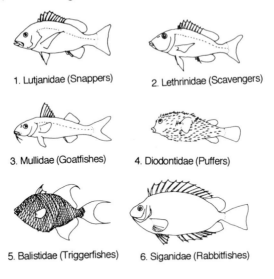

1. Lutjanidae (Snappers)   2. Lethrinidae (Scavengers)

3. Mullidae (Goatfishes)   4. Diodontidae (Puffers)

5. Balistidae (Triggerfishes)   6. Siganidae (Rabbitfishes)

b). Common seagrasses

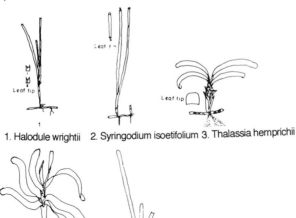

1. Halodule wrightii   2. Syringodium isoetifolium   3. Thalassia hemprichii

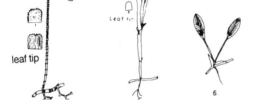

4. Thalassodendron ciliatum   5. Cymodocea rotundata   6. Halophila ovalis

Figure 5.10. Diagram of (a) the common seagrass fishes and (b) the common seagrasses found in the seagrass ecosystem.

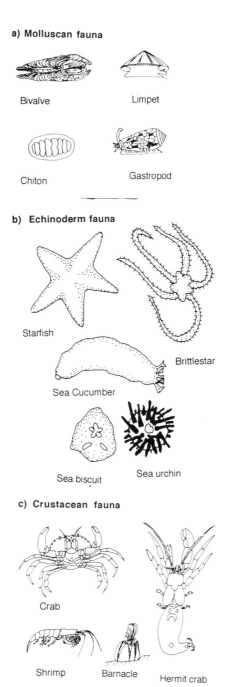

a) Molluscan fauna

Bivalve

Limpet

Chiton

Gastropod

b) Echinoderm fauna

Starfish

Brittlestar

Sea Cucumber

Sea biscuit

Sea urchin

c) Crustacean fauna

Crab

Shrimp

Barnacle

Hermit crab

Figure 5.11. Common invertebrates (a) molluscs, (b) echinoderms, and (c) crustaceans found in seagrass beds.

(2)   Excessive human and industrial waste disposal. Coastal towns continue to grow faster than sanitation facilities. Waste disposal has become a problem as waste is frequently dumped in the intertidal zones and can kill animals and plants at high concentrations.

(3)   Improper siting of houses, hotels, industries and aquaculture farms increases coastal soil erosion and reduces coastal productivity. For example, in many countries salt industries have cleared whole mangrove forest ecosystems and drastically reduced associated timber and fishing activities. Hotels are often built too close to the beaches without any regard for a buffer zones for protecting sand dune plants that stabilize beaches. Dune plants are replaced by unsightly sea walls that can increase beach erosion in front of the sea wall and on adjacent properties. Bird and turtle nesting sites in dune vegetation have rarely been given consideration for protection. Wastes from hotels are frequently discharged in pits but some waste reaches the sea by groundwater seepage.

(4)   There is indiscriminate removal of invertebrate organisms by beach combing tourists and local beach hawkers. Because intertidal zones are easily accessible this collection may be a threat to intertidal invertebrates communities.

(5)   Deviation or damming of rivers and estuaries or the extraction of water from underground aquifers causes changes in sediment deposition and erosion patterns plus salinity changes that cause various ecological stresses for intertidal organisms.

(6)   Unsuitable farming methods will enhance soil erosion and cause high silt loads in rivers thereby threatening various intertidal and shallow-water organisms.

(7)   Tar balls from oil tankers are frequently encountered in the intertidal zones as waves and currents cause them to accumulate in the intertidal zone. Tar balls can stick on shells of hermit crabs making the latter almost immobile besides being a potential health hazard.

Integrated Coastal Zone Management

Environmental problems effecting the intertidal zone occur from many activities undertaken by diverse economic sectors frequently operating on unilateral and short-term economic decisions. Consequently, sustainable use of resources is unlikely without some coordination between sectors. To minimize these conflicts integrated coastal zone management (ICZM) has been proposed and implemented in a few countries. Integrated coastal zone management is a resource management system for controlling development and human activities that affect the condition of economic resources and environmental quality by coordinating activities such that the capacity of the environment to sustain economic activities for the future is considered. ICZM is an attempt at developing a comprehensive and integrated framework for research, policy, planning and management of coastal resources. The goals are to insure the optimum use of coastal resources while maintaining high levels of biological diversity, conservation of critical habitats, maintenance of fisheries, the attractiveness of the environment to locals and tourists and to promote public health (5).

Boundaries of the coastal zone depend on political, administrative, legal, ecological and pragmatic considerations. Shorelines and intertidal areas are certainly included in the coastal zone but also the inclusion of watersheds is essential for coastal protection. Whole small islands are usually considered to be coastal zones. Because the sea is effected by distant events that occur far inland and far offshore it is difficult to protect coastal zones from unsustainable development without including watersheds and open-ocean environments. Key elements in the implementation ICZM are: planning, research, education, coordination, legislation and enforcement. Because of the level of coordination, sophistication and associated costs required for ICZM its actual implementation has largely been restricted to developed countries such as Australia and the United States. Nonetheless, some of the principles can be applied to developing countries and forms a goal for future efforts.

Guidelines for Coastal Zone Management

Planning. Planning lays the foundation for sustainable economic use of resources. In most cases the planning is sectoral whereby resource use and development are considered in different sectors even when the same resources are shared. In contrast ICZM insists on integrated development and management of coastal resource and planning is multisectoral involving all the stakeholders. Key stakeholders vary from country to country but should include the artisanal fishery communities, mangrove foresters, local villages, farmers in watershed areas, public health and tourism concerns, port and harbors, urban councils, wildlife and parks departments, fisheries department, and forest department.

During the planning stage, boundaries of the ICZM are defined and delineated based on the concerns and uses of the above groups and the ecosystems themselves. Boundaries must encompass natural ecosystems and critical habitats for biological diversity preservation but also towns, industrial centers, local rural communities, ports and stakeholder interaction need to be integrated within these ecological boundaries and the full set of interactions between man and nature must be considered in the planning process. Based on this planning process the key issues and conflicts need to be recognized and addressed through interactions between the planners and the stakeholders. It is noteworthy that priorities will vary from country to country or within a single country's coastline. This is especially true when considering the practical realities of political, economic, and social life.

Research. Scientific and socio-economic research is necessary for providing information for decision making. Information is important in project reviews of development projects and for determining detrimental environmental impacts of planned projects. Frequently, research can suggest alternatives for minimizing environmental threats or conflicts. Consequently, an Environmental Impact Assessment (EIA) is mandatory for implementation of a new project. In general, an EIA can be defined as the process of identifying, interpreting, predicting and communicating the potential effects that a proposed project may have on the environment. A very important ingredient in the EIA process is a knowledge of the facts and understanding of the natural environment. This usually requires long-term research that is often not part of the frequently short-term EIA process. Consequently, basic and long-term research must be the foundation for EIA and not economic, political or social motives. Unfortunately, research can often become a tool for supporting the views of a particular interest group rather than the search for an objective truth. This can be avoided if research is conducted independently of the stakeholders interests. Even researchers, however, are often part of a "world view" that can effect the way they approach problems and their interpretation of results. Therefore, most scientific findings only become facts after they have been reviewed and tested by scientists working independently of each other. It is obvious that arriving at the facts is only accomplished through a significant expenditure of time and money. This can often be a point of contention between developers, scientists and planners.

After the project is reviewed and the opportunities and constraints in the natural systems are identified, project are accepted, rejected or modified. The EIA process should continue from project initiation so that information is gathered throughout the project cycle to effectively monitor the environment for corrective measures. In essence an information feedback loop needs to be developed between researchers, managers and users so that changes can be made as circumstances require. Without adaptation to changing circumstances and new findings the ICZM concept and EIA process will fail.

Education. Continuous education and awareness of the activities of ICZM to stakeholders is an integral part of the success of ICZM. Target groups should include local communities, planners, managers, administrators, researchers, politicians, hoteliers, fishermen, mangrove cutters, dive and boat operators, forest guards, fisheries scouts, and wildlife rangers among others. Specific education programs tailored for the different target groups will facilitate the build up of knowledge and understanding that can lead to positive attitudes and wise decision making in conflict resolution. The purpose of the education is to create a forum for discussion of issues and to inform users of research findings and changes in policy and not as propaganda agenda for enforcing the world view of policy makers. Without the consensus of users the enforcement of policies becomes difficult or impossible and therefore education is one of the surest ways to achieve enforcement.

Coordination. This is a very crucial administrative task in the ICZM that involves networking with governmental and non-governmental organizations, business sectors and the public. The main objectives in the coordination is to reduce fragmentation and duplication of efforts by ensuring that different groups know what each other does and to coordinate joint efforts. Coordination can share expertise and reduce budgets.

Legislation and Enforcement. Successful implementation of ICZM has to be backed by law. In most developing countries there are no specific laws for management of coastal zone resources except for sectoral acts (such as the Water

Act, Forest Act, Fisheries Act, Wildlife Act, Health Act and others) that are rarely integrated or specific about the limits of resource use. Permits are often granted but without any specific limits on the user (except for the need to purchase the permit on a regular basis). The above acts are not effective enough to resolve the complex conflicts in the coastal zone areas because the Acts are either vague, poorly distributed, understood or simply not enforced. Enforcement requires significant expenditures to insure that the research and planning stages are successfully implemented.

Conclusions

Information for conservation purposes is mostly generated from ecological studies because such studies are directly involved with the study of the structure and function of nature and the consequences of human use of resources. Such information is only the beginning of conservation endeavors. For the effective achievement of conservation goals, ecological information needs to be integrated with ideas, desires, needs, and information from other stakeholders because they play an equally important role in conservation.

The institutional framework for executing these processes is at its infancy in this region. The capacity for undertaking research and the development of research facilities largely depends on bilateral and multilateral donor support programs. The practices of ICZM need to be initiated in this region because of the multiple uses of most marine resources. ICZM has the potential to avert:

(1)   overexploitation of both living and non-living resources,
(2)   pollution that will cause low biological diversity, health problems and cause a decline in economic activities such as tourism which is a major foreign exchange earner for many countries in the region,
(3)   physical destruction or alteration of critical habitats which will cause elimination of the species that require such habitat refuges.

With sustainable use of intertidal wetland resources, the economic productivity in these areas can further be enhanced.

References

1.   Barnard, K.H. 1950. Description catalogue of south african decapod crustacea. *Annals of the South African Museum* 38: 1-837
2.   Brakel, W.H. 1982. Tidal patterns of the East African Coast and their implications for littoral biota. UNESCO/ALESCO Symposium on Coastal and Marine Environment of the Red Sea, Gulf of Aden and Tropical Western Indian Ocean. The Red Sea and Gulf of Aden Environmental Programme 2: 403-418
3.   Broekhuysen, G.J., Taylor, H. 1959. The ecology of south Africa estuaries, Part 8, Kosi Bay estuary system. *Annals of the Natal Museum* 44: 279-296
4.   Chapman, V.J. 1977. Introduction: In *Ecosystems of the World: Wet Coastal Ecosystems*, ed. Chapman, V.J., pp. 1-29. Amsterdam: Elsevier Publishing Company
5.   Clark, J.R. 1992. Integrated management of coastal zones. *FAO Fisheries Technical Paper No. 327*. Rome: FAO

6. Chelazzi, G., Vannini, M. 1980. Zonation of intertidal molluscs on rocky shores of Southern Somalia. *Estuarine and Coastal Marine Science* 10: 569-583

7. Crosnier, A. 1962. Crustaces decapodes: *Portunidae Faune de Madagascar* 16: 1-154

8. Crosnier, A. 1965. Crustaces decapodes: Gapsidae et Ocypodidae. *Faune de Madagascar* 18: 1-143

9. Day, J.H. 1974. *A Guide to Marine Life on South African Shores.* Capetown: A.A. Balkema

10. Day, J.H., Morgans, J.F.C. 1956. The ecology of South African estuaries, Part 7, The biology of Durban Bay. *Annals of Natal Museum.* 13: 259-312

11. Hartnoll, R.G. 1975. The Grapsidae and Ocypodidae (Decapoda: Brachyura) of Tanzania. *Journal of Zoology* (London). 177: 305-328

12. Hartnoll, R.G. 1976. The ecology of some rocky shores in tropical East Africa. *Estuarine and Coastal Marine Science* 4: 1-21

13. Isaac, W.E., Isaac, F.M. 1968. Marine botany of Kenya coast. 3. General account of environment, flora and vegetation. *Journal of East African Natural History Society* 27(1): 2-28

14. Jasuund, E. 1976. *Intertidal Seaweeds in Tanzania.* Tromso: University of Tromso

15. Lewis, J.R. 1964. *The Ecology of Rocky Shores.* London: English University Press

16. Lockwood, A.P.M. 1976. Physiological adaptation to life in estuaries. In *Adaptation to Environment: Essays on the Physiology of Marine Animals,* ed. Newell, R.C., pp. 315-392. London: Butterworth Company

17. Lugo, A.E., Snedaker, S.C. 1975. The ecology of mangroves. *Annual Review of Ecology and Systematics* 5: 39-64

18. Macnae, W. 1968. A general account of the fauna and flora of mangrove swamps and forests in the Indo-West Pacific region. *Advance Marine Biology* 73-270

19. Macnae, W., Kalk, M. 1962. The fauna and flora of sand flats at Inhaca Island, Mozambique. *Journal of Animal Ecology* 31: 93-128

20. McClanahan, T.R., Nugues, M., Mwachireya, S. 1994. Fish and sea urchin herbivory and competition in Kenyan coral reef lagoons: the role of reef management. *Journal of Experimental Marine Biology and Ecology.* 184: 237-254

21. Menge, B.A., Lubchenco, J. 1981. Community organization in temperate and tropical rocky intertidal habitats: prey refuges in relation to consumer pressure gradients. *Ecological Monographs* 51: 429-450

22. Menge, B.A., Lubechenco, J., Ashkenas, L.R. 1985. Diversity, heterogeneity and consumer pressure in a tropical intertidal community. *Oecologia* 65: 394-405

23. Millard, N.A.H., Harrison, A.D. 1953. The ecology of South African estuaries, Part 5: Richards Bay. *Transactions of the Royal Society of South Africa* 34: 157-179

24. Newell, B.S. 1957. A preliminary survey of the hydrography of the British East African coastal waters. *Fishery Publications,* Colonial Office (London) 9: 1-21

25. Newell, R.C. 1976. *Adaptation to Environment: Essays on the Physiology of Marine Animals.* London: Butterworth and Company

26. Odum, E.P. 1971. *Fundamentals of Ecology.* Philadelphia: W.B. Saunders Company.

27. Oyieke, H.A., Ruwa, R.K. 1987. Non-encrusting, macroalgal zonation on rocky cliffs around Mombasa, Kenya. *Kenya Journal of Science and Technology* 8: 77-96

28. Reay, P.J., Haig, J. 1990. Coastal hermit crabs (Decapoda: Anomura) from Kenya, with a review and key to East African species. *Bulletin of Marine Science* 46: 578-589

29. Richards, D. 1984. *South African Shells: A Collectors Guide.* Cape Town: C. Struik Publishers

30. Ruwa, R.K. 1983. Field demonstration of a tidal component in the rhythmic vertical migrations of *Grapsus maculatus* (Brachyura: Grapsidae) on a rocky shore in Kenya. *Kenya Journal of Science and Technology* 4: 49-52

31. Ruwa, R.K. 1984. Invertebrate faunal zonation on rocky shores around Mombasa, Kenya. *Kenya Journal of Science and Technology* Series B 5: 49-65

32. Ruwa, R.K. 1989. Macrofaunal composition and zonation on sandy beaches at Gazi, Kanamai and Malindi Bay, Kenya. *Kenya Journal of Science and Technology* Series B 10(1-2): 31-45

33. Ruwa, R.K. 1990. The effects of habitat complexities created by mangroves on macrofaunal composition in brackish water intertidal zones at the Kenyan coast. *Discovery and Innovation* 2: 49-55

34. Ruwa, R.K. 1993. Zonation and distribution of creek and fringe mangroves in the semi-arid Kenya Coast. In *Towards Rational Use of High Salinity Tolerance Plants Deliberations about High Salinity Tolerance Plants and Ecosystems,* eds. Lieth, L.H., Al Massoum, A. Vol. 1, pp. 97-105. Dordrecht: Kluwer Academic Publishers

35. Ruwa, R.K. 1993. Disturbances in mangrove vegetation and their possible influence on other biota. SAREC Documentation: Conference Reports 1993. 1: 104-126. Stockholm, Sweden

36. Ruwa, R.K. 1994. Halophytic coastal marsh vegetation in East Africa. In *Halophytes as a Resource for Livestock and for Rehabilitation of Degraded Lands.* eds. Squires, V.R., Ayoub, A.T., pp. 201-210. Dordrecht: Kluwer Academic Publishers

37. Ruwa, R.K., Brakel, W.H. 1981. Tidal periodicity and size-related variation in the zonation of the gastropod *Nerita plicata* on an East African rocky shore. *Kenyan Journal Science and Technology* 2: 61-67

38. Ruwa, R.K., Jaccarini, V. 1986. Dynamic zonation of *Nerita plicata, N. undata* and *N. textilis* (Prosobranchia: Neritacea) populations on a rocky shore in Kenya. *Marine Biology* 92: 425-430

39. Ruwa, R.K., Jaccarini, V. 1988. Nocturnal feeding migrations of *Nerita plicata, N. undata* and *N. textilis* (Prosobranchia: Neritacea) on the rocky shores at Mkomani, Mombasa Kenya. *Marine Biology* 99: 229-234

40. Ruwa, R.K., Polk, P. 1986. Additional information on mangrove distribution in Kenya: Some observations and remarks. *Kenya Journal of Sciences (Series B).* 7: 41-45.

41. Ruwa, R.K., Polk, P. 1994. Patterns of spat settlement recorded for the tropical oyster *Crassostrea cucullata* (Born 1778) and the barnacle, *Balanus amphitrite* (Darwin, 1854) in a mangrove creek. *Tropical Zoology* 7: 121-130

42. Ryther, J.H. 1969. Photosynthesis and fish production in the sea. *Science* 166: 72-76

43. Segal, E. 1956. Adaptive differences in waterholding capacity of an intertidal gastropod. *Ecology* 37: 174-178

44. Semesi, A.K. 1988. Seasonal changes of macro-epiphytes on seagrass *Thalassodendron ciliatum* (Forssk.) den Hartog at oysterbay, Dar-es-Salaam, Tanzania. In *Proceedings of Workshop on Ecology and Bioproductivity of the Marine Coastal Waters of Eastern Africa,* ed. Mainoya, J.R. pp. 51-58. Dar-es-Salaam University Press: Tanzania

45. Sikes, H.L. 1930. The drowned valleys of Kenya. *Journal of East Africa Natural Historical Society* 38-39: 1-7

46. Smith III T.J., Chan, H.T., Mclvor, C.C., Robblee, M.B. 1989. Composition of seed predation in tropical, tidal forests from three continents. *Ecology* 70(1): 146-151

47. Snedaker, S.C., Brown, M.S. 1982. Primary productivity of mangroves. In *CRC Handbook on Biological Resources*, Vol. 1. eds. Black, C.C., Mistu, A., pp. 477-485. Baton Rouge, Florida: CRC Press

48. Southwood, A.J. 1965. *Life on the Seashore*. London: Heinemann Educational Books

49. Steele, J.H., Baird, I.E. 1968. Production ecology of a sand beach. *Limnology and Oceanography* 12: 14-25

50. Stoddart, D.R. 1971. Environment and history in Indian Ocean reef morphology. In *Regional Variation in Indian Coral Reefs*, Symposia of the Zoological Society of London, eds. Stoddart, D.R., Yonge, M., pp 3-38. London: Academic Press

51. Taylor, J.D. 1968. Coral reef associated invertebrate communities (mainly molluscan) around Mahe, Seychelles. *Philosophical Transactions of the Royal Society London* Series B 254: 129-206

52. Taylor, J.D. 1970. Intertidal zonation at Aldabra Atoll. *Philosophical Transactions of the Royal Society of London* 260: 173-213

53. Taylor, J.D. 1971. Reef associated molluscan assemblages in the Western Indian Ocean. In *Regional Variation in Indian Coral Reefs*, Symposia of the Zoological Society of London, eds. Stoddart, D.R., Yonge, M., pp 501-534. London: Academic Press

54. Taylor, J.D. 1976. Habitats, abundance and diets of muricacean gastropods at Aldabra atoll. *Zoological Journal of Linnean Society* 59: 155-193

55. Taylor, W.R. 1960. *Marine Algae of the Eastern Tropical and Subtropical Coasts of the Americas*. Ann Arbor: The University of Michigan Press

56. Vannini, M. 1975. Researchers on the Coast of Somalia. The shore and dune of Sar Uanle 4. Orientation and anemotaxis in the land hermit crab, *Coenobita rugosus* Milne Edwards. *Monitore Zoologica Italiano* 6: 57-90

57. Vannini, M. 1976. Researches on the coast of Somalia - The shore and dune of Sar Uanle. Sandy beach decapods. *Monitore Zoologico Italiano* 8: 255-286

58. Vannini, M., Chelazzi, G. 1978. Field observations on the rhythmic behaviour of *Nerita textilis* (Gastropoda: Prosobranchia). *Marine Biology* 10: 308-314

59. Vannini, M., Valmori, P. 1981. Researches on the coast of Somali. The shore and the dune of Sar Uanle. 30: Grapsidae (Derapoda Brachyura). *Monitore Zoologico Italiano* 6: 57-101

60. Vermeij, G.J. 1971a. Temperature relationships of some tropical Pacific intertidal gastropods. *Marine Biology* 10: 308-314

61. Vermeij, G.J. 1971b. Substratum relationships of some tropical Pacific intertidal gastropods. *Marine Biology* 10: 315-320

62. Warburton, K. 1973. Solar orientation in the snail *Nerita plicata* (Prosobranchia Neritacea) on a beach near Watamu, Kenya. *Marine Biology* 23: 93-100

63. Whittaker, R.H. 1975. *Communities and Ecosystems*. New York: MacMillan Press

Section III:

Inland-Water Ecosystems

Chapter 6

Rivers and Streams

Scott D. Cooper

Rivers and streams constitute a valuable resource for the people of East Africa. Rivers and streams are a major source of water for agricultural, industrial, and domestic use, produce fish, are important habitats for wildlife, and act as highways for commerce and trade. Rivers, streams, and their associated wetlands also harbor vectors of disease, and act as conduits for the disposal of waste generated by human beings. Population expansion and development projects have resulted in widespread land-use changes with far-reaching consequences for catchments and their drainage streams. Both direct use of, and indirect influences on, freshwater resources by human beings can result in the pollution of water making it unfit for domestic, agricultural, or industrial use. The management of riverine systems often requires balancing the goals of fish production, hydroelectric generation, wildlife habitat, drinking water, and irrigation use while controlling contamination and the spread of water-borne disease.

Running-water (=lotic) ecosystems are home to a variety of species with economic, nutritional, aesthetic, and scientific value, and conservation of these species requires an understanding of the ecology of their habitats. Because streams and rivers reflect the physical, chemical, and biological conditions in their catchments, changes in the biota of lotic ecosystems also frequently provide indices of human influence on adjacent land and water resources. In the following chapter I review the physical, chemical, and biological characteristics of streams and rivers in eastern Africa, discuss human influences on running-water ecosystems, and conclude with recommendations for conserving their biodiversity.

Geography and Biogeography

Eastern Africa is bounded on the north by the Ethiopian Highlands and Somalian Desert, on the west by the western Rift, and on the south by the lower Zambezi River. This area includes rivers flowing into and out of the lakes in the upper Nile and Zaire, and lower Zambezi, basins; the eastward-flowing rivers of Somalia, Kenya, Tanzania, and northern Mozambique; the rivers feeding the lakes and swamps of the Rift Valleys; and isolated drainage systems in arid and semi-arid zones (such as the northern Uaso Nyiro flowing into the Lorian Swamp in Kenya, inlet streams of Lake Eyasi in Tanzania; Fig. 6.1). The geological history of this area has encompassed the isolation of some river systems, the capture of some systems by other drainage networks, hydrological connections during wet periods, changes in river direction owing to highland development through seismic or volcanic activity, and blockage of rivers by lakes formed in or between rift basins, all with implications for the composition and diversity of riverine faunas and floras (9,55,79).

The location and direction of flow of East African rivers and streams have been affected profoundly by the formation of the Rift Valleys and associated volcanic activity. The northern part of the eastern Rift was once connected to the upper White Nile, as recently as 7500 years ago, and, therefore, the fish faunas of Lakes Turkana and Albert are similar. Uplifting of the eastern ridge of the western Rift (30,000 years ago) blocked and reversed the flow of large rivers which formerly flowed from east to west across the Victoria plateau to the upper Nile River. When coupled with subsidence in the plateau between the two Rift Valleys, this flow reversal resulted in the formation of Lake Victoria, the blockage of many streams by lake waters, and connections among the Victoria, George, and Edward basins. The rivers connecting Lakes Edward, Kyoga, and Victoria with the Nile possess rapids or waterfalls that act as important barriers to fish migration. These barriers coupled with the desiccation of these lakes during past geological eras resulted in the development of a depauperate non-cichlid fish fauna with species affinities with the Nile fauna. By contrast, the haplochromine cichlids in these lakes underwent a rapid, recent radiation resulting in diverse, endemic flocks of cichlid species (see Chapter 8). The Malagarasi River, that now flows into eastern Lake Tanganyika, was part of the upper Zaire system before the formation of the western Rift, which began at the end of the Miocene (> 12 million years ago). The Malagarasi River, today, still possesses some species found elsewhere only in the Zaire basin.

Lake Kivu was once part of the Lake Edward basin, but the eruption of the Virunga Volcanoes (20,000 years ago) blocked its north-flowing outlet causing it to rise and spill into the Ruzizi River and, thence, into Lake Tanganyika. Waterfalls on the Ruzizi River have prevented all but one Tanganyika fish species (*Barilius moorei*) from invading the upper Ruzizi and Lake Kivu, and Lake Kivu's fauna is

Figure 6.1. Facing page. Map of eastern Africa, showing the major rivers. The fine dashed lines delineate the Rift Valleys. For the Nile and Zambezi River systems, only the Upper Nile and Lower Zambezi Rivers are shown. Middle and lower sections of the Nile and upper and middle sections of the Zambezi are not depicted. Lake Turkana is fed by the Omo River, which drains the southern Ethiopian highlands to the north, and the Turkwel and Kerio Rivers, which originate in the Cherangani Hills and the slopes of Mt. Elgon to the south. Until recently, the lakes of the southern Ethiopian Rift (Chew Bahir, Chamo, Abaya) drained into Lake Turkana. The upper Nile basin includes Lakes Edward, George, Victoria, Kyoga, and Albert, and their influent and effluent streams and rivers. Lake Tanganyika is fed by streams and rivers in western Tanzania (such as the Malagarasi River), as well as by the Ruzizi River draining Lake Kivu. It is connected to the Zaire River system by the Lukuga River. A number of small streams and rivers flow from highlands lining the Western Rift into Lake Malawi, and Lake Malawi is connected to the lower Zambezi by the Shire River. Most of the remaining lakes associated with the Rift Valleys (such as Lakes Baringo, Nakuru, Naivasha, Magadi, Natron, and Manyara in the eastern Rift; Lakes Rukwa and Chilwa in or near the western Rift) lack outlets and are fed by small rivers or streams draining nearby highlands. Most of the remaining rivers in East Africa flow eastward from higher inland areas to the Indian Ocean. Redrawn from published maps (9).

most similar to that in Lake Edward. At different times in its history, the Lake Malawi basin probably was connected to the Upper Zambezi River via the Bangweulu swamps and drained into the Indian Ocean via the Rovuma River, but became disconnected from these water bodies during formation of the western Rift. Lake Malawi and the upper Shire River are isolated from the lower Shire and Zambezi systems by 80 km of rough water, the Murchison Rapids, and the upper Shire has few fish species found in the lower Zambezi.

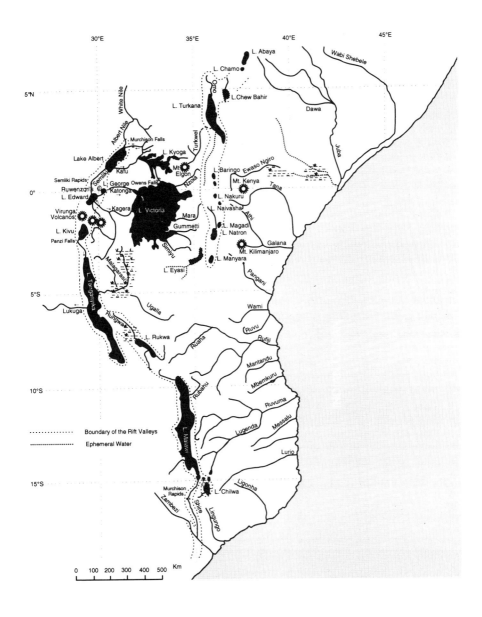

Eastward - flowing rivers of eastern Africa drain the largest mountains in the area, including Mt. Kenya and Mt. Kilimanjaro, and their fish faunas indicate past exchanges with the Lake Victoria system, in the north, and the Malagarasi and Malawi systems, in the south. Because many river systems are separated by low-lying swamps, it is probable that faunal exchanges occurred among some river basins, such as between the Malagarasi River and Lake Victoria basins or among the Malagarasi, Ruaha, Rufiji, and Rovuma rivers, during floods in wet geological periods.

The River Ecosystem

Unidirectional currents and a linear form distinguish streams and rivers from other ecological systems. The linear form of streams insures an intimate relationship between a stream and its drainage basin. The riparian vegetation that lines the banks of most unaltered streams contributes leaves, branches, fruits, seeds, flowers, pollen, and trunks to streams and this dead plant material (detritus) provides an important energy source for stream microbes (bacteria, fungi), and food and cover for invertebrates and fish (38). The riparian vegetation also filters nutrients and sediment carried into streams from upland areas, the roots of riparian vegetation stabilize banks minimizing erosion, and the riparian canopy can completely shade small headwater streams, thereby limiting the production of aquatic plants and algae, and lowering stream temperatures. Light levels in streams are also dependent on water depth and turbidity, and many East African rivers and streams are very turbid owing to the transport of large sediment loads.

River current speeds vary greatly, being low near the bottom and sides of channels owing to friction with the bottom and banks, and being lower on the downstream than upstream sides of obstructions. Many rivers and streams have well-developed riffles (fast-flowing shallow areas) interspersed with pools (slow-flowing deep areas). In meandering streams the outsides of meander bends are zones of fast currents and erosion, the insides of bends are characterized by slower currents and sediment deposition, and shallow, rapidly-flowing water occurs between bends. There is an intimate relationship between current speeds and the particle size composition of substrata, because the erosion, transport, and deposition of sediments are functions of current velocity. Riffles usually contain coarse substrata (such as rocks and boulders), whereas pools and floodplain habitats are depositional zones for fine sediments (silt and sand). Similarly, headwater streams are often narrow, torrential and underlain by coarse substrata, whereas large, lowland rivers are broad, meandering, and underlain by fine substrata. Many stream organisms reach their highest abundance under particular current and substratum conditions (63).

The River Continuum Concept

The River Continuum Concept (RCC) provides a comprehensive description of community and ecosystem responses to habitat changes proceeding from the headwaters of rivers to their mouths, emphasizing linkages between upstream and downstream areas via the longitudinal transport of dissolved substances, particulates, and organisms (65,84; Box 6.1). The RCC predicts that shading by the riparian tree canopy keeps autochthonous (in-stream) production by algae and plants at low levels in headwater areas. Inputs of CPOM (coarse particulate organic matter, detritus fragments with diameters > 1 mm; that is leaves, twigs, branches, and fruit) encourage microbial growth and provide a rich food source for organisms, called shredders, that eat CPOM. Fragmentation of CPOM by microbes and shredders produces fine particulate organic matter (FPOM, diameter < 1 mm) which is extensively utilized by invertebrates called collectors. Because autochthonous production is low and microbial respiration is high, ratios of photosynthesis to respiration (P/R) are less than one. Fish in these shaded headwater streams eat primarily aquatic or terrestrial invertebrates.

In mid-reach sections, the riparian canopy opens up allowing light to reach stream bottoms; consequently, autochthonous (in-stream) production by periphyton (attached algae) and, where fine substrata are available, higher plants increase. Invertebrate grazers, which eat algae growing on the bottom (= periphyton), become a dominant component of the mid-section river fauna. Because FPOM is exported from upstream to downstream areas, and is produced by the decomposition of algae and plants, collector biomass remains high in mid-reach sections. Because of increased in-stream photosynthesis, P/R ratios are > 1. Common fish species eat grazers and collectors, and some algivorous fish may be present.

Near the mouth of the river, light penetration is limited by the abundance of suspended sediments and depth. Some phytoplankton production may occur in slow-flowing waters near the surface, and limited periphyton and plant growth may occur at river margins. Autochthonous production, however, is generally low owing to light limitation below the water's surface. FPOM from upstream areas is suspended in the water column or deposited on the river bottom, and forms the primary food source for most invertebrate collectors. Because in-stream photosynthesis is low, P/R ratios are < 1. Fish species include planktivores and a variety of benthic feeders.

Although the RCC has been useful for synthesizing information on longitudinal changes in rivers in temperate regions, it is not clear that the RCC applies to tropical rivers. Although the RCC assumes gradual, continuous changes in abiotic conditions proceeding downstream, East African rivers show alternating placid and torrential sections owing to changes in gradient. Although it is likely that consumers will be most abundant where their resources are abundant, actual patterns observed in river systems have not always fit the patterns predicted by the RCC (2,80,105). Many headwater streams in East Africa and elsewhere, such as those at high elevations, those in arid regions, and those extensively modified by human activity lack a dense, riparian tree canopy. In addition, riparian inputs may be washed out of steep, headwater streams by unpredictable or seasonal floods, resulting in low CPOM and shredder abundance. The RCC also does not consider the important role that disturbances, such as floods and droughts, play in stream systems, or the successional sequences of species that occur after disturbances. Also, many of the generalist consumers present in East African streams do not fit neatly into feeding categories (such as shredders, collectors, grazers, predators; see also 49).

Of particular importance is the high turbidity or suspended sediment concentrations found in many East African stream and river systems. Because of high turbidity, even mid-reach sections of rivers lack substantial light penetration; consequently, production of algae and plants in stream channels may be always low.

Box 6.1. River Ecology from Headwaters to the Ocean

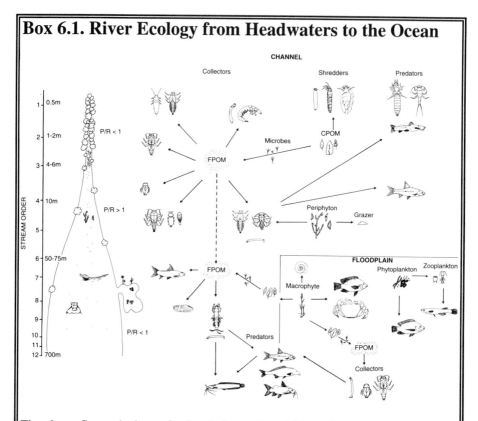

The above figure depicts a food web for an East African river system, proceeding from its headwaters to its mouth (8,10,26,29,32,37,40,51,61,65,75,84,104). Numbers on the left side of the left-hand scale are stream orders and numbers on the right-hand side of the scale are stream widths. Headwater streams are first order streams, and a second order stream is formed by the confluence of two first order streams. Stream order only changes when two streams of equal order flow together, whereas a stream or river's order does not change when a stream of lower order flows into it. The stream map indicates major food sources for stream consumers, and P/R (community photosynthesis to respiration) ratios, for different points in the stream proceeding from its headwater to the mouth.

Arrows (trophic links) point from resources to the consumers that use them. The trophic links next to the map emphasize the importance of fine detritus (FPOM) and collectors to stream ecosystems in East Africa. Most collectors have widespread distributions, but a few are associated with particular elevations (such as helodid larvae which are found at high elevations, and the mayfly *Ephemerythus* and elmid beetles which are found at lower elevations; 104). Subsidiary trophic links include shredders which eat coarse detritus (CPOM) in headwater, shaded streams and grazers which eat attached algae (periphyton) in sunny streams. Macroinvertebrates, particularly collectors, are eaten by a variety of invertebrate predators and fish, which are eaten, in turn, by other fish.

The lower right-hand portion of the diagram depicts some of the major characteristics of food webs in floodplain pools and lakes, including the importance of higher plants as structure and food in these habits. Phytoplankton and zooplankton are well-developed in these habitats, and form an important food source for some fish. Higher plants provide food directly to some consumers (such as crabs, some fish), provide substrata for periphyton that are eaten by grazers (such as snails), or die and form detritus (CPOM, FPOM) which is eaten by a variety of macroinvertebrates, particularly collectors, and some fish. Arrows indicate some of the links between floodplain and channel habitats. Specific identities of invertebrates and fish can be found in Figures 6.4 and 6.5.

S.D. Cooper & S.W. Wiseman

Extensive silt deposition on stream bottoms will also inhibit periphyton production. Given the generalized food habits, the high potential growth and colonization rates of many river consumers, and the variability and unpredictably of hydrological regimes, it appears that East African river communities are composed of widespread, opportunistic species, rather than specialized species that occur predictably at particular times and places.

At least in its original conception the RCC neglected the role of floodplains in the ecology of river systems. Considerable exchange of dissolved substances, particulates, and organisms occurs between rivers and their floodplains. Although the RCC emphasizes longitudinal linkages in river channels, lateral linkages between the channel and floodplain may be as, or more, important to the economy of river systems (32,89,93,96,97).

In general, then, the distributions and abundances of lotic organisms are determined by a hierarchy of factors which operate at different spatial and temporal scales (31,38,64). Climatic patterns determine seasonal variation and overall levels of river discharge at broad geographic scales. The topography, geology, and vegetation of individual river basins determine run-off patterns within rainfall regions, as well as the water chemistry of individual rivers (62). Within the stream, macro and microscale variation in current, substratum, food, and light conditions will further restrict the distributions of organisms. In subsequent sections I describe specific physical, chemical, and biological conditions found in East African rivers and streams.

Hydrology and Physical Factors

The composition and diversity of river communities are determined, to a large degree, by variation in river discharge, and concurrent effects on current speeds, substratum composition, and disturbance intensity (76). Rivers and streams in eastern Africa vary greatly in the amounts of water that they carry, ranging from headwater and seasonal streams that can have negligible flows to the lower Zambezi River, having an average discharge of 3800 m^3/second. The Victoria Nile has an average discharge of approximately 1000 m^3/second, and average discharges for the eastward-flowing rivers of eastern Africa range from around 24 m^3/second for the Athi-Galana-Sabaki to 150 to 200 m^3/second for the Tana to 1000 m^3/second for the Rufiji River. Large Kenyan rivers flowing into Lake Victoria (such as the Nzoia, Migori, Mara, Sondu, Yala) have discharges ranging from 30 to 100 m^3/second. Rivers in the central and southern part of the eastern Rift have lower discharges (approximately 1 to 20 m³/second; Molo, Gilgil, Perkerra, Malewa, southern Uaso Nyiro), reflecting low rainfall levels.

Rivers draining large lakes and reservoirs, such as the Victoria Nile, have fairly constant discharges (6,83). In contrast, discharge in most rivers and streams in eastern Africa fluctuate over an order of magnitude throughout the year, reflecting seasonal changes in rainfall (Fig. 6.2). Rivers draining areas near the equator usually have discharge peaks at two times of the year, shortly after the spring and autumnal equinoxes. Rivers further from the equator tend to show more unimodal peaks occurring shortly after the summer solstice, in July to September in northern East Africa and January to April in southern East Africa.

Africa is a dry continent and a relatively small proportion (1 to 12 %) of rainfall becomes river run-off (6,83). Many arid and semi-arid areas of eastern Africa have streams or rivers that are ephemeral or intermittent, consisting of isolated pools during the dry season. Most perennial rivers or streams have headwaters in well-watered highland areas, but even these rivers have seasonal fluctuations in discharge. Small headwater streams can experience scouring floods that reduce the abundance of resident plants and animals (85). In contrast, flood waters in large rivers usually overflow into adjacent floodplain areas composed of a mosaic of channels, temporary and permanent lakes and ponds, levees, islands, and upland areas. Eastern Africa is also noted for large interannual variations in rainfall, and droughts are common. As a consequence, river discharges in this part of the world often vary greatly from year-to-year, and even moderately large rivers may dry up during droughts.

Many stream organisms can tolerate only a fairly narrow range of temperatures, and the geographic distributions of many riverine plants and animals are related to latitude, elevation, and degree of geothermal influence because of the effects of the above factors on water temperature (9). Glacier-fed streams on Mt. Kenya, Mt. Kilimanjaro, or in the Ruwenzori Mountains have temperatures near freezing, and many of the highland streams (elevations >1800 m) are cool (< 18° C). Typical temperatures for streams at elevations less than approximately 1500 m range from 18 to 30° C, but higher temperatures have been recorded from streams on the floor of the eastern Rift (such as 33° C for the southern Uaso Nyiro; 18,83). High temperatures (>40° C) can be found in streams fed by geothermal springs, which are abundant in the Rift Valleys.

Within a stream or river, temperatures vary with seasonal changes in discharge, with higher temperatures in the dry than wet season, and in drying river beds or floodplain pools (up to 35° C) than perennial streams. Large rivers show less seasonal variation in temperature than small streams.

Light levels play a critical role in determining the levels of in-stream primary production and, consequently, the abundance of grazers (see above on RCC). Although rivers in arid and semi-arid areas probably were naturally turbid owing to the transport of large amounts of sediment, erosion has been exacerbated in recent times and sediment loads have increased in these and more mesic areas by the reduction of native vegetation by human activities, including livestock grazing, burning, farming, road construction, and urban development (28; Chapters 4; 16).

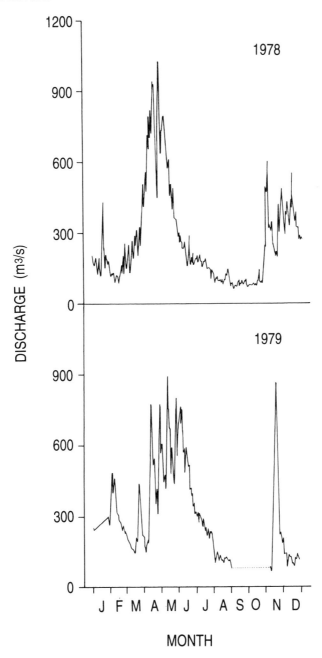

Figure 6.2. A graph of daily discharge (m³/s) versus time (a hydrograph) for the Tana River, Kenya, 1978 (top) and 1979 (bottom).

Chemistry

Most stream organisms require high levels of oxygen to sustain normal levels of activity. In most streams and rivers, dissolved oxygen is near saturation levels. Stagnant floodplain habitats, and the remnant pools of intermittent streams, may have low oxygen levels (< 2 mg/l) owing to their high temperatures, lack of water exchange, and decomposition of accumulated organic debris. Oxygen can also reach low levels under the canopies of dense floating plants. Fish and invertebrates that survive in these habitats may possess a variety of physiological and behavioral adaptations for coping with low oxygen (9).

Most freshwater organisms cannot tolerate high salt concentrations and require circumneutral or slightly alkaline pHs (5.6 to 8.5). African rivers usually show low salt concentrations (< 100 mg/l), and intermediate pHs (6.5 to 8.2); however, pHs as high as 10 have been recorded for streams flowing into the alkaline lakes of the eastern Rift. Typical water conductivities range from 30 to 300 microSiemens per centimeter (mS/cm), with lower values from forested, montane streams and higher values from lowland, savanna or desert streams (72,74,83). Higher conductivity values (> 500 mS/cm) have been recorded from rivers draining lakes with high salinities or alkalinities, such as the Ruzizi River draining Lake Kivu and the Semliki River draining Lake Edward.

In unshaded, clear rivers and streams, plant and algal growth is often limited by nutrients, such as nitrogen, phosphorous, or, in the case of diatoms, silicon. African rivers have high silicate concentrations, and phosphate concentrations are often high relative to nitrogen (60,74,87). Concentrations of major solutes and nutrients in rivers and streams are influenced by climate, soils, and vegetation. Soils are well-leached in areas of high rainfall (such as the Ruwenzoris, Aberdares, Mt. Elgon, Mt. Kenya) leading to low solute concentrations in receiving waters. Native vegetation covering drainage basins can also reduce erosion and store nutrients, leading to low solute and nutrient concentrations in drainage streams.

One of the few detailed studies of water chemistry in rivers was conducted in the Naivasha basin, Kenya (33; see Chapters 7 and 9 for information on Lake Naivasha). The Malewa River, the primary inflow to Lake Naivasha, drains volcanic highlands, and its chemical composition is similar to rivers draining other volcanic areas in the region. Sodium and calcium, the major cations, are derived primarily from chemical weathering of the rocks in the catchment, whereas bicarbonate, the major anion, is derived from gaseous CO_2. In contrast most of the sulfate in the Malewa River originates from atmospheric deposition. The chemistry of streams draining non-volcanic regions, such as the Ruwenzori Mountains and the Precambrian Plateau of Uganda, are determined primarily by rock weathering (48).

Chemical conditions in floodplain and swamp habitats often differ from those in the main river channel. High temperatures, acidic pHs, and low dissolved oxygen in floodplain and swamp waters promote the release of iron, manganese, and phosphorous from sediments. Vigorous plant growth in such habitats can result in the removal of substantial quantities of nutrients, but decomposition of detritus can produce high concentrations of ammonia and phosphate. Although there is little information on the roles of fish, birds, crocodilians, and large mammals in nutrient cycling in East African rivers, calculations indicate that hippos contribute only a minor component to river nutrient budgets (72).

The Food Web

Biota

Detritus, plants, and algae are primary food sources for invertebrate consumers in streams and rivers (Box 6.1). In East Africa, food webs are often dominated by trophic interactions involving fine detritus (FPOM) and collectors. These invertebrates are eaten, in turn, by predatory invertebrates and fish. The apex of riverine food webs is often occupied by large fish, crocodilians, predatory birds (such as fish eagles and kingfishers), or mammals, such as otters and, increasingly, humans.

Aquatic Microbes. Aquatic microbes, particularly suspended and benthic bacteria, play an important role in nutrient cycling and energy flow in rivers and streams. Bacteria and fungi decompose dead vegetation, thereby improving the nutritional value of detritus for many of the larger aquatic consumers. Although few leaf-decomposition studies have been conducted in tropical streams, high water temperatures indicate that decomposition of plant detritus will be rapid. Leaf decomposition rates will depend also on the toughness and chemical composition of leaf material and the chemical composition of the water (62,91).

Plankton. Plankton assemblages are poorly developed in many small to moderately-sized streams, because planktonic organisms cannot maintain their position in currents and are washed out of the system. Plankton, however, are abundant in floodplain lakes and ponds and constitute an important food source for fish, and are transported subsequently into main channels where they are eaten by fish and filter-feeding invertebrates living downstream. Resident plankton assemblages can develop in large rivers. Diatoms, rotifers, and copepods are common in some free-flowing rivers (for example the upper White Nile), whereas blue-green algae (cyanobacteria) and cladocerans often dominate in and below reservoirs or lakes (24,81).

Periphyton. In clear streams and floodplain habitats an extensive community of microscopic algae, attached to the surfaces of sediment, rocks, and higher plants, develops. These algae, called periphyton, are dominated by diatoms and colonial blue-green and green algae, but the red alga, *Lemanea*, is sometimes found in small, undisturbed streams (104). After scouring floods, diatoms are often the first algae to colonize stream surfaces, but can be displaced later by filamentous or colonial green or blue-green algae. Similarly, intensive grazing by invertebrates, such as snails and herbivorous insects, often favors the dominance of tightly-adherent (adnate) diatoms over more loosely-attached, erect taxa. Although numerous temperate studies have shown large effects of grazers on periphyton, few grazing studies have been conducted in tropical waters. There is some indication, however, that cichlids and snails that graze on periphyton associated with plant surfaces can reduce periphyton standing stocks in African waters (27,81). Filamentous green (such as *Cladophora*) and blue-green (such as *Oscillatoria*) algae are also favored by high nutrient and light conditions, and may clog waterways.

Aquatic Macrophytes. Although macroscopic vascular plants are not common in shaded, headwater streams, mosses, liverworts, and plants belonging to the family Podostemonaceae, can be found growing attached to rocks in small East African streams. Eastern Africa's rivers are often turbid and have shifting sand substrata; consequently, higher rooted plants are absent or rare in their channels. Aquatic macrophytes, however, do occupy the margins of many river systems and reach their maximum development in floodplain and swamp habitats (26). Macrophytes, here, refers to higher plants, mosses, ferns, and macroscopic algae (Characeae: stoneworts) with submergent, emergent, and floating growth forms (Fig. 6.3).

Many macrophyte populations wax and wane with seasonal water levels, reaching their greatest development during high water and, in some cases, become concentrated and die when their habitats become desiccated. Most aquatic macrophytes are not used directly by herbivores, perhaps owing to high concentrations of secondary defense compounds and their high fiber content. Some decapods (crabs, crayfish) and fish (tilapia), a few aquatic insects (Lepidoptera, Trichoptera, Coleoptera, Diptera:Chironomidae), and terrestrial insects and mammals, however, will eat these plants (52,68). Most macrophyte production apparently enters food webs as detritus; however, data on the fate of organic compounds fixed by aquatic macrophytes are scanty (26). Macrophytes provide surfaces for attachment, cover from predators, and attached algae as food for invertebrates and fish. Additionally, macrophytes take up and store large quantities of nutrients, with important effects on nutrient cycling and water chemistry, as well as act as conduits for the release of some gases (such as methane).

Dry parts of riparian and floodplain zones are occupied by a complex mosaic of grassland, scrubland, woodland, and forest habitats. The distribution and diversity of plant formations in these habitats are related to soil and moisture conditions, and disturbances associated with fires, flooding and drying (32,45).

Aquatic Invertebrates. Minute animal taxa (< 1 mm in length), including protozoans, microturbellarians, rotifers, nematodes, oligochaetes, gastrotrichs, tardigrades, chydorid cladocerans, ostracods, copepods, and mites are associated with interstices and surfaces of bottom sediments and higher plants. Most of these taxa graze on bacteria, fungi, fine detritus, and periphyton associated with these surfaces, but they have been little-studied in African waters and their role in rivers is unknown.

River and stream macroinvertebrates, primarily aquatic insects, crustaceans, and molluscs, are associated with the bottom (benthic), and the insects tend to be short-lived (generation time < 1 year) and small (< 2 cm in length; Fig. 6.4; 8,29,40,75,103). Most aquatic insects spend their larval stages in the water but are terrestrial as adults; however, both the larvae and adults of most aquatic beetles (Order: Coleoptera) and true bugs (Order: Hemiptera) are aquatic. Some macroinvertebrates play a role in the transmission of human diseases. Some blackflies (*Simulium*) occurring in East and West Africa act as vectors for the filarial worm that causes river blindness (onchocerciasis); some mosquitoes (*Anopheles*)

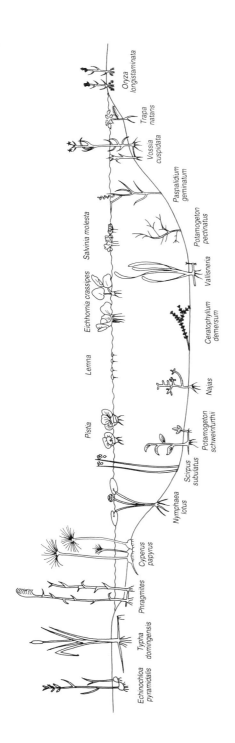

Figure 6.3. Common aquatic macrophytes found in eastern African rivers and associated floodplains. Submerged macrophytes are those completely covered by water and include taxa with (*Vallisneria*, *Najas*, *Potamogeton*) and without (*Ceratophyllum*) roots. Floating-leafed macrophytes are rooted in bottom sediments but have leaves floating at the surface and include the water lilies and some potamogetons (*Nymphaea*, *Potamogeton thunbergii*). Surface-floating macrophytes float on the water surface and are not rooted to the bottom, and include water ferns (*Salvinia*, *Azolla*), water cabbage (*Pistia*), water hyacinth (*Eichhornia*), and duckweeds (such as *Lemna*). Finally, emergents are rooted or anchored to the bottom, but their stems extend through the air-water boundary and their vegetative structures are exposed to the air. Common emergents in East Africa include grasses occupying seasonal wetlands (*Oryza*, *Echinochloa*), as well as reeds (*Phragmites*), sedges (*Scirpus*, *Cyperus*, especially *C. papyrus*), cattails (*Typha*), and other grasses (*Paspalidium*, *Vossia*) occupying permanent wetlands. Some plants, such as the water chestnut (*Trapa natans*), have both submerged and floating leaves. Submerged and floating-leafed macrophytes, as well as plants that have the characteristics of both, are sometimes called euhydrophytes (26). Redrawn from figures in 26.

Plate A

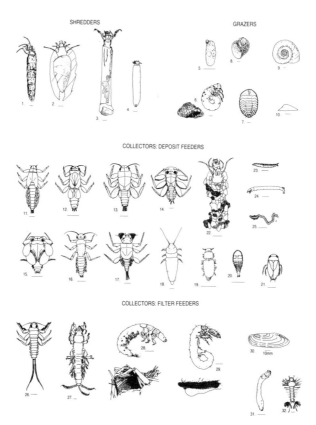

Figure 6.4. Common aquatic macroinvertebrates found in eastern African rivers and streams. Drawn from personal specimens or redrawn from diagrams (29,34,61,73,102). Length of scale = 1 mm, unless otherwise noted. In general, only the basal parts of the caudal filaments of mayfly nymphs are shown. **Plate A:** Functional groups include **shredders**: (1.) *Goerodes* (Trichoptera: Lepidostomatidae), (2) *Anisocentropus* (Trichoptera: Calamoceratidae), (3) Leptoceridae (Trichoptera), and (4) *Tipula* , cranefly (Diptera: Tipulidae); **grazers**: (5) Hydroptilidae, microcaddisfly (Trichoptera), (6) *Helicopsyche* and case (Trichoptera: Helicopsychidae), (7) *Eubrianix* , water penny (Coleoptera: Psephenidae), (8) *Bulinus* (Gastropoda), (9) *Biomphalaria* (Gastropoda), and (10) *Burnupia* , limpet (Gastropoda); **collectors, deposit feeders**: (11) *Baetis* (Ephemeroptera: Baetidae), (12) *Caenis* (Ephemeroptera: Caenidae), (13) *Ephemerythus* (Ephemeroptera: Tricorythidae), (14) *Afronurus* (Ephemeroptera: Heptageniidae), (15) *Dicercomyzon* (Ephemeroptera: Tricorythidae), (16) *Choroterpes* (Ephemeroptera: Leptophlebiidae), (17) *Tricorythus* (Ephemeroptera: Tricorythidae), (18) Helodidae (Coleoptera), (19) Elmidae adult (Coleoptera), (20) *Pseudancyronyx* larva (Coleoptera: Elmidae), (21) *Micronecta*, water boatman (Hemiptera: Corixidae), (22) Leptoceridae (Trichoptera), (23) *Pericoma* (Diptera: Psychodidae), (24) Chironomidae, "midge" (Diptera), and (25) *Branchiura* (Annelida: Oligochaeta: Tubificidae); **collectors, filter-feeders**: (26) *Elassoneuria* (Ephemeroptera: Oligoneuriidae), (27) *Eatonica* (Ephemeroptera: Ephemeridae), (28) *Cheumatopsyche* and net, net-spinning caddis (Trichoptera: Hydropsychidae), (29) *Chimarra* and net (Trichoptera: Philopotamidae), (30) *Mutela* (Pelecypoda), (31) *Simulium* , blackfly (Diptera: Simuliidae), (32)*Culex*, mosquito (Diptera: Culicidae).

Plate B

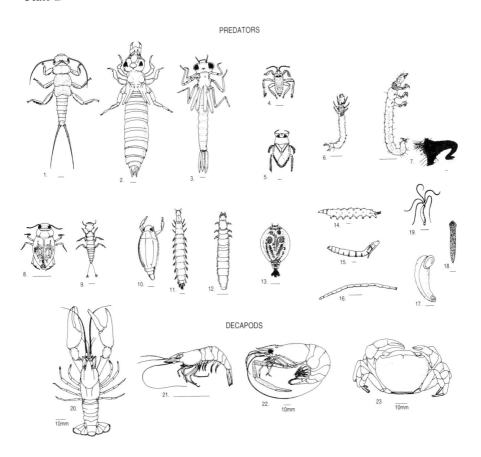

Figure 6.4. Common aquatic macroinvertebrates found in eastern African rivers and streams. Drawn from personal specimens or redrawn from diagrams (29,34,61,73,102). Length of scale = 1 mm, unless otherwise noted. **Plate B: Predators**: (1) *Neoperla* (Plecoptera: Perlidae), (2) *Progomphus* , dragonfly (Odonata: Gomphidae), (3) Coenagrionidae, damselfly (Odonata), (4) Veliidae (Hemiptera), (5) *Naucoris* (Hemiptera: Naucoridae), (6) Ecnomidae (Trichoptera), (7) *Polycentropus* and net (Trichoptera: Polycentropidae), (8) *Yola* adult (Coleoptera: Dytiscidae), (9) *Laccophilus* larva (Coleoptera: Dytiscidae), (10) Gyrinidae adult, whirligig beetle (Coleoptera), (11) Gyrinidae larva (Coleoptera), (12) Hydrophilidae larva (Coleoptera), (13) *Prosopistoma* (Ephemeroptera: Prosopistomidae), (14) *Atherix* (Diptera: Rhagionidae), (15) Tipulidae (Diptera), (16) Ceratopogonidae (Diptera), (17) Hirudinea, leech (Annelida), (18) *Dugesia*, flatworm (Turbellaria: Tricladidae), (19) *Hydra* (Coelenterata: Hydridae). **Decapoda**: (20) *Procamburus clarkii* , Louisiana red swamp crayfish (Introduced), (21) *Caridina* (Atyidae), (22) *Macrobrachium* (Palaemonidae), (23) *Potamonautes* (Potamidae). Feeding group designations are provisional and based on preliminary dietary analysis on personal specimens, as well as published information on these or closely-related forms (29,30,56,61,73,102).

are vectors for the pathogens causing malaria; and some snails (*Bulinus* and *Biomphalaria*) act as intermediate hosts for the trematode that causes schistosomiasis.

The relative abundance of different feeding groups (shredders, collectors, grazers, predators) depends on the relative availability of different types of food (CPOM, FPOM, algae, or other invertebrates), whereas species composition within each group is related to elevation and longitudinal position within the system (8,40, 75,103). The shredder fauna is not well-developed in eastern Africa, and consists primarily of case-building caddis (Order Trichoptera: Families Calamoceratidae, Lepidostomatidae, some Leptoceridae) and cranefly larvae (Order Diptera: Family Tipulidae). Shredders preferentially feed and show highest growth rates on leaf litter with the highest nutritional value (low C:N ratio).

Collectors are divided into filter-feeders, which remove fine particles being carried downstream in the water column, and deposit-feeders, which eat fine detritus after it has settled on the bottom. The dominant filter-feeders in fast-flowing waters in eastern Africa are the larvae of blackflies (Order Diptera, Family Simuliidae) and net-spinning caddisflies (Order Trichoptera, Families Hydropsychidae and Philopotamidae; 21,40,83,104). Blackfly larvae filter particles from the water column using fans of setae attached to their heads, and net-spinning caddis larvae capture particles using fine nets made from silk secreted by glands located in their heads. In slower flowing waters, clams (Order Pelecypoda) and some mayflies (Order Ephemeroptera, Family Ephemeridae) filter particles out of the water column by creating their own feeding currents. Mosquito larvae are associated with the water's surface in stagnant, backwater areas and filter fine particles from the water using bristly mouthparts. Other common filter-feeders include other mayfly larvae (Order Oligoneuriidae), sponges, and bryozoans.

Fine-particle, deposit feeders are often the most common taxa found in East African streams, and include many mayflies (particularly Family Baetidae, Family Leptophlebiidae, Family Heptageniidae, Family Caenidae), dipterans (Family Chironomidae), caddis larvae (Family Leptoceridae), and adult and larval elmid beetles (Family Elmidae; 8,40,58,59,75,85,104). Some taxa, including a variety of snails, some caddis larvae, and water pennies (Order Coleoptera, Family Psephenidae) have mouthparts adapted for scraping attached algae from hard surfaces, and hydroptilid caddis larvae pierce cell walls and remove the cell contents of filamentous algae with sucking mouthparts (Fig. 6.4). Because many of the deposit-feeders will also eat loosely-attached algae, the distinction between deposit-feeders and grazers is often blurred.

Invertebrate predators include stoneflies (Family Perlidae), a mayfly (*Prosopistoma*), dragonfly and damselfly naiads (Order Odonata), true bugs (Order Hemiptera), some dipterans (Family Athericidae), some caddis larvae (Families Polycentropodidae, Ecnomidae), some beetles (Family Dytiscidae, larvae of Family Hydrophilidae), hydroids, leeches, and flatworms (8,30,40,56,58,85). These predators show a variety of cruising and ambush foraging tactics, and engulf, entrap, or pierce small invertebrates. Flatworms produce mucus which allow them to glide along substratum surfaces, and they engulf prey that have become trapped in their mucus trails or that they encounter while moving along the bottom. Hydroids possess stinging cells (nematocysts) in their tentacles for entangling prey and, contrary to popular myth, most leech species are not sucking ectoparasites but, rather, carnivores preying on small, aquatic invertebrates. There are also invertebrate predators associated with the water's surface, including water striders

(Order Hemiptera: Families Veliidae, Gerridae) and whirligig beetles (Family Gyrinidae), which prey on insects caught in the surface film.

Most stream invertebrates develop throughout the year in East Africa and have a number of generations each year. Stream invertebrates are fairly vagile and quickly colonize newly-created habitats. During floods or droughts, invertebrates may occupy undisturbed refuges along stream banks, behind boulders, deep within the substratum, or in headwater springs, then recolonize denuded areas from these refuges. Many aquatic insects have a terrestrial adult stage that acts as a repository for the insect population during extreme disturbance events, and which oviposits in streams when these disturbances have passed. Many benthic invertebrates will occasionally enter and be carried downstream in the water column, and this "drift" is a major route for the recolonization of new or disturbed habitats. Invertebrates in a stream near Mt. Kenya reached peak abundance on newly-introduced substrata within 10 days (59). Baetid mayfly and blackfly larvae in stream sections in Uganda recovered to natural levels within a month after being dosed with DDT; however, large, long-lived taxa, such as larvae of predatory stoneflies and omnivorous hydropsychid caddisflies, took longer to recover (47).

A prominent difference between the invertebrate faunas of temperate versus tropical streams is the conspicuous presence of a variety of decapods (crabs, prawns, shrimp) in tropical streams. River crabs (Family Potamidae) are commonly found in streams and rivers throughout eastern Africa, and prawns (*Macrobrachium* spp.) are found in large rivers, particularly those flowing into the Indian Ocean, where they form part of the catch of indigenous fisheries (101,103). Among the river crabs, the species of *Potamonautes* (particularly *P. niloticus*) inhabiting western Kenya and eastern Uganda have attracted considerable scientific attention because a local vector of onchocerciasis, *Simulium neavi*, was always found attached to these crabs, a phenomenon called phoresis ("hitch-hiking") (21). Owing to control measures directed at *Simulium neavi*, including application of DDT, onchocerciasis has been eliminated from this area. In addition to these native decapods, the Louisiana red swamp crayfish, *Procambarus clarkii*, has been introduced to parts of eastern Africa and is spreading through rivers, streams, ponds, and lakes in Kenya and Uganda (44).

Fish and Fisheries. The fish faunas of eastern African rivers are dominated by representatives of the minnow family (Family Cyprinidae) with important contributions by the Cichlidae, Characidae, Mormyridae, and various catfish families (Clariidae, Bagridae, Mochokidae, Amphilidae, Synodontidae, Schilbeidae) (Table 6.1, Fig. 6.5; 6,10,19,37,53,79). Of particular interest are the presence of two ancient lineages whose representatives possess lungs (lungfish, Family Protopteridae; bichirs, Family Polypteridae). A variety of estuarine or marine species are found at the mouths of rivers flowing into the Indian Ocean, including gobies, marine catfish (*Arius*), milkfish, pipefish, tarpon, juvenile mullets, and sharks. Several eel species (*Anguilla* spp.) spawn in the Indian Ocean and ascend vast distances up East African rivers where they grow and mature (86).

Many of the fish species in East Africa are generalized predators on invertebrates, or are omnivores that eat a variety of invertebrates as well as algae and plant fragments (Table 6.1). The most common genus in East Africa, *Barbus* (Family Cyprinidae), is a generalized predator on invertebrates. *Barbus* has numerous species in East Africa, including a number that have localized

Plate A

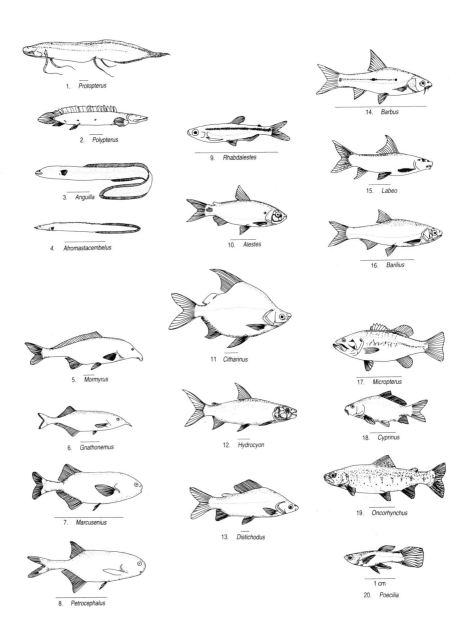

1. *Protopterus*
2. *Polypterus*
3. *Anguilla*
4. *Afromastacembelus*
5. *Mormyrus*
6. *Gnathonemus*
7. *Marcusenius*
8. *Petrocephalus*
9. *Rhabdalestes*
10. *Alestes*
11 *Citharinus*
12. *Hydrocyon*
13. *Distichodus*
14. *Barbus*
15. *Labeo*
16. *Barilius*
17. *Micropterus*
18. *Cyprinus*
19. *Oncorhynchus*
20. *Poecilia*

1 cm

Plate B

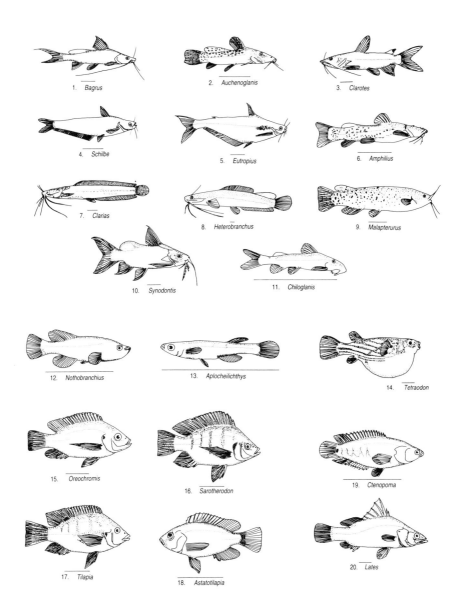

Figure 6.5. Representative fish found in eastern African rivers and streams. Drawn from personal specimens or redrawn from figures or photographs (10,19,37,53,67). Length of scale = 5 cm, unless otherwise noted. Families, common names, and dominant food types eaten by the adults of each genus are noted in Table 6.1.

Table 6.1. Families, common names, and food habits of freshwater fish found in East African streams and rivers. Plate no. refers to the figure number of fish pictured in Fig. 6.5. Some fish families with restricted distributions in East Africa are not pictured in Fig. 6.5. The East African genera and distributions of these families, which are not shown, are listed in the "common name" column.

Family	Common name	Plate no.	Food
Protopteridae	lungfish	5A:1	primarily aquatic invertebrates, esp. molluscs and insects; small fish, frogs
Polypteridae	bichir	5A:2	primarily fish, also insects, decapods, frogs
Anguillidae	eel	5A:3	fish, crabs; some insects
Mastacembelidae	spiny eel	5A:4	primarily aquatic insects, some crustaceans
Mormyridae	electric fish	5A:5 to 8	primarily aquatic insects; some small crustaceans and molluscs
Characidae	tigerfish	5A:9	small insects and crustaceans
		5A:10	omnivore, incl. terrestrial & aquatic inverts, algae, seeds, plants, detritus small fish
		5A:12	primarily fish
Citharinidae		5A:11	mud eater; FPOM
Distochodontidae		5A:13	plants, detritus
Cyprinidae	Introduced common carp	5A:14	omnivore, primarily aquatic insects, crustaceans, molluscs, but some filamentous algae, plants, seeds, and detritus
		5A:15	mud eater; FPOM
		5A:16	small invertebrates, including zooplankton, fish fry
		5A:18	omnivore: algae, plants, FPOM, aquatic invertebrates, especially chironomids
Centrarchidae	Introduced, largemouth black bass	5A:17	primarily fish, some decapods, insects, frogs
Salmonidae	Introduced, rainbow trout	5A:19	aquatic invertebrates, esp. insects, crustaceans, molluscs; terrestrial invertebrates; occasional fish
Poeciliidae	Introduced, guppy	5A:20	zooplankton, small insects
Bagridae		5B:1	primarily fish, some molluscs and insects
		5B:2	primarily aquatic insects; also, molluscs, oligochaetes, crustaceans, fish fry
		5B:3	primarily fish, some molluscs and aquatic insects
Schilbeidae		5B:4 and 5	omnivores, including aquatic and terrestrial insects, crustaceans, molluscs, seeds, fruit, frogs, fish
Amphiliidae		5B:6	aquatic insects

Family	Common name	Plate no.	Food
Clariidae	walking catfish	5B:7 and 8	omnivores: fish, aquatic invertebrates, detritus, seeds, birds
Malapteruridae	electric catfish	5B:9	primarily small fish, some decapods and aquatic insects
Mochokidae		5B:10	omnivore: primarily aquatic invertebrates, but also fish, detritus, seeds, plants, algae
		5B:11	aquatic invertebrates, especially insects; periphyton
Aplocheilidae	annual killifish	5B:12	small crustaceans, including zooplankton; small insects
Poeciliidae Subfamily Aplocheilichthyinae	top minnow	5B:13	small crustaceans, including zooplankton; small insects
Tetraodontidae	pufferfish	5B:14	molluscs
Cichlidae		5B:15	primarily phytoplankton, occasionally invertebrates, detritus, periphyton
		5B:16	primarily phytoplankton, some zooplankton, detritus
		5B:17	primarily plants, some invertebrates, periphyton, detritus
		5B:18	aquatic invertebrates
Anabantidae		5B:19	small terrestrial and aquatic insects, crustaceans
Centropomidae	Nile perch	5B:20	primarily fish
Osteoglossidae	*Heterotis*, In: Nile, Omo River - Lake Turkana	Not shown	omnivore: invertebrates, fish, amphibians, plankton, seeds, plants
Notopteridae	*Xenomystus*, In: Nile	Not shown	
Gymnarchidae	*Gymnarchus* In: Nile, rivers affluent to L. Turkana	Not shown	fish, frogs, aquatic invertebrates
Cromeriidae	*Cromeria,* In:Nile	Not shown	
Kneriidae	*Kneria,* In: Tanzanian rivers	Not shown	aquatic insects
Channidae	*Parachanna,* snake head, In: Nile	Not shown	
Gobiidae	gobies, estuarine	Not shown	aquatic invertebrates, small fish
Eleotridae	*Kribia,* others, primarily estuarine, but *Kribia* is in the Nile	Not shown	

A wide variety of marine fishes may enter the lower ends of coastal rivers.

distributions. The minnow family (Cyprinidae) also includes the widespread genus *Labeo*, which possesses a ventral mouth adapted for ingesting bottom sediments and associated detritus and organisms.

The Cichlidae include generalized haplochromine species (such as *Astatotilapia* spp.) that feed on invertebrates in streams, as well as a variety of *Oreochromis*, *Sarotherodon*, or *Tilapia* species that ingest phytoplankton, periphyton, or aquatic macrophytes. The cichlid genus *Serranochromis* includes many species in the Zambezi River system which are predators on aquatic invertebrates or fish (10). Many of the cichlids thrive in lentic (standing water) habitats associated with floodplains and often become the dominant species in reservoirs formed by dam construction. Some of the cichlids are endemic to particular river systems; however, the widespread introduction of many cichlid species (*Oreochromis mossambicus, O. niloticus, Tilapia zillii, T. rendalli*) has created some uniformity in the composition of cichlid assemblages.

Large catfish (such as *Clarias gariepinus, Bagrus, Schilbe*), tigerfish (*Hydrocynus*), and Nile perch (*Lates*) eat other fish. The commercially-important *Distochodus* (Family Distochodidae) is primarily herbivorous, whereas the closely-related *Citharinus* (Family Citharinidae) eats fine detritus in bottom sediments (10,19,51). Catfish belonging to the family Schilbeidae and some characins (*Alestes*) are omnivores that will eat algae, plants, seeds, fruits, zooplankton, terrestrial and aquatic invertebrates, and fish, depending on the availability of different food types.

Some fish show seasonal and ontogenetic changes in their food habits. For example, some of the characins eat seeds and invertebrates that have fallen out of flooded vegetation when they move onto the floodplain at high water, but switch to aquatic invertebrates and fish as water levels decline and terrestrial invertebrate and seed availability decreases (51). Many of the tilapiines (Family Cichlidae) feed on microcrustaceans (copepods, cladocerans) or small insects when they are fry, but specialize on algal or plant foods as they mature. Juvenile stages of the pufferfish (*Tetraodon*) feed on small insects and crustaceans whereas adults eat primarily molluscs. Many piscivorous species feed on invertebrates in their juvenile stages.

Many headwater streams in eastern Africa possess a depauperate fish fauna characterized by small cyprinid species, usually *Barbus* or, in Ethiopia, *Garra*. In some small, fast-flowing, stony streams the small catfish *Amphilius* or *Chiloglanis* are present, and *Amphilius* was apparently the only fish originally inhabiting many Kenyan highland streams before they were stocked with trout (19). These catfish possess expanded pectoral fins or suckers for attaching to, or lodging in, substrata in torrents. Provided upstream reaches are not blocked by waterfalls, small streams may also be visited by migratory *Barbus* or *Labeo* species (86).

Fish species diversity tends to increase as one proceeds down the course of river systems (98). Terrestrial and aquatic invertebrates are the major food source for headwater fish; however, a great variety of algal, detrital, plant, invertebrate, and vertebrate food sources become available in lowland rivers with their complex floodplain habitats (54,97). Although many fish in savanna rivers are generalists, there is some spatial segregation of fish species by depth and current speed, with species with ventral mouths feeding on the bottom (such as *Labeo* and some catfish), species with upturned mouths feeding at the surface (top minnows, *Ctenopoma*), and streamlined species with terminal mouths feeding in mid-water (such as some *Barbus, Alestes*).

In addition to the well-known introduction and re-distribution of commercially important cichlids and Nile perch in East Africa (see Chapter 8), a variety of exotic fish species have been intentionally or inadvertently introduced into eastern Africa including black bass (*Micropterus salmoides*), bluegill (*Lepomis macrochirus*), rainbow trout (*Oncorhynchus mykiss*), and mosquitofish (*Gambusia affinis*) from North America; guppies (*Poecilia reticulata*) from the Caribbean; and common carp (*Cyprinus carpio*), silver carp (*Hypophthamichthys molitrix*), grass carp (*Ctenopharyngodon idella*), and brown trout (*Salmo trutta*) from Eurasia (66). Grass carp were introduced to control water weeds and silver carp were imported to crop phytoplankton; however, these introductions were largely unsuccessful and these species have largely disappeared. In contrast, common carp are now a prominent component of the fisheries catch from reservoirs and farm ponds, and rainbow and brown trout are commonly found in cool, highland rivers. Trout form the basis of an important sport fishery in the central highlands of Kenya, and trout farming is practiced on Mt. Kenya and Kilimanjaro.

Many riverine species, including a variety of cyprinids (*Barbus*, *Labeo*), mormyrids, characins (*Alestes*), and catfish (*Clarias*, *Schilbe*, *Synodontis*), undergo spawning migrations when floodplains are inundated (86,101). Spawning movements are associated with rising water at the onset of rains and spawning occurs in floodplain habitats, small tributary streams, or headwater areas during high water. Adults return to the main channel leaving their young to grow and develop near spawning habitats, usually on the floodplain (97). There are also fish species that feed, grow, and mature in lakes but ascend inflowing rivers to spawn (called potamodromous species; 14,92,99,100). In some cases, these potamodromous species form the basis for important fisheries. For example, historically the Luo people used a variety of traps and fish fences to capture fish, particularly the fwani (*Barbus altianilis radcliffi*) and ningu (*Labeo victorianus*), ascending or descending Lake Victoria's affluent streams during spawning migrations. Similarly, the salmon-like mpaca (*Opsaridium microlepis*) ascends the northern streams affluent to Lake Malawi in large numbers and is heavily fished on these spawning migrations.

Although most migrating species broadcast large numbers of eggs on spawning substrata, many of the fish resident in floodplain habitats produce fewer eggs but provide offspring with some parental care. Tilapiines (Family Cichlidae) build and protect nests on lake or pond bottoms, lungfish guard nests built in vegetation, and haplochromines (Family Cichlidae) brood eggs and fry in their mouths. Populations of many riverine species peak at high water when the area of inundated floodplain is at its maximum and after eggs have hatched. Populations quickly decline during falling water, however, as habitat shrinks, food supplies dwindle, and predation pressure, from fish, birds, and humans, increases. Some fish predators, however, may show maximum feeding rates during this period because their prey become more concentrated during this time (93). Many fish trapped in floodplain environments perish owing to predation or desiccation as their habitats dry, and population levels are lowest during low or early rising water (97).

Given the general aridity of Africa and its long history of dry and wet periods, it is not surprising that a variety of fish species can deal with high temperatures, low oxygen levels, and the desiccation of floodplain habitats. Many of these species can tolerate high water temperatures, and they possess a variety of adaptations for dealing with low oxygen. Top minnows (Family Poecilidae, Subfamily

Aplocheilichthyinae, *Aplocheilichthys*) have upturned mouths and often gulp the minute layer of oxygenated water found at the air-water interface during low water periods (98). Lungfish, bichirs, clariid catfish, and anabantids (*Ctenopoma*) have lungs or accessory respiratory structures that allow them to use atmospheric oxygen. Lungfish, clariid catfish and anabantids, that often dominate in swamps or isolated wetlands (such as oxbows), can walk on land, and leave de-oxygenated ponds and travel to new habitats at night. Some lungfish species survive through dry periods by constructing a burrow and surrounding their bodies with a mucus cocoon to prevent desiccation. Annual killifishes (Family Aplocheilidae, *Nothobranchius*) occur in temporary ponds and streams, and survive through dry periods by producing drought-resistant eggs.

Many riverine environments in eastern Africa are characterized by low water transparency; consequently, many of the fish in these habitats have mechanisms for dealing with low light levels and a paucity of visual cues. Almost all fish have lateral lines that are sensitive to water movements, and the catfishes and *Barbus* have barbels that are probably sensitive to chemical and mechanical cues. The mormyrids produce weak electric currents, and respond to distortions in their electric fields produced by obstructions, prey, and other fish. Different mormyrid individuals produce different electrical signals and there are consistent differences in electrical signals between the sexes, suggesting the importance of electrical discharges to social or mating behavior (20). In addition, mormyrids will change their electrical signals in response to changes in their chemical environment; consequently, it has been suggested that mormyrid electrical signals can be used as indicators of water quality (36). The electric catfish, *Malapterurus*, produces strong electrical discharges (450 V) which may be used in stunning prey or thwarting predators. Because they produce a nasty shock, fishermen will induce these fish to produce an electrical discharge before handling them. It takes several hours after discharge before these fish are able to produce another strong shock.

Riverine and potamodromous fish have provided an important protein source for many of the people of eastern Africa. Approximately 40% of the freshwater fisheries catch in Africa comes from rivers and their floodplains, amounting to around 500,000 tons of fish per year (94,95). In eastern Africa a large proportion of the catch comes from the large Rift lakes; however, river fisheries have developed along the spawning routes of some migratory species, and in floodplain and swamp waters. Fishing activity often supplements floodplain farming and is highly seasonal. Fishermen concentrate on migratory species during flooding periods and on floodplain species during falling and low-water periods. Fish are caught using traps, baskets, fences, hook-and-line, spears, poisons, and nets, and, historically, most fish were used locally, although small markets for dried, salted, or smoked fish existed. In Africa, the total yield of fish is a direct function of river and catchment size, and floodplain rivers with extensive floodplains have a much higher yield than similarly-sized rivers with "normal" floodplains (94,95,96,97; Fig. 6.6). Because floods provide nursery habitat for riverine fishes, there is a direct relationship between flooding intensity and duration, on one hand, and the fisheries catch in one or two years after the fish have grown to a catchable size. The introduction of modern fishing equipment, such as monofilament gill nets and boats powered by outboard motors, has resulted in the wholesale exploitation of some stocks, and modern distribution systems have encouraged the export of fish to urban markets.

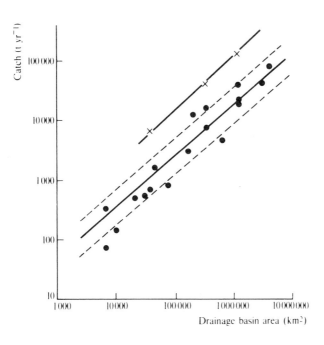

Figure 6.6. The relationship between fisheries catch (C) and drainage basin area (A) for African rivers. The dots indicate lakes with "normal" floodplain development, and the "x's" indicate lakes with extensive floodplains. Note the log-log scale of axes. The equation for the "normal" river line: $C = 0.13A^{0.85}$; the equation for the floodplain river line: $C = 0.44 A^{0.90}$. (Taken from 96, with permission).

Other Vertebrates. Riverine environments are also important habitats for a variety of amphibian, reptile, bird, and mammal species. Amphibian populations are well-developed in wetland habitats that lack predatory fish. Among reptiles, omnivorous turtles, such as the side-necked turtle, and predatory crocodilians are common inhabitants of river and floodplain habitats. Although crocodiles have been exterminated from many parts of their range, they reach high population densities and act as top predators in unperturbed areas, such as the Mara River system in the northern Serengeti. Monitor lizards are commonly found in riparian zones and can be important predators on the eggs of birds and crocodiles. A variety of snakes, including the large pythons, prey on other reptiles, mammals, and birds. A variety of birds, including herons, hammerkops, ibises, ducks, coots, storks, grebes, cormorants, pelicans, geese, spoonbills, jacanas, cranes, kingfishers, plovers, finfoots, sandpipers, and fishing eagles make extensive use of aquatic habitats and foods (3). Among large mammals, conspicuous inhabitants of riverine and floodplain environments include hippopotami, buffaloes, and a variety of antelopes (kob, sitatunga, reedbuck, water buck), and, among smaller mammals, cane rats and marsh mongooses are never found far from water. Needless to say, almost all large herbivorous and predatory mammals, and many small species, use river wetlands as sources of food, water, and cover, including riparian species which are rare and endangered (such as Tana River red colobus and mangabey). Two otter species are found in East Africa and prey on freshwater decapods, molluscs, and fish, and an insectivorous otter shrew occurs in rainforest streams draining the Ruwenzoris (39,50).

Human Influences on Streams and Rivers

Humans have had large effects on the streams and rivers of eastern Africa through land and water development, the introduction of sediments, pollutants and exotic species, and the overexploitation of fish stocks (1,25). These effects have resulted in the demise of some species that cannot tolerate conditions created by human activity, and an increase in those species cultured by human beings or favored by human-induced alterations in the aquatic environment. Studies of the viability of development projects require a clear recognition of the value of indigenous farming, grazing, and fishing practices as well as the value of development benefits. Certainly, the conservation of native species and life styles should play a role in development issues, because of their potential value to this and future generations.

Habitat Alteration

Water Development. Humans have had large effects on river systems, either directly, by damming and channelizing rivers and building water distribution facilities, and indirectly, by removing native vegetation in drainage basins, replacing it with agricultural, urban, industrial, or mining developments. Eastern African rivers show considerable seasonal fluctuations in discharge; consequently, many dams have been built for the storage of floodwaters for use in the dry seasons. Dams have far-reaching effects on riverine environments by converting lotic to lentic systems above dams, and by changing hydrological regimes, temperature conditions, nutrient levels, and sediment transport below dams. Typical riverine species of invertebrates (baetid mayflies, blackflies, net-spinning caddis) and fish (mormyrids, some cyprinids (such as *Labeo*) and siluroids) decline and are replaced by more lentic taxa (oligochaetes, chironomids, cichlids, carp) in reservoirs (22,23,24,81). Sections of river downstream from dams usually show a more constant discharge and less temperature variation through the year compared to the original river. Exact temperature and nutrient levels in this section will depend on whether water is released from the upper or lower layer of the reservoir. Because of the release of large quantities of plankton to downstream areas, filter-feeding invertebrates (simuliids, hydropsychids) can reach high levels in these areas.

Downstream floodplain areas are deprived of normal flooding and drying schedules, and normal levels of sediment deposition, and the dams act as barriers to migrating fish species. Dams may also disrupt the hydrological or chemical cues needed to induce migratory and spawning behavior in fish. Because flooding durations are often reduced, there is a general desiccation of the floodplain environment, with drying lakes and ponds becoming more saline and floodplain plant species being replaced by more xeric forms. Because extensive flooding is prevented in downstream floodplains, old floodplains are subjected to constant grazing pressure from livestock and wildlife leading, in some cases, to overgrazing (4). Although dry season farming is a common activity on floodplains, dam construction promotes the expansion of intensive, year-round cultivation, with the attendant destruction of native plants. Aseasonal floods produced by controlled dry season releases from dams can destroy the larval habitats of many frogs and toads, and destroy nesting sites for crocodiles, monitor lizards, terrapins, and aquatic birds in downstream areas. Some land development schemes associated with dam construction, such as the expansion of rice farming, provide increased habitat for pathogens and their vectors (such as *Anopheles* mosquitoes that carry the protozoan, *Plasmodium*, that causes malaria). In general, then, the construction of dams has far-reaching consequences for the economies and ecologies of surrounding areas.

The construction of canals for distributing and transporting water, and channelization of rivers and streams, often accompanies water development and flood control projects. River and stream channelization reduces habitat diversity and, consequently, aquatic biodiversity, whereas canals can act as pathways for the dispersion of exotic species and vectors of pathogens.

Land Development. The indirect effects of human activity on river systems are usually mediated through the removal of native vegetation. Reductions in native vegetation in drainage basins and riparian zones through overgrazing; burning; deforestation; road construction; urban, industrial, and agricultural expansion; and mining results in increased erosion and sediment inputs into streams, and decreases infiltration and groundwater recharge. As a consequence, streams and rivers in developed basins have higher floods and lower dry season discharge than those in unperturbed basins (see Chapter 16). Sediments can cover benthic organisms, abrade the integuments of aquatic animals, inhibit oxygen penetration into sediments, and reduce light penetration and photosynthesis. Removal of riparian vegetation will also reduce bankside shading and detrital inputs, resulting in more extreme water temperatures and higher light levels, but lower daytime oxygen levels, standing stocks of detritus, and cover for fish and invertebrates.

Pollution

Human activities introduce contaminants to river and stream systems that reduce native biodiversity and the quality of water for human use (41,42,46,57,77,90). One of the most obvious forms of pollution is the introduction of domestic sewage to running-water systems. Extensive work in Europe and North America has documented the effects of organic sewage discharged into streams. In general, introduction of large quantities of sewage into streams will cause a large increase in microbial respiration (Biochemical Oxygen Demand, BOD), a drastic decline in oxygen levels, the appearance of "sewage fungus", and the disappearance of most macroinvertebrates. Provided there are no additional sewage inputs to a river, the river will recover as it flows downstream and may achieve the original complement of organisms found above the sewage discharge. In zones with extremely low oxygen the invertebrate fauna is dominated by tubificid worms and *Chironomus* midges. As oxygen levels increase, additional oligochaete and chironomid taxa, and flatworms, appear. At the highest oxygen levels, a diverse fauna of mayfly, stonefly, caddisfly, blackfly, beetle, and hemipteran taxa are present.

Results from studies of East African streams have verified this pattern, emphasizing the utility of macroinvertebrate assemblages as indicators of pollution stress. For example, several studies have examined chemical, floral, and faunal changes in the Nairobi River as it flows through the city of Nairobi (35,69,70,71). Above Nairobi, the Nairobi River's invertebrate fauna is very diverse and includes many invertebrate taxa. As it flows through the Nairobi suburbs, this stream receives sediment and nutrient inputs from roads, dwellings, and gardens, and only a few tolerant taxa persist (such as introduced crayfish, baetid mayflies, chironomids, oligochaetes, hydropsychid caddis larvae). Within Nairobi this stream becomes an open sewer receiving human waste directly as well as waste from vegetable kiosks, garages, and latrines. Chironomini midges reach maximum population levels at the interface between suburban and urban areas of the city, in areas where oxygen levels are sufficient for life processes and where organic inputs provide a rich food source. Even chironomids, however, are absent from stream sections draining most urban areas, and the only aquatic insects commonly collected in these sections are air-breathing psychodid larvae (Fig. 6.4). The invertebrate fauna of the Nairobi River recovers below Nairobi until a fairly diverse fauna is present in the Athi River at Fourteen Falls, around 50 km below the city limits; however, many species present in the upper Nairobi River are absent at Fourteen Falls and vice-versa.

Pesticides have been used extensively throughout Africa to control the vectors of onchocerciasis, malaria, schistosomiasis, and rypanasomiasis. These pesticides have been introduced directly into aquatic systems, or have drifted into aquatic systems from land. These pesticides can have direct toxic effects on non-target organisms, depending on the exact pesticide and dose used, and can also have indirect effects on other organisms by affecting their food base (25). Similarly, herbicides used to control water weeds have had direct and indirect effects on other aquatic organisms by removing habitat and food resources. Industrial and mining processes can release organic toxins, acids, metals, or heated water into streams and rivers, sometimes exceeding the lethal limits of riverine organisms. Many factories processing agricultural or forestry products, such as paper mills and sugar and coffee factories, introduce large amounts of organic matter and, in the case of paper mills, chlorinated waste, directly into rivers and streams (7,11,25). These toxins and organic waste not only affect the aquatic biota, but also impair the quality of water for human use. Proposed oil development projects in eastern Africa require attention because of their potential for polluting water bodies with oil residues and toxic hydrocarbons.

Introduction of Exotic Species

Although a number of exotic species have been introduced to East Africa, their effects on native species are unclear. The North American snail, *Physa acuta*, was accidentally introduced into Africa from Europe and is now common in polluted waters. The South American snail, *Marisa cornuarietis*, and North American snail, *Helisoma*, were introduced to act as competitors for schistosome-carrying snails (13). Experimental trials have indicated that these snails can control the abundance of *Bulinus* and *Biomphalaria*, but there is concern about their effects on other components of the biota, such as other snails. They have not spread, however, from sites of experimental introduction.

The Louisiana red swamp crayfish is spreading throughout eastern Africa and is being considered as a control agent for the snails that carry the schistosomes causing schistosomiasis (43). Because crayfish are voracious omnivores taking a variety of algal, plant, and animal material, they have had profound effects on stream and lake systems in temperate areas, suggesting that introductions of crayfish should be approached with caution.

Because trout have been introduced into high-elevation rivers that were originally nearly devoid of fish, it would appear that trout introduction effects on river ecosystems have been minimal. Some investigators, however, have suggested that trout have reduced native crab and amphilid catfish populations, and trout effects on native insects are unknown (19,85). In South Africa, trout introductions have resulted in the demise of several fish species (2). Common carp have become abundant in some of the large reservoirs and rivers of eastern Africa, and form an important part of the catch in these areas. Observations in temperate areas suggest that carp may have a large effect on aquatic systems by increasing turbidity by stirring up bottom sediments. They also reduce native fish populations by eating the eggs of native species, or by covering eggs with sediments. Their interactions with native bottom-grubbing taxa in Africa, such as *Labeo*, are unknown. Mosquitofish and guppies were introduced into eastern Africa for mosquito control, and have spread to a number of streams. Although effects of these species on native species are unrecorded, the similarity of these fish to native topminnows (such as *Aplocheilichthys*) underlines the potential for interactions between them.

A number of fish native to Africa have been introduced from native to new habitats. These introductions may represent the greatest threat to the genetic integrity and persistence of native fish populations. The effects of Nile perch introductions on Lakes Victoria and Kyoga are well-known; however, there is little information on the effects of these introductions on fish faunas in the connecting Victoria Nile. Commonly raised cichlids have been widely introduced throughout eastern Africa and may interbreed, compete with, or modify the habitat of endemic species. Similarly, some *Barbus* species have been introduced into non-native waters with possible implications for endemic *Barbus* populations. In general, then, many exotic species have been introduced to eastern African fresh waters; however, their effects on native species is largely unknown. Given the documented, profound effects of introduced on native species in rivers and lakes in other parts of the world, the spread of already-introduced exotic species should be monitored and, where possible, checked, and the introduction of additional species should be approached with extreme caution.

Overfishing

The introduction of modern fishing equipment, particularly the monofilament gill net, has resulted in the overexploitation of some native fish stocks. A large, indigenous fishery was once centered on the *Barbus altianilis radcliffi* and *Labeo victorianus* populations that ascended Lake Victoria's tributaries to spawn. The widespread use of gill nets in this fishery resulted in some early large catches followed by large drops in fish yields (7,15). Similarly, fisheries officers noted the decimation of fish stocks in floodplain waters of the lower Sabaki River associated with intense fishing and postulated that fish stocks were maintained by refuge areas located upstream (101). Overfishing has also been implicated in the declines of potamodromous *Labeo mesops* and catfish stocks in Lake Malawi, and may have contributed to declines in the *Opsaridium* fishery (82). Although alternative explanations for fishery declines have been forwarded, overfishing appears to have played a role in the decline of some fish stocks, particularly those of migratory species.

Problems and Recommendations

Rivers and streams in eastern Africa contain a variety of aquatic species with food, aesthetic, recreational, and scientific value. In addition, rivers are parts of complex floodplain habitats that support a diversity of plants and animals adapted to seasonal changes in water level. Because streams and rivers drain and receive inputs from their catchments, they are quite sensitive to human-induced disturbances in their catchments. Agricultural, industrial, domestic, and urban expansion affects hydrological regimes, erosion and sedimentation, and the input of toxins and nutrients to streams and rivers. Dam construction, in particular, affects the behavior, demographics, diversity and distribution of organisms in upstream areas inundated by reservoir waters and downstream areas deprived of natural flooding cycles. The introduction of exotic species, pollution, and overfishing can also reduce the diversity of native species.

The previous analysis suggests key species that should be considered when evaluating future development projects. Because of the ecological importance of riparian and floodplain forests and herbaceous vegetation, the impacts of planned projects on these plant communities should be considered. Similarly, migratory fish are sensitive indicators of the health of whole river systems, because they use a variety of floodplain and channel habitats. Environmental degradation of only one migratory fish habitat can disrupt life cycles, resulting in population declines and the loss of an important meat resource for local people.

Endemic species should also receive special consideration because disturbances in their home habitats can result in extinctions. Although many aquatic invertebrates and fish in eastern African waters are widespread, there are taxa with localized distributions, including some cichlid and *Barbus* species, and some mammals. Although the invertebrate fauna of streams and rivers is not well-known, species with restricted distributions have been found in most groups that have been examined extensively (such as snails and decapods; 13,103). It is probable that more endemic species will be found as other invertebrate groups are studied more intensively. Small streams are quite vulnerable to human disturbance because their low discharge does not effectively dilute contaminant or sediment inputs. As a consequence, even small, local disturbances can have large effects on the biota of these ecosystems.

Some of the deleterious effects of land and water development can be mitigated. Migratory fish populations can be maintained by the construction of fish ladders and hatcheries; however, these measures are not always effective in allowing fish to pass around dams or in maintaining native fish stocks. The loss of indigenous fisheries can be compensated for by the development of fish farming and reservoir fisheries activities. This will do little, of course, to conserve native fish stocks. Dam releases can be timed to coincide with natural floods, insuring that downstream areas experience similar flooding cycles, albeit reduced, as before dam construction (5;

Chapter 14). This will greatly reduce many of the negative effects of dams on downstream ecosystems. These procedures, however, may not be followed if they contradict flood control and agricultural expansion policies based on complete water storage in wet seasons and controlled water releases during dry seasons. The judicious timing of dam releases can also be used as an effective means of controlling nuisance macrophyte growth in reservoirs and disease vectors in downstream and flooded areas (such as *Simulium, Anopheles*; 2,23,24).

Soil erosion from agricultural activity can be reduced through restrictions on land use (such as prohibiting farming on slopes that are too steep), by planting ground cover or trees, by terracing, by controlling herd size and grazing intensity, and by contour or strip farming. Sediment and nutrient inputs can be greatly reduced by maintenance of riparian vegetation along streams. Although many of these structural solutions to environmental problems are too expensive to be widely used, some, such as tree planting and preservation of riparian bufferstrips, are relatively inexpensive and do not require sophisticated technology.

Similarly, there are a variety of methods available for reducing the discharge of pollutants into receiving waters. Urban and rural areas need adequate sewage treatment facilities to reduce water quality degradation. At present, many African cities do not possess plumbing networks for collecting all storm and sewage run-off, and sewage treatment plants are often non-operational or overburdened. The result is that expanding cities have gone farther and farther afield to find potable water sources, because more proximal sources have become too contaminated for human use. Rural and slum areas often lack the most rudimentary sewage facilities and waste is often dumped directly into streams or rivers. The use of settling ponds in mining operations and treatment facilities for industrial operations can reduce contaminant inputs to streams and rivers. In addition, operational modifications, such as reducing the use of chlorine in bleaching processes in paper mills, can greatly reduce the input of toxic waste to water ways.

Many of these pollution abatement techniques are costly. The environmental costs of reduced water quality on human health, fisheries, wildlife, and environmental quality are seldom considered in East Africa because the costs of mitigation frequently affect profit margins. Procedures to evaluate the efficacy and environmental effects of pesticides and herbicides on aquatic systems are required. In place of pesticide and herbicide use, which often affects a broad spectrum of animal and plant species, authorities could put increased emphasis on the biological control of pest organisms. Such control could involve the use of pathogens, such as *Bacillus thuringiensis*, which only affect pest species, as well as the use of predators for nuisance invertebrates (such as fish that eat mosquito larvae) and herbivores for nuisance macrophytes (such as herbivorous tilapiines). Care should be taken, however, in introducing new organisms for biological control because of the potential impact of these introduced species on native plant and animal communities.

With a few exceptions, many of the introductions of exotic species to East African rivers have been haphazard or inadvertent. Such introductions may have caused the disappearance of native species (see above and Chapters 7 and 8). There should be controls on the introduction of exotic species, and a careful consideration of the influence of introduced species on native communities, particularly because exotic species, once successfully introduced, are almost impossible to remove.

Effective management of riverine fisheries may depend on the use of standard management options (see Chapter 3) including restrictions on fishing gear (such as mesh size, net length), closed seasons or areas, habitat preservation or improvement, licensing systems, and the maintenance of adequate and fluctuating water flow. Some of these options are already being used in eastern Africa (for example in Malawi), but it is too early to tell if these management policies have effectively preserved fish stocks while maintaining fish yields. By prohibiting fishing activities in national parks, some nations (for example Tanzania) have effectively preserved native fish but have removed an important meat resource from the diets of local peoples. Clearly, the conservation of native species will depend critically on the participation and cooperation of local people. To be effective, some conservation measures will require compromises between the human utilization of natural resources and the preservation of natural ecosystems.

Inadequate attention has been paid to the influence of dam construction on native species, indigenous agricultural activities, and traditional life styles. Rivers and their floodplains support a variety of fish species, and woody and herbaceous plants, that are extensively utilized by indigenous peoples. The fishing, farming, and grazing activities of these people are geared to normal flood cycles. Construction of dams disrupts flooding cycles, blocks fish migrations, disrupts nutrient subsidies to floodplains and encourages the removal of native vegetation by intensive farming or grazing. Flooded areas upstream from dams are lost to future farming or grazing activity. Water and land development disrupt natural hydrological cycles and degrade riverine habitats, all with deleterious effects on native species diversity and indigenous human activities.

Development projects are often geared to produce export products for urban markets. Dam construction produces hydroelectricity, and provides flood protection and a constant water source for irrigating crops and watering livestock. All of these benefits produce subsidized products for the industrial, governmental, and service sectors of urban areas.   These developments and population expansion insure the replacement of traditional life styles by intensive agricultural activity for export markets, and the disproportionate growth of urban relative to rural areas. The value of traditional activities are almost never considered, and peoples in target areas seldom included in planning development projects.

The optimal use of natural resources requires an ecosystem perspective that balances local resource use, wildlife habitat, water quality, soil conservation, and indigenous fisheries with the production of export products for urban markets. Many eastern African countries have set up Water Basin authorities for the effective management of river catchments; however, wholesale water and land development geared towards production of export products has continued to have priority in resource use decisions. This situation can be traced to centralized control of the decisions of Water Basin authorities, the reliance on engineering solutions to development problems, interference from international aid agencies or donor nations, and the lack of rural participation in management decisions (78). Water basin authorities that include local representation and extensive local participation, as well as representatives from environmental and natural resources agencies, would allow for a more comprehensive, broader-based approach to natural resources development.

It is apparent that much of our knowledge of plant and animal populations in the riverine habitats of eastern Africa is fragmentary, but such knowledge will be necessary in carrying out conservation and development that consider native diversity. A key element in future conservation activities should be an inventory of river and floodplain resources in eastern Africa, including information on the composition, diversity, and abundance of resident plant and animal communities. Such an inventory would include detailed surveys of aquatic, riparian, and floodplain habitats; taxonomic and systematic work on collected specimens; the development of useful keys to East African material; the determination of trophic relationships and delineation of food webs; and correlational studies of relationships between biotic and abiotic conditions, on one hand, and the distribution and abundance of riverine species, on the other. Comparisons of undisturbed systems with similar systems that have been affected by human activity can also provide some indication of the effects of anthropogenic perturbations on river systems. Such comparisons should go hand-in-hand with long-term monitoring programs that follow plant and animal populations before, during, and after development projects are in place. The results of such monitoring studies can provide a guide to the effects of human activities and indicate ways to reduce the negative effects of human activity on riverine systems. Furthermore, development activities can be modified based on monitoring results to achieve a more desirable outcome (adaptive management; Chapter 3; 88). Such survey and monitoring efforts will require the co-operation and participation of personnel from museums, universities, colleges, and governmental agencies, and residents in target or affected areas. Such information will be necessary to put management decisions on a rigorous scientific basis (12,16,17).

References

1. Allan, J.D., Flecker, A.S. 1993. Biodiversity conservation in running waters. *BioScience* 32: 32-43
2. Allanson, B.R., Hart, R.C., O'Keeffe, J.H., Robarts, R.D. 1990. *Inland Waters of Southern Africa: An Ecological Perspective.* Dordrecht: Kluwer Academic Publishers
3. Andrews, P., Groves, C.P., Horne, J.F.M. 1975. Ecology of the lower Tana River food plain (Kenya). *Journal of the East Africa Natural History Society and National Museum* 151: 1-31
4. Attwell, R.I.G. 1970. Some effects of Lake Kariba on the ecology of a floodplain of the Mid-Zambezi Valley of Rhodesia. *Biological Conservation* 2: 190-196
5. Awachie, J.B.E. 1981. Some general considerations on river basins in Africa and their management and development in relation to fisheries. In *Seminar on River Basin Management and Development*, Blantyre, Malawi, 8 - 10 December, 1980. ed. Kapetsky, J.M., pp. 2-23. Food and Agriculture Organization of the United Nations, Committee for Inland Fisheries of Africa. Technical Paper No. 8
6. Balek, J. 1983. *Hydrology and Water Resources in Tropical Regions.* Amsterdam: Elsevier
7. Balirwa, J.S., Bugenyi, F.W.B. 1980. Notes on the fisheries of the River Nzoia, Kenya. *Biological Conservation* 18: 53-58
8. Barnard, P.C., Biggs, J. 1988. Macroinvertebrates in the catchment streams of Lake Naivasha, Kenya. *Revue d'Hydrobiologie Tropicale* 21(2):127-134
9. Beadle, L.C. 1981. *The Inland Waters of Tropical Africa. Second edition.* London: Longman
10. Bell-Cross, G., Minshull, J.L. 1988. *The Fishes of Zimbabwe.* Harare: National Museums and Monuments of Zimbabwe
11. Bernacsek, G.M. 1981. Freshwater fisheries and industry in the Rufiji River basin, Tanzania: The prospects for coexistence. In *Seminar on River Basin Management*

and Development, Blantyre, Malawi, 8-10 December, 1980. ed. Kapetsky, J.M., pp. 69-88. Rome: Food and Agriculture Organization, Committee for Inland Fisheries of Africa. Technical Paper No. 8

12. Boon, P.J. ed. 1992. *River Conservation and Management.* Chichester, England: John Wiley and Sons

13. Brown, D.S. 1980. *Freshwater Snails of Africa and their Medical Importance.* pp. 487, London: Taylor and Francis Limited

14. Cadwalladr, D.A. 1965. Notes on the breeding biology and ecology of *Labeo victorianus* Boulenger (Pisces: Cyprinidae) of Lake Victoria. *Revue de Zoologie et de Botanique Africaines* 72: 109-134

15. Cadwalladr, D.A. 1965. The decline in the *Labeo victorianus* Blgr. (Pisces: Cyprinidae) fishery of Lake Victoria and an associated deterioration in some indigenous fishing methods in the Nzoia River, Kenya. *East African Agricultural and Forestry Journal* 30: 249-256

16. Calow, P., Petts, G.E. eds. 1992. *The Rivers Handbook. Volume I: Hydrological and Ecological Principles.* Cambridge, MA: Blackwell Scientific Publications

17. Calow, P., Petts, G.E. eds. 1994. *The Rivers Handbook. Volume II: Problems, Diagnosis, and Management.* Cambridge, MA: Blackwell Scientific Publications

18. Coe, M.J. 1969. Observations on *Tilapia alcalica* in Lake Natron on the Kenya-Tanzania border. *Revue de Zoologie et de Botanique Africaines* 80: 1-14

19. Copley, H. 1958. *Freshwater Fishes of East Africa.* London: H.F. & G. Witherby Ltd

20. Crawford, J.D. 1992. Individual and sex specificity in the electric organ discharges of breeding mormyrid fish (*Pollimyrus isidori*). *Journal of Experimental Biology* 164: 79-103

21. Crosskey, R.W. 1969. A re-classification of the Simuliidae (Diptera) of Africa and its islands. *Bulletin of the British Museum of Natural History. (Entomology), (Supplement)* 14:1-195

22. Dadzie, S. 1980. Recent changes in the fishery of a new man-made lake, Lake Kamburu (Kenya). *Journal of Fish Biology* 16:361-367

23. Davies, B.R. 1979. Stream regulation in Africa: A review. In *The Ecology of Regulated Streams.* eds. Ward, J.V., Stanford, J.A., pp. 113-142. New York: Plenum Press

24. Davies, B.R., Walker, K.F. eds. 1986. *The Ecology of River Systems.* Dordrecht: Dr W. Junk Pub. [Chaps. 3 (Nile - Dumont) and 7 (Zambezi - Davies)]

25. Dejoux, C. 1988. *La Pollution des Eaux Continentales Africaines.* Paris: Editions de l'ORSTOM, Travaux et Documents No. 213

26. Denny, P. ed. 1985. *The Ecology and Management of African Wetland Vegetation.* Dordrecht: Dr. W. Junk

27. Denny, P., Bowker, D.W., Bailey, R.G. 1978. The importance of the littoral epiphyton as food for commercial fish in the recent African man-made lake, Nyumba ya Mungu Reservoir, Tanzania. *Biological Journal of the Linnaean Society, London* 10:139-150

28. Dunne, T. 1979. Sediment yield and land use in tropical catchments. *Journal of Hydrology* 42: 281-300

29. Durand, J.R., Lévêque, C. eds. 1980 and 1981. *Flore et Faune Aquatiques de l'Afrique Sahelo-Soudanienne, Vols. 1 and 2.* Paris: O.R.S.T.O.M

30. Fontaine, J. 1980. Regime alimentaire des larves de deux genres d'Ephéméroptères *Raptobaetopus* Müller-Liebenau, 1978 et *Prosopistoma* Latreille, 1883. In *Advances in Ephemeroptera Biology.* eds. Flannagan, J.F., Marshall, K.D., pp. 201-210. New York: Plenum Press

31. Frissell, C.A., Liss, W.J., Warren, C.E., Hurley, M.D. 1986. A hierarchical framework for stream habitat classification: viewing streams in a watershed context. *Environmental Management* 10: 199-214

32. Gaudet, J.J. 1992. Structure and function of African floodplains. *Journal of the East Africa Natural History Society and National Museum* 82 (199): 1-32

33. Gaudet, J.J., Melack, J.M. 1981. Major ion chemistry in a tropical African lake basin. *Freshwater Biology* 11:309-333

34. Gillies, M.T. 1954. The adult stages of *Prosopistoma* Latreille (Ephemeroptera), with descriptions of two new species from Africa. *Transactions of the Royal Entomological Society of London* 105: 355-372

35. Githunguri, E.W. 1991. *Ecological Impact of Urban Sewage and Waste Water on the Nairobi-Athi River*. Nairobi: MSc Thesis, University of Nairobi

36. Gould, R. 1990. Electric elephant fish are shocked by pollution (using electric fish to detect water pollution).*New Scientist* 128 (1738, Oct. 13): 22

37. Greenwood, P.H. 1958. *The Fishes of Uganda*. Kampala: The Uganda Society

38. Gregory, S.V., Swanson, F.J., McKee, W.A., Cummins, K.W. 1991. An ecosystem perspective of riparian zones. *Bioscience* 41: 540-551

39. Haltenorth, T., Diller, H. 1980. *A Field Guide to the Mammals of Africa, including Madagascar*. (Translated by R.W. Hayman) London: Collins

40. Harrison, A.D., Hynes, H.B.N. 1988. Benthic fauna of Ethiopian mountain streams and rivers. *Archiv fur Hydrobiologie Supplement* 81(1):1-36

41. Haslam, S.M. 1990. *River Pollution: an Ecological Perspective*. London: Belhaven Press.

42. Hellawell, J.M. 1986. *Biological Indicators of Freshwater Pollution and Environmental Management*. London: Elsevier

43. Hofkin, B.V., Koech, D.K., Ouma, J., Loker, E.S. 1991. The North American crayfish *Procamburus clarkii* and the biological control of schistosome-transmitting snails in Kenya: Laboratory and field investigations. *Biological Control* 1: 183-187

44. Hofkin, B.V., Mkoji, G.M., Koech, D.K., Loker E.S. 1991. Control of schistosome-transmitting snails in Kenya by the North American crayfish *Procamburus clarkii*. *American Journal of Tropical Medicine and Hygiene* 45(3):339-344

45. Hughes, F.M.R. 1990. The influence of flooding regimes on forest distribution and composition in the Tana River floodplain, Kenya. *Journal of Applied Ecology* 27:475-491

46. Hynes, H.B.N. 1960. *The Biology of Polluted Waters*. Liverpool: Liverpool University Press

47. Hynes, H.B.N, Williams, T.R. 1962. The effect of DDT on the fauna of a central African stream. *Annals of Tropical Medicine and Parasitology* 56:78-91

48. Kilham, P. 1984. Sulfate in African inland waters: sulfate to chloride ratios. *Verhandlungen, Internationale Vereingung fur Theoretische und Angewandte Limnologie* 22:296-302

49. King, J.M., Day, J.A., Hurly, P.R., Henshall-Howard, M-P, Davies, B.R. 1988. Macroinvertebrate communities and environment in a southern African mountain stream. *Canadian Journal of Fisheries and Aquatic Sciences* 45:2168-2181

50. Kingdon, J. 1989. *Island Africa*. Princeton: Princeton University Press

51. Lauzanne, L. 1988. Chapter 10: Les habitudes alimentaires des poissons d'eau douce africains. In *Biology and Ecology of African Freshwater Fishes*. eds. Lévêque, Bruton, C.M.N., Ssentongo, G.W., pp. 221-242. Paris: Editions de l'ORSTOM, Travaux and Documents No. 216

52. Lodge, D.M. 1991. Herbivory on freshwater macrophytes. *Aquatic Botany* 41: 195-224

53. Lowe-McConnell, R.H. 1975. *Fish Communities in Tropical Freshwater*. pp. 337, London: Longman

54. Lowe-McConnell, R.H. 1987. *Ecological Studies in Tropical Fish Communities*. Cambridge: Cambridge

55. Lowe-McConnell, R.H. 1991. Ecology of cichlids in South American and African waters, excluding the African Great Lakes. In *Cichlid Fishes: Behaviour, Ecology and Evolution*. ed. Keenleyside, M.H.A., pp. 60 - 85. (Chap. 3). London: Chapman and Hall

56.   Marlier, G. 1954. Recherches hydrobiologiques dans les rivieres du Congo Oriental. II. Etude ecologique. *Hydrobiologia* 6:225-264
57.   Mason, C.F. 1991. *Biology of Freshwater Pollution. 2nd edition.* New York: John Wiley and Sons
58.   Mathooko, J.M. 1988. *Downstream Drift of Invertebrates in Naro Moru River, a Tropical River in Central Kenya.* M.Sc. Thesis, University of Nairobi
59.   Mathooko, J.M., Mavuti, K.M. 1992. Composition and seasonality of benthic invertebrates, and drift in the Naro Moru River, Kenya. *Hydrobiologia* 232:47-56
60.   Melack, J.M., MacIntyre, S. 1991. Phosphorous concentrations, supply and limitation in tropical African lakes and rivers. In *Phosphorous Cycles in Terrestrial and Aquatic Ecosystems.* SCOPE/UNEP Regional Workshop 4: Africa. eds. Tiessen, H. Frossard, E., pp. 1-18. Saskatoon, Canada: Saskatchewan Inst. of Pedology
61.   Merritt, R.W., Cummins, K.W. 1984. *An Introduction to the Aquatic Insects of North America, second edition.* Dubuque, Iowa: Kendall/ Hunt Publishing Company
62.   Meyer, J.L., McDowell, W.H., Bott, T.L., Elwood, J.W., Ishizaki, C., Melack, J.M., Peckarsky, B.L., Peterson, B.J., Rublee, P.A. 1988. Elemental dynamics in streams. *Journal of the North American Benthological Society* 7: 410-432
63.   Minshall, G.W. 1984. Aquatic insect-substrate relationships. In *The Ecology of Aquatic Insects.* eds. Resh, V.H., Rosenberg, D.M., pp. 358-400. New York: Praeger Publishers
64.   Minshall, G.W. 1988. Stream ecosystem theory: a global perspective. *Journal of the North American Benthological Society* 7:263-288
65.   Minshall, G.W., Petersen, R.C., Cummins, K.W., Bott, T.L., Sedell, J.R., Cushing, C.E., Vannote, R.L. 1983. Interbiome comparison of stream ecosystem dynamics. *Ecological Monographs* 53:1-25
66.   Moreau, J., Arrignon, J., Jubb, R.A. 1988. Chapter 19: Les introductions d'espèces etrangeres dans les eaux continentales africaines. Interets et limites. In *Biology and ecology of African freshwater fishes.* eds. Lévêque, C., Bruton, M.N., Ssentongo, G.W., pp. 395-426. Paris: Editions de l'ORSTOM, Travaux and Documents No. 216
67.   Moyle, P.B. 1976. *Inland Fishes of California.* Berkeley: University of California Press
68.   Newman, R.M. 1991. Herbivory and detritivory on freshwater macrophytes by invertebrates: a review. *Journal of the North American Benthological Society* 10: 89-114
69.   Njuguna, S.G. 1978. *A Study of the Effects of Pollution on a Tropical River in Kenya.* M.Sc Thesis, University of Nairobi
70.   Odipo, R.W. 1987. *The Effects of Industrial and Domestic Effluents on the Quality of the Receiving Waters of Nairobi, Ngong, and Ruiruaka Rivers.* M.Sc Thesis, University of Nairobi
71.   Pacini, N. 1989. *The Ecology of the Nairobi-Athi River and the Impact of Polluting Discharges.* M.Sc. Thesis, University of Leicester
72.   Payne, A.I. 1986. *The Ecology of Tropical Lakes and Rivers.* Chichester: John Wiley & Sons
73.   Pennak, R.W. 1989. *Freshwater Invertebrates of the United States, 3rd ed .* New York: John Wiley & Sons
74.   Petr, T. 1977. Limnology of the Nzoia River in western Kenya. I. Physical and chemical characteristics. *Revue de Zoologie Africaine* 91: 31-44
75.   Petr, T., Paperna, I. 1979. Limnology of the Nzoia River in western Kenya. II. Fish, fish parasites, and the benthic fauna. *Revue de Zoologie Africaine* 93: 539-567
76.   Poff, N.L., Ward, J.V. 1989. Implications of streamflow variability and predictability for lotic community structure: a regional analysis of streamflow patterns. *Canadian Journal of Fisheries and Aquatic Sciences* 46: 1805-1818
77.   Rosenberg, D.M., Resh, V.H. 1992. *Freshwater Biomonitoring and Benthic Macroinvertebrates.* New York: Chapman and Hall

78.   Rowntree, K. 1990. Political and administrative constraints on integrated river basin development: an evaluation of the Tana and Athi Rivers Development Authority, Kenya. *Applied Geography* 10:21-41

79.   Skelton, P.H. 1988. Chapter 4: The distribution of African freshwater fishes. In *Biology and Ecology of African Freshwater Fishes.* eds. Lévêque, C., Bruton, M.N., Ssentongo, G.W., pp. 65-91. Paris: Editions de l'ORSTOM, Travaux and Documents No. 216

80.   Statzner, B., Higler, B. 1985. Questions and comments on the River Continuum Concept. *Canadian Journal of Fisheries and Aquatic Sciences* 42:1038-1044

81.   Symoens, J.J., Burgis, M., Gaudet, J.J. eds. 1981. *The Ecology and Utilization of African Inland Waters.* Nairobi: UNEP Reports and Proceedings, Series I. [Chaps. 4 (Invertebrates) and 6 (Rivers)]

82.   Tweddle, D. 1981. The importance of long-term data collection on river fisheries, with particular reference to the cyprinid (*Opsaridium microlepis*, Gunther, 1864) fisheries of the affluent rivers of Lake Malawi. In *Seminar on River Basin Management and Development,* Blantyre, Malawi. 8-10 December, 1980. ed. Kapetsky, J.M., pp. 145-163, Rome: Food and Agriculture Organization, Committee for Inland Fisheries of Africa Technical Paper No. 8

83.   Vanden Bossche, J.P., Bernacsek, G.M. 1990. *Source Book for the Inland Fishery Resources of Africa. Vol. 1.* Rome: Food and Agriculture Organization, Committee for Inland Fisheries of Africa, Technical Paper 18/1

84.   Vannote, R.L., Minshall, G.W., Cummins, K.W., Sedell, J.R., Cushing, C.E. 1980. The river continuum concept. *Canadian Journal of Fisheries and Aquatic Sciences* 37:130-137

85.   van Someren, V.D. 1952. *The Biology of Trout in Kenya Colony.* Nairobi: Govt. Printer

86.   van Someren, V.D. 1962. The migration of fish in a small Kenya river. *Revue de Zoologie et de Botanique Africaines* 66: 375-393

87.   Viner, A.B. 1975. The Supply of Minerals to Tropical Rivers and Lakes (Uganda). In *Coupling of Land and Water Systems.* ed. Hasler, A.D., pp. 227-261. New York: Springer-Verlag

88.   Walters, C.J. 1986. *Adaptive Management of Renewable Resources.* New York: Macmillan

89.   Ward, J.V. 1989. Riverine-wetland interactions. In *Freshwater Wetlands and Wildlife, CONF-8603101, DOE Symposium Series No. 61.* eds. Sharitz, R.R., Gibbons, J.W., pp. 385-400. Oak Ridge, TE: USDOE Office of Scientific and Technical Information

90.   Warren, C.E. 1971. *Biology and Water Pollution Control.* Philadelphia: Saunders

91.   Webster, J.R., Benfield, E.F. 1986. Vascular plant breakdown in freshwater ecosystems. *Annual Review of Ecology and Systematics* 17: 567-594

92.   Welcomme, R.L. 1969. The biology and ecology of the fishes of a small tropical stream. *Journal of Zoology, London* 158: 485-529

93.   Welcomme, R.L. 1975. *The Fisheries Ecology of African Floodplains.* Rome: Food and Agriculture Organization, Committee for Inland Fisheries of Africa, Technical Paper No. 3

94.   Welcomme, R.L. 1976. Some general and theoretical considerations on the fish yield of African rivers. *Journal of Fish Biology* 8: 351-364

95.   Welcommme, R.L. 1978. Some factors affecting the catch of tropical river fisheries. In *Symposium on River and Floodplain Fisheries in Africa,* Bujumbura, Burundi. 21-23 November, 1977. Review and experience papers. ed. Welcomme, R.L., pp. 266-275. Food and Agriculture Organization, Committee for Inland Fisheries of Africa, Technical Paper No. 5

96.   Welcomme, R.L. 1979. *Fisheries Ecology of Floodplain Rivers.* London: Longman

97.   Welcomme, R.L. 1985. *River Fisheries.* Rome: Food and Agriculture Organization, Fisheries Technical Paper 262

98.   Welcomme, R.L., de Merona, B. 1988. Chapter 12: Fish communities of rivers. In
       Biology and Ecology of African Freshwater Fishes. eds. Lévêque, C., Bruton,
       M.N., Ssentongo, G.W., pp. 251-276. Paris: Editions de l'ORSTOM, Travaux and
       Documents No. 216
99.   Whitehead, P.J.P. 1959. The anadromous fishes of Lake Victoria. *Revue Zoologie et
       de Botanique Africaines* 59: 329-363
100.   Whitehead, P.J.P. 1959. The river fisheries of Kenya. I - Nyanza Province. *East
       African Agricultural and Forestry Journal* 24:274-278
101.   Whitehead, P.J.P. 1960. The river fisheries of Kenya. Part II - The lower Athi (Sabaki)
       River. *East African Agricultural and Forestry Journal* 25:259-265
102.   Wiggins, G.B. 1977. *Larvae of the North American Caddisfly Genera (Trichoptera).*
       Toronto: University of Toronto Press
103.   Williams, T.R. 1991. Freshwater crabs and *Simulium neavei* in East Africa. III.
       Morphological variation in *Potamonautes loveni* (Decapoda: Potamidae). *Annals of
       Tropical Medicine and Parasitology* 85: 181-188
104.   Williams, T.R., Hynes, H.B.N. 1971. A survey of the fauna of streams on Mount
       Elgon, East Africa, with special reference to the Simuliidae (Diptera). *Freshwater
       Biology* 1: 227-248
105.   Winterbourn, M.J., Rounik, J.S., Cowie, B. 1981. Are New Zealand stream
       ecosystems really different? *New Zealand Journal of Marine and Freshwater
       Research* 15:321-328

Chapter 7

Saline and Freshwater Lakes of the Kenyan Rift Valley

J.M. Melack

Lakes are a conspicuous feature of the eastern African Rift Valley which stretches from the Afar depression of northeastern Ethiopia through the highlands of Ethiopia, Kenya and Tanzania to southern Tanzania (29). Because no rivers flow to the ocean from the eastern African Rift Valley, the whole region is a basin of internal drainage in which lie closed-basin or endorheic lakes. Tectonic and volcanic activity associated with the formation of the Rift Valley created the lake basins. The lakes that are in the rift valley that bisects the central highlands of Kenya from Lake Magadi in the south to Lake Bogoria in the north are emphasized in this chapter (Fig. 7.1, Table 7.1).

Evaporation is the only route of water loss from endorheic lakes which, therefore, concentrate solutes carried by their inflowing streams and springs and often become very salty. Many of the lakes in the Kenyan Rift Valley are alkaline and saline (34), though notable exceptions are fresh, such as Lake Naivasha (9). Because the water balance of closed-basin lakes is especially sensitive to seasonal and longer term fluctuations in rainfall and evaporation, water levels, areas and salinities can vary widely. Some of the shallow lakes of today were much larger and deeper 5000 to 10,000 years ago when the climate was wetter (28).

One of the first scientific investigations in eastern Africa was J. W. Gregory's geological expedition to the Kenyan Rift Valley in 1893. He noted raised beaches around some of the lakes and attributed these features to prior periods having higher lake levels and a wetter climate. L. S. B. Leakey made further observations on these old shore lines during his archeological expeditions to Kenya in the 1920s and sponsored the first limnological study of lakes in the Kenyan or Gregory Rift Valley in 1929. P. M. Jenkin made pioneering measurements of physical and chemical conditions and collections of flora and fauna in lakes Naivasha, Sonachi (called the Crater Lake by her), Elmenteita, Nakuru and Baringo (15). The following year E. B. Worthington lead the Cambridge Expedition to East African lakes during which L. C. Beadle conducted his limnological studies of lakes Naivasha, Sonachi (formerly called Naivasha Crater Lake), Bogoria (formerly called Lake Hannington), Baringo and Turkana (formerly called Lake Rudolf) as well as other lakes in Uganda (2).

Table 7.1. Major lakes in Rift Valley of central Kenya (9,34,48). Lakes change in area and depth; values shown for period in 1970s emphasized in text. Lake Magadi is only intermittently inundated and is not listed.

Lake name	Altitude (m asl)	Area (km^2)	Mean depth (m)
Bogoria	963	33	5.4
Nakuru	1760	44	2
Elmenteita	1776	20	1
Sonachi	1891	0.18	4
Oloidien	1887	5.5	5.6
Crescent Island Crater	1887	2.1	11
Naivasha	1887	145	4.7

Occasional ecological investigations of these lakes occurred during the next forty years, but not until the 1970s was concerted research in well-equipped laboratories conducted.

Tropical Limnology

Limnology is the study of physical, chemical and biological processes and interactions among organisms in non-marine aquatic environments. While much of the scientific understanding of aquatic ecosystems relies on studies done in the mid-latitudes, tropical limnology has matured considerably in recent decades and now provides a rich source of information about lakes in the tropics (3,27). Several significant studies have been done in lakes of the Rift Valley of Kenya. The soda lakes have attracted the most attention, and a discussion of their ecology is the heart of this chapter. Freshwater Lake Naivasha with its rich communities of aquatic plants has been significantly altered by introductions of exotic species, and the discussion of the ecology of Lake Naivasha provides striking contrast to the neighboring soda lakes.

Equatorial lakes of low to moderate altitude such as those in the Kenyan Rift Valley receive high insolation and maintain warm temperatures year round. The solar radiation evaporates and stratifies the water, and provides the energy for photosynthesis of organic matter by aquatic plants. While equitable sunlight and temperatures are conducive to consistently high rates of photosynthesis, marked seasonal and interannual variations in rainfall, river discharge and mixing of the water can occur and modify the supply of nutrients and light to aquatic plants.

The seasonal range and variability of phytoplankton photosynthesis is now known for a wide variety of tropical lakes. Synthesizing these studies J. M. Melack (32) demonstrated the lack of a latitudinal trend in temporal variability within the tropics (the region between the Tropics of Cancer and Capricorn) but instead identified three patterns of seasonal change. All three patterns occur in lakes of the

Figure 7.1. Landsat image of Kenyan Rift Valley from Lake Baringo in north to Lake Magadi in south. Figure 7.1 is a composite of two multispectral scanner scenes acquired 21 December 1973 and displayed as band 7. Lakes appear white, and from north (top) to south (bottom) are as follows: Baringo, Bogoria, Ol Bolassat, Nakuru, Elmenteita, Naivasha, Magadi. The highlands bordering the rift valley have scattered cumulus clouds and darker patches of forest.

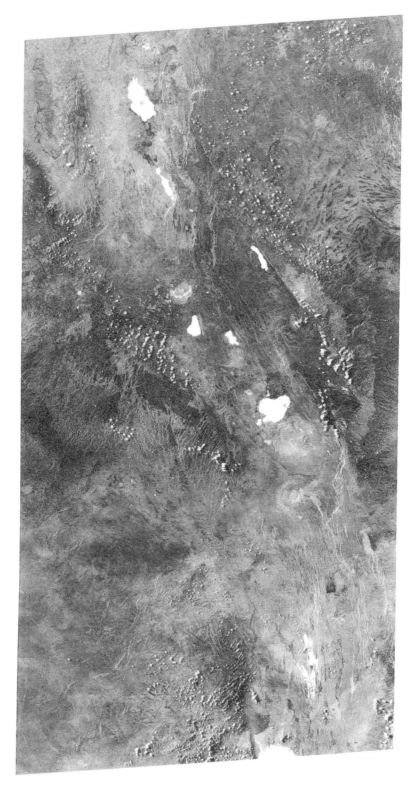

Kenyan Rift Valley and are explained further later in this chapter. These patterns include: 1) most tropical lakes have pronounced seasonal fluctuations in biological composition and processes that usually correspond to variations in rainfall, runoff or vertical mixing within the lake (such as lakes Sonachi and Oloidien). 2) a second pattern occurs in lakes with muted seasonal fluctuations in which 24-hour variations often exceed month to month changes (such as Lake Naivasha and Lake Nakuru from 1971 to 1974). 3) an abrupt switch from one persistent condition to another may occur (such as found for Lake Nakuru in 1974 and Lake Elmenteita in 1973 to 1974).

Geochemistry

The amounts, proportions, sources and sinks of the major dissolved inorganic constituents define the principle geochemical conditions in lakes. Most eastern African waters are solutions of sodium (Na^+) and bicarbonate (HCO_3^-) plus carbonate (CO_3^{2-}) derived largely from weathering of sodium silicate rocks (23,29). Other major solutes, common worldwide, are calcium (Ca^{2+}), magnesium (Mg^{2+}), potassium (K^+), chloride (Cl^-) and sulfate (SO_4^{2-}) with silicate (SiO_2) and fluoride (F^-) also being important. pHs can exceed 10 in the soda lakes because of their high alkalinity.

The total concentration of solutes, also known as salinity, is expressed as the total weight of dissolved solids per weight or volume of solution. Electrical conductivity is often used as an easily made measurement of the salinity, but the relation between conductance and salinity becomes non-linear as the salinity increases. Within the Kenyan Rift Valley the surface waters range from dilute freshwater to saline brines, and these great differences in chemical composition and concentration can strongly influence the distribution and abundance of organisms (13,25). Based in part on the biological communities, or species present, the distinction between fresh and saline water is made at between 3 and 5 grams/liter of total dissolved solutes.

The geochemical and biogeochemical processes that control the chemical evolution of the rift valley lakes are like those that operate elsewhere in closed-basin lakes: evaporative concentration, intermittent flooding and desiccation, sorption and ion exchange with sediments, reactions of cations with alumino-silicate precursors to form clay minerals (= reverse weathering), precipitation primarily of carbonates and sulfides from moderately saline waters and of other salts from hypersaline brines, CO_2 and sulfide loss by degassing and sulfate reduction (6,23). Although the geochemical evolution of closed-basin lakes undergoing evaporative concentration is usually considered a gradual process, the change can be rapid and discontinuous if associated with long-term stratification followed by abrupt mixing. Rapid change occurs when deep vertical mixing permits H_2S gas to be lost to the atmosphere. The chemistry of the lake is changed because loss of every equivalent of sulfide produces an equivalent of alkalinity that remains in the lake (23).

The extensive surveys of many eastern African lakes done by J.F. Talling (45,49) and P. Kilham (20), and the detailed analyses by H.P. Eugster and his colleagues at Lake Magadi (6) provide valuable information about the chemical conditions in the waters of the Kenyan Rift Valley. These data indicate that sodium, chloride, sulfate and bicarbonate plus carbonate all increase more or less linearly as salinity increases from fresh to saline water. In contrast, precipitation of calcium as

$CaCO_3$ and precipitation or sorption of magnesium begins to occur at salinities as low as about 0.2 grams/liter. As salinity continues to rise, saturation of sodium carbonates occurs. Exploitation of trona ($Na_2CO_3 \cdot NaHCO_3 \cdot 2H_2O$) precipitating at Lake Magadi supports the large mining operation there. Eventually, NaCl and other rarer salts will form as evaporative concentration proceeds.

Fluoride concentrations so high as to be inhibitory to aquatic organisms occur in Kenyan soda lakes (22). Although some local rocks do contribute to these high values, it is the scarcity of calcium, because of its precipitation as $CaCO_3$ so early in the evaporative evolution, that leads to these high fluoride levels. Calcium is too low in concentration to form fluorite (CaF) as is common in most places. Instead, the concentration of dissolved fluoride is controlled by the much more soluble villiaumite (NaF).

Nutrient Limitation

Of the factors regulating primary productivity in lakes, the supply of nutrients, in particular phosphorus and to a lesser extent nitrogen, is often of major importance (44). In general, three factors determine the relative role of phosphorus or nitrogen in limiting the abundance or productivity of phytoplankton (12): (1) the ratio of nitrogen to phosphorus in hydrologic and meteorologic inputs to a lake, (2) preferential losses from the euphotic zone by processes such as denitrification, sorption of phosphorus to particles and differential settling of particles with different N:P ratios, and (3) the contribution from nitrogen fixation. Unfortunately, these processes have not been measured in a coordinated manner in the lakes in the eastern African Rift Valley. Instead, inferences from several indicators of nutrient limitation must be made.

The ambient concentrations of dissolved nutrients can provide an indication of the likelihood of limitation. Most lakes in the eastern African Rift Valley sampled for soluble reactive phosphorus (SRP) have moderate to high concentrations, although freshwater and saline lakes with very low concentrations of SRP do occur (such as in lakes Naivasha, Sonachi and Elmenteita; 37). Some of the saline lakes have exceptionally high SRP values, and assays with radioactive phosphate in Lake Nakuru indicated that most of the SRP was biologically available orthophosphate (42). Few reliable data with which to assess the concentrations of dissolved inorganic nitrogen (primarily nitrate and ammonium) in eastern African lakes have been published (36,45,49). Nitrate is very low in almost all the sampled lakes. Ammonium is usually low in the euphotic zone but can be high toward the bottom, and it can reach moderate concentrations in surface waters. In general, the ambient concentrations of dissolved nitrogen and phosphorus in eastern African lakes would indicate a greater likelihood of nitrogen than phosphorus limitation.

N:P ratios in suspended particulate matter are another indicator of the nutritional status of phytoplankton. Healthy algae have a N:P ratio of about 16, the Redfield ratio (12). Ratios less than 10 often indicate nitrogen deficiency and ratios greater than 20 can indicate phosphorus deficiency. N:P ratios of suspended particulates indicate limitation by phosphorus in lakes Elmenteita and Sonachi, but possible nitrogen deficiency in Lake Bogoria (36). Ratios of total nitrogen to total phosphorus in samples spanning about a year indicated that lakes Oloidien and Sonachi were highly phosphorus deficient and that Lake Naivasha was moderately phosphorus limited (17). The total nitrogen to total phosphorus ratios throughout

S. Njuguna's year long study of Lake Sonachi adds further support to the importance of phosphorus limitation in that soda lake (41).

The rate of uptake of radioactive phosphate by particulate matter suspended in lakes is a widely used index of phosphorus demand by the plankton. Turnover times are typically rapid when phosphorus is in short supply and are slow when supply is adequate. In mid-1974 uptake was rapid in samples from lakes Elmenteita, Naivasha, Sonachi and slow in samples from Lake Nakuru and the Crescent Island Crater basin of Lake Naivasha (42) Measurements from late 1979 through 1980 documented very rapid uptake throughout the year in lakes Oloidien and Sonachi, and a wide seasonal range in Lake Naivasha with average values indicative of phosphorus limitation (17).

Phytoplankton growth depends on the rate of nutrient supply not concentrations. Consequently, measurements of nutrient inputs, outputs and recycling via biological, chemical and physical processes are required to provide a comprehensive understanding of the nutritional status of a lake. Unfortunately, all such measurements and, therefore, a biogeochemical model are not available for any Kenyan lake. Valuable future studies should include investigations of nitrogen fixation and denitrification, exchanges of nitrogen and phosphorus with sediments, regeneration of nutrients by animals and nutrient loading to lakes via rain and rivers.

Nutrient limitation is best assessed by experimental manipulation of nutrient levels. Although experiments can be done on scales ranging from small flasks to outdoor enclosures to whole lakes, larger volume experiments provide more realistic conditions than small containers. In the first and only experimental enrichment of a large enclosure suspended in an eastern African lake, the phytoplankton in Lake Sonachi showed a strong positive response to phosphorus addition and a lesser response to ammonium enrichment (36).

The available information about Kenyan Rift Valley lakes provides some evidence for the important role of phosphorus and less evidence for limitation by nitrogen, but needs improvement for guiding management of these lakes. Long-term records required to document trends in the degree of eutrophication and comprehensive investigations of the biogeochemical cycles of phosphorus and nitrogen are lacking. With increased human population size, and associated agricultural and industrial activities in the catchments of the lakes, further understanding of the lakes' responses to altered nutrient supplies is critical to planning wise economic development in the region.

Lake Types

Soda Lakes

Soda lakes in the Rift Valley of eastern Africa are among the world's most productive, natural ecosystems. A conspicuous feature of these lakes are enormous flocks of lesser flamingos grazing on thick green suspensions of algae (47). In apparent contrast to such prolific biological activity are the harsh physical and chemical conditions and a depauperate biota.

Ecological Setting and Biota. Soda lakes are susceptible to marked fluctuations in water level and salinity because they lie in closed catchments with a semi-arid climate. Intervals of near dryness alternate with periods of high water. Salinity ranges over two orders of magnitude based on measurements during the last few decades. Major shifts in diatom assemblages are recorded in the lacustrine sediments and indicate large changes in salinity during the last few thousands of years in these lakes (13,28).

Soda lakes in broad, shallow basins such as lakes Elmenteita and Nakuru usually become thermally stratified during the morning. However, afternoon winds and nocturnal cooling typically mix the lakes to the bottom each day. Many of the small lakes in volcanic craters, such as Lake Sonachi, have saltier water underlying fresher water and are chemically stratified; they may not mix from top to bottom for many years (30).

Abundant phytoplankton or occasionally high concentrations of suspended mineral particles occur in soda lakes. Sunlight is absorbed and scattered by these particles and does not penetrate far into the water. If a black and white disk, called a Secchi disk by limnologists, is lowered into these turbid waters, it can disappear from view within only a few tens of centimeters. The scarcity of light in much of the lake can strongly reduce photosynthesis and growth of algae.

A characteristic feature of soda lakes throughout tropical Africa is persistent, almost monospecific algal blooms of the blue-green alga, *Spirulina platensis.* This alga is the primary food for the lesser flamingo (47), and has been a protein-rich food for people in Chad for many years (26). It has become a health food fad in the United States in recent years, and is now mass cultured in large outdoor facilities. Under natural conditions in soda lakes of eastern Africa, such as lakes Simbi and Nakuru, *Spirulina platensis* attains some of the highest rates of photosynthesis ever recorded (33,48).

Very few species of phytoplankton occur in African soda lakes. Besides *Spirulina platensis*, there are other blue-green alga such as *Anabaenopsis arnoldii*, which can convert atmospheric nitrogen into organic nitrogen. Small spherical green and blue-green algae can also be abundant. Several diatoms occur whose siliceous cell walls can remain in the lake's sediments for thousands of years and provide a record of ecological conditions (see Chapter 1).

Benthic or bottom-living algae cover the sediments if sufficient light reaches the lake bottom (35). These algal communities are a mixture of diatoms and blue-green algae; they can reach standing crops similar to planktonic abundance and at times are an important source of food for the flamingos.

Heterotrophic bacteria can attain extremely high numbers in soda lakes, but little is known about their changes through time or their ecological importance (21).

Few aquatic invertebrates live in African soda lakes (29). Protozoa are the most diverse group; 21 species of ciliates have been collected from Lake Nakuru (7). The zooplankton is typically composed of a couple of species of rotifers and one copepod (*Paradiaptomus africanus*). Several water boatmen or corixids (Hemiptera), a backswimmer or notonectid (Hemiptera) and lake flies or chironomids (Diptera) are common insects. A fairy shrimp (*Branchinella spinosa*) can occasionally be abundant in fishless lakes such as Lake Elmenteita.

In contrast to large and fresh Rift Valley lakes (see Chapter 8) only a few species from one genus of cichlid fish inhabit eastern African soda lakes. One species is endemic to Lake Natron, another is endemic to Lake Manyara, and a third, found naturally in Lake Magadi (*Sarotherodon alcalicus grahami*) has been introduced into Lake Nakuru. This mouth-brooding cichlid can survive temperatures up to 41°C and live over a wide range of salinities (4).

Bird life on and near African soda lakes is world renowned. Especially striking is the multitude of lesser flamingos wading along the shore and swimming offshore. lesser flamingos are filter-feeders with a bill and mouth specially designed to extract large algae such as *Spirulina platensis* from the surface of the lakes (16). Accurate estimates of the number of flamingos in a particular lake and of the population throughout the region are difficult to obtain. By electronic counting of aerial photographs taken monthly at Lake Nakuru, E. Vareschi calculated a mean population size of about 900,000 in 1972 and 1973; a maximum of 1.5 million was observed (47). In 1974 the population dropped to a mean of about 100,000 with a minimum of about 30,000. A decrease at one lake may indicate increases at neighboring lakes, and regional aerial surveys done from 1972 to 1974 in Kenya and northern Tanzania recorded an average total of 0.5 million flamingos among all the lakes in the area (46). The major nesting place of the lesser flamingo is the salt flats in Lake Natron. Protected from most predators by kilometers of caustic evaporates and thick mud, the adult and young flamingos must endure surface temperatures over 50° C. The mounds of mud that form the nests offer the young flamingos a slightly cooler roost than the surface of the salt flats.

The greater flamingo also frequents African soda lakes and can number in the tens of thousands. Many species of ducks, geese, sandpipers, plovers, stilts and avocets abound. Fish-eating birds such as great white and pink-backed pelicans, white-necked cormorants, fish eagles and a variety of herons, egrets, terns, gulls, grebes and kingfishers are common in lakes where fish occur. Islands in Lake Elmenteita provide protected breeding grounds for greater glamingos and great white pelicans. Hippopotamus live in many soda lakes, although their numbers are seldom high.

The variations in depth, so typical of shallow lakes in semi-arid climates, can dramatically influence birds that depend on the lakes for reproduction and food. For example, in 1962 lake levels rose abruptly in eastern African soda lakes due to abnormally high rainfall. With Lake Natron too deep for lesser flamingos to breed on the salt flats, about 1.2 million birds moved north to Lake Magadi. Unfortunately, many young birds were lost to predation by hyenas or because salt precipitated around their ankles and hindered their movements. With Lake Natron enlarged and freshened, the cichlid fish temporarily increased to exceptionally large numbers and many great white pelicans came to breed on newly formed islands and to eat the fish. Unfortunately, the fish suddenly died and the pelicans were forced to abandon their breeding.

Ecosystem Dynamics.   Low species diversity and abundant resident populations make soda lakes especially appealing environments in which to conduct investigations of trophic dynamics and ecosystem processes. Lake Nakuru has been the focus of a comprehensive study of production and energy flow (48). Primary producer dynamics have been examined in neighboring lakes (34,35).

When intensively studied in the early 1970s, Lake Nakuru was shallow (average depth about 2 m) and moderate-sized (about 40 km²). A bloom of *Spirulina platensis minor* characterized Lake Nakuru in 1972 and 1973 and accounted for an average of 98% of the biomass of primary producers and about 94% of the total biomass of the lake (Fig. 7.2). During a typical day, however, only about 10% of the lake's volume received sufficient light for algal photosynthesis to occur. Hence, the daily difference between photosynthetic production of organic carbon and

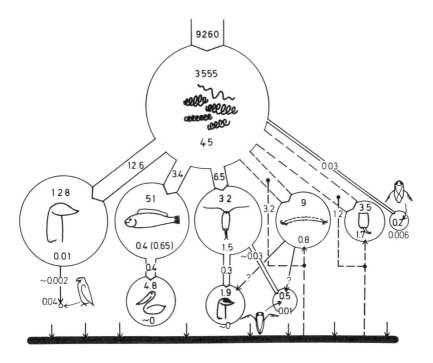

Figure 7.2. Energy flow in Lake Nakuru during the *Spirulina platensis* bloom in 1972 and 1973 (48). Primary producers: almost exclusively the cyanophyte *Spirulina platensis minor*. Primary consumers: The lesser flamingo *Phoeniconaias minor*; the cichlid fish *Sarotherodon alcalicus grahami*; the copepod *Lovenula* (= *Paradiaptomus*) *africana*; the midge larva *Leptochironomus deribae*; the rotifer *Brachionus dimidatus* (in 1972) and *B. dimidiatus* and *B. plicatilis* (in 1973); the water bug *Micronecta jenkinae* (*Corixidae*). Secondary consumers: the African Fish Eagle *Haliaeetus vocifer*, the Marabu Stork *Leptoptilos crummeniferus*, pelicans (*Pelecanus onocrotalus, P. rufescens*), Cormorants (*Phalacrocorax carbo*) and about 50 other fish-eating birds, the greater flamingo *Phoenicopterus ruber*; the water bug *Anisops varia* (Notonectidae). Decomposers and detritus are represented by the bottom bar. Area of the circles reflects the logarithm of the biomass, the upper numbers in the circles = biomass of organisms (kilojoules/m³), lower numbers = production (kilojoules/m³/ day). Arrow width reflects logarithm of the consumption rate, numbers within the arrows consumption rates (kilojoules/m³/day); top arrow is total incident radiation (also in kilojoules/m³/day to facilitate comparison). Broken lines to chironomids and rotifers indicate that the proportions contributed by bacteria (+detritus) and algae to the diet are unknown. Fish production rate in brackets indicates results from laboratory growth rates extrapolated to mean lake temperature.

respiratory use of the stored carbon (net primary production) was only modest, and the turnover of algal biomass was about 1% per day. Consequently, if consumption by herbivores were to about balance net production by algae, only a small fraction of the standing stock of algae could be consumed each day.

Primary consumers constituted 6% of the total biomass of the lake and were represented by five major taxa: the lesser flamingo, a cichlid fish, a calanoid copepod, a chironomid larva and a rotifer (see caption of Fig. 7.2 for species names).

Average ingestion by primary consumers was about 80% of the daily net algal production but only 1% of the standing stock. The two taxa with the highest proportion of the total primary consumer biomass (flamingos, 58% and fish, 22%) ingested 34% and 9%, respectively, of the algal production. In contrast, the copepods and rotifers contributed 14% and 2% to the primary consumer biomass but ingested 18% and 30% of the algal production. The net growth or production of the consumers was about 10% of net primary production, a value typical of nutrient-rich lakes with short food chains.

The two secondary consumers with greatest abundance, the great white pelican and greater flamingo, constituted only 0.2% of the total biomass of the lake. The fish-eating birds, over 90% of which were pelicans, ate a weight of fish each day about equivalent to the production of fish. Greater flamingos fed primarily on copepods and some water bugs and chironomid larvae, but consumed only a very small portion of these animals' daily productivity. Overall, about 20% of the net production by primary consumers was taken by secondary consumers. Essentially no production of pelicans or flamingoes occurs at Lake Nakuru because neither species breeds there and almost all birds at the lake are adults.

The ecological conditions observed at Lake Nakuru in 1972 and 1973 may have persisted for a decade (29), but changed abruptly in early 1974 (48). During a dry period from December 1973 to March 1974, the level of the lake fell 50 centimeters and salinity rose about 25% to 25 grams/liter. In January and February 1974 *Spirulina platensis minor* declined to 0.5% of the total algal biomass. During this period the biomass of primary producers fell over 50%, and net primary productivity declined 70%. The copepods and water bugs disappeared in the first two months, and the rotifers increased about 40-fold to become 25% of the primary consumer biomass. Peak algal ingestion rates by rotifers at this time were 20 times greater than the ingestion rates of the lesser flamingoes. By May 1974 the rotifer population had returned to earlier levels, but the phytoplankton had shifted to a mixture of small green algae, diatoms and a nitrogen-fixing blue-green alga. Total primary producer biomass was about half of previous levels.

In Lake Elmenteita biological changes similar to those observed in Lake Nakuru occurred during a period of evaporative concentration in 1973 (35; Table 7.2). The blue-green algae, *Spirulina platensis, Spirulina laxissima* and *Anabaenopsis arnoldii*, and the copepod, *Paradiaptomus africanus,* decreased dramatically in abundance. Chlorophyll concentration, a measure of algal abundance, fell 20-fold. Photosynthetic rates of phytoplankton declined five-fold. As the phytoplankton abundance diminished, the transparency of water increased and the productivity of the benthic algae increased.

Several hypotheses have been proposed to explain the abrupt changes in lakes Nakuru and Elmenteita (35). The factors responsible are likely to have included salinity tolerance of the biota and the nutrient and light supply to the primary producers. The most likely suggestion is that the rapid rate of change of salinity and a salinity exceeding about 25 grams/liter could not be tolerated by the copepod, *Paradiaptomus africanus* , and adversely affected the nitrogen-fixing blue-green alga, *Anabaenopsis arnoldii.* Furthermore, the loss of the copepod and increased oxygen concentrations at the sediment-water interface reduced the rate of nutrient recycling to the water column which gave small green and blue-green algae a competitive advantage and supported a lower total biomass of phytoplankton.

Evidence of varying strength supports this hypothesis. Based on a survey of 48 African lakes (25), the copepod *P. africanus* appeared to occur in lakes with

Table 7.2. Characteristics of Lakes Elmenteita (E) and Nakuru (N) before and after major changes in early 1970s (35).

Characteristics	Before	After
Spirulina platensis	Abundant (E, N)	Undetectable(E, N)
Chlorophyll	High (E, N)	Low (E, N)
Phytoplankton photosynthesis	High (E, N)	Low (E) Moderate (N)
Benthic algal photosynthesis	Trivial (E, N)	Important (E) Minor (N)
Nannoplankton	Minor (E, N)	Major (E, N)
Paradiaptomus africanus	Abundant (E, N)	Undetectable (E, N)
Anoxic water over sediments	Common (E, N)	Absent (E) Rare (N)
Salinity	Below 21 g l^{-1} (E, N)	Above 21 g l^{-1} (E, N)

salinities up to about 17 grams/liter; its presence in lakes Nakuru and Elmenteita in 1973 extends that range to about 21 grams/liter. Hence, the salinity tolerance of the copepod seems to have been exceeded by the salinities above 24 grams/liter that occurred in lakes Nakuru and Elementeita. Studies of the tolerance of *A. arnoldii* and other nitrogen - fixing algae to salinity have not yet been done.

Scanty evidence from soda lakes with *Spirulina platensis* blooms indicates that high concentrations of orthophosphate usually co-occur with the algal blooms, but that dissolved inorganic nitrogen ranges widely from very low to high concentration. Even after the crash of *S. platensis* in Lake Nakuru in 1974, slow turnover times of radioactively labeled phosphorus indicated that the supply of orthophosphate exceeded the demand. However, in Lake Elmenteita, subsequent to the crash of the phytoplankton, orthophosphate concentrations were below the limit of detection and the turnover time of radioactively labeled phosphorus was very fast. Perhaps the differences in the availability of phosphorus between lakes Nakuru and Elmenteita contributed to the higher biomass of phytoplankton in Lake Nakuru than in Lake Elmenteita following the declines in *S. platensis* .

While zooplankton are known to be important recyclers of nitrogen and phosphorus, no measurements are available specifically for the copepod *P. africanus*. Sediments can supply nutrients to the overlying water and can sequester nutrients. Experiments done with sediments from Lake Nakuru demonstrated that the net flux of orthophosphate was into the sediments when oxygenated water overlaid the mud, and that the net flux was out of the sediments when water low in dissolved oxygen overlaid the mud (37).

Clearly, further observational and experimental research is needed to understand the fascinating ecological variations and rapid shifts that can occur in soda lakes. Long-term measurements should be combined with well-designed experiments conducted in the field and in the laboratory and with mathematical modeling. While of value to specific lakes, research on ecosystems with few species, such as soda lakes, may provide important insights into ecosystem dynamics in general. For example, recent results from Mono Lake, a large saline lake in North America, have demonstrated the interplay between reductions in water column mixing and nitrogen supply, zooplankton excretion and algal primary productivity (14).

Conservation and Exploitation. Lake Nakuru is the heart of a popular National Park world renowned for its bird life, which is the apex of its aquatic food web. Unfortunately, Lake Nakuru may be endangered by pollution because it lies in a closed hydrologic basin with runoff from expanding agricultural lands and from a rapidly expanding town which releases domestic sewage and industrial wastes into the lake. Investigations in the early 1970s detected high concentrations of potentially toxic metals (copper and zinc) in tissues of birds and fishes (24). Construction of a factory producing copper oxychloride and other pesticides in the mid-1970s caused further concern, especially after experimental toxicity tests found that copper reduced photosynthesis and growth of *S. platensis* and the abundance of rotifers from Lake Nakuru (19). Domestic sewage from the town of Nakuru is treated in oxidation ponds before the effluents are discharged into the lake. Although historically a highly eutrophic lake, continued inputs of nutrients and organic matter may alter ecosystem processes or the relative abundance of the species in Lake Nakuru. Only regular monitoring and remedial action will ensure that human endeavors do not harm the spectacular bird life and aquatic ecosystem on which the birds depend.

As the flights by flamingos, pelicans and other birds from one soda lake to another so conspicuously show, lakes Natron, Magadi, Elmenteita, Nakuru, Bogoria and other lakes in the region form an integrated group of ecosystems. Lake Bogoria is now a National Park, Lake Elmenteita lies within large agricultural estates and cooperatives, and other lakes are protected partially by their harsh environments and remoteness. Still human activities influence even the distant lakes. Lake Magadi is the site of a large facility which extracts various salts from a series of evaporating pans into which water is pumped. Plans to harvest fish and *Spirulina platensis* from eastern Africa soda lakes have been proposed. It is unlikely, however, that natural nutrient supplies to the lakes are sufficient to sustain intensive removal of the alga.

As is evident from the pronounced shifts in species composition and abundance that have occurred in response to natural variations in water level, the ecology of shallow soda lakes is especially sensitive to hydrologic changes. Hence, one of their greatest risks is diversion of inflowing rivers. Western North America provides ample examples of how diversion of water supplies can modify saline lakes (38), and these examples should offer instructive lessons to resource managers in eastern Africa. Clearly, complete desiccation or prolonged periods with very high salinities will destroy the aquatic ecosystems. Conversely, increased inputs of fresh water will dilute the soda lakes sufficiently to allow invasion of many more species with concomitant changes in the food webs and productivity. Hence, protection of saline lakes requires variations in water level and salinity to be maintained within bounds dictated by the ecological characteristics of the resident species.

Freshwater Lakes

Lake Naivasha is distinctive among the lakes lying in the Rift Valley of Kenya because its water is fresh, an anomaly for an endorheic basin. The fresh water provides a habitat suitable for swamps of *Cyperus papyrus*, a varied association of littoral aquatic macrophytes, wonderfully diverse birdlife, and a commercial and sport fishery. It encourages recreational uses and helps irrigate the fertile farmlands bordering the lake. Shallow, freshwater lakes fringed by papyrus swamps, such as Lake Naivasha, are common and ecologically important in Africa, and the wetlands, if left intact, offer a buffer separating the lakes' open water from the surrounding terrestrial habitats that are increasingly being developed by human enterprises (see Chapter 9).

The Naivasha basin contains four topographically distinct bodies of water. The deepest water lies within a volcanic crater bounded in part by a portion of the crater's rim called Crescent Island. To the southwest of the main lake is Oloidien Lake. Both of these basins have been connected to the main lake in recent years when lake levels were higher. About 4 km west of Lake Naivasha is Lake Sonachi, a soda lake mentioned above.

The fresh waters of the Naivasha basin are a slightly alkaline solution of predominately sodium and bicarbonate. Salinity increases from the Malewa River (1) to the papyrus swamp (2 and 3) to the main lake (4) to the western lagoon between the shore and the floating papyrus reef (5) to Crescent Island Crater (6), and to Oloidien Lake (Fig. 7.3; 9). The Malewa River drains the Aberdare Mountains and the Kinangop Plateau and supplies about 90% of the river discharge to Lake Naivasha. Rainfall directly onto the lake and seepage into the lake bottom account for almost all the remainder. Evaporation from the lake accounts for about 80% of the loss of water from the lake; seepage out of the lake and water extracted for irrigation make up the balance.

Several hydrological and biogeochemical factors combine to keep Lake Naivasha's water fresh (9). These include 1) a large fraction of the water is supplied to the lake from dilute rivers and rain, 2) the lake does not actually lie in a hydrologically closed basin but loses water and solutes via groundwater seepage, and 3) exchanges with near and off-shore sediments and sedimentation of particles removes some solutes.

Water levels in the Naivasha basin have fluctuated a great deal during both recorded history and in the more distant past. Evidence derived from surveying elevations of ancient shorelines and from examination of sediment cores (43) concurs that a much larger (around 610 km^2) and deeper (around 100 meters) lake existed around 10,000 and 12,000 years BP (=Before Present) and overflowed through Njorowa gorge (Hell's Gate) to the south. By about 9200 years BP the gorge had been cut to its present level, and a 400 km^2 lake occupied the Naivasha basin until 5700 years BP. A drier climate resulted in still lower lake levels until the lake dried up completely for perhaps 100 years about 3000 years BP. During the last three millennia a small lake with variable level has existed with perhaps several occasions of complete desiccation. Writings of European travelers describe a rising lake which reached 1896 m above sea level (= m asl), 4 to 5 m higher than any level since, at the end of the last century (1).

Continuous records began in 1908 at 1890 m asl. Following a decline to an elevation of about 1882 m asl and an area of about 100 km^2, which persisted until the late 1950s, the water level rose rapidly in the early 1960s to just over 1887 m asl. Two fluctuations of 2 to 3 meters have occurred over the last two decades with the lowest level of the century being reached in early 1988. During these last two decades the combination of changes in water level and introductions of species alien to the lake have had noticeable effects on the ecology of Lake Naivasha.

Lake Naivasha is ecologically unusual because it combines freshness with fluctuating water levels (much as a reservoir does) and because its native fish fauna was composed of only one species. Desiccation of the lake in the recent and more distant past coupled with the remoteness of the basin from waters with suitable

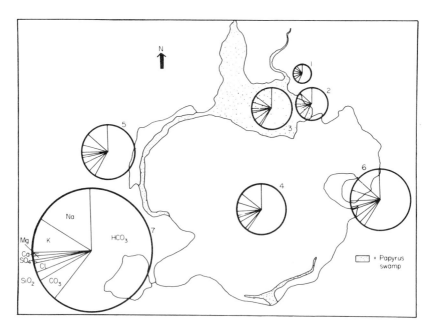

Figure 7.3. Diagram of chemical differences in various parts of Naivasha basin (9): Malewa River(1), northern papyrus swamps (2 and 3), Lake Naivasha (4),western lagoon between shore and papyrus reef (5), Crescent Island Crater (6) and Oloidien Lake (7). Areas of circles are proportional to total dissolved solutes. CO_3 is lacking in compartments 1 and 2.

freshwater-adapted species may explain the depauperate fish fauna. In contrast, the higher plant (8) and algal flora (18), and the zooplankton (31) and macro-invertebrate fauna (5) and the waterfowl are diverse.

To balance the conservation and the development of Lake Naivasha and its catchment in light of the natural fluctuations in level, the many introduced species and the intensifying agricultural uses of the lake shore and uplands are a major challenge. An examination of the ecological conditions in Lake Naivasha during the last couple of decades provides an excellent perspective on the extent of the challenge. A wide range of research has been done on the lake, and this offers a rich source of information; the studies by D.M. Harper and his associates (11) are especially pertinent to understanding recent changes in the lake's ecology (see Chapter 9).

Prior to 1925 the only fish living in Lake Naivasha was the endemic zooplanktivorous small-tooth carp (*Aplocheilichthyes antinorii*) (39). In 1925 the cichlid *Oreochromis spilurus niger* was successfully introduced, and a few years later the American large-mouthed bass (*Micropterus salmoides*) was added. These two species fell to very low abundance or disappeared during the low lake levels in the late 1940s and early 1950s. In the subsequent few years Large-mouthed bass was re-introduced together with two cichlids, *Tilapia zillii* and *Oreochromis leucostictus*. Other introductions during the last five decades have had limited success; however,

Box 7.1. Predator-Prey Interactions and Defenses

With the exception of plants, most species must eat or parasitize living and dead organisms to survive and reproduce. Plants must avoid being eaten but in some cases, like pollen or seeds that need to be dispersed away from adults to mate or survive, being eaten at the right time is an advantage. Predators and prey have evolved numerous behavioral, morphological, and physiological adaptations to eating and being eaten (1). Species of predators and prey inhabiting the same area frequently simultaneously change their genetic composition or adapt over evolutionary time (they coevolve) to improve their ability to capture prey or escape from predators. There is little evidence that either prey or predators become so effective in their capture or defense mechanisms as to eliminate the other over evolutionary time (humans being an exception). Instead, species tend to coevolve, that is, the genetic composition of both species changes, resulting in escalating evolution or an "arms race" that often produces complex adaptations unique to particular combinations of predators and prey. Complex adaptations, however, are often not adaptive in low energy or stressful environments because the energetic costs of these adaptations are greater than the benefits - especially for those species competing with other species that are not using as much acquired energy on these adaptations. Escalating coevolution can be stopped, retarded, or eliminated by disturbances or environmental stresses that cause species to go locally extinct.

In most cases, evolutionary adaptations restrict species to particular locations where they are adapted to their predators and/or prey. If experimentally or accidentally transplanted to another environment, species are often poorly adapted to the new environment, and are extirpated. In fewer cases (but frequently on islands) the introduced species may be more effective at predation or avoiding predators than the local inhabitants (2). In such cases the introduction of a foreign species can cause dramatic changes in the species composition of the ecosystem. In some cases these changes are short-lived because the species is so successful that it eventually eliminates itself by eliminating its resources. Alternatively, the species composition of the ecosystem will change and become stable with a new species composition and food web configuration. This change may be associated with mass species extinctions and changing ecological processes.

The increased movement and abundance of people in the last few centuries has caused many introductions of species into foreign ecosystems. The results of these species introductions are often surprising, hard to predict, and different from the outcome expected by the people introducing the species. In many cases the changes are detrimental, not only to local species but also to humans, and management is required to rectify the conditions. In some cases the condition is unrectifiable or deemed to costly, and people and the surviving indigenous species must adapt to the new ecological conditions.

T.R. McClanahan

References

1. Vermeij, G.J. 1987. *Evolution and Escalation: An Ecological History of Life.* Princeton: Princeton University Press

2.   Pimm, S.L. 1991. *The Balance of Nature: Ecological Issues in the Conservation of Species and Communities*. Chicago: The University of Chicago Press

the migration of a riverine catfish, *Barbus amphigramma*, from the Malewa River during high water in 1982 established a population in the lake.

The fishery in Lake Naivasha is based almost exclusively on the artificial introductions of the bass which are caught by rod and line for sport and the cichlids since the commercial gill-net fishery opened in 1959. The tonnage of fish harvested from the lake peaked in late 1960s but declined abruptly in the early 1970s, perhaps as a consequence of over-fishing with smaller and smaller gill-net sizes. Although recovering after the mid-1970s, the harvests have fluctuated during periods of rising and falling water levels.

Lake Naivasha may have the potential for a greater harvest of fish than its average catch has been. The maximum sustainable yield is estimated to be about 420 metric tons per year. Theoretical yields derived from the morpho-edaphic index (a measure of the waters' solute concentration) and from regression models related to primary productivity suggest a maximum yield between 500 and 700 metric tons per year (39). One recommendation for an enhanced fishery is more introductions that would diversify the food web and convert more of the lake's productivity to harvestable fish (40). Such a proposal requires very careful scrutiny in light of the still incomplete understanding of Lake Naivasha's ecological dynamics.

Introductions other than fish have further modified the ecology of Lake Naivasha (11). The Louisiana red crayfish (*Procambarus clarkii*) was added in 1970 and has been harvested since 1975 with catches ranging from several hundred to about forty metric tons per year. The coypu (*Myocastor coypus*), a large rodent, imported to the Kinangop Plateau for fur-farming in 1950, escaped and entered Lake Naivasha in 1965. Attempts to control the coypu with introductions of the Python (*Python rebae*) were not successful, but, after attaining a large population in the early 1970s, the coypu appear to have disappeared by 1984.

In 1961 the floating Water-fern *Salvinia molesta* was observed in Lake Naivasha and caused alarm because of its well-known ability to cover large areas quickly. Spraying of herbicide and biological control with the grasshopper *Paulinia ecuminata* had only modest success, and the fern became widespread in the lake, covered much of the near-shore water and hindered access to the open water for fishing, recreation and pumping of water for irrigation. Recent biological control with the introduced weevil *Cyrtobagous salviniae* appears to be working. However, another aquatic weed, the Water Hyacinth (*Eichhornia crassipes*), first appeared in 1988 and is now ubiquitous around the lake 's edge.

J. J. Gaudet's studies of the aquatic vegetation in Lake Naivasha during the 1970s characterized the higher plant communities of the lake as an emergent papyrus swamp and littoral lagoons with Water-lilies (*Nymphaea caerula*), *Salvinia molesta* and several submerged macrophytes (8). Conditions were very different by the early 1980s. Floating islands of papyrus and *S. molesta* covered up to 25% of the lake's surface, and Water-lilies and the submerged plants were missing from the main lake. As water levels fell during the 1980s, large areas of papyrus were cleared to expand agricultural lands on the northern side of the lake, and Water-lilies and submerged plants reappeared. Furthermore, from 1982 to 1989, the abundance of phytoplankton, estimated from chlorophyll measurements, increased five-fold in the off-shore waters, whose transparency then declined three-fold (10).

The ecological causes for the multifaceted changes that have occurred in Lake Naivasha during the past two decades can only be inferred from the partial evidence (11) and are ripe for more intensive monitoring and experimental research. Improvements in the cichlid yields in the late 1970s probably reflected effective enforcement of mesh-sizes for gill-nets and increased productivity of littoral regions with rising lake levels. The bass consumed the crayfish which thrived during the 1970s, and bass catches rose. Disappearance of the Water-lilies is most likely owed to grazing by the coypu and crayfish; the crayfish then turned to the submerged plants and caused their disappearance by 1983. With the loss of the aquatic macrophytes as a primary food, the crayfish could not sustain so high a productivity, and the yields to fishermen and the bass fell. In turn, the catches of bass dropped. Moreover, the disappearance of the coypu and decline of the crayfish probably permitted the regrowth of the Water-lilies and submerged plants.

The loss of the northern papyrus swamp and conversion of this fertile land to agriculture probably contributed to increased inputs of nitrogen and phosphorus to Lake Naivasha with concomitant growth of phytoplankton during the 1980s. The key to modulating the input of nutrients to the lake is maintenance of the fringing papyrus and associated littoral vegetation. These wetlands are also an important habitat for many of the birds and mammals that make Lake Naivasha so spectacular to naturalists.

The continued vitality of Lake Naivasha depends on an adequate supply of fresh and non-polluted water, a resource in jeopardy. Water losses are occurring because surface and ground water inflows to the lake are being used to irrigate farmlands and as cooling water for the Hell's Gate geothermal power station, and, most seriously, are being diverted out of the catchment to supply the needs of the expanding town of Nakuru.

Four steps towards ecologically sound management of the resources in the Naivasha basin were proposed in 1990 (11). These are 1) maintain a buffer of natural vegetation around the lake, 2) develop a comprehensive management plan that balances needs of the fishery, wildlife, agriculture, domestic water consumption and the ecosystem of the lake, 3) implement an ecological and hydrological monitoring scheme and 4) establish an education program on environmental and water conservation. Many of these recommendations appear to have been incorporated into a recent environmental impact assessment for Lake Naivasha prepared as part of an effort to identify the lake as a world-class ecosystem worth conserving. While motivated by the acute problems at Lake Naivasha, the sound conservation and management of lakes throughout eastern Africa would benefit from similar programs.

References

1. Ase, L.E. 1987. A note on the water budget of Lake Naivasha, Kenya - especially the role of *Salvinia molesta* Mitch. and *Cyperus papyrus* L. *Geografiska Annaler* 69A: 415-429
2. Beadle, L.C. 1932. Scientific results of the Cambridge Expedition to the East African lakes 1930-31. 4. The waters of some East African lakes in relation to their fauna and flora. *Journal of the Linnean Society of Zoology* 38: 157-211
3. Beadle, L.C. 1981. *The Inland Waters of Tropical Africa*. London: Longman
4. Coe, M.J. 1966. The biology of *Tilapia grahami* in Lake Magadi, Kenya. *Acta Tropica* 23: 146-177

5.   Clark, F., Beeby, A., Kirby, P. 1989. A study of the macro-invertebrates of Lakes Naivasha, Oloidien and Sonachi, Kenya. *Revue d'Hydrobiologie Tropicale* 22: 21-33

6.   Eugster, H.P., Hardie, L.A. 1978. Saline lakes. In *Lakes*, ed. Lerman, A., pp. 237-293, New York: Springer-Verlag

7.   Finlay, B.J., Curds, C. R., Bamforth, S.S., Bafort, J.M. 1987. Ciliated Protozoa and other microorganisms from two African soda lakes (Lake Nakuru and Lake Simbi, Kenya). *Archiv für Protistenkund* 133: 81-91

8.   Gaudet, J.J. 1977. Natural drawdown on Lake Naivasha (Kenya) and the formation of papyrus swamps. *Aquatic Botany* 3: 1-47

9.   Gaudet, J.J., Melack, J.M. 1981. Major ion chemistry in a tropical African lake basin. *Freshwater Biology* 11: 309-333

10. Harper, D.M. 1992. The ecological relationships of aquatic plants at Lake Naivasha, Kenya. *Hydrobiologia* 232: 65-71

11. Harper, D.M., Mavuti, K.M., Muchiri, S.M. 1990. Ecology and management of Lake Naivasha, Kenya, in relation to climatic change, alien species' introductions, and agricultural development. *Environmental Conservation* 17: 328-336

12. Howarth, R.W. 1988. Nutrient limitation of net primary production in marine ecosystems. *Annual Review of Ecology and Systematics* 19: 89-110

13. Hecky, R.E., Kilham, P. 1973. Diatoms in alkaline, saline lakes: ecology and geochemical implications. *Limnology and Oceanography*. 18: 53-71

14. Jellison, R.S., Melack, J.M. 1993. Algal photosynthetic activity and its response to meromixis in hypersaline Mono Lake, California. *Limnology and Oceanography* 38: 818-837

15. Jenkin, P.M. 1936. Reports on the Percy Sladen Expedition to some Rift Valley lakes in Kenya in 1929. VII. Summary of the ecological results with special reference to the alkaline lakes. *Annuals and Magazine of Natural History, Series 101*. 18: 133-181

16. Jenkin, P.M. 1957. The filter-feeding and food of flamingos (Phoenicopteri). *Philosophical Transactions of the Royal Society, London, Series B*. 240: 401-493

17. Kalff, J. 1983. Phosphorus limitation in some tropical African lakes. *Hydrobiologia* 100: 101-112

18. Kalff, J., Watson, S. 1986. Phytoplankton and its dynamics in two tropical lakes: a tropical and temperate zone comparison. *Hydrobiologia* 138: 161-176

19. Kallqvist, T., Meadows, B.S. 1978. The toxic effect of copper on algae and rotifers from a soda lake (Lake Nakuru, East Africa). *Water Research* 12: 314-327

20. Kilham, P. 1971. *Biogeochemistry of African Lakes and Rivers*. Ph.D. Dissertation, Duke University, Durham, N.C., USA

21. Kilham, P. 1981. Pelagic bacteria: extreme abundances in African saline lakes. *Naturwissenschaften* 67: 380-381

22. Kilham, P., Hecky, R.E. 1973. Fluoride: geochemical and ecological significance in East African waters and sediments. *Limnology and Oceanography* 18: 932-945

23. Kilham, P., Cloke, P.L. 1990. The evolution of saline lake waters: gradual and rapid biogeochemical pathways in the Basotu lake district, Tanzania. *Hydrobiologia* 197: 35-50

24. Koeman, J.H., Pennings, J.H., de Goeij, J.J.M., Tjioe, P.S, Olindo, P.M., Hopcraft, J. 1972. A preliminary survey of the possible contamination of Lake Nakuru in Kenya, with some metals and chlorinated hydrocarbon pesticides. *Journal of Applied Ecology* 9: 771-775

25. LaBarbera, M.C., Kilham, P. 1974. The chemical ecology of copepod distribution in the lakes of East and Central Africa. *Limnology and Oceanography* 19: 459-465

26. Léonard, J., Compère, P. 1967. *Spirulina platensis* (Gom.) Geitl., algue bleue de grande valeur alimentaire par sa richesse en proteines. *Bulletin du Jardin Naturel de Belge* 37, Supplement 1

27. Lewis, W.M. Jr. 1987. Tropical limnology. *Annual Review of Ecology and Systematics* 18: 159-184

28. Livingstone, D.A. 1975. Late quaternary climatic change in Africa. *Annual Review of Ecology and Systematics* 6: 249-280

29. Livingstone, D.A., Melack, J.M. 1984. Some lakes of subsaharan Africa. In *Lakes and Reservoirs*, ed. Taub, F.B., pp. 467-497. Amsterdam: Elsevier Science Publishers

30. MacIntyre, S., Melack J.M. 1982. Meromixis in an equatorial African soda lake. *Limnology and Oceanography* 27: 595-609

31. Mavuti, K.M., Litterick, M.R. 1981. Species composition and distribution of zooplankton in a tropical lake, Lake Naivasha, Kenya. *Archiv für Hydrobiologie* 93: 52-58

32. Melack, J.M. 1979. Temporal variability of phytoplankton in tropical lakes. *Oecologia* 44: 1-7

33. Melack, J.M. 1979. Photosynthesis and growth of *Spirulina platensis* (Cyanophyta) in an equatorial lake (Lake Simbi, Kenya). *Limnology and Oceanography* 24: 753-760

34. Melack, J.M. 1981. Photosynthetic activity of phytoplankton in tropical African soda lakes. *Hydrobiologia* 81: 71-85

35. Melack, J.M. 1988. Primary producer dynamics associated with evaporative concentration in a shallow, equatorial soda lake (Lake Elmenteita, Kenya). *Hydrobiologia* 158: 1-14

36. Melack, J.M., Kilham, P., Fisher, T.R. 1982. Responses of phytoplankton to experimental fertilization with ammonium and phosphate in an African soda lake. *Oecologia* 52: 321-326

37. Melack, J.M., MacIntyre, S. 1991. Phosphorus concentrations, supply and limitation in tropical African lakes and rivers. In *Phosphorus Cycles in Terrestrial and Aquatic Ecosystems*, Regional workshop 4: Africa. eds. Tiessen, H., Frossard, E., pp. 1-18. Saskatoon, Canada: Saskatchewan Institute of Pedology

38. Mono Basin Ecosystem Study Committee. 1987. *The Mono Basin Ecosystem - Effects of Changing Lake Level*. Washington, D. C.: National Academy Press

39. Muchiri, S.M., Hickley, P. 1990. The fishery of Lake Naivasha, Kenya. In *Catch-effort Sampling Techniques and their Application in Freshwater Fisheries Management*, ed. Cowx, I., Chapter 34. Oxford, England: Blackwells

40. Muchiri, S.M., Hickley, P., Harper, D.M., North, E. 1992. The potential for enhancing the fishery of Lake Naivasha, Kenya. In *Proceedings of the International Symposium on Rehabilitation of Inland Fisheries*, ed. Cowx, I. Oxford, England: Blackwells

41. Njuguna, S.G. 1988. Nutrient-phytoplankton relationships in a tropical meromictic soda lake. *Hydrobiologia* 158: 15-28

42. Peters, R.H., MacIntrye, S. 1976. Orthophosphate turnover in East African lakes. *Oecologia* 25: 313-319

43. Richardson, J.L., Richardson, A.E. 1972. History of an African rift lake and its climatic implications. *Ecological Monographs* 42: 499-534

44. Schindler, D.W. 1978. Factors regulating phytoplankton production and standing crop in the world's freshwaters. *Limnology and Oceanography* 23: 478-486

45. Talling, J.F., Talling, I.B. 1965. The chemical composition of African lake water. *Internationale Revue der gesamten Hydrobiologie* 50: 421-463

46. Tuite, C.H. 1979. Population size, distribution and biomass density of the Lesser Flamingo in the Eastern Rift Valley, 1974-1976. *Journal of Applied Ecology* 16: 765-775

47.   Vareschi, E. 1978. The ecology of Lake Nakuru (Kenya). I. Abundance and feeding of the Lesser Flamingo. *Oecologia* 32: 11-35

48   Vareschi, E., Jacobs, J. 1985. The ecology of Lake Nakuru. 6. Synopsis of production and energy flow. *Oecologia* 65: 412-424

49.   Wood, R.B., Talling, J.F. 1988. Chemical and algal relationships in a salinity series of Ethiopian inland waters. *Hydrobiologia* 158: 29-67

Chapter 8

The Great Lakes

Les Kaufman, Lauren J. Chapman & Colin A. Chapman

Lake Superior is greater in surface area than Lake Victoria. Lake Baikal is deeper than Lake Tanganyika. Nonetheless, the Great Lakes of East Africa are second to none, when it comes to the wealth of native fishes and the number of people dependent on these lakes. Lake Malawi is thought to host more than 500 species of fishes, nearly all endemic (27). Endemism is also >99% among the haplochromine cichlids of Lake Victoria, where more than 400 species appear to have evolved in less than 200,000 years (29) - geologically and evolutionarily little more than an instant. The most diverse faunas occur in Lake Tanganyika, the deepest and oldest lake. Fishes and snails in Lake Tanganyika are so varied in form that the occupants of this one lake might easily be mistaken for marine organisms hailing from dozens of unrelated taxa. Through the fish protein that they produce, the Great Lakes fuel a human society with currently one of the earth's highest rates of population growth.

Despite their impressive faunal diversity, large size, and fisheries production, these lakes are all threatened by changes resulting from human modification of the landscape and the effects of human activities on water quality, biological diversity, and fish stocks. In this chapter we review the distinguishing attributes of the major East African Lakes, the faunas they presently support, and current significant perturbations affecting the aquatic systems. Where possible, we identify factors underlying current patterns of change. Since Lake Victoria has experienced the most rapid and dramatic changes in the last century, we consider Lake Victoria in detail, outlining how changes in this lake may foretell the future for other lakes, unless careful management plans are constructed. We will end by briefly discussing potential conservation programs that may aid in mitigating the ecological changes that are occurring in the Lake Victoria ecosystem.

Lake Geography

East Africa is rich in aquatic resources, including nine large lakes. Lakes Albert, Kivu, Tanganyika, Malawi, and Turkana are steep-sided true Rift Valley lakes; while Lakes Edward, George, Kyoga, and Victoria are shallow-basin lakes (Fig. 8.1).

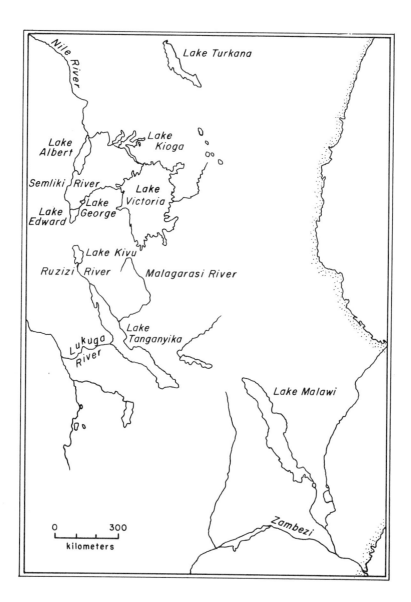

Figure 8.1. A map of the major lakes and rivers in East Africa.

Comparison of the Great Lakes of East Africa to other large lakes around the world highlights their significance from a number of perspectives (Table 8.1). Lakes Victoria (68,800 km^2), Tanganyika (32,900 km^2), and Malawi (22,490 km^2) are among the largest lakes in the world. As such these lakes play particularly vital roles in the regulation and maintenance of tropical ecosystems on local, regional and potentially global scales.

Table 8.1. The area and maximum depth of freshwater lakes of the world (greater than 8000 km² in area).

Lake	Area, km²	Maximum depth, m
Africa		
L. Victoria	68800	79
L. Tanganyika	32900	1435
L. Malawi	22490	706
North and Central America		
L. Superior	83270	393
L. Huron	60700	229
L. Ontario	19230	237
L. Michigan	58020	281
L. Erie	25680	64
L. Athabasca	8080	91
L. Winnipeg	24510	21
Great Bear Lake	31790	319
Great Slave Lake	8270	140
L. de Nicarague	8270	70
South America		
L. Titicaca	8340	304
Asia		
L. Baikal	30500	1741
L. Balkhash	17400	26
Europe		
L. Ladoga	18390	230
L. Onega	9600	124

The Lakes

Lakes Victoria and Kyoga. Lake Victoria is an immense inland sea, the second largest freshwater body in the world, with a catchment area of approximately 184,00 km² (Table 8.1, Fig. 8.1). Situated across the equator at an altitude of 1134 m, the lake is set in a large shallow basin. Although Lake Victoria is enormous in area (300 km from north to south and 280 km from east to west, covering an area of 68,800 km²), it is relatively shallow with a maximum depth of only 79 m and a mean depth of only 40 m. The waters of Lake Victoria are shared by three countries: Tanzania with 51% of the lake area, Uganda with 43%, and Kenya with only 6% of the lake area within its boundaries.

Lake Victoria is fed by a number of rivers; the most significant in terms of volume is the Kagera River. The major source of water, however, is from direct rainfall on the surface (2). Lake Victoria discharges into the Nile river system, through the damn built at the Owen Falls. Subsequently, the waters flow through Lake Kyoga and then over the Murchison Falls and into Lake Albert. The Murchison Falls form a barrier to the Nile fish ascending from Lake Albert towards Lake Victoria (26). As a result of its shallow basin form, the edges of the lake are often fringed with extensive papyrus swamps (*Cyperus papyrus*). This swampy margin has potentially been very significant in the evolution of the fauna (see below) and may, in the future, play a particularly significant role in conservation of many of the fish species endemic to the lake.

The origin and past history of the lake has been an area of considerable debate (2,23). It is thought that the Lake Victoria basin started to form in the Miocene (approximately 20 million years ago) with the faulting of the Western Rift Valley and uplifting of land to the west of the basin (Table 8.2, Fig. 8.2). This process eventually resulted in the reversal of flow of major rivers such as the Katonga and Kagera. These rivers had previously drained the Lake Victoria region and flowed into Lake George. However, as a result of the uplifting, instead of flowing to the west, the water in these rivers drained to the east into the basin that was to become Lake Victoria. The uplifting process was, however, very slow and Lake Victoria may have drained into Lake George well into the late Pleistocene - possibly as late as 25,000 to 35,000 years ago (2; Table 8.2).

Sediment cores near the mouth of the Nile suggest that the level of the lake was too low to reach the outlet from about 14,500 to 12,000 years ago and again about 10,000 years ago (2). These were glacial times when the region as a whole was much cooler and drier (Chapter 1). Thus, it is likely that during one or more periods in the last 15,000 years Lake Victoria was much reduced in size, and became very shallow, and perhaps saline. It is probable that such water-level fluctuations repeatedly isolated and reconnected bays and depressions. Thus, the emerging picture is that Lake Victoria formed as a result of the uplifting and lake depression, but climatic fluctuations may have had a tremendous effect on the biological components of the lake through periods of isolation and reconnection. From a comparative perspective, climatic changes would have had an even more dramatic effect on the shallow lake basin of Victoria than on the deeper rift lakes of Tanganyika and Malawi.

Lake Kyoga lies just to the north of Lake Victoria. It receives the outflow from Lake Victoria and drains to the north, over the Murchison Falls and into Lake Albert. The lake was formed through the same tectonic changes and uplifting of the escarpment that led to the formation of Lake Victoria, but Lake Kyoga's formation may have been slightly later than that of Victoria (2).

Lake Tanganyika. Both Lakes Tanganyika and Malawi differ radically in physical or geologic structure from Lake Victoria. Unlike the shallow basin of Lake Victoria, these lakes sit in mountainous terrain with numerous precipitous drops past rocky shores and sandy beaches to the bottom of a freshwater abyss. Lake Tanganyika is 650 km long with an average width of about 50 km (2; Fig. 8.1). In general, the water depth drops off very rapidly as one moves out from the shore (maximum depth = 1470 m only 4 km from the shore, mean depth 570 m; 2). Although there is a shallow gradient in the very north and south ends of the lake (the 100 m contour is less than 10 km from shore), there are a few areas of shallow swampy water. Sections of the lake are 1470 m in depth. The majority of the deeper areas of the lake are permanently without oxygen (below 200 m) and support little life except bacteria. The lake's main inflow are the Ruzizi River from Lake Kivu (the faunas are separated by the Panzi Falls) and the Malagarasi River flowing in from a swampy region to the east. In comparison to Lake Victoria, the catchment area is very small, but again the major source of water to the lake is through direct rainfall onto the surface. The outflow is on the western side in the middle of the lake, the Lukuga River. Historical evidence suggests that this outflow was intermittent and even presently, it is not strong (2; Table 8.2). Thus, the major loss of water to the lake is through evaporation from the water surface,

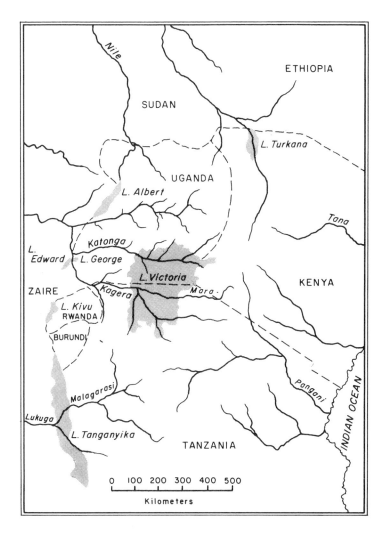

Figure 8.2. The pre-Pleistocene drainage in the East African Region, as proposed by (2).

a situation that is also true for both Lakes Victoria and Malawi, but not to the extent as observed in Lake Tanganyika (2).

Lake Malawi. Lake Malawi is physically similar to Lake Tanganyika. It is long and narrow, occupies a mountainous region (especially in the north) and drops off very rapidly as one travels out from the shore. The lake is 600 km long, up to 75 km wide, and has a maximum depth of 706 m with only one basin (Fig. 8.1). Like Lake Tanganyika, Malawi has an intermittent outlet, the Shire River (2). The catchment area of the lake is very small, and the lake is largely dependent on direct

Table 8.2. The major geological events that occurred in East Africa and the associated changes in the aquatic systems (Adapted from 3). The evidence for the approximate dating of some the events is controversial and further research is required to verify the time when each of these events occurred.

Geological Period (Years Before Present)	Event
25 Million Miocene separate Lakes Rivers	Lake Karunga covering the North East corner of the future Lake Victoria Beginning of the uplifting (by more than 1000 m) that will form the Rift Lake Basin, Edward and George from Lake Victoria and reverse the flow of the Kagara and Katonga Rivers
12 Million Pliocene 7 million	Eruption of isolated volcanoes such as Kilimanjaro, Kenya, Elgon
2-3 Million Pleistocene 2 million - 35,000	Rwenzori Mountains rise Further rising and formation of more rift lakes 'Kaiso' lake in Albert and Edward Basins Upwarping of the western side of the future Lake Victoria Basin Eruption of the Virunga Volcanoes
30,000 - 25,000	Uplifing on the western side of the present Lake Victoria causes the reversal of the Kagea and Katonga Rivers, and the formation of Lake Victoria, which eventually overflows into the Nile Basin to the north
25,000 - 20,000	The water level of Lake Victoria was approximately 30 m above the present lake Glacial period associated with cooler temperature and lower water levels in the lakes Major eruptions of the Virunga Volcanoes leads to the formation of Lake Kivu Lake Tanganyika drains out the Lukuga outlet

20,000 - 15,000	Maximum extension of the glaciers on the Rwenzori Mountains
14,500	Lake Victoria falls from previous high level

Holocene

14,500 - 12,000	Lake Victoria is too low to reach the outlet
12,000 - 10,000	Rising temperatures and higher rainfall levels
10,000	Lake Victoria is again too low to reach the outlet
10,000 - 8,000	Eruption of the Katwe and Banyuruguru craters near Lakes Edward and George occurred during this period and likely earlier and may have resulted in the extinction of a large number of fish, NileCrocodile, and mollusk species
7500	The period when Lake Turkana was last connected to the Nile
3,700	The date of the last drop in Lake Victoria's water level likely associated with the cutting down of the outflow at Jinja and climatic changes
840 - 940	The southern arms of Lake Malawi dried out.
490 - 140	The southern arms of Lake Malawi again dried out.
30	Lake Victoria rises 2 m as a result of high rainfall
10	Nile crocodile again reported from Lakes Edward and George.

rainfall and small inflow streams that come directly off the mountains. The lake's level exhibits marked seasonal changes. For example, the lake fluctuates just over a meter between the wet and dry seasons (25). The area of the lake below 250 m in depth is anoxic (permanently without oxygen; 26). Rapid response to both seasonal and long-term changes in rainfall repeatedly isolate and reconnect inshore habitats. Trewavas (43) suggested that such changes facilitated the speciation of the cichlids found in the lake. One of the features that distinguishes Lake Malawi from the other Great Lakes is that it is much further from the equator than the other lakes, resulting in cooler surface water temperatures than Lake Tanganyika and Victoria (27).

The geologic history of Lake Malawi is poorly understood. The lake is younger than Lake Tanganyika, but has been isolated long enough to produce a remarkable endemic fauna (2). Geological evidence suggests that at earlier times, the lake drained to the east coast, and the fauna has ties to east coast river fishes and fishes in the Upper Zambezi. It is evident that none of the Malawi fish have come directly from the north. The lake level has fluctuated with climatic factors. For example, during drier periods of the last glacial cycle, the lake was 250 to 500 m lower than that of its present level. During a drier cycle between 1150 and 1250 A.D., and again sometime between 1500 and 1850 A.D., the southern arms of the lake dried up (2). These areas are now inhabited by many endemic rock-dwelling cichlids (26,37) that evolved since these drier periods.

Lake Albert. Lake Albert is a typical Rift Valley lake, 150 km long with an average width of about 35 km. Its maximum depth is only 56 m, which occurs 7 km from the mid-western shore. The lake has two major inflow rivers, the Semliki to the south, and the Victoria Nile, that enter the lake over Murchison Falls at the north end. Because the waters from the Victoria Nile have less effect on the lake as a whole than the waters from the Semliki, they enter immediately prior to the lake's outflow. As with the other lakes, there is evidence that its water level has changed in response to changing climatic conditions associated with glacial cycles (2; Chapter 1). For example, during the last glacial period, between 14,500 and 12,000 years ago, it seems likely that there was a stoppage of the Victoria Nile, that resulted in a drop in water level. The history of the lake corresponds to events described for Lake Victoria, involving the uplifting of the Western Rift (2).

The lake can become stratified during calm periods of the year, although dissolved oxygen in deeper waters is not greatly reduced. Consequently, a number of the species from this lake (such as *Lates niloticus*) show little tolerance to low oxygen waters (2,8).

Lake Turkana. Lake Turkana, located in the arid hot region of northern Kenya, is another large rift lake with a surface area of 4350 km². It is 265 km in length and averages approximately 30 km in width. It has a maximum depth of 120 m in a small depression in the lake's southern end, and a maximum depth excluding this depression of 80 m (2). The River Omo flowing from the Ethiopia highlands supplies 98% of the inflow to Lake Turkana. The level of the lake has fluctuated extensively, and there are fossil fish that can be found several km from the shore, indicating that it was once much larger than it is today. The geological events that lead to the formation of the lake commenced in the early Miocene (approximately 20 million years ago), and since then, the lake has had a dynamic history. Early in its

Table 8.3. Physical and biological characteristics of the major East African Lakes (biological refers to cichlid radiations only). Taken from (2,27, and the Times Atlas).

	Victoria	Tanganyika	Malawi	Albert	Turkana	Edward	George	Kivu	Kyoga
Physical									
Size	68800	32900	22490	6410	4250	2250	270	...	...
Max Depth(m)	100	1435	706	53	73	112	3	>500	...
Altitude (m)	1134	773	474	619	380	912	916	1460	1033
Mean Lattitude	10S	70S	120S	1.50N	3.50N	0	0	20S	1.50N
Conductivity	96	610	210	735	3300	925	200	1240	...
Salinity %(g/l)	0.093	0.530	0.192	0.597	2.482	0.789	0.139	1.115	...
pH range	7.1-8.5	8.0-9.0	8.2-8.9	8.9-9.5	9.5-9.7	8.8-9.1	8.5-9.8	9.1-9.5	...
Biological									
Endemic Cichlids	400+	170+	500+	Few	Few	60+	60+	Few	
Trophic Diversity	High	Med	High	Low	Low	High	High	Low	High
Reproductive Divversity	Low	High	Low	Low	Low	Low	Low	Low	Low
Phyletic Diversity	Low	High	Med	Low	Low	Low	Low	Low	Low

history the lake exhibited rapid speciation of its molluscan fauna (2). Nonetheless, this endemic fauna did not persist, presumably the result of low water levels, high levels of salinity, or lake desiccation.

Recent history has also been dynamic. The lake experienced very high water levels about 20,000 years ago, but was again very low during the glacial dry periods around 12,000 years ago. Much of the fauna died out during this period. Subsequently, the lake rose to more than 80 m above its present level, and was connected to the Nile river system and its fauna. About 7500 years ago, the water level dropped below the outlet and the lake became isolated (2). Not surprisingly, the lake is more saline than any other of the East African Great Lakes, due to falling water levels and no outflow (Table 8.3). The lake, however, still harbors a rich fauna (2). A remarkable feature of the Lake Turkana is the absence of papyrus (*Cyperus papyrus*) which is the dominant vegetation type throughout much of the Nile basin. Beadle (2) speculates that the high salinity or alkalinity accounts for its absence.

Lakes Edward and George. Lakes Edward and George lie in a rather peculiar depression in the western rift valley, surrounded on three sides by escarpments; the Kichwamba Escarpment to the east, the Zaire Escarpment to the west, and Ruwenzori Mountains to the north (Fig. 8.1). Lake George, which is bisected by the equator in western Uganda, and lies to the west of Lake Edward, is 290 km² in size, with a maximum depth of only 3 m. The lake is fed from rivers flowing off the Ruwenzori Mountains (Nonge, Mbuku, Bumlikwesi, Dura, and Mpanga Rivers), but before they reach the lake proper, their waters filter through a large papyrus swamp at the northern edge of the lake. Lake George has higher productivity and is more saline than Lake Edward. Lake George is connected to Lake Edward by the Kazinga Channel, which is 40 km long and less than 1 km wide (2).

Lake Edward (2250 km²) is much larger than Lake George (270 km²), and has a maximum depth of 112 m. The lake's major inflows are from the Nyamugasani River, which drains off of the Ruwenzori Mountains, and the Ishasha, Rutshuru, and Rwindi Rivers, which drain the Rwandan highlands. According to Beadle (2) the annual contribution of the Kazinga Channel is probably much less than the input from these rivers. The lake has one outflow, the Semliki River, that flows to the North, eventually draining into Lake Albert. Hydrological data on Lake Edward have demonstrated prolonged stratification and water without oxygen below about 50 m (2). The faunas of the Lakes Edward and Albert are separated by a series of rapids on the Semliki. During its 250 km course from Lake Edward to Lake Albert, the river descends 300 m, the majority of this is in one section of rapids.

Like many East African lakes, geological and paleontological evidence suggest that Lakes Edward and George have had a very dynamic history. In the early Pleistocene (2 million years ago), there may have been passable connections between the Lake Edward and Lake Albert basins. The faunas of the lakes were very similar: fossil remains of *Lates*, *Hydrocynus*, and *Crocodilus niloticus* in the Edward basin, link the two faunas. Subsequently, Lakes Edward and George may have gone through a number of mass extinction and recolonization events (2). The last of these events probably occurred between 8000 and 10,000 years ago and was associated with the eruption of volcanoes that formed the Katwe and Banyuruguru craters. Beadle (2) speculates that these volcanoes may have deposited enormous amounts of toxic ash on the lake, that would have killed species such as *Lates*, which may require well-oxygenated water. Similarly, the Nile crocodiles (*Crocodilus niloticus*) were no longer present in the areas of Lakes Edward and George after the eruption of these neighboring volcanoes. They were presumably kept from colonizing the lake by difficulties in passing the Semliki falls. However, recently with forest clearing on the sides of the falls, crocodiles have reinvaded Lake Edward.

Lake Kivu. Volcanic activity in the late Pleistocene in the Virungas created a barrier about 100 km south of Lake Edward. These eruptions completely blocked the flow of the upper Rutshuru River and the water flooded the valley forming the present day Lake Kivu. As the water levels rose, they eventually overflowed to the south, over the Panzi Falls and into Lake Tanganyika (2).

Lake Kivu is a medium-sized lake, with a maximum depth greater than 500 m. The thickness of sediments in the deep northern basin suggests that there may have been a lake in the area of the present-day Lake Kivu, prior to the volcanic activity that blocked the river (2). The surface of this lake would have been only 60 m higher than Lake Edward. This shallow grade between this older Lake Kivu and Lake Edward may have allowed exchange of species between the two lakes. This lends credence to the claim of Poll (2) who suggested in 1939 that the fauna of Lake Kivu is the remains of a richer fauna that was destroyed as a result of ash and chemical deposited on the surface of the lake during the period of volcanic activity.

The Faunas

The faunas of the Lakes Victoria, Tanganyika, and Malawi are all dominated and shaped by members of a single fish family, the Cichlidae (Table 8.3). These Great Lakes, plus Lakes Edward, George, and Kivu all have many endemic cichlids species (10,11,13,14). All are primarily evolutionary radiations of feeding-adaptation. The assemblages in the Great Lakes offer many striking examples of evolutionary convergence. Cichlids species with similar morphologies and functional characteristics have evolved independently in each of the lakes.

Lake Victoria and Kyoga. The fish fauna of Lake Victoria was dominated by a diversified, presumably monophyletic species flock of haplochromine cichlids (29). The explosive speciation and adaptive radiation displayed by these cichlids remains unrivaled among vertebrates. No other group of vertebrates is known in which such an extensive and rapid adaptive radiation was realized in so little time and with so little overall body form diversity. Lake Victoria probably had more than 400 species of endemic haplochromine cichlid fish until the early 1980s when a mass extinction occurred.

The haplochromines are a very interesting group of fish. All defend their young until self-sufficient. The large eggs and developing young are held in their mouth. They are behaviorally aggressive, physiologically adaptable, and phenotypically very plastic. Many haplochromine cichlids can individually alter tooth and skull morphology in response to diet change (11). They exhibit extraordinary evolutionary diversification that is represented to some degree by the number of feeding methods they exhibit. There are detritivores, phytoplanktivores, zooplanktivores, epiphytic algae grazers, fish that crush mollusks with their oral jaws, fish that crush mollusks with their pharyngeal jaws, prawn eaters, piscivores, fish scale eaters, crab eaters, and paedophages ("child eaters"; 11). Paedophages are predators with large mouths, thick rubbery lips, and outward pointed anterior teeth. They take the head end of brooding females into their mouth and suck out the eggs or young.

It is likely that the fishes of the Lake Victoria are descendants of fishes living in pre-Pleistocene rivers that flowed through the area that is now Lake Victoria, west towards Zaire (2). As in Lakes Tanganyika and Malawi, Victoria cichlid fishes are the product of rapid speciation from a few ancestors (27). However, in Lake Victoria, evolution has been more rapid than in either Lakes Malawi or Tanganyika. A variety of mechanisms may have operated to produce the difference in the rates of speciation between Lake Victoria and the deep rift lakes.

It seems likely that four different sorts of speciation mechanisms could be operating in Lake Victoria; all are associated with the relatively shallow morphometry of the lake. First, the shallow morphometry of the Lake Victoria Basin allows for speciation in satellite lakes with only modest fluctuations in lake level (11). During periods of low lake level, species can diverge from the parent stock in isolated satellite lakes. When lake level rises and satellite lakes are once again connected to the main lake, populations mix, and, if some mechanism prevents interbreeding, two distinct species may be recognized from one common ancestor. Secondly, the abundance of islands, rocky headlands, and pinnacles in Lake Victoria affords ample opportunity for speciation on these islands when lake levels are high. These offshore islands can separate populations of fish that will not cross large stretches of open water. Consequently, when water levels fall and populations come back into contact they may have diverged enough to prevent interbreeding (11). Thirdly, the shallow basin creates a situation where there are segments of vegetated swamp margins (typically papyrus) separating rocky or sandy segments of shoreline. Papyrus swamps, with their chronic low oxygen conditions, can serve as barriers to the movement of species intolerant of low oxygen conditions (5). Finally, during periods of desiccation, fish may retreat up rivers to avoid saline or hypoxic conditions. This creates the potential for a population with subpopulations now isolated in different rivers to diverge (23). Thus, speciation dynamics in Lake Victoria should be more rapid, more continuous, and more geographically diffuse than in either Lakes Malawi or Tanganyika.

The fish fauna of Lake Kyoga is extremely similar to that of Lake Victoria. Prior to the construction of the hydroelectric dam on the Victoria Nile, Owen Falls was little more than a series of rapids that was thought to be no barrier to the movement of fish from Lake Victoria and Kyoga (2). The swampy margins and lagoons of Lake Kyoga may play an important role in protecting the haplochromine fauna from Nile perch predation serving as refugia for haplochromines that are endangered or extinct in the larger lake (discussed below).

Lake Tanganyika. The cichlid fauna of Lake Tanganyika is, like the lake, very old. Several well represented lineages hail from origins that lie millions of years in the past. Unlike Lakes Malawi and Victoria, that are dominated by the mouthbrooding haplochromine and tilapiine cichlids, a large proportion of the Lake Tanganyikan cichlids are substratum-spawning "lamprologines", derived from ancestors in the Zaire River. Representative species in Lake Tanganyika of any given feeding adaptation are more highly specialized morphologically than their look-alikes in Lakes Malawi or Victoria. Two exceptions to this pattern stand out: Lake Tanganyika is poor in species of snail-crushing cichlids, and apparently lacks true cichlids paedophages.

Lake Tanganyika has by far the richest non-cichlid lacustrine fish assemblage in East Africa (27). Included among these are the world's largest tigerfish (*Hydrocynus goliath*), four species of *Lates* (relatives of the Nile Perch, *Lates niloticus*), and two pelagic sardines, which together with the *Lates*, comprise the bulk of the fishery. The lake is also richly endowed with catfishes, carp, characoids, spiny eels, and many others (26,27).

Lake Tanganyika's long period of isolation is reflected in a high level of endemism: 220 out of the 287 species (76%) in the lake are endemic; 172 of these endemic fishes are cichlids (2,25,27). Of the 115 noncichlid species, 46% are endemic. What illustrates the greater age of Lake Tanganyika and distinguishes its fauna from that of Lake Victoria, is the wider divergence in the non-cichlids. Eight genera are endemic to Lake Tanganyika. Apart from *Xenoclarias* in Lake Victoria, none of the other lakes have endemic genera (27). Lake Tanganyika also has more fish families than the other great lakes; reflecting its ties to the species-rich Zaire River system. Of the 24 families of fishes found in the Zaire river, 18 are found in tributaries and marshes around the lake, 12 occur in the littoral and sublittoral zones, 7 families occur in the benthic zone, and 4 in the pelagic zones (27). Interestingly, there are some species presently found in Malagarasi river draining the country to the east of the lake (and south of Lake Victoria) that are evolutionarily related to species in the Zaire River system, but are no longer found in the lake itself.

The level of the Lake Tanganyika has fluctuated with changing geological and climatic patterns. Evidence suggests that 200,000 years ago, the different basins of the lake were totally separated (27). It would have required a drop in water level of 700 m to completely isolate the two basins (24). Approximately 50,000 years ago the lake was a unified body, but the north and south basins were connected only by the Kalemie Strait (27). It has been suggested that repeated separation of the two halves of the lake may have fueled a process of rapid speciation (27). There is some taxonomic evidence to support this idea. For example, there are two subspecies of the cichlid *Opthalmochromis ventralis*, with a northern and a southern variety. However, climatic changes sufficient to cause lake-level variation of the required magnitude for basin isolation have not occurred since the last glacial period (>12,000 years ago). Thus, it seems probable that other processes are also contributing to the rapid rate of speciation.

Lowe-McConnell (27) suggests that the many rocky segments of Lake Tanganyika shoreline separated by stretches of sand or swamp have created a patchy situation that facilitates allopatric speciation particularly among mouth brooders were juvenile dispersal is minimal. Fluctuating lake levels alone do not occur at a high enough frequency to account for the present number of endemic species, but certainly may have contributed to the high endemism above the species level. These speculations are supported by the fact that the rocky habitats support the most diverse species assemblages (19,27).

Lake Malawi. Lake Malawi, like Lakes Victoria and Tanganyika, harbors an impressive endemic flock of cichlids. Like the other species flocks of cichlids, estimates of the numbers of species have changed over the years, reflecting the difficulties of cichlid systematics and the lack of extensive collections. In 1981 Beadle (2) reported that there were 245 species known from Lake Malawi from only 7 families, which is about half the number in the other Great Lakes (25). In contrast, with further investigations and more collections, Ribbink (39) in 1988 and Lowe-McConnell (27) in 1993 report 400 and greater than 500 species, respectively, of which all but 4 are endemic.

Lake Malawi's fish fauna is most closely related to species found in the upper Zambezi, although only 9 out of the 14 families in the Zambezi are found in the lake. In addition to the species flock of cichlids found in the lake, Malawi harbors a unique flock of 12 species of clariid catfishes (25), and 4 endemic tilapias.

Lake-level events in the last 500 years have permitted us to re-evaluate the estimated speed of cichlid speciation in Lake Malawi. An early study of Lake Nabugabo (a satellite lake of Lake Victoria) by Greenwood (11) suggested that haplochromines could speciate in as little as 4000 years, since this was when Lake Nabugabo is estimated to have separated from Lake Victoria. Evidence suggests, however, that the southern end of Lake Malawi was dry 500 years ago. Presently, there are endemic species living around islands in areas that were dry only 500 years ago (27,37).

Lake Albert. Lake Albert has a diverse fish fauna but is strikingly different from other East African Great Lakes, in having very few cichlid species. Considering the present-day mass extinction of cichlids in Lake Victoria associated with the eruption of the introduced Nile perch (*Lates niloticus*), it is tempting to speculate that the scarcity of the cichlid fauna in Lake Albert might be related to the presence of predators such as *Lates* and *Hydrocynus*. However, the cichlid species flock in Lake Tanganyika where a number of species of *Lates* are abundant argues against this speculation. Fryer and Iles (10) suggest that Lakes Albert and Tanganyika received fully differentiated riverine faunas that occupied available niches after catastrophic lake-level disturbances. This may have limited opportunities for cichlid species radiation. The fauna of Lake Albert is separated from that of Lake Victoria by Murchison Falls. Commercially valuable fish are *Oreochromis niloticus, Lates niloticus, Alestes baremose*, and *Citharinus citharus* (2).

Lake Turkana. The fish fauna of Lake Turkana is derived from the Nile drainage, particularly Lake Albert and, like Lake Albert, has few cichlids. Hopson and Hopson (2) documented 48 fish species in the lake, 30 species were found throughout the Nile Drainage, 22 species also occurred in Lake Albert, and 10 species were endemic to Lake Turkana. Fish faunal diversity is relatively low, concurring with the geological evidence that the lake may have recently dried up completely, or may have become too saline to permit most fish to persist. A number of species found in Lake Albert are absent from Lake Turkana (26).

Lakes Edward and George. The fish fauna of the Lakes Edward and George is as diverse as its geological history. Some species in Lakes Edward and George are also found in Lake Albert (such as *Bagrus docmac, Oreochromis niloticus*), while other species (such as *Polypterus senegalus, Hydrocynus* spp.), and even families (such as Mastacembelidae, Characidae, Schilbeidae) that typify the Nilotic fauna from Lake Albert are absent. Many of the cichlids from Lakes Edward and George are closely related to species in Lake Victoria (2,12,13). In conjunction with geological evidence, this supports the idea that a very recent connection existed by which species moved from Lake Victoria to Lake Edward (2). As discussed previously, it is now generally agreed that Lake Victoria was formed primarily by the Kagera and Katonga Rivers when they were back-ponded 760,000 years ago (23). It is believed that the lake continued to flow westward until further uprising finally stopped and reversed the river flow (2). Connections between Lakes Victoria and Edward may have existed as recently as a few thousand years ago, but more likely during the early Pleistocene, approximately 1 million years ago (2).

It is not surprising that the fish fauna of the very productive Lake George is dominated by the herbivorous *Oreochromis niloticus* and *Enterochromis nigripinnis* (30). Commercially important fish in both lakes all originate from Lake Albert ancestors and include *Oreochromis niloticus, Barbus altianalis, Protopterus aethiopicus, Bagrus docmac*, and *Clarias lazera*. The number of endemic cichlids in the lakes is uncertain and requires further study. Until recently only 23 species of cichlids were reported from these lakes, five of which were considered to be common to Lake Victoria (12,13). More recently, estimates have increased to over 60 endemic cichlid species (27). Three of the five species thought to be shared with Lake Victoria have been redescribed as endemic to Lakes Edward and George (the other two are widely distributed throughout the basin - *Astatotilapia nubilis* and *Astatoreochromis alluaudi*). Occasionally massive fish kills of fish, without accessory respiratory organs, have been reported from Lake George (26) associated with the loss of water-column oxygen.

Lake Kivu. Lake Kivu has a relatively impoverished fish fauna and is considered ecologically immature due to its recent formation (2). The fish fauna is most closely aligned with Lake Edward, but a number of lineages present in Edward are lacking in Lake Kivu (families such as Lepidosirenidae, Mormyridae, Bagridae, Cyprinodontidae; 2). This is to be expected from the geological evidence that suggests that Lake Kivu was recently formed when volcanic actions damned a valley previously draining into Lake Edward (2). Families well represented in Lake Tanganyika (Characidae, Centropomidae, Mochokidae, Mastacembelidae) are absent both from Lakes Kivu and Edward. However, during the period that Lake Kivu has been isolated from Lake Edward, there has been some cichlid speciation - all six *Haplochromis* found in the lake are endemic (2). As with Lakes Edward and George, the fauna of Lake Kivu may have been reduced during periods of volcanic activity.

Freshwater herrings (*Stolothrissa tanganikae* and *Limnothrissa miodon*) were introduced to Lake Kivu from Lake Tanganyika in 1963 to fill a vacant niche for zooplanktivorous fishes. One species (*L. miodon*) became established, but was of low food value because it picked up a strong sulfur smell and taste from the volcanic waters of the lake (1).

Human Influences on African Great Lake Ecosystems

The Great Lakes do not end at the water's edge. These lakes are parts of a landscape that include terrestrial ecosystems as varied as tropical rainforest, savanna, desert, and human cities. The effects of human activities on the composition of the drainage waters hundreds of miles away directly influence the lakes' ecology. Human influences on these ecosystems have been increasing for a variety of reasons, all linked one way or the other to increasing the human population. The countries bordering these lakes have some of the highest population growth rates in the world (for example Uganda = 3.4% per annum, Kenya = 4.1%, Tanzania = 3.6%; Table 8.4).

Human activities have influenced the Great Lakes in a variety of ways. In some cases, the short term effects can, from one perspective, be viewed as very positive for the people involved. The natural resources of the basin have been used, and the capital reinvested in modernized infrastructure. Ecosystem changes and mechanization have produced vast expansions in fisheries and agricultural yields. The increase in food supplies and access to modern medicines facilitated by economic growth has allowed some people to live longer and have more productive lives.

Some would argue that some of these benefits, while desirable, have come at a high price, particularly since the development is probably unsustainable. In the Great Lakes Basins, nature is in retreat. Crocodile, whale-headed stork, python, monitor lizards, otter, hippopotamus, sitatunga, and a variety of other animal populations have been greatly reduced from precolonial levels. The forests that we see today are only remnants of much more extensive forest blocks that existed prior to clearance by man. East African forests have been affected by agriculturists for a considerable period. Evidence from pollen analyses, archeological digs, and linguistic studies suggest that widespread forest destruction occurred at least as far back as 4800 years ago (17). Little regard has been paid to the fact that resources in such forests are small, and deforestation in East Africa has been extensive (Uganda = 1.3% per annum, Kenya = 1.7%, Tanzania 0.7%; 28 and also see Chapters 1 and 2).

All of the lakes are threatened by pollution (7), increased sedimentation resulting from deforestation (6), and overfishing (7), and some lakes are threatened by potential oil exploration (Lakes Tanganyika, Malawi, and Albert). Perhaps one of the greatest threats to these ecosystems is the introduction of exotic species. The potential effect of species introductions is most dramatically illustrated in the lake that has experienced the greatest level of disturbance, the highest rate of species loss, and which is in the greatest danger of ecosystem collapse: Lake Victoria. Therefore, we focus this next section on the recent history of Lake Victoria.

Table 8.4. Vital statistics for selected African countries: their size, population characteristics, forests, and deforestation rates.

	Uganda	Kenya	Tanzania	Rwanda	Burundi	Malawi	Mozambique
Country Size (km^2)[1]	236,578	582,645	939,762	26,330	27,835	94,080	784,755
Population (1989), millions[1]	17	24.1	26.3	7.0	4.8	7.3	13.4
Annual Growth Rate[1]	3.4%	4.1%	3.6%	3.4%	3.3%	3.3%	2.6%
Doubling Time, years[1]	20	17	19	20	21	21	27
Pop. Dens. per km^2	71.9	41.4	28.0	265	196	93	19
Percent Urban	9%	19%	19%	6%	5%	13%	20%
Land Use							
Percent Cultivated[1]	28	4	5	38	47	20	4
Percent Pasture[1]	21	6	37	16	33	16	55
Original Forest (km^2)[2]	103,400	81,200	176,200	9,400	10,600	10,700	246,900
Est. of Remaining Forest	7500	6900	14,400	1554	424	320	9350
Defor. Rate (%/yr)[5]	1.3	1.7	0.7	2.7	2.7	....	1.1
Forest Rem. 1994[3]	6245	5428	13050	1057	288	....	8005
Prot. Areas(km^2)[4]	6084	13,148	77,008	150	379	....	....

[1] Stuart et al. 1990; [2] Groombridge 1992; [3] Based on FAO estimated Remaining Forest (1980) and on Deforestation Rates; [4] Tanzania and Kenya are from Groombridge 1992 and probably include many parks with little area of forest, Uganda is from the data in Howard 1991, and includes only forested areas; [5] McNeely et al. 1990 and FAO 1981

Lake Victoria - Brief History of Human Disturbance

In the last one hundred years Lake Victoria has experienced dramatic changes in both its limnological parameters (such as dissolved oxygen, productivity) and native fishery stock (22,23,34,44). Overfishing was the cause of the earliest recorded anthropogenic changes in the fish fauna. In the early part of the century the British introduced gill nets and other modern fishing methods to the area. This led to the rapid depletion of important anadromous fishes (*Barbus altianalis* and *Labeo victorianus*) that once made spectacular spawning runs up rivers that flow into Lake Victoria, and *Oreochromis esculentus* that was a major food fish in the lake (34). To replace these devastated fish stocks, several exotic species of tilapia were introduced into Lake Victoria in the 1950s (27,33). These introductions were slow to take hold, but eventually *Oreochromis niloticus* established itself and became an important part of the fishery. Also around this time, Nile perch, *Lates niloticus*, was introduced into Lake Victoria (33). Nile perch was slow to take hold and persisted as a minor component of the fauna for decades until the early 1980s, when their populations increased dramatically (22,34,44). The Nile perch population explosion is suggested to be the leading cause for the demise of the native fishes (1,34). The decline in native fishes, and the increase in Nile perch populations are almost perfectly reciprocal (34,44), and the replacement occurred very rapidly. In 1978, the haplochromine fauna were intact and contributed about 80% of the yield and Nile perch less than 2%, with the remainder consisting of the introduced *Oreochromis niloticus* and native non-cichlids (44). By the mid 1980s (22,34,44), the haplochromine fish community had been virtually destroyed, and the Nile perch comprised better than 80% of the catch (34). The remaining 20% consisted of *Oreochromis niloticus*, the tiny native cyprinid, *Rastrineobola argentea*, and a small remnant of other native fishes.

The Nile perch is a top predator that grows up to 2 m in length and to a weight of more than 100 kilograms and feeds on a wide range of fish and invertebrate species. It has been recorded to eat mormyrids, *Alestes*, *Barbus*, *Clarias*, *Protopterus*, *Rastineobola*, *Tilapia* sp. *Oreochromis* sp. and a variety of other fish species including its own young (33). It ingests large volumes of fish; in the stomach of a 110 cm long Nile perch, Okedi (36) recorded 57 *Haplochromis*.

The above situation is, however, not so clear cut. Coincident with the Nile perch explosion, was a dramatic change in the physical environment of the lake. In the early 1980s, the regular appearance of dense algal blooms was associated with low oxygen levels, and fish kills (31,32). Prior to 1978, the lake was well-mixed, though oxygen fell to lower levels in deeper waters (> 60 m) for brief periods during the rainy seasons (41). Now, Lake Victoria appears to be stratified for the entire year (3,18,22). Water depths between 25 and 50 m have recently experienced increases in the abundance of *Cardina niloticus* (the lake's small, native detritivorous shrimp, that appears to be highly tolerant of low oxygen conditions; 34).

Recently, water hyacinth (*Eichhornia crassipes*) was introduced into the lake. Water hyacinth occurs naturally in the floodplains of South America, where its spread is limited by naturally occurring predators, fungi, and diseases. It was introduced to Africa at the turn of the century and was first reported in Lake Victoria in 1990, where it is believed to have invaded from the Kagera River flowing into the lake from Rwanda. Unrestricted by nutrients, and natural enemies, a single plant can produce 140 million daughter plants per year, enough to cover 140 hectares with a wet weight of 28,000 tons. The plant also produces seeds that can survive in the mud for 30 years, creating severe problems of reinfestation after elimination programs (World Bank Unpublished Report).

The water hyacinth has been established in Lake Victoria for only a few years, but it has proliferated at a great rate. It is now a common feature along shores in Uganda and Tanzania and has made its appearance in Kenya. Hyacinth floats on the water surface, often forming dense floating mats of vegetation. Hyacinth has a number of effects on the lake by reducing light penetration, limiting water-column, mixing, and increasing detrital inputs. By shading out the sun it is likely to provide concealment to ambush predators that feed on indigenous species, such as the black bass (*Micropterus salmoides*) - now established in a Kenyan river that feeds into Lake Victoria. Further it may shade out other forms of bottom-dwelling vegetation that often is located just outside of the papyrus beds (such as *Ceratophyllum*). If this is the case, it may reduce the value of the shore margins as refugia from predation. At this time the direction of these relationships is speculative; data verifying the effect of hyacinth on the ecosystem are required.

A number of other aspects of the lake's ecology also appear to have changed: productivity and turbidity have both increased, papyrus swamps are on the decline, aquatic snails are increasing in abundance, the land surrounding the lake has been extensively deforested, and human populations bordering the lake have risen dramatically.

Arguments concerning the introduction of Nile perch and Nile tilapia into Lake Victoria are in many ways moot, since their introduction is irreversible. From a fisheries perspective, the introductions may have been beneficial, at least on the short term. With the establishment of the Nile perch and Nile tilapia fisheries, the total quantity of fish landed from the waters of all three countries has increased between two and six fold (35). In Kenya the catch increased from 20,000 tons in 1977 to 123,000 tons in 1988 (35). In Uganda the catch increased from 17,000 tons in 1983 to 132,000 tons in 1988; and in Tanzania the catch increased from 64,000 tons in 1982 to 89,000 tons in 1985 (35).

Recent statistics indicate that in Lake Victoria over 100,000 artisanal fishermen using more than 20,000 small craft caught an estimated 550,000 tons of fish in 1990. The estimated market value of this catch is approximately US $280 million (World Bank Report). This situation has led to the rapid, and virtually unchecked, development of fish processing capacity along the lake shore, including large-scale trawling, and fish freezing plants. This has occurred despite variable growth rates in the fisheries industry. There is, however, good reason to believe that the Nile perch fishery will crash in the near future. It is believed that the Nile perch stocks are being over-fished. This has been intensified by the use of small mesh sizes that capture immature fish (New Vision, Sept. 4, 1992). There is limited quantitative data since the last comprehensive assessment of fish stocks dates back to 1971, when the lake was a multispecies fishery based on indigenous fishes. Accelerated efforts in faunal survey and fisheries stock assessment are required to provide information necessary for the development of informed conservation and management plans for the lake and region as a whole.

The most likely scenario for the future of the Lake Victoria fisheries seems to be one of instability. If the Nile perch fishery crashes, there will be a period of extremely low catch. Whether fish shortage will cause the governments of these three countries or other agencies to advocate new introductions is not known. Such introductions should be discouraged, since their long-term effects on the ecosystem are unknown. Nonetheless, even if introductions are not advocated by the appropriate governmental bodies, introductions may still be made since the control of such a large body of water will be difficult.

In any event, virtually all stakeholders would regard a stable market as better than an irruptive one. Thus, the more that is known about the present ecosystem, the better we can predict the future course of the lake, and potentially limit future instability and low catches.

Conservation of East Africa's Great Lakes

The East African Great Lakes provide many goods and services to the people who live around them including: food, water, waste dilution, transport, hydroelectric power, building materials (such as papyrus), recreation, tourism, and aquarium fishes as an export commodity. The countries in East Africa are striving for sustainable economic growth, while at the same time advocating the conservation of existing habitats. These goals are in many ways incompatible, creating tension between different factions. For example, to preserve the watershed and the quality of the water draining into the lakes, advocates for conservation of the aquatic systems favor the halt of forestry activities and the reforestation of habitats that ring the lake to prevent erosion, retard nutrient inputs, and to restore the populations of terrestrial organisms, like hippopotamus, that probably played a significant role in the ecosystems. Opposed to such plans are the powerful economic sectors in favor of agricultural and industrial development.

Opposing sectors represent differing philosophies with respect to the wise use of resources. Conservationists might argue against development schemes, on the basis that the costs in terms of ecological degradation are too high for the returns and that the development will not be sustainable. If the Nile perch fishery in Lake Victoria does crash, it could be heralded as an example of such a sequence. It seems, however, that advocating non-development is inappropriate, if not hypocritical. Thus, we must look for alternatives that are based on appropriate technologies and indigenous species. For example, there could be development of commodities based on native products like papyrus and aquaculture of indigenous food species. It is theoretically possible to draw human sustenance from intact, native ecosystems without bringing about their demise. Unfortunately, it seems probable that such an approach will not work at the present time, for two reasons; 1) the knowledge and experience to conduct projects to the economic scale that is necessary does not presently exist, and 2) the volume of goods and services derived from the lakes will, by necessity, increase as a result of human population growth.

There is, however, good reason to believe that his situation will not continue. Population growth rates are declining, and technological systems are being developed to more efficiently use existing resources. The term sustainable development is, however, an oxymoron that will remain so until demand can be brought into line with ecological production. Thus, we suggest that agencies advocating "sustainable development" must first employ "ecological first-aid" to maintain or restore ecological systems and both conserve and develop options for future economic growth. Parallel efforts are the key to success in this endeavor.

One priority is the preservation of remnant fragments of intact natural ecosystem in protected areas. This is a temporary step, since such fragments are rarely large or self-sufficient enough to remain intact for long. A second defense is the development of extractive reserves or "managed exploitation areas". These areas should be regulated to provide materials for local use as best as the ecosystems can provide. Protected areas are often created for their long-term benefits. However, they also have the immediate value of providing a source of colonizers for managed and restored habitats. Third, the deployment of extractive reserves must be planned within the context of a desired end product, and an ecological landscape that has been recast so as to meet the needs of a large human population. Defining and managing this landscape constitutes the most challenging of all the tasks currently faced by environmental scientists and decision makers working in the Lake Victoria region.

In subsequent discussions we will attempt to illustrate ways in which such a management system could aid in long-term conservation goals, provide future options for sustainable development, and contribute to immediate needs. Let us begin by examining two examples of potential management systems that could be employed in "Managed Exploitation Areas" for Lake Victoria.

Sustainable Resources

Fish Ponds. The development of an extensive fish aquaculture industry would seem to be a prudent advancement. Conducted on a local level, this development could provide a much needed protein supply for local villagers, and potentially reduce the pressure on the existing lake fisheries. Aquaculture conducted on a larger scale has the potential of generating significant revenues. Such a development may be particularly timely, if the Nile perch fishery declines and leaves behind an extensive processing and export infrastructure. For aquaculture to be successful on the scale needed to reduce fishing pressure, wide-ranging development, education, and technological assistance are required.

Oreochromis niloticus is an ideal fish for aquaculture. An advantage of farming this species is that it is already in Lake Victoria, so there is no danger of accidental introduction. Local people are familiar with the species, and it is already in local markets. In addition *O. niloticus* breeds quickly and easily, thrives on cheap, readily available plant foods, and produces high yields. This species is both tolerant of wide temperature and salinity ranges, and low oxygen.

Aquaculture development endeavors could be situated to assist conservation efforts. It has recently been advocated (Lake Victoria Workshop, August 1992) that "Fish Parks" be established in some areas to preserve the existing haplochromines and in other areas to maintain breeding stock for commercially important species (21). If these parks were established, the economies of some villages may be negatively affected, at least on the short term until tourism or other options could be developed. Fish ponds could replace the income of villagers who relied on areas designated as fish parks.

Papyrus Swamps. Papyrus (*Cyperus papyrus*) swamps are extensively distributed around the shores of Lake Victoria and throughout wetter regions of East Africa (2,42). In Uganda alone, it is estimated that papyrus swamps occupy more than 10% of Uganda's land surface (16) with only 2.25% of this area protected in National Parks or Reserves (see Chapter 9). The swamps provide numerous benefits by maintaining microclimate, preventing flooding, purifying water, reducing sediment loading into lakes, supplying fish, and producing reeds for mats, roofing, and export. The commercial use of papyrus need not disrupt the ecosystem.

Papyrus is the fastest growing herb in the world (42), and unlike forested areas, nutrients are continuously supplied to harvested areas by flowing water.

In areas where there is high population pressure and thus high demand for agricultural land, swamps have been drained (4). However, the long-term fertility of drained swamps is questionable (42), and in many cases the reclaimed soils are not very productive in the first place (42). Drainage schemes have led to soil erosion, decreased water table levels, loss of water supplies for rural communities, and soil acidification (42). One solution to excessive swamp drainage is to increase the economic use and sustainable management of papyrus, thereby ensuring preservation of swamp ecosystems (42). Presently the technology exists to use papyrus for commercial specialty paper production and for fuel briquettes that would decrease the pressure placed on existing forests.

Interim Solutions, the Role of Scientists, and Refugia. It has been extremely difficult to develop scientifically rigorous predictions concerning ecosystem function, particularly predictions accurate over a short time and a large geographical area. However, to be able to understand and potentially manage the Great Lakes, this is exactly what must be done. We propose that the scientific community can most directly contribute to the conservation of the Great Lakes by resolving key unknowns in the ecosystem. For example, how will pollution or increased sedimentation affect limnological parameters, water quality, and species interactions. Additionally, what are the social implications of changes in the fisheries industry, and at what level of exploitation is the fisheries industry sustainable? For the remainder of the paper we focus on how scientists can contribute to the survival of endangered or threatened faunas.

The most direct means of conserving an endangered fauna is to identify and protect areas where the fauna can persist (refugia) even if perturbations to the ecosystem are not brought under control. Such an approach has been advocated for the Great Lakes a number of times. For Lake Tanganyika, rocky shores in sections of erosion-resistant watershed would be most suitable for protection. Rocky areas would have low levels of sedimentation even if deforestation was to occur (6). The extension of existing terrestrial park boundaries into the water is also feasible, since the watershed is also protected (7). In Lake Malawi, emphasis may be most appropriately placed on protecting areas harboring speciose and endemic faunas (38). Expansion of the Lake Malawi National Park boundaries to include such areas as the Chisumulu Islands and additional mainland rocky areas are good conservation possibilities (38).

In Lake Victoria, the situation becomes very complex, since Nile perch, the major threat to the fauna, is mobile and can very easily invade protected areas. As a result, identification of refugia will necessitate determining what sorts of areas Nile perch will not invade. Considering what is known about Nile perch and the changes that have occurred in the lake since their introduction, the following assumptions are tenable.

(1) Some haplochromines can tolerate oxygen levels as low as 1 ppm for an unknown period of time (5).
(2) Adult Nile perch require water with high dissolved oxygen since their blood has a low affinity for oxygen (8).

(3) Nile perch of sufficient size to take large, inshore species inhabiting-shallow rocky shores or swamps cannot prey effectively because of the structural complexity of these areas.

(4) Nile perch are less effective predators in riverine habitats due to the high ratio of marginal refugia to open water.

Based on these assumption, escape from Nile perch can be achieved by the restriction of activities to structural refugia, or (for species tolerant of low oxygen) restriction to portions of the water column where dissolved oxygen concentration is low, or by migration and colonization of riverine habitat. In consideration of these factors, we predict the existence of four refugia: 1) structural refugia, 2) low oxygen refugia, 3) peripheral refugia behind anoxic/hypoxic or geographical barriers, and 4) riverine refugia.

There is accumulating evidence that papyrus swamps can form significant barriers to the dispersal of fish species intolerant of low oxygen conditions. For example, recent collections identified remnant populations of *Oreochromis esculentus*, as well as endangered haplochromines, in Lake Kanyaboli (a small satellite lake joined to Lake Victoria by the Yala River). It is likely that these populations exist because the Nile perch has not been able to disperse through the Yala swamp. We predict that shallow-water prey species tolerant to low oxygen conditions may find refuge from predation in the enormous area of peripheral swamps. Laboratory experiments on the responses of some species Lake Victorian cichlids to progressive and acute hypoxia have demonstrated that some of these fishes can tolerate very low oxygen levels (5). We also predict that fishes tolerant of chronic and/or acute hypoxia may disperse through peripheral swamps to find refuge in hypoxic lagoon areas behind the fringing swamp. Prey species tolerant of low oxygen may also find refuge from predation by inhabiting the deeper waters of the lake.

Areas that have complex spatial elements may be serving as refugia from the Nile Perch. Previous studies have found evidence that the more tolerant haplochromines and the Nile perch are mainly rock-dwelling species and species inhabiting inshore areas close to vegetation (34,44).

Many of the non-cichlids fishes once made spawning runs up the rivers that flow into Lake Victoria (12). Recently there is evidence that some of these species may not be descending back into the lake. It seems reasonable to speculate that rivers are serving as refugia for these species from Nile perch predation, since Nile perch continue to show restricted use of riverine habitat. Examples may include *Labeo victorianus*, *Barbus* spp., *Schilbe mystus*, and various mormyrids (34). It is possible that viable populations of such species exist in the rivers. Nonetheless, they are being prevented from re-establishing in Lake Victoria by the Nile perch that prey on individuals that disperse back into the lake.

Conclusions

The Great Lakes of East Africa are windows on the relationship between biological diversity and the welfare of humans. The lakes are some of the most biologically diverse ecosystems known. For example, Lake Malawi hosts more than 500 fish species, and Lake Victoria harbors more than 400 cichlid species that have evolved in less than 200,000 years. Similarly, the fishes and snails in Lake Tanganyika are so varied in form that the members of a single family might easily be mistaken for unrelated taxa. But this rich biological diversity is threatened by human activities. The countries in East Africa are striving for sustainable economic growth, while at the same time advocating the conservation of existing habitats. These goals are in many ways incompatible, thus creating tension between different factions. We suggest that agencies advocating sustainable development must first repair ecological damage to both conserve and develop options for future economic activity. Such a process should involve the development of 1) a series of "Protected Areas" that will serve as source of species diversity, preserving options for the future, 2) a series of "Managed Exploitation Areas" that would be regulated to provide materials for local use as best they can, and 3) a vision of the long-term goals for the ecological landscape: including patterns of land use, watershed changes, and incorporation of human needs and activities as an integral component of the landscape model. Scientists can contribute most directly to these goals by quickly and responsively resolving key unknowns in the ecosystem. In particular, the identification and protection of faunal refugia are critical to the persistence of threatened faunas.

References

1.   Balon, E.K., Bruton, M.N. 1986. Introduction of alien species or why scientific advice is not heeded. *Environmental Biology of Fishes* 16:225-230
2.   Beadle, L.C. 1981. *The Inland Waters of Tropical Africa. An Introduction to Tropical Limnology*. London: Longman
3.   Bootsma, H.A., and Hecky, R.E. 1993. Conservation of the African Great Lakes: A limnological perspective. *Conservation Biology* 7:644-656
4.   Bugenyi, F.W.B. 1991. Ecotones in a changing environment: Management of adjacent wetlands for fisheries production in the tropics. *Verhandlungen Internationale Vereinigung Limnology* 24:2547-2551
5.   Chapman, L.J., Liem, K.F. in press. Papyrus swamps and the respiratory ecology of *Barbus neumayeri*. *Environmental Biology of Fishes*
6.   Cohen, A.S., R. Bills, C.Z. Cocquyt, Caljon, A.G. 1993. The impact of sediment pollution on biodiversity in Lake Tanganyika. *Conservation Biology* 7:667-677
7.   Coulter, G.W., Mubamba, R. 1993. Conservation in Lake Tanganyika with special reference to underwater parks. *Conservation Biology* 7:678-685
8.   Fish, G.R. 1956. Some aspects of the respiration of six species of fish from Uganda. *Journal of Experimental Biology* 33:186-195
9.   Fryer, G. 1960. Concerning the proposed introduction of Nile perch into Lake Victoria. *East African Journal of Agriculture* 25:267-270
10.   Fryer, G., Iles, T.D. 1972. *The Cichlid Fishes of the Great Lakes of Africa: Their Biology and Evolution*. London: Oliver and Boyd
11.   Greenwood, P.H. 1965. The cichlid fishes of Lake Nabugabo, Uganda. *Bulletin of the British Museum Natural History (Zoology)* 12:315-357
12.   Greenwood, P.H. 1966. *The Fishes of Uganda*. Kampala: Uganda. The Uganda Society
13.   Greenwood, P.H. 1973. A revision of the *Haplochromis* and related species (Pisces: Cichlidae) from Lake George, Uganda. *Bulletin of the British Museum Natural History (Zoology)* 25:141-242
14.   Greenwood, P.H. 1974. The cichlid fishes of Lake Victoria, East Africa: The biology and evolution of a species flock. *Bulletin of the British Museum Natural History (Zoology)* 26:1-134

15. Groombridge, B. 1992. *Global Biodiversity: Status of the Earth's Living Resources.* London: Chapman and Hall

16. Gumonyi-Mafabi, P. 1992. *Conservation of Biodiversity in Uganda.* Makerere Kampala: University Press

17. Hamilton, A.C. 1974. Distribution patterns of forest trees in Uganda and their historical significance. *Vegetatio* 29:21-35

18. Hecky, R.E., Bugenyi, P.W.B. 1991. Hydrology and chemistry of the African Great Lakes and water quality issues: Problems and solutions. *Mitteilungen Internationale Vereinigug fur Theoretische und Angewandte Limnologie* 23:45-54

19. Hori, M., M.M. Gashagaza, Nshombo, M., Kawanabe, H. 1993. Littoral fish communities in Lake Tanganyika: Irreplaceable diversity supported by intricate interactions among species. *Conservation Biology* 7:657-666

20. Howard, P.C. 1991. *Nature Conservation in Uganda's Tropical Forest Reserves.* IUCN, Switzerland: Gland

21. Kaufman, L., Ochumba, P., Ogutu-Ohwayo, R. 1993. *People, Fisheries, Biodiversity and the Future of Lake Victoria.* Boston: New England Aquarium

22. Kaufman, L.S. 1992. The lessons of Lake Victoria: Catastrophic change in species rich freshwater ecosystems. *Bioscience* 42:846-858

23. Kendall, R.L. 1969. An ecological history of the Lake Victoria Basin. *Ecological Monographs* 39:121-176

24. Livingstone, D.A. 1965. Sedimentation and the history of water level change in Lake Tanganyika. *Limnology and Oceanography* 10:607-609

25. Lowe-McConnell, R.H. 1969. Speciation in tropical freshwater fishes. *Biological Journal of Linnean Society* 1:51-75

26. Lowe-McConnell, R.H. 1975. *Fish Communities in Tropical Freshwaters.* London: Longman

27. Lowe-McConnell, R.H. 1993. Fish faunas of the African Great Lakes: Origins, diversity, and vulnerability. *Conservation Biology.* 7:634-643

28. McNeely, J., Miller, K., Reid, W., Mittermeier, R., Werner, T. 1990. *Conserving the World's Biological Diversity.* Gland: IUCN Publishers

29. Meyer, A., Kocher, T.D., Bassasibwaki, P., Wilson, A.C. 1990. Monophyletic origin of Lake Victoria cichlid fishes suggested by mitochondrial DNA sequences. *Nature* 347:550-553

30. Moriarty, D.J.W., Darlington, J.P.E.G., Dunn, I.G., Moriarty, C.M., Tevlin, M.P. 1973. Feeding and grazing in Lake George, Uganda. *Proceedings of the Royal Society of London (B)* 184:299-319

31. Ochumba, P.B.O. 1987. Periodic massive fish kills in the Kenyan part of Lake Victoria. *Water Quality Bulletin* 12:119-122

32. Ochumba, P.B.O., Kibaara, D.I. 1989. Observations on blue-green algal blooms in the open waters of Lake Victoria, Kenya. *African Journal of Ecology* 27:23-34

33. Ogutu-Ohwayo, R. 1990. Changes in the prey ingested and the variation in the Nile perch and other fish stocks of Lake Kyoga and the northern waters of Lake Victoria (Uganda). *Journal of Fish Biology* 37:55-63

34. Ogutu-Ohwayo, R. 1990. The decline of the native fishes of Lake Victoria and Kyoga (East Africa) and the impact of introduced species, especially the Nile perch, *Lates niloticus,* and Nile tilapia, *Oreochromis niloticus. Environmental Biology of Fishes* 27:81-96

35. Ogutu-Ohwayo, R., R.E. Hecky. 1991. Fish introductions in Africa and some of their implications. *Canadian Journal of Fisheries Aquatic Sciences* 48:8-12

36. Okedi, J. 1971. *Further Observations on the Ecology of the Nile Perch (Lates niloticus Linne) in Lakes Victoria and Kyoga.* East African Freshwater Fisheries Research Organization, Annual Report 1970

37. Owen, R.B., Crossley, R., Johnson, T.C., Tweddle, D., Kornfield, I., Davidson, S., Eccles, D.H., Engstrom, D. 1990. Major low levels of Lake Malawi and their

          implications for speciation rates in cichlid fishes. *Proceeding of Royal Society
          of London* 240:519-553

38.   Reinthal, P. 1993. Evaluating biodiversity and conserving Lake Malawi's cichlid
          fish fauna. *Conservation Biology* 7:712-718

39   Ribbink, A.J. 1988. Evolution and speciation of African Cichlids. In *Biology and
          Ecology of African Freshwater Fishes*, eds. Leveque, C., Bruton, M.N.,
          Ssentongo, G.W., pp.35-51. Institut Francais de Recherche Scientifique pour le
          Developpement en Paris: Cooperation

40.   Stuart, S.N., R.J. Adams, Jenkins, M.D. 1990. *Biodiversity in Sub-saharan African
          and its Islands: Conservation, Management, and Sustainable Use.* Occasional
          Papers of the IUCN Species Survival Commission

41.   Talling, J.F. 1966. The annual cycle of stratification and phytoplankton growth in
          Lake Victoria, East Africa. *Internationale Revue der gesamten Hydrobiologie*
          51:545-621

42.   Thompson, K. 1976. Swamp development in the head waters of the While Nile. In
          The Nile: Biology of an Ancient River, ed. Rzoska, J. pp.177-196. Hague: Junk

43.   Trewavas, E. 1947. Speciation of cichlid fishes of East African Lakes. *Nature*
          160:96

44.   Witte, F.T. Goldschmidt, J. Wanink, M.V. Oijen, K. Goudswaard, E. Witte-Mass,
          Bouton, N. 1992. The destruction of an endemic species flock: Quantitative data
          on the decline of the haplochromine cichlids of Lake Victoria. *Environmental
          Biology of Fishes* 34:1-28

Chapter 9

Freshwater Wetlands and Marshes

David Harper & Kenneth Mavuti

The term 'wetland' encompasses a wide range of vegetation types characterized by soils permanently or temporarily waterlogged, by either salt (such as mangroves and marshes) or fresh water (such as swamps and pet bogs). Several of these types of wetland are the focus of separate chapters in this book (inter-tidal, riverine forests, shallow lakes) and so the definition used by this chapter is that freshwater wetlands are 'temporarily or permanently wet ecosystems dominated by emergent grasses and reeds or swamp vegetation'. The wetlands of this chapter are closest to the ecosystem defined by the International Biological Program (which promoted ecosystem research in the decade 1964 to 1974) rather than the broader definition used by the Ramsar Convention on Wetlands of International Importance in 1971 which is the basis of modern wetland conservation. The Ramsar definition would encompass the full range of wet ecosystems described in this book (Chapters 4,5,6,14).

Waterlogged soils that encourage wetland vegetation occur because their drainage is impeded. In the most obvious case impeded drainage occurs at the margins of lakes and rivers, and is responsible for the zonation of plants that follows the water table and water depth. Less obviously, there may be a surfacing of ground water or blockage of stream water flow that leads to the formation of bogs at higher altitudes (such as on the mountain blocks the Aberdares and Mt Kenya) or swamps in valley floors, as were once common in Uganda. Seasonal riverine swamps occur when water inflow exceeds outflow as in alluvial floodplains or inland deltas. Whatever the type of wetland, they are communities at the edge (or ecotone) between dry land and open water (see Chapter 5). Ecotones are often highly productive and rich in unique species, drawing species from both adjacent ecosystems, and wetlands world-wide are no exception. In the oldest agricultural cultures of Europe, Asia, and Africa artificial wetlands, in the form of shallow ponds, played an important role in fish protein production, often in rotation with crop production on a deliberate filling and emptying cycle.

Wetlands have come to the forefront of the conservation agenda in the last two decades because of their rapid rate of destruction caused by three main activities:

(1)   drainage for agriculture (their soils are usually very fertile because of their past retention of nutrients and vegetation detritus and they provide a source of irrigation water),

(2)   development (their topography is usually flat and lowering of their outlet height can easily drain them),

(3)   flood control (the construction of headwater reservoirs often drowns important wetland sites and the consequent downstream regulation of discharge removes flood peaks hence reduces seasonal wetlands).

The very real sustainable benefits of wetlands have often not been readily quantified economically (such as natural storage for drinking-water, the seasonal agriculture on floodplains, seasonal subsistence fisheries, maintenance of migrating fish, bird, mammal and human communities and even purification of effluents).

In the past decade, and particularly since 1991 (The Rio Convention), attitudes to natural ecosystems have changed, moving towards the concept of sustainable development rather than merely economic development. This means that, for wetlands which have survived the already massive destruction, goals are now sustainable multi-purpose uses. Such goals should retain the integrity of the basic wetland ecology and functioning while recognizing the need for an economic output.

The increasing interest in wetland ecology and their conservation has produced a numbers of recent books that provide detailed information about the structure, functioning, and conservation of such ecosystems in eastern Africa (32) with emphasis on vegetation (6), wetland fisheries (33) and to their wider ecology (4,30). A recent volume on wetlands of the earth contains much of interest for East Africa (25). The International Union for the Conservation of Nature (IUCN, the union of the world's conservation bodies), has been highly active in promoting regional conferences in eastern and southern Africa dealing with issues of wetland conservation (3,19,22,23). Consequently, at all educational levels material concerning wetland ecology and conservation is widely available from international and regional agencies such as IUCN, WWF (World Wide Fund for Nature) and IWRB (International Waterfowl Research Bureau).

Wetland Classification and Distribution

Wetland Types

The wetland types found in East Africa are formed by the same environmental gradients that prevail in other parts of the world – altitude, hydrological patterns, geology and soil mineral content. The predominant abiotic influence is usually local topography and discharge pattern rather than rainfall, since most wetlands are dependent for their water supply upon distant rather than local rainfall. At higher altitudes the predominant abiotic factors are rainfall intensity, topography and temperature. This makes montane wetlands distinct from those of lower altitude. There are six main types of freshwater wetlands in eastern Africa conveniently distinguishable without detailed taxonomic or hydrologic classification. These are: 1) lake-edge swamps, 2) valley swamps, 3) seasonal floodplains and inland deltas 4) small ponds and dams, 5) swamp forests, and 6) high-altitude swamps and mires. Riverine forests would form a seventh category covered in a later chapter (Chapter 14).

Lake and Riverine Edge Swamps. Shallow lake edges, particularly along the sheltered deltas and shores of inflowing rivers, often present the widest zonation of different wetland plants. This is because lakes have the greatest hydrological gradients, ranging from shallow water with floating and submerged plants (submerged plants may grow to depths of 15 m but more usually to around 5 m) to wet soils where the water table depth progressively decreases below the surface (a 1 m water table is approximately the depth limit for characteristically wetland plants). Variation in water level (and hence water table) is seasonal and also changes between years but the change is usually not very great (less than 1 m vertical change in a year). Most wetland plants can not cope with greater fluctuations in abiotic conditions associated with water-level changes (such as oxygen and nutrient concentration changes). The paucity of aquatic vegetation associated with reservoirs is attributed to excessive fluctuations in water level.

The main vegetation zones of lake margins are illustrated in Figure 9.1. From the lakeward side they are: submerged macro-algae, submerged (both rooted and non-rooted) angiosperms (flowering plants), floating-leafed (but rooted) angiosperms, free-floating (non-rooted) angiosperms and ferns, emergent angiosperms (nearly all monocotyledons), water-tolerant grasses and herbs, water-dependent trees and shrubs.

Submerged macro-algae, members of the family Characeae or stoneworts, are generally found in moderately hard waters (their English name derives from their stony feel that is caused by their incorporating of calcium carbonate in their tissue). In Lake Naivasha, Kenya, two genera occur, *Nitella* and *Chara*. Stoneworts occur in depths below 2 to 3 m presumably because they have no rooting structure and are thus susceptible to wave action (although calcium carbonate makes them heavy and helps them resist currents). Such species often occur associated with beds of the angiosperm *Ceratophyllum demersum*, itself characterized by the lack of a root system and living in deep water (up to 8 m in Lake Bunyoni, Uganda; 5). *C. demersum* together with *Utricularia spp.* (bladderwort, so-named because of small vesicles that passively trap and then digest zooplankton) have also been recorded in rapidly-changing water regimes, such as newly-flooded reservoirs, precisely because they are not confined by roots (for example Kainji Lake, Nigeria; 7).

Species (and hybrids) of the genus *Potamogeton* are the dominant submerged plants in East African shallow lake edges. The exact species composition of submerged plants is determined by factors that are not fully understood, although ionic strength and composition are probably the most important abiotic factors. Denny (7) compared ionic gradient (conductivity from 0.05 to 20 x 10^3 µS/cm) in a series of east and southern African lakes with species distribution and demonstrated the tolerance of *C. demersum* and *P. pectinatus* to a wide range of ionic strengths. In contrast, *Ruppia* species were confined to brackish water and *Myriophyllum* species were confined to the fresher water. More recently, the increasing salinity of Oloidien Lake, an enclosed basin at Lake Naivasha, from around 300 µS/cm ten years ago to 2000 at the present time, has been marked by the progressive loss of first *Potamogeton octandrus*, then *Najas pectinata*, and then *P. schweinfurthii* until only *P. pectinatus* remains at the current ionic composition (14). Table 9.1 gives an indication of the ionic tolerances of different submerged and floating-leafed macrophyte species in East African lake-edge communities.

Most submerged species have flower heads that project above the water surface for pollination and seed dispersal (although many can also reproduce vegetatively). Some species, such as *Potamogeton octandrus*, support their flower heads on floating leaves and others such as *P. thunbergii* are fully floating-leafed. The most widely known floating-leafed species are the water lilies that, in many lakes, form a dominant zone in depths of 1 to 2 m between submerged beds and emergent swamp. In low-gradient littoral zones, where shallow open-water lagoons and swamp are intermingled, they may cover the lagoon surface. Associated with these littoral zones are free-floating plants that range in size from *Wolffia spp.*, a duckweed 1 mm across, to the alien *Eichhornia crassipes* (water hyacinth; a native of South American river systems) which, in favorable environments, can produce clumps several meters across. Many African lakes and reservoirs have been dominated by free-floating plants, mainly a consequence of two successful alien species invasions; the free-floating *Salvinia molesta* and *Eichhornia* (24). *S. molesta* acquired the nick-name 'Kariba weed' because of its effect on that newly-flooded Zambian reservoir in the 1960s. Both species evolved in river systems rather than lakes, where dispersal was by seasonal flooding, and so both have adaptations for the rapid colonization of newly created water bodies and of inter-locking plants to form a dense mat that counteracts the absence of sediment anchoring by roots. Each species does. however, have limited ability to root in the substrate and in lakes with water level fluctuations and draw-down zones, such as Naivasha, this ensures survival of plants from one dry spell to the next. *S. molesta* is limited to wet mud, but *E. crassipes* has a more extensive, true root system, and can survive on apparently dry soils until overgrown by semi-terrestrial grasses.

Emergent plants grow from around 1 m in water depth to saturated soils above the lake edges. The most well-known species in eastern Africa is papyrus, *Cyperus papyrus*, that achieves heights of around 5 m when fully mature (9 m under optimal growing conditions; 37) and forms impenetrable stands of swamp. Several other species may also occur under different conditions of hydrology or chemistry*Phragmites* (the cosmopolitan reed *P. australis*) and *Typha* (reedmace or cattail) approach papyrus in height. *Miscanthidium* and *Vossia* are smaller grasses. *Miscanthidium spp.* are characteristic of more acid, low-nutrient waters, while *Vossia cuspidata* is particularly well adapted to unstable hydrologies. It has a rapid growth rate combined with a long, flexible stem that allows it to follow the rise and fall of water levels; it is more common than other species in fluctuating lakes such as Chad and in river channel edges. It tends to be displaced or restricted to the water's edge under more stable hydrological regimes. Most emergents tend, particularly when mature, to occur in single-species stands that exclude other species. On the swamp's outer edges they may be associated with climbers such as bindweeds and smaller herbs and grasses that germinate in patches of soil and organic debris at the base of emergent plants.

Many water edges, particularly with fluctuating levels, contain emergent swamp plants forming a mat that lies on the ground at low water and floats when the water rises. Parts of the mat may then break off and form floating islands, moving across the water surface for several months until stranded. This is a characteristic feature of shallow lakes such as Naivasha, or large riverine wetlands such as those of the Nile in the Sudd, southern Sudan (18). A large number of amphibious plant species may become associated with such floating swamps and moving islands because of the greater swamp edges and the creation of shallow lagoons: Gaudet (10) catalogued the diverse flora of Lake Naivasha during a draw-down phase, and was able to provide a complete cross-section and species list of lake edge vegetation zonation at that time (see Fig. 9.1).

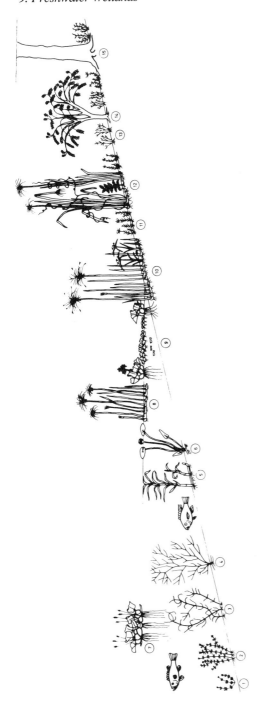

Figure 9.1. Typical succession from 4-5 meters below to 4-5 meters above the water surface in a relatively stable wetland ecotone. The example used is Lake Naivasha. The deepest water is colonised by rootless charophyte algae (1) and *Ceratophyllum demersum* (2) and is a feeding station of open-water fish such as the large-mouthed bass *Micropterus salmoides*. The submerged macrophyte zone proper is colonised by plants which reach the surface to flower such as *Naias pectinata* (3), *Potamogeton pectinatus* (4), and *Potamogeton schweinfurthii* (5); within the plant beds occur bottom-feeding tilapia species. Inside the submerged plants is a zone of floating-leaved plants lilies (6) intermingled with mats of floating hyacinth (7) on which marsh grasses and herbs germinate. Floating or partially-grounded islands of papyrus (8) are trapped among the shallow water vegetation, forming sheltered lagoons which become densely covered by *Eichhornia* and *Salvinia* (9) and these warm, rich habitats become rapidly colonised by juvenile fish and amphibians. At the lake edge, a rooted swamp of *Cyperus papyrus* and *Cyperus dives* (10) has been present since 1988; the latter seems to germinate more rapidly after recent inundation of wet soils but is progressively replaced by the former. On the inside of lake-edge swamps is a pasture of water-dependent grasses such as *Cynodon dactylum* (11) with clumps of old papyrus (12) stranded by former lake rises. These survive for at least ten years if not cleared or grazed, progressively colonised by climbing plants such as *Ipomea* and by the seedings of shrubs such as *Acacia* and *Sesbania*. On dryer ground *Cynodon* gives way to rain-dependent grassland species such as *Hyparrhenia* (13) underlying shrubs such as *Cassia* (14) or *Sesbania*. The whose succession is bounded by a characteristic fringing woodland of 'fever trees' - *Acacia xanthophloea* (15) which are capable of rooting to the permanent water table depth.

Table 9.1. Examples of the salinity tolerance range of submerged and floating-leafed macrophyte species in East African freshwater wetlands (from 2).

Species	Conductivity range, $\mu S \ cm^{-1}$
Ruppia spp	$8 - 20 \times 10^3$
Potamogeton pectinatus	$0.2 - 10 \times 10^3$
Ceratophyllum demersum	$0.05 - 1.5 \times 10^3$
Potamogeton schweinfurthii	$0.05 - 1 \times 10^3$
Nymphaea caerulea	$0.05 - 0.85 \times 10^3$
Potamogeton thunbergii	$0.05 - 0.4 \times 10^3$

Valley Swamps. Characteristic of Uganda, particularly the regions to the north and west of Lake Victoria including the Yala Swamp in western Kenya, valley swamps are similar in species composition to lake-edge swamps but different in size and hence their influence on the landscape. Valley swamps may also be more species-poor because of their uniform nature. They were formed in the Pleistocene by water level changes and river flow reversal, leading to deposition of silts, infilling of former lakes and colonization by rooted species: this has often led to complex mosaics of plant communities (20). Where papyrus has dominated in wetlands with more stable hydrology, the swamp may sit on water several meters in depth. Valley swamps are very susceptible to draining by man and many have been converted to agriculture in the last four decades, losing some of the unique vegetation types (9). Initial successes of agriculture on these swamps have not always been sustained because problems arise from soil drying and oxidation of organic matter leading to a loss of fertility. Also increased flooding occurs due to the loss of the wetlands' capacity as a 'sponge' or buffer for inflowing waters during the wet seasons.

Seasonal Floodplains. Seasonal floodplains occupy large areas of parts of Africa, and some of the largest occur in eastern Africa. In the Sudd of southern Sudan for example, seasonally-flooded grasslands occupy about 17,000 km², a similar quantity to the permanent swamps and open water (20). In Tanzania the major inland wetlands are all seasonal floodplains up to 7500 km² in size (the Malagarasi; 16). The absolute size of such grasslands may be less important to their ecological value than their relative size and the role that they play as seasonal refuge for migratory species. The most well-known and complex effects of seasonal wetland changes are probably in Amboseli National Park, Kenya, where wildlife and pastoralists both rely on the seasonal expansion of the grasslands around a permanent swamp and careful management of pastoralists, wildlife and water is necessary to avoid conflict (21,36).

The vegetation of seasonally flooded areas is complex, but often dominated by grasses (Fig. 9.2). Plant structure is believed to be maintained, in recent times, by grazing from both livestock and wildlife. *Echinochloa* and *Oryza* are common genera dominating grasslands on the landward side of *Vossia* swamps. These and several other prominent genera are highly nutritious to herbivores. The boundary between those grasslands seasonally flooded and those seasonally waterlogged by rain is gradual, and many grass species are found in both environments. Usually, however, the dominants change to less palatable species, such as *Hyparrhenia* (found in both the Sudd and the Kafue flats of Zambia) in seasonally rain-inundated soils. The species composition of floodplain grasslands is influenced by many factors such as the duration of flooding, soil structure and chemistry, and particularly the soil salinity.

Within seasonally rain-fed grasslands there may be areas of water logging which, if sufficiently prolonged, allow the development of aquatic communities in the dry season. Such temporary pools often have local names (vlei, dambos, mbuga, wadi) reflecting their importance to the traditional agricultural and fisheries economies (2). Other important forms of temporary pools are those which remain on the courses of ephemeral rivers and streams when flow ceases, providing refuges for some vertebrates (such as fish), breeding environments for others (particularly amphibians) and at the same time important feeding stations for others (birds). The communities of such temporary ponds have been well-studied in other arid areas (34) and recent surveys of such wetlands in Kenya have been initiated (Virani, Pacini, Cooper, unpublished data). The importance of such water bodies to local and regional biodiversity needs further study.

Ponds and Dams. Small, sometimes seasonal wetlands, that may be natural depressions but more frequently are artificial farm dams for dry season irrigation or drinking water, can play an important role in maintaining biodiversity on the landscape scale. Found in these pools are a unique assemblage of invertebrates whose eggs can survive desiccation (mainly crustacea) and fish that can aestivate (such as lungfish *Protopterus*; 28). Ponds also support a high diversity of plants and birds (27) because their small size allows for considerable heterogeneity from pond to pond. Farm ponds are usually less important for plants as they are heavily affected by domestic live stock, but are important for birds, not least because the dung-fertilized and trampled littoral edge provides excellent foraging for waders.

New reservoirs and irrigation schemes must also be considered as wetlands, albeit artificial ones in which plant communities are highly modified. Their ecological importance lies in segments of the natural wetland communities that they support. For example, irrigated rice schemes such as Mwea on the Tana river in Kenya contain a diverse bird fauna. A cascade of five hydroelectric dams on the upper Tana of Kenya traverses several vegetation zones themselves too arid to support permanent natural water bodies, and therefore they achieve a regional importance, again for their bird fauna. Part of the riverine community eliminated by the conversion to lentic conditions also find refuge in the smaller and shallower dams, such as Tana river crocodiles and hippopotamus which thrive in the margins of Kindaruma Reservoir.

Swamp Forests. Swamp forests are uncommon wetlands in comparison with the temperate zone, where unchecked succession in emergent swamps frequently leads to woody plant dominance by plant genera such as *Salix* (willow). Palms (*Raphia*) occur in swamp forest in Uganda and westwards in Zaire, possibly relicts of a former more common vegetation. At higher altitudes, woody species of *Erica* and *Myrica* create a shrubby forest type.

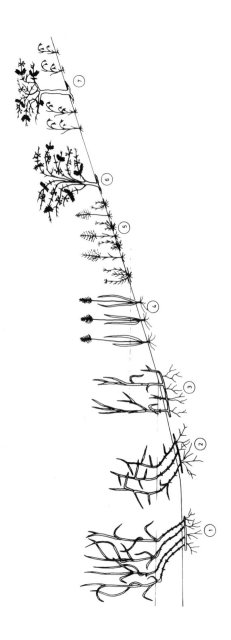

Figure 9.2. Typical succession in a wetland with unstable water levels, such as seasonally rain-flooded depressions or seasonal floodplains: a general pattern described from several examples after [15]. Growing within the zone flooded for several months each year are species characterized by extensible rooting stems; *Vossia cuspidata* (1) and *Echinochloa scabra* (2). Genera such as *Oryza* (3) and *Miscanthidium* (4) are typical of wet grasslands, flooded for shorter periods. Rain-fed grasslands close to the upper limit of flooding are dominated by grasses such as *Cynodon dactylis* and *Panicum spp.* (5). At the upper edge, grading into what is termed 'rain-fed savannah' is grassland dominated by characteristic shrubs such as the 'whistling thorn' - *Acacia drepanolobia* (6) with a mixed group of grass species depending upon precise soil and topography, but including *Themeda triandra* (7).

High Altitude Swamps and Mires. Vegetation differences in wetlands become apparent at higher altitude, as wetlands accumulate organic debris rather than decompose it under the higher rainfall and lower temperatures which prevail there. This debris accumulates, leading to more acidic soil conditions. Papyrus disappears above an altitude of around 2300 m and swamps tend to be dominated by acid-tolerant grasses such as *Miscanthidium* and by *Carex spp.* sedges. Characteristic high-altitude plants such as 'red-hot poker' *Kniphophia* and tussock-forming sedges also occur.

Eastern Africa contains many of the most important peat-forming wetlands in the African continent. This is due to the predominance of high-altitude mountain blocks in the Aberdares (=Nyandaruas), Mounts Kenya and Elgon, the Ruwenzoris (Chapter 16) and the highlands of Rwanda and Burundi. High altitude wetlands also tend to be the only wetlands where protection of the water supply is secured by inclusion of the entire catchments within national parks (32).

Wetland Distribution

The largest wetlands of Africa are just outside the area generally known as 'eastern' – the Sudd swamps of Sudan, the Okavango delta of Botswana, the Kafue/Bangweulu floodplains of Zambia, and the swamp forests of Zaire. East Africa contains wetlands that are smaller in size but nevertheless important in their variety and hence diversity (Fig. 9.3): lowland valley swamps on the fringes of Lake Victoria (mostly in Uganda but a few on the east shores of Kenya); high altitude peatlands and wetlands on the mountain ranges; inland deltas and floodplains of the Malagarasi (western Tanzania), the Lorian swamp (northern Kenya) and Amboseli (southern Kenya); small lacustrine wetlands in the eastern rift valley such as lakes Naivasha and Baringo; and new wetlands associated with reservoirs such as Nyumba ya Munga (Tanzania) and the upper Tana cascade (Kenya). Riparian fringes and small ponds are too small and widespread to show at this scale.

Wetland Community Structure and Functioning

The ecological value of wetlands is a consequence of both their structure and function, which are interlinked. The structure of plant communities is controlled by the adaptations of individual species to water availability and depth, combined with competitive interactions between species that share similar adaptations. The lack of uniformity of water supply in most wetlands, especially small ones, means that they are much more heterogeneous than terrestrial systems of equivalent size (the ecotone consequence referred to above). Each vegetation zone supports not just a different assemblage of plants, but a different assemblage of animals as well, so that the entire ecotone achieves an astonishing diversity in a very small area. At Lake Naivasha for example, in July to August 1987, approximately 50% more species of birds were recorded in the wetland ecotone than in an equivalent-sized adjacent terrestrial area (15). Figure 9.4 compares the species richness of ecotone habitats at Lake Naivasha and shows how each contributes a varying proportion of unique species to the overall list. The most rich is the shallow flooded lagoon habitat, because it contains a greater mosaic of mud, floating plants, detritus and drowned aerial insects. These same characteristics account for its low contribution to

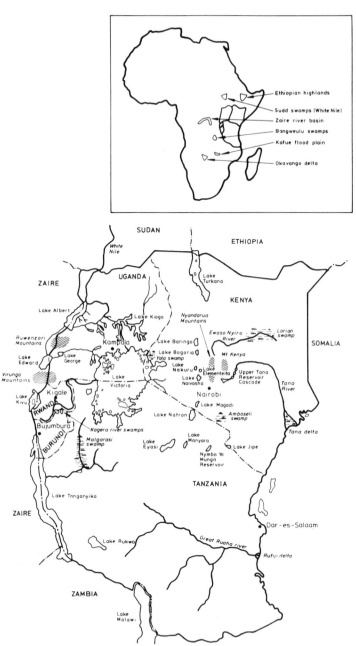

Figure 9.3. The distribution of wetlands in eastern Africa. The inset map of Africa indicates the location of the largest wetlands of continental importance in a band around the five countries of east Africa. The main map shows the wetlands of regional importance. They fall into the following groups: (a) montane wetlands, in Kenya and Uganda/Rwanda, (b) eastern rift valley shallow lakes and reservoirs in Kenya, (c) deltaic riverine wetlands, in Kenya and Tanzania, (d) the shallow edge wetlands of Lake Victoria, in Uganda, Kenya and Tanzania, and (e) river valley wetlands in Uganda, Rwanda and Burundi.

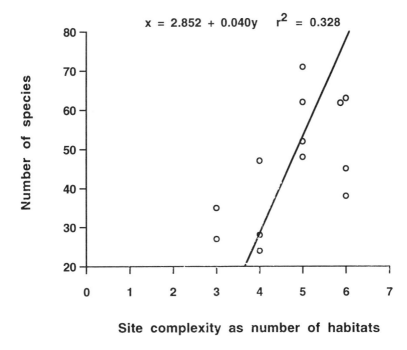

$$x = 2.852 + 0.040y \quad r^2 = 0.328$$

Site complexity as number of habitats

Figure 9.4. Number of bird species in different of the wetland ecotone at Lake Naivasha as a function of the number of habitats.

uniqueness. Open shoreline on the other hand, is used almost entirely by waders - birds specialized for probing in the mud (including many migrants) and hence its richness is low but its uniqueness is the highest. Figure 9.5 confirms that sites with greater diversity of habitats tend to have a higher species richness. High numbers of certain European migrants such as ducks, geese, pelicans and storks also use east African wetlands as winter refuge areas, accentuating their value.

Wetlands of all types are important for fish, both in supporting unique species and in the sustenance of open-water species. Swamp species adapted to low oxygen levels and desiccation (such as species in the Protopteridae and Clariidae) tend to be characteristic of wetlands and shallow benthic environments, while many small fishes (such as in the families Poecillidae and Cichlidae) frequent vegetated shallow ponds and littoral environments. Probably the most important quantitative roles of wetlands to fishes is their role as breeding grounds for lacustrine and riverine species and for food. Most fish species use some kind of surface for spawning and hence are dependent upon the architecture or structure of the littoral zone substrate and vegetation. In the permanent littoral zone many species such as *Tilapia* and *Heterotis* are nest-builders; many other species such as cichlids are mouth-brooders or live-bearers. In the spatially larger but temporally smaller floodplain grasslands and pools many species are adapted to reproduction with high fecundity coincident with annual or semi-annual floods. In such floodplain rivers many species migrate outwards to spawn as the water rises and inundates shallow grasslands, releasing nutrients from dry soils and stimulating an explosion of algal and invertebrate (usually crustacea) production. Large numbers of rapidly-growing fish fry inhabit

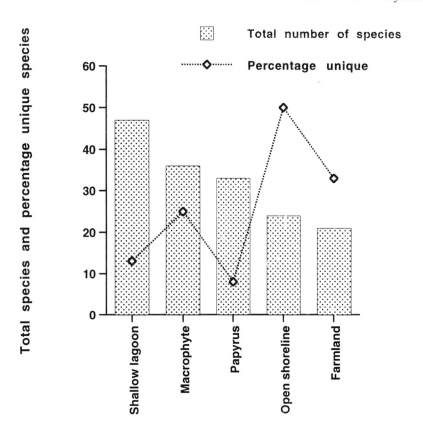

Figure 9.5. Species richness at twelve sites on the shoreline of Lake Naivasha plotted against the number of vegetation zones at that site as a measure of habitat complexity.
such as shrimps and beetles more prominently in the food webs, whereas the seasonally-flooded wetlands are more dominated by microinvertebrates, typically crustacea hatching from resting eggs in the mud, and insects dispersed by aerial adults.

the vegetated shallows and new ponds, sustained by aquatic plant and invertebrate production (such as insects). As flood waters recede fish become concentrated into smaller areas; some find their way back into the permanent channels but the majority become prey for birds and humans.

Other higher vertebrates characteristic of swamps include all amphibia and some reptiles such as crocodiles and turtles (29). Many mammals are associated with wetlands because they need drinking water and also the relatively high wetland plant production in an otherwise barren dry season environment (for example elephants and buffalos using Lake Amboseli). A number of ungulates have highly-developed seasonal migration patterns based on the production of floodplain grasses after flood recession. Species such as *Kobus* (lechwe) in the Nile and Zambezi river systems, *Damaliscus* (tiang) and gazelles have been well-studied (18). Other species are totally dependent on permanent wetlands and are, therefore, non-migratory. These species include *Hippopotamus amphibius* (common to Murchison Falls, Uganda, and Naivasha, Kenya) and the sitatunga (*Tragelaphus*; Saiwa swamp, Kenya and

Bangweulu flats, Zambia). Each species has adaptations for spreading its weight in mud or waterlogged soil and vegetation. These species are completely dependent on the wetland, either for keeping cool (*Hippopotamus*) or for food (*Sitatunga*). The habitat requirements of small and medium-sized mammals are not so well-known, but several are so closely associated with wetland or riparian ecotones (for example otter and mongoose species) as to be effectively dependent. Others are inhabitants of the woodland fringe associated with the landward edge of semi-natural wetlands. These species would become regionally extinct without these 'islands' (for example black and white colobus *(Colobus)* in lacustrine *Acacia xanthophloea* woodlands in the Gregory Rift Valley).

Wetland Function

The structure of wetlands cannot be separated from aspects of their ecological function, as indicated from the above discussion. The exact function is driven by the pattern of hydrology and degree of change between dry-season and wet-season state, and details are beyond this review (for river systems see 35 and Chapter 6 and for lacustrine wetlands see 17). Nevertheless, there are important common factors to their function that increasingly forms the basis for the economic evaluation of wetlands. Economic evaluation is necessary to ensure wetland protection.

The first common function is hydrological buffering. Most riverine and valley wetlands act like sponges or buffers, receiving water at times of high flow in larger rates and slowly releasing it, but releasing a more constant volume all year round hence sustaining the flow during dry periods. Downstream ecosystems are thus more consistently sustained. This has important consequences for human activities such as irrigation and water supply as well as the availability of watering points for large herbivores through dry seasons. The upland wetlands of the Aberdares mountains have been utilized in this way to supply Nairobi with drinking water.

The second function is high primary productivity of wetlands. Almost all ecological textbooks, when discussing ecosystem or biome types, highlight wetlands as the most productive ecosystems in the world (37), and the basis for this is the combined supply of the essential requirements for photosynthesis - water itself and carbon dioxide from air – together with nutrients for cellular growth from either water or sediment. Swamp plants have considerable adaptations for nutrient conservation and most possess the C_4 pathway of photosynthesis, which provides a more efficient water and nutrient use than the conventional C_3 pathway.

The third function, linked to photosynthetic efficiency, is mineral and nutrient buffering. Nutrients are not readily available all year round from inflowing sources but are trapped and recycled by the vegetation. Gaudet (11) has shown that emergent plants in lacustrine swamp plants sequester inorganic nutrients from rainy season inflows and utilize them during production. Much of this production is stored (translocated downwards into the rhizome). A small fraction of the shoot biomass is consumed by animals but the bulk dies, accumulates in the swamp as debris and is slowly decomposed. Leaching downwards and outwards from the swamp and into the lake littoral zone is a continuous stream of fine particulate organic matter and organic dissolved nutrients. This particulate organic material provides food for littoral heterotrophs - from bacteria through filter-feeding zooplankton to detritivorous fish. The dissolved nutrients also sustain littoral plant growth. A further buffering occurs within the littoral zone, through complex interactions between macrophytes, their epiphytic algae, and small grazing invertebrates such as crustacea and hemiptera (Figs. 9.6 and 9.7). Open water receives only a small fraction of incoming inorganic nutrients, again as a steady rather than a pulsed supply. Available inorganic nutrients are tightly recycled in plankton production rather than retained as algal biomass (this is why lakes where the fringing vegetation has been removed suffer from outbreaks of algal 'blooms'). Sinking and sedimenting organic matter is utilized by benthic invertebrates and recycled by bottom-feeding fish. The similarity of pathways but difference in detail of Figures 9.6 and 9.7 are both derived from studies of wetlands in the Sudd, southern Sudan. The permanent littoral zone with aquatic plants contains larger, macroinvertebrates

Wetland Economic Uses

The economic uses of wetlands hinge entirely on their relative constancy and high productivity. Earliest human uses of floodplain wetlands were based on seasonal changes of fishing in the wet season and grazing floodplain grasses in the dry season, supplemented by the use of swamp grasses (papyrus and reeds) for roofing and other aquatic species for food (*Nymphaea* for tubers and seeds). Such uses were still common among peoples of the Sudd until recently (18) and small swamp areas in Kenya such as Loboi/Kesubo swamps are still used in this way by the Njemps/Tugen people. More constant-sized wetlands, such as Naivasha, were originally used by migrating pastoralists alongside migrating wildlife as a dry season water and food source.

Fisheries exploitation of wetlands continues to be of great importance. For example, over 50% of Zambia's fish production is from freshwater wetland swamp or floodplain habitats. Kenya, Uganda and Tanzania fisheries are dominated by Lake Victoria or marine catches but with a quarter of Tanzania's fish catch still coming from small ponds and wetlands. The littoral wetland component of catches in lake Victoria (shoreline and rivers) as opposed to the open water catches are also substantial. For Lake Naivasha, fishes caught in the open water have both fed and bred in the littoral wetlands and so the entire lake catch is dependent on the littoral wetland (33).

An extension of wetland fish harvesting is fish farming, typically in small ponds. Developments of 'polyculture' pioneered in Asia are progressively being introduced on a small scale in eastern Africa. Polyculture combines fish ponds with rice and vegetable farming along the swamp/land ecotone. This use preserves the integrity of the swamp while maximizing yields.

Many wetlands have been rapidly converted to large-scale agricultural production since national independence, driven by the twin pressures of rapid rates of population increase and development initiatives in all eastern African countries. Some wetlands have been converted to wetland agriculture, such as rice cultivation, while other wetlands have been drained and fully converted to arable agriculture. Some of these drainage schemes have been successful, but many have had negative effects on both humans and wildlife (9). Loss of the hydrological buffering effect of wetlands has led to flooding in the wet season, and to loss of spring and well-water sources in the dry season; increased use of fertilizers and pesticides has led to downstream pollution and agricultural yields have fallen as soils have dried and organic matter oxidized. The irrigation necessary in the dry season has led to salinization of soils due to high evaporation.

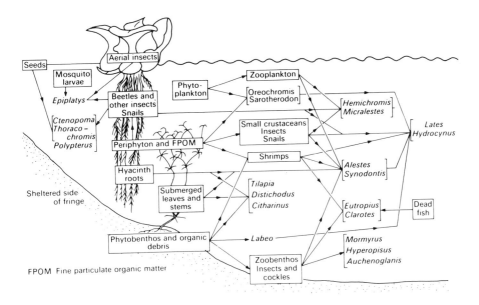

Figure 9.6. Invertebrate/fish feeding relationships in the submerged and floating macrophyte zone of wetlands; an example from the Sudd lakes of southern Sudan. From [16], copyright Cambridge University Press, with permission.

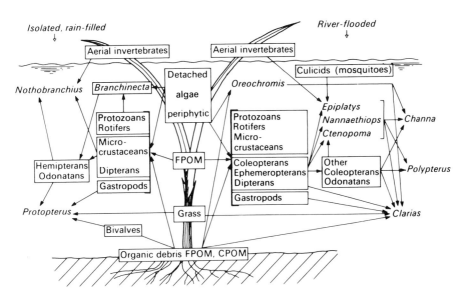

Figure 9.7. Invertebrate/fish feeding relationships in the seasonally-flooded zone typical of riverine wetlands; an example from the Nile floodplain of southern Sudan. From (18), copyright Cambridge University Press, with permission.

Subsistence hunting once undoubtedly took place alongside grazing and fishing in wetlands, and this became supplanted by organized game hunting with European colonization. Species such as buffalo, hippopotamus and lechwe have been utilized both in hunting and controlled cropping schemes in the past in many countries, but with little long-term success. Recreational wildfowl shooting has also taken place in many wetlands. Nowadays, however, wetlands offer modern African states much greater potential for tourist development than hunting, with National Parks such as Lake Nakuru, Kenya; Lochinvar, Zambia; Okavango, Botswana and Murchison Falls, Uganda, combining the spectacular viewing of wetlands wildfowl against a backdrop of terrestrial large mammals.

Sustainable Management of Wetlands

Two trends are beginning to affect the future of east African wetlands. The first is that authorities are becoming increasingly aware that special effort is needed to protect a representative selection of wetland habitats. With this realization has come the awareness that there is a considerable dearth of knowledge about how many wetlands a country has, of what size and vegetation type they are, and the means necessary to secure their future. Most countries have begun to conduct wetland inventories, through government agencies such as the wildlife management authorities and through voluntary bodies such as 'wetland groups'; assisted by international agencies such as WWF, IUCN, and the Ramsar Committee.

The second trend is towards integrated uses of wetlands, along lines laid out in agreed management plans, and incorporating wetland catchments as well as the wetlands themselves (see Chapter 5). This follows policies recently established in developed countries, that integrated use allows easily-quantifiable exploitative economic values to be set alongside ecosystem and global values that cannot be so easily quantified, and that integrated rather than single use maximizes the possibility for sustainable management in the long-term (25).

One wetland in Kenya, Lake Naivasha, has been commercially exploited for over 50 years yet still retains its core ecological value. This lake can be used to illustrate the characteristics of east African wetlands, their hydrological and biological exploitation and the needs for sustainable management. The general ecology of the open water system of Lake Naivasha has been covered in Chapter 7 so the focus here will be on the interactions and human use of the littoral wetlands.

Lake Naivasha

Location and History. Naivasha lies in the floor of the Gregory Rift Valley at an altitude of 1888 m. The climate of the area is warm (monthly mean air temperatures range from 15.9 to 18.5° C) and semi-arid. Rainfall near the lake is bimodal with the main rains in April and May and short rains in November. Naivasha town, an agricultural trading center with a population estimated at 15,000 and serving another 100,000 in its hinterland, lies on the lake shores.

The geological history of the lake basin determined from core samples shows substantial changes in climate during the past 30,000 years. From 12,000 to 9200 years ago a large (612 km^2) lake including the present-day area of Elementeita and Nakuru overflowed southward cutting the Njorowa gorge in what is now Hell's Gate

National Park. A smaller lake of around 400 km² lake occupied the basin around 5700 years ago, and decrease in rainfall reduced the size of the lake and closed the basin. By 3000 years ago the lake dried out completely and remained dry for about 100 years. During the past 3000 years a small fluctuating lake has existed in the basin which may have dried up on more than one occasion; the most recent about 150 years ago. In the last few decades the lake area has varied between 100 to 150 km².

Hydrology and Chemistry. The majority of the water entering the basin comes from the Malewa River (estimated at a mean of around 140 x 10⁶ m³ /y) that drains the Aberdare Range and Kinangop Plateau. Direct precipitation (70 x 10⁶ m³/y) and seepage-in (50 x 10⁶ m³/y) account for the rest. Evaporation accounts for the bulk of the water loss (170 x 10⁶ m³/y), seepage-out approximately matches seepage-in and agricultural and industrial abstraction accounts for an uncertain amount of water estimated at between 45 and 80 x 10⁶ m³/y. A new water supply off-take for Nakuru and Gilgil towns in the upper catchment threatens to reduce the inflow by 25%. There is thus now a net deficit, except in wet years, which is the most unstable current feature of the lake. No evidence of discrete underground outflows has been found; such outflows were once proposed to explain the anomalous freshness of Naivasha. The dilute nature of Lake Naivasha water is now explained by a combination of factors, principally the inflow of dilute water coupled with the outflow of more concentrated seepage water, and precipitation of solutes. Chemically, the lake is dominated by sodium and potassium cations, and bicarbonate anions.

Ecology. Ecologically Lake Naivasha is an unusual and incomplete ecosystem. This is because it combines freshness with fluctuating water levels - a situation more usual of man-made impoundments – and it is incomplete because its impoverished native fauna leaves several important niches unoccupied. Only one species of fish existed in the Naivasha basin prior to further introductions by man. It is probable that the desiccation of the lake in the past, coupled with the relative isolation of the lake from suitable fresh water lakes that contain potential replacement species (all but one of the other lakes in the Gregory Rift are saline), is responsible for the impoverished nature of the fish fauna (see Chapter 7).

Other components of the ecosystem whose species have better powers of dispersal are not impoverished. For example, Gaudet (10) found a high number of families (43) and species (108) of aquatic plants. The most common type of emergent vegetation is papyrus swamp. Once occupying several tens of square kilometers, this is now largely confined to a fringe varying in thickness from a few meters on steep shores to several hundred meters on shallow-gradient shores. *Cyperus papyrus* is the dominant species, although when new swamp germinates *C. dives* is initially successful, being replaced by *C. papyrus* over a period of about three years. Occasional stands of the cattail *Typha* occur in sheltered areas. Swamp germination occurs during periods of water-level decline if a rapid seasonal rise occurs, inundating bare, nutrient-rich soil. Gaudet showed how this had occurred four times in the first half of this century, creating the mosaic of papyrus reefs and islands which were characteristic of the wetland up to the 1970s. A similar event occurred in 1988, but by then the previous generations of papyrus had been cleared or burnt for agricultural development and thus the present wetland is characterized by a much more even-aged papyrus in stands and islands.

In the littoral zone, beyond the papyrus, and with the exception of the inflowing areas where substrate is flocculent and deoxygenated, a zone of submerged plants dominated by *Potamogeton schweinfurthii* and *Najas pectinata* with *P. pectinatus* and *P. octandrus* occurs to a varying extent down to 4 to 5 m in depth. Floating vegetation is also common. Ten years ago this included *Pistia stratiodes* but currently *Eichhornia crassipes* is the dominant species with scattered *Salvinia molesta*. The water lily, *Nymphaea caerulea* dominated the floating-leafed community in shallow water up to the mid 1970s but has now almost completely disappeared, confined to sheltered backwaters and isolated ponds. Free-floating duckweed species such as *Lemna* and *Wolffia* occur on the inner edges of papyrus reefs or the lake edges of the papyrus fringe.

The natural vegetation of the shoreline consists of scrub dominated by shrubs such as *Cassia* and *Sesbania* species capable of withstanding waterlogging, which progressively colonize the clumps of stranded papyrus. These give way to seedlings of *Acacia xanthophloea* that eventually produce a closed canopy with a mixed understory of shrubs. As the water-table depth increases with distance away from the lake shore, *Acacia drepanolobium* and *Euphorbia spp.* come to dominate the open savanna environment.

In the open water of the lake, shallow depth and eutrophic conditions support a phytoplankton community whose density has risen from around 20 mg/m^3 chlorophyll a concentration in the 1970s to around 50 mg/m^3 in the 1990s. This is a consequence of generally declining water levels concentrating minerals and nutrients, coupled with occasional short periods of water level rise which re-dissolve those nutrients mineralized from the dried soils. Nevertheless, the degree of biological recycling is such that dissolved nitrate and phosphate is rarely detected in the lake water. The algae are sufficiently dense to color the water distinctively green or brown depending upon species, and restrict photosynthesis to the top meter.

The phytoplankton community of the lake was formerly dominated by cyanobacteria such as *Lyngbya*, *Microcystis* and *Aphanocapsa* but more recently diatoms and such as *Melosira* and chlorophytes such as *Closterium* have become more abundant. Those studies which have investigated nutrient flux thoroughly concluded that algae were generally phosphorus-limited, as is also typical of temperate eutrophic waters. The phytoplankton are grazed by a diverse community of zooplankton, the most important of which are cladocerans. The species commonly found are *Simocephalus vetulus, Daphnia pulex, D. laevis, Diaphanosoma excisum* and species of Ostracods. Within the submerged plant beds these crustacea reach high densities and create very clear water through their grazing impact upon algae; a consequence of the refuge for zooplankton which the submerged plant beds provide from fish predation (13).

The benthic fauna of the open water is poor in both species richness (only about a dozen species of chironomids and oligochaetes) and in abundance. The reasons for this are not clear, but may be associated with the highly flocculent detrital substrate and possible interactions with crayfish. The dominant species by biomass is the swamp worm *Alma emeni*, both in the bottom muds of the littoral zone and colonizing floating vegetation wherever sediment accumulates - such as the roots of *Eichhornia* and the mat of papyrus. It has specific morphological adaptations for low oxygen conditions (1) and is an important link in the food chain to both fish and probing birds such as lily trotter (*Actophilornus africanus*) and long-toed lapwing (*Vanellus crassirostris*)

The littoral invertebrate community is dominated by insects in the water column – chiefly the hemipteran *Micronecta scutellaris* – but with approximately thirty other species of beetles and bugs. The littoral edge is dominated by the crayfish *Procambarus clarkii*, an introduced voracious catholic omnivore blamed for fluctuations in plant populations in Naivasha and implicated in some of the changes in invertebrate composition.

Several hundred hippopotami (*H. amphibius*) live in groups of 8 to 20 in shallow and isolated areas of the lake and its inflow. They play an important role in creating paths through submerged plant beds that then provide small areas of open water colonized by different species of macrophytes. The channels have different conditions of oxygen and temperature and are used by fish and invertebrates. Similar paths through papyrus open up the swamp to water movement, again facilitating fish passage and oxygenation of swamp water. Hippopotomi feed predominantly on land at night but quite extensively within macrophyte beds by day.

Bird species recorded at the lake and its wetland environs have exceeded 300, but few studies have characterized the bird community of the wetland and its associated riparian vegetation fringe exclusively, leaving out casual recordings. Over two months in 1987, a total of 218 species were recorded in the area (during the north temperate summer, so the count excludes migrants). Of these, 161 species used the lake and associated vegetation and 79 were essentially aquatic or semi-aquatic. Aquatic species are the most abundant; the numbers of coot and duck can at times exceed several tens of thousands, associated with an extensive submerged plant zone (in 1982, 1987 and 1991). Pelicans occasionally occur in large numbers, associated with high fish reproductive success, usually after water level rise (over 4000 were counted in early 1983). Probably the most characteristic species of the wetland is the African fish eagle (*Haliaeetus vocifer*) whose population of around 60 pairs represents the densest concentration in the continent.

Many of the birds, and a varying population of humans, depend on fish. Fish are made up almost entirely from introduced species. Only *Barbus amphigramma*, a riverine species that appeared in the past ten years, is natural. The original native species disappeared in the 1960s. Over a dozen species of fish have been introduced in the past seventy years, some more than once, but at present only three are common. *Tilapia zillii* is the dominant species but over the past two decades its population and catch have fluctuated wildly. Both *T. zillii* and the other main species, *Oreochromis leucosticta*, are predominantly detritivores and do not appear to be food-limited. A third commercially-exploited species is the crayfish *Procambarus clarkii*. All three are consumed by the only predatory fish, the large-mouthed bass *Micropterus salmoides*, as well as by humans and piscivorous birds. Bass are sport-fished by rod and line and commercially using gill nets. Bass spawn off the small areas of rocky shoreline at the southern end of the lake whereas *Tilapia* is a soft-substrate nest-builder and *Oreochromis* a mouth-brooder; both spawn in the swamp edges.

Exploitation of the Naivasha Wetland.
Natural fluctuations in water level of the lake this century have exceeded four vertical meters; and the receding lake bed has been used for agriculture by legal agreement between the Government and

riparian owners for over sixty years. The lake is now the center of Kenya's horticultural industry with approximately 100 km² of former lake bed and adjacent land under intensive cultivation. This uses an uncertain quantity of irrigation water perhaps equivalent to one third of the average inflow of the major river, and half this amount again is abstracted for Olkaria geothermal power station to the south of the lake. There is thus an average net annual loss of water from the lake of at least 15 to 20 x 10⁶ m³. This is currently roughly equivalent to a decline of 15 to 20 cm in lake level. The implications are that the wetland size and integrity will be sustained only if river inflow is maintained at least 12 to 15% above its long-term average but that it will decline if the inflow is less. Any additional catchment offtakes from inflow rivers, such as a proposed second-phase water supply for Nakuru town, will increase this deficit threefold. The consequences, then, are that the wetland will inevitably shrink to a fraction of its current size. Riparian owners concerned about their own collective use of lake water together with international conservation agencies concerned about the newly-declared status of the lake as a 'Ramsar' site, are thus confronted with the prospect of catchment-scale events, controlled by international aid programs and national priorities in energy and water supply, which deprive the ecosystem of its fundamental means of support.

Even if this resource-use problem is solved by judicious allocation around the lake and use of alternative supplies for Nakuru, there remain two linked problems, of resource quality and ecosystem quality. Water quality has deteriorated in ways that cannot be fully explained. The lake has become enriched with nutrients, leading to greater phytoplankton crops and decreased transparency. Enrichment has been variously attributed to decreased lake levels, increased fertilization from agriculture, increased sewage inflow, loss of the buffering of submerged plants due to crayfish, loss of the buffering of fringing papyrus by burning, clearance for grazing or cultivation, and erosion in the catchment leading to greater silt loads. The truth is probably a mixture of all these explanations. There is also an unknown threat from other agricultural chemicals, such as pesticides and herbicides, with some evidence that their on-farm use around the lake is not always safe. Resource quality has also deteriorated in terms of fish yields. The cause is overexploitation due to insufficient resources for regulating the fishery. Licensed fishermen are using illegal gill-mesh sizes and poachers are seine-netting in the shallows; the result is an ever-decreasing mean size of the fish caught.

Ecosystem quality (biological diversity) has also deteriorated, in ways that cannot be fully explained. Disappearance of the original native (and endemic) fish species was attributed to predation by bass. Disappearance of certain plant and invertebrate species has been attributed to the effects of crayfish. Dietary studies of the crayfish species in other environments support this contention. Disappearance of certain bird species, such as great-crested grebe (*Podiceps cristatus*) and African darter (*Anhinga rufa*), is attributed to disturbance by overfishing and use of small-meshed gill nets in which they become trapped. Loss of large uninhabited areas of natural vegetation zones around the lake fringe is almost total, with only two areas of the lake unsettled or uncultivated, one which is currently threatened by development. The *Acacia xanthophloea* woodland that fringes the lake had remained intact in most parts of the system until the end of the 1980s. Building development had taken place among the trees with shrub clearance on the edge of the woodland. A disturbing trend of most recent agricultural developments however, has been the complete clearance of both woodland as well as fringing papyrus. Unchecked agriculture will put both the water quality and the ecosystem quality at risk of permanent and irreversible damage.

Sustainable Management. Sustainable management requires a reversal of these deteriorating trends. Water use at present is managed as a 'commons' where everybody can draw water as they need (but based upon their financial ability to invest in equipment to draw and distribute it). Many people argue that it is a fundamental right of a modern democratic and capitalistic society to provide water without restriction. In this wetland, water is drawn by various agencies and people from three sources: groundwater, river, and lake. Each source may appear unconnected to the individual water-user. In any future allocation however, water and its needs should be regarded as connected to the whole watershed. Principles of the commons need to give way to the principle of resource allocation, if the 'tragedy of the commons' (12) is to be avoided.

Water quality needs to be protected. The most effective way is to use existing wetland vegetation to purify urban effluents and the re-creation of continuous buffer zones around the wetland edge. Most of this can only occur through individual initiative, and so needs to be associated with a widespread education program. Artificial wetlands can be constructed, following principles successfully developed in the temperate zones. One example would be utilization of *Eichhornia* ponds to purify effluents both from Naivasha town's sewage and un-sewered drainage; another the re-creation of continuous buffer zones around the wetland edge by engineering and replanting.

Biological diversity needs to be protected and, wherever possible, enhanced. This is the most difficult task (although not the most pressing), as agreement on goals may not be easy. Suggestions have been made that new (East African) species of fish should be introduced to diversify the fishery and the food web. But these species might have negative effects on diversity, as the introduction of crayfish is supposed to have done twenty years before. Suggestions have been made that *Eichhornia* should be controlled by the introduction of a control weevil, as happened with *Salvinia* three years ago, apparently successfully. But this ignores the potential benefits of *Eichhornia* as a wetland buffer. Perhaps the ecosystem quality, and ultimately the water quality, might best be protected by lower-intensity agriculture and higher-intensity tourism and public access: there are both pros and cons for this. The most pressing need now is for a management plan that addresses each resource issue and stimulates debate about the options for each management decision.

The problems confronting the wetland of Lake Naivasha are similar to many problems facing wetlands across East Africa. Wetlands will survive only through multi-purpose uses, with each use respecting the environmental needs of the other users. This will require a scientific understanding of the resource base, its fragility and interconnectedness, supporting a management plan implemented by consent among potentially conflicting users. Most developed countries find these objectives hard to achieve for wetland management. East African societies have the opportunity to lead the world with sustainable management and utilization of wetlands such as Lake Naivasha.

References

1.   Beadle, L. C. 1981. *The Inland Waters of Tropical Africa*. Harlow, Essex: Longman
2.   Chidumayo, E. N. 1992. The utilisation and status of dambos in southern Africa: a Zambia case study. In *Wetlands Conservation Conference for Southern Africa*, eds. Matiza, T. Chabwela, H.N., pp. 105-108. Gland, Switzerland: IUCN
3.   Crafter, S. A., Njuguna, S. G., Howard, G. W. 1992. *Wetlands of Kenya*. IUCN: Gland Switzerland
4.   Davies, B., Gasse, F. 1988. *African Wetlands and Shallow Water Bodies*. Collection Travaux et Documents no. 211, Institut Francais de Recherche Scientifique pour le Development en Cooperation: Paris
5.   Denny, P. 1973. Lakes of south-western Uganda II: vegetation studies on Lake Bunyoni *Freshwater Biology* 3: 123-135
6.   Denny, P. 1985a. *The Ecology and Management of African Wetland Vegetation*. Dordrecht Dr. W. Junk Publishers
7.   Denny, P. 1985b. Submerged and floating-leaved aquatic macrophytes (euhydrophytes). In *The Ecology and Management of African Wetland Vegetation*, ed. Denny, pp. 19-42. P Dortrecht: Dr. W. Junk Publishers
8.   Denny. P. 1985c. The structure and functioning of African euhydrophyte communities: the floating-leaved and submerged vegetation. In *The Ecology and Management of African Wetland Vegetation*, ed. Denny, P., pp. 125-152. Dortrecht: Dr W. Junk Publishers
9.   Denny, P., Turyatunga, F. 1992. Ugandan wetlands and their management. In *Conservation and Development: The Sustainable Uses of Wetland Resources*, eds. Maltby, E., Dugan P.J., Lefevre, J.C., pp. 77-86. Gland, Switzerland: IUCN
10. Gaudet, J. J. 1977. Natural drawdown on Lake Naivasha and the formation of papyrus swamps. *Aquatic Botany* 3: 1-47
11. Gaudet, J. J. 1979. Seasonal changes in nutrients in a tropical swamp: north swamp, lake Naivasha. *Journal of Ecology* 67: 953-981
12. Hardin, G. 1968. The tragedy of the commons. *Science* 162: 1243-1248
13. Harper, D. M. 1992. The ecological relationships of aquatic plants at Lake Naivasha Kenya. *Hydrobiologia* 232: 65-71
14. Harper, D. M., Adams, C., Mavuti, K. M. 1995. The aquatic plant communities of the Lake Naivasha wetland, Kenya: pattern, dynamics and conservation. *Wetlands Ecology and Management* 3: 111-123
15. Henderson, I. G. 1987. *Lake Naivasha Vertebrate Survey*. Leicester: University of Leicester, Zoology Department
16. Howard, G. W. 1992. Introduction. In *Wetlands of Tanzania*, eds. Kamukala, G.L., Crafter S.A., pp. 1-6. Gland, Switzerland: IUCN
17. Howard-Williams, C., Gaudet, J. J. 1985. The structure and functioning of African swamps. In *The Ecology and Management of African Wetland Vegetation*, ed. Denny, P., pp 153-176. Dortrecht: Dr W. Junk, Publishers
18. Howell, P., Lock, M., Cobb, S. 1988. *The Jonglei Canal: Impact and Opportunity*. Cambridge: Cambridge University Press
19. Kamukala, G. L., Crafter, S. A. eds. 1992. *Wetlands of Tanzania*. Gland, Switzerland: IUCN
20. Lind, E. A., Morrison, M. E. S. 1974. *East African Vegetation*. London: Longman
21. Lindsay, W. K. 1987. Integrating parks and pastoralists: some lesons from Amboseli. In *Conservation in Africa: People, Policies and Practice*, eds. Anderson, D., Grove, R., pp. 149-168. Cambridge: Cambridge University Press
22. Maltby, E., Dugan, P. J., Lefeuvre, J. C. 1988. *Conservation and Development: The Sustainable Use of Wetland Resources*. Gland, Switzerland: IUCN
23. Matiza, T., Chabwela, H. N. eds. 1992. *Wetlands Conservation Conference for Southern Africa*. Gland, Switzerland: IUCN
24. Mitchell, D. S. 1985. African aquatic weeds and their management. In *The Ecology and Management of African Wetland Vegetation*, ed. Denny, P., pp. 177-202. Dortrecht: Dr W. Junk Publishers

25. Mitsch, W. J. ed. 1994. *Global Wetlands, Old World and New.* Amsterdam: Elsevier
26. Muchiri, S. M., Hart, P. J. B., Harper, D. M. 1995. The persistence of two introduced tilapia species in Lake Naivasha, Kenya, in the face of environmental variability and fishing pressure. In *The Impact of Species Changes in African Lakes*, eds. Pitcher, T.J., Hart, P.J.B., pp. 299-319. London: Chapman and Hall
27. Ng'weno, H. 1992. Seasonal wetlands in Nairobi. In *Wetlands of Kenya*, eds. Crafter, S.A., Njuguna, S.G., Howard, G.W., pp. 55-64. Gland, Switzerland: IUCN
28. Okeyo, D. O. 1992. Wetland fish of Kenya. In *Wetlands of Kenya,* eds.Crafter, S.A., Njuguna, S.G., Howard, G.W., pp. 47-54. Gland, Switzerland: IUCN
29. Simbotwe, M. P. 1992. Economic value of the herpetofaunal resource in wetland areas. In *Managing the Wetlands of the Kafue Flats and Bangweulu Basin*, eds. Jeffrey, R.C.V., Chabwela, H.N., Howard, G. W., pp.65-70. Gland, Switzerland: IUCN
30. Symoens, J. J., Burgis, M., Gaudet, J. J. eds. 1981. *The Ecology and Utilization of African Inland Waters.* Reports and Proceedings Series, Nairobi: UNEP
31. Thompson, K. 1985. Emergent plants of permanent and seasonally-flooded wetlands. In *The Ecology and Management of African Wetland Vegetation*, ed. Denny, P., pp. 43-108. Dortrecht: Dr W. Junk Publisher
32. Thompson, K., Hamilton, A. C. 1983. Peatlands and swamps of the African continent. In *Mires: Swamp, Bog, Fen and Moor*, ed. Gore, A.J.P., pp. 331-374. Amsterdam: Elsevier
33. Vanden Bossche, J.-P., Bernacsek, G. M. 1990. *Source Book for the Inland Fishery Resources of Africa.* CIFA Technical Paper, Rome: FAO
34. Watson, G. F., Davies, M., Tyler, M. J. 1995. Observations on temporary waters in northwestern Australia. *Hydrobiologia* 299: 53-73
35. Welcomme, R. L. 1985. *River Fisheries.* Technical Paper 262, Rome: FAO
36. Western, D. 1975. Water availability and its influence on the structure and dynamics of a savannah large mammal community. *East African Wildlife Journal* 13: 265-286
37. Whittaker, R. L. 1975. *Communities and Ecosystems.* New York: Macmillan Publishing Company

Section IV:

Grass, Shrub, and Woodland Ecosystems

Chapter 10

Arid and Semi-arid Ecosystems

David M. Swift, Michael B. Coughenour & Menwyelet Atsedu

The ecology of the arid and semi-arid regions of East Africa is essentially the ecology of subsistence pastoralism. Pastoralists and their herds are ubiquitous in the drier parts of East Africa. Pastoralists need these areas for their survival and, in turn, significantly affect them. Pastoralists are participants in the dynamics of intact ecosystems in ways that agriculturists and most people living in moister ecosystems are not. Consequently, the ecology of arid ecosystems can not be understood in isolation from an understanding of pastoral society and the management of their livestock. In tropical areas where annual precipitation is 600 millimeters or less, and is divided between two rainy seasons, crop agriculture is not possible. Nonetheless, these arid areas have been inhabited by people and their livestock for thousands of years. The result, in many cases, is a human-dominated ecosystem in which livestock have partially taken over the role of wildlife as consumers of grass and shrubs.

Neither humans nor livestock could survive in arid areas without the other. Without livestock, the human population would be limited to the few who could survive from hunting and gathering - a tenuous mode of life that supports lower human population numbers than pastoralism. Without the herders to care for them, livestock would fall victim to predation or be unable to find drinking water during the long dry periods. Living together in a symbiotic relationship, both can survive and even flourish in these harsh environments.

Although arid ecosystems have persisted for very long periods, these ecosystems are far from being stable. Extended periods, too dry to support the growth of plants, occur every year, and droughts of a year or longer are common. Cessation or failure of plant growth triggers a sequence of events that is felt throughout the ecosystem at every feeding level, including humans. Thus the environment is generally harsh, and, during unpredictable droughts, very harsh. Only organisms adapted to dry and unpredictable conditions can survive; and even populations of well-adapted organisms experience wide fluctuations in numbers and levels of productivity.

From the perspective of the pastoral people adaptation consists of coping with drought - of finding ways to organize themselves as families and as societies to deal with life in an inherently risky environment. Much of pastoral management and social organization can be interpreted as adaptations for insuring survival in the face

Box 10.1.   East African Climate and its Classification

East Africa is fortunate in having numerous meteorological stations that have collected rainfall and climate data for over 50 years. An analysis of the long-term seasonal and annual rainfall from 90 stations in Kenya, Tanzania and Uganda were used to create a map of the rainfall patterns dependent on the degree of similarity of the rainfall stations (5,8). The maps displayed below show the results of this analysis for the (left) seasonal (4 seasons) and (right) annual rainfall records.

Both maps indicate a number of distinct rainfall regions and some geographic gradients in rainfall that are largely predictable dependent on the annual migration of the Intertropical Convergence Zone, elevation and closeness to water bodies (see Chapter 3). The analysis by season, however, produces more groupings particularly around Lake Victoria and the Kenyan highlands. The maps suggest a number of distinct high-rainfall regions that include the coast, Mt. Kilimanjaro, the area around large lakes such as Lake Victoria and Tanganyika, and the Kenyan highlands. The existence and predictability of single or double rainfall peaks also distinguish areas in the region (9) where northeastern Uganda and southern Tanzania have single rainfall peaks while southern Uganda and northern Kenya have two rainfall peaks of equal intensity. The remaining areas have a dominant southeast monsoon and weaker northeast monsoon peak in rainfall. There is also a pattern of decreasing rainfall from the west to the east that reflects that decreasing influence of the expansion of the low-pressure wet air from central Africa. This expansion produces much of the northeast monsoon rains in the western part of this region. Because rain is frequently the most important resource for plant production, many plant communities and their productivity should reflect these rainfall patterns and clusters.

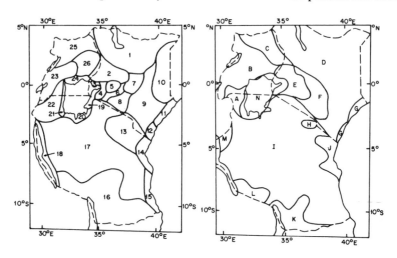

The maps above are based on long-term averages and it should be appreciated that rainfall patterns change (Chapter 1; 1,5,10). Analyses of the trends suggest that there are weak oscillations in rainfall that occur at periods of around 2 to 3, 5 to 6, and 10 to 11 years (4,5,9). Some of these changes are probably related to the intensity of the Southern Oscillation with strong winds associated with warm periods in the central Pacific ('El Nino') and vice versa (1,2,3,4,6,7). Warm periods in the central Pacific may produce low rainfall particularly in the coastal and northeast portions of East Africa during the northeast monsoons because strong Westerlies that move towards the warm pool in the central Pacific results in extensive upwelling of cold water off of Somalia during the southeast monsoons. This cold seawater will reduce evaporation and the rainfall that reaches East Africa once the monsoons switch. The northeast monsoons, particularly at the coast and the north-eastern part of this region, can, therefore, be highly unpredictable and may often produce the drought conditions that frequently plague the region. The opposite pattern appears to apply to the Kalahari in southern Africa and the Sahel and Sudan deserts appear to less affected by the Southern Oscillation (4).

T.R. McClanahan

References

1. Cadet, D.L. 1985. The Southern Oscillation over the Indian Ocean. *Journal of Climatology* 5: 189-212
2. Farmer, G. 1988. Seasonal forecasting of the Kenya coast short rains. *Journal of Climatology* 8: 489-497
3. Hutchinson, P. 1992. The Southern Oscillation and prediction of 'der' season rainfall in Somalia. *Journal of Climate* 5: 525-531
4. Nicholson, S.E., Entekhabi, D. 1986. The quasi-periodic behavior of rainfall variability in Africa and its relationship to the Southern Oscillation. *Archives for Meteorology, Geophysics, Bioclimatology* 34: 311-348
5. Ogallo, L.J. 1980. Regional classification of the East African rainfall stations into homogeneous groups using the method of principal component analysis. *Statistical Climatology and Developments in Atmospheric Sciences* 13: 255-266
6. Ogallo, L.J. 1984. Temporal fluctuations of seasonal rainfall patterns in east Africa. *Mausam* 35: 175-180
7. Ogallo, L.J. 1987. Relationships between seasonal rainfall in East Africa and the Southern Oscillation. *Journal of Climatology* 7: 1-13
8. Ogallo, L.J. 1989. The spatial and temporal patterns of East African seasonal rainfall derived from principal component analysis. *International Journal of Climatology* 9: 145-167
9. Potts, A.S. 1971. Application of harmonic analysis to the study of East African rainfall data. *The Journal of Tropical Geography* 34: 31-42
10. Rodhe, H. and Virji, H. 1975. Trends and periodicities of East African rainfall. *Monthly Weather Review* 104: 307-315

of recurring drought. One of these drought coping strategies, the accumulation of livestock, has given rise to the widely held belief that pastoralism is a destructive mode of resource exploitation that inevitably leads to degradation of rangelands at best and desertification at worst. The fact that pastoralists have persisted here for thousands of years forces us to examine that belief carefully.

In this chapter, we examine the structure and dynamics of arid and semi-arid grazing ecosystems in East Africa. We discuss pastoral management strategies and attempt to understand them in light of the long-term goals of pastoralists and as responses to the stresses that these hard and unpredictable environments generate. We address the issue of degradation of these dry grazing ecosystems and consider the roles of both climate and pastoral management on their long-term persistence. We also look at the history of development efforts in these areas and try to explain why success has been so elusive.

Climate and Ecology

Climate

East African arid lands can be defined as those areas having climates that are marginal or totally unsuitable for crop-based agriculture. These lands include the driest portions of the "semi-arid" eco-climatic zone, and all of the "arid" eco-climatic zone as defined by Pratt and Gwynne (26). Temperatures in the dry semi-arid and arid zones are continuously warm to hot and little variation occurs throughout the year (Fig. 10.1). In Lodwar, Kenya mean daily temperature ranges between 29° and 30° C. Lowest temperatures occur in July and August and highest temperatures occur in March and October. Mean maximum temperatures range from to 33° in July and August to 37° in February and March. The absolute monthly maximum in Lodwar is 40° in April. Mean minimum temperature is 23°. Temperatures in Wajir, Kenya are on average about 2° cooler than those in Lodwar, but display the same annual pattern. Temperatures are influenced by elevation, decreasing with elevation at a rate of about 6° C per 1000 m (26). Potential evaporation (PE) rates in excess of 2500 mm/y are typical for arid regions of East Africa and rates of nearly 4000 mm/y can be observed locally. PE rate is strongly influenced by temperature. Since temperature is affected by elevation, PE rate decreases with increasing elevation (36).

Mean annual rainfall varies greatly throughout the region, from less than 150 mm to 400 mm per year or more at higher elevations; always far less than potential evaporation. Rainfall through most of the arid zones of northern Kenya and southern Ethiopia is bimodal, with peaks occurring around April and November (Fig. 10.2). These peaks are produced by the movements of the Intertropical Convergence Zone (ITCZ), a band of low pressure where the trade winds of the northern and southern hemispheres converge (see Chapter 3). The ITCZ moves north and south over the equator, reaching a northernmost position in late July and a southernmost position in late January. The "short rains" of October to early December and the "long rains" of March to May occur when the ITCZ passes overhead. Parts of western Kenya are also influenced by the moist Congo air mass and by Lake Victoria, which tend to reduce dryness from June through September. In extreme western Kenya and in neighboring Uganda, rainfall is unimodal, with the peak rains occurring from March to May (Fig. 10.2). Thus, parts of western Turkana district may also show little increase in monthly rainfall during October and November relative to rainfall in July through September.

Year-to-year variation in rainfall is high, more so in drier parts of the region. For instance, in the driest year on record in Lodwar only 2 mm of rain fell, while in the wettest year 498 mm fell (Fig. 10.3). The mean rainfall between 1926 and 1988 was 182 mm, with a standard deviation of 107 mm. Thus it is not uncommon to have a "poor" rainfall year or even two "poor" years in a row. There are also longer term periods of good and poor rainfall years. The decade of 1926 to 1936 was very dry (mean 133 mm/y). The 1930s and 1940s were relatively dry (145 and 152 mm/y) while the 1960s and 1970s were relatively moist (258 and 221 mm/y). The early to mid 1980s were also drought years followed by increased rainfall into the 1990s.

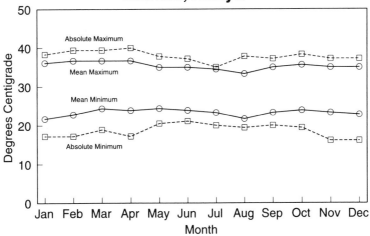

Figure 10.1. Average monthly maximum (day time) and minimum (night time) air temperatures in an arid zone of northern Kenya.

Spatial variability of rainfall is a key feature of the regional climate. Spatial variations in rainfall are obviously significant to animals or pastoralists who move about in response to changing spatial distributions of forage. Rainfall is spatially patchy, particularly over short time scales. Convective storms and highly localized rain showers are quite common. In general, rainfall increases with elevation (see 7). The driest areas of East Africa are those located at low elevations, such as the floor of the Rift Valley. Topography produces rainfall gradients within landscapes and regions. For example, within southern Turkana, rainfall is over 600 mm/y at high elevations, but only 200 mm/y on adjacent lowlands.

Ecology

Vegetation of the dry semi-arid and arid zones is physiognomically diverse, defying simple characterization as "desert" or "savanna", and the physiognomic classification system recommended by the East African Range Classification Committee (25,26) aptly describes the wide range of this variation. "Grasslands" are grass-dominated ecosystems with shrub and tree covers of less than 2%. Dwarf shrubs (< 70 cm tall) are often mixed with the herbaceous plants to form "dwarf shrub grasslands". "Shrub grasslands" and "wooded grasslands" have a woody covers of less than 20%. In a "shrubland" the woody plants do not exceed 6 m in height but their cover exceeds 20%. A "woodland" is comprised of plants > 6m tall with cover in excess of 20%. A "forest" has a closed (100% cover), multi-layered canopy of trees 7 to 40 m tall. All of these plant-community physiognomies can be found in the dry climatic zones of East Africa.

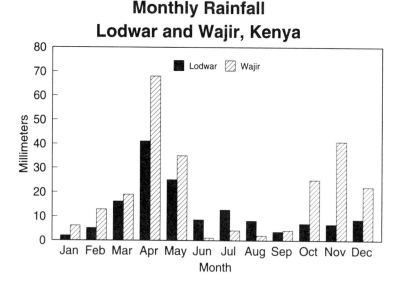

Figure 10.2. Average monthly rainfall in an arid zone of northern Kenya.

Vegetation physiognomy is often related to water availability, which is affected by soil texture, direct rainfall and water redistribution on the landscape. Woody canopy cover on sites not subsidized by run-on is roughly correlated with mean annual rainfall (5). Water-subsidized sites are likely to support more woody cover and larger trees. Where water is concentrated into ephemeral stream beds it infiltrates to a depth that is beyond the rooting zone of herbaceous plants. Large, partially phreatophytic, trees such as *Acacia tortilis* are supported on these sites. Woodlands develop in the alluvial zones of larger stream beds. Riparian forests composed of *A. elatior* and *A. tortilis* occur on massively water-subsidized sites in the alluvial zones of permanent water courses (such as the Turkwel River; see Chapter 14). The doum palm (*Hyphaene coriacea*) is fairly common along larger ephemeral streams as well as permanent water courses. The fruits of doum palm are edible and the young leaf buds are harvested for basket making.

In southern Turkana, the three most abundant tree species are *A. reficiens*, *A. tortilis* and *A. senegal* (5,24). *A. reficiens* tends to dominate upland interfluvial sites while *A. tortilis* dominates the ephemeral stream channels, with *Delonix elatior* found less commonly. The evergreen shrub *Salvadora persica* is often found along these channels. *A. senegal* is the dominant species of tree found on hillslopes and mountain sides, along with species of *Commiphora* and *Boswellia hildebrandtii*. A similar pattern has been described in Samburu District by Barkham and Rainy (1). Barkham and Rainy hypothesized that *A. senegal* and *Commiphora* occur on well-drained sites because they cannot tolerate dissolved salts and minerals that accumulate in alluvial sites. Diverse mixtures of *Acacia* and *Commiphora* and other woody species can be found near and on mountain-foot slopes. Other important *Acacias* found in these ecoclimatic zones include *A. nubica*, *A. mellifera*, *A. paolii* and *A. etbaica*. Other species of trees such as *Balanites orbicularis*, *Maerua crassifolia*, and *Boscia coriacea* occur on sites not dominated by the *Acacia* species.

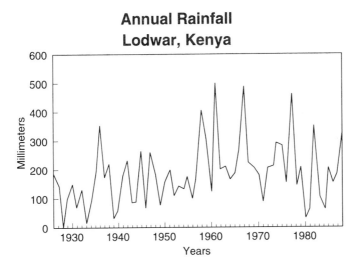

Figure 10.3. Total annual rainfall in an arid zone of northern Kenya collected over a 60-year period.

The herbaceous layer is dominated by annual grasses below about 300 to 350 mm/y of rainfall. Annual grasses are best able to withstand prolonged droughts as seeds. *Aristida adscensionis*, *A. mutabilis* and *Eragrostis cilianensis* are the most commonly encountered annual grasses. Clumps of *Cenchrus ciliaris*, *Stipagrostis uniplumis* and *Leptothrium senegalense* perennial bunchgrasses and *Kyllinga weltwitschii*, a perennial dryland sedge, can be found interspersed among the annual grasses at some locations. "Weak perennials" (such as *Tetrapogon cenchriformes* and *Perotis patens*) persist vegetatively through short dry seasons, but also produce numerous seeds. On sites with good soils about 7 kg/ha/y of forage is produced for each mm of rainfall (7). The most common dwarf shrub is the spiny legume *Indigofera spinosa*, a very important forage for livestock (6,8) and native browsers like the Grant's gazelle (*Gazella granti*). A similar species, *I. cliffordiana*, is less spiny and tends to occur on slightly moister microsites. The dwarf shrubs *Sericocomopsis* and *Duosperma* are important in many areas, especially in the Marsabit District of Kenya.

Interestingly, perennial grass species (*Dactyloctenium Brachiaria* and *Sporobolus*) often occur in the shade of large *A. tortilis* trees (32), even below 300 mm/y rainfall. Reduced evaporation rates in the shade, along with nutrient inputs from tree leaf litter, create favorable microsites for these species. These patches are often the first to green-up in response to early rains.

Above the 300 to 350 mm/y rainfall level, perennial grasses prevail. These commonly include *Stipagrostis* and *Leptothrium*, as well as *Sporobolus*, *Cynodon* and *Eragrostis*. Grass-like sedges (*Carex* and *Kyllinga spp.*) are also locally common.

Herbaceous biomass dynamics are extreme. Depending on the season, a landscape may appear quite barren or quite lush. Areas covered by ancient lava flows appear particularly barren in the dry season. The seasonal contrast is most apparent in annual grasslands. Grasses respond quickly to good rains, germinating and producing green ground-cover canopies. Peak greenness usually lasts for a relatively short period of time (around 1 month). Then, leaves rapidly dry and senesce. With the ensuing extreme dryness and occasional strong winds, leaf tissues disintegrate. Much of the dead tissue is consumed beneath the soil surface by termites, which use the crop of litter as a substrate for "fungal gardens". Within four to eight months after the rains, sites take on a "barren" appearance once again.

As a consequence of the rapid decline in forage quality, and the disappearance of forage due to consumption by termites and abiotic weathering, large mammalian herbivores are able to consume but a small fraction (less than 10%) of total forage production. This low "foraging efficiency" applies to species of grazing livestock as well as the native ungulates. Since only a small fraction of this annual plant growth can be consumed during the time it is available, the effects of herbivory on fast-growing annual plants is greatly diminished. Herbivores fill up on the abundant resources during good times and then compete for the more resilient perennials during drier times. Competition for forage must also be reduced during the annual flush, because there are few herbivores relative to available forage production when good forage is available. However, weak competition for this good forage apparently does not weaken the relationship between rainfall (and presumed forage production) and herbivore population density (3). The extent to which herbivore numbers are regulated through competition for forage or die-offs due to a lack of water in drought years requires further field study.

The extent to which pastoralists and their livestock have influenced the abundance of native wildlife species is poorly known, since pastoralists have been present in most locations for many generations. Grant's gazelles are abundant, apparently competing little with livestock for forage. While Grant's gazelles are primarily grazers on open grasslands, in Northern Kenya they are mixed feeders, relying heavily upon browse from dwarf shrubs and larger woody plants, particularly during the dry season. Other large herbivore species are present, such as elephants (*Loxodonta africana*), cape buffaloes (*Syncerus caffer*), giraffes (*Giraffe camelopardis*) and Burchell's and Grevy's zebras *Equus burchelli, E. grevyi*). Elephants tend to move along permanent and ephemeral water courses, often over great distances. Buffaloes can be a hazard for pastoralists in bushlands. Beisa Oryx (*Oryx beisa*) are prevalent in this ecoclimatic zone. The gerunuk (*Litocranius walleri*) is another classic dryland ungulate species. They are well adapted to browse on *Acacia* and *Commiphora* trees and shrubs with their long necks and rear leg foraging stance. They also do not require any drinking water. Competition with domestic goats, which forage similarly, must be significant. While black rhinoceroses (*Diceros bicornis*) were undoubtedly once fairly common, they are now absent from this region. Decreases in megafaunal species such as rhino and elephant due to hunting by humans in the last few decades may have unforeseen consequences for vegetation structure, given the ability of these species to reduce shrub and tree covers.

Predator species include spotted hyenas (*Crocuta crocuta*), black-backed jackals (*Canis mesomelas*) as well as occasional cheetahs (*Acinonyx jubatus*) and lions (*Panthera leo*). Predators tend to be more common in higher rainfall zones, probably in response to higher densities of native prey species.

The ubiquitous vervet monkeys (*Cercopithecus aethiops*) and the Anubis or olive-backed baboons (*Papio anubis*) can be found in this zone, particularly in the higher rainfall areas.

East African Pastoralism

History

The history of pastoralism in East Africa goes back 4000 or 5000 years. Sometime during that period, people who combined livestock husbandry with hunting and gathering and possibly some agriculture moved to the south from the Sahara and the Sahel (19). At first, pastoralists may have kept only sheep and goats, but bones of domesticated cattle have been recovered from sites inhabited 4000 years ago, and bones of the currently dominant cattle species *Bos indicus* dating back 2000 years have been discovered. Archeological evidence suggests that the relative importance of livestock, hunting and gathering, and agriculture has varied over time as environmental conditions changed, favoring first one strategy and then another. Still, pastoralism has successfully remained an important part of the mix of human lifestyles since it first arrived in East Africa. How have these pastoral people managed to be so persistently successful in such a difficult and unpredictable environment?

Goals and Strategies of Subsistence Pastoralism

To understand subsistence pastoralism and its ecological role we need to understand how the goals of subsistence pastoralists differ from those of market-oriented pastoralists. Once the goals are identified, we can interpret the strategies of subsistence pastoralists as a means of achieving those goals. The orientation of the two types of management systems are enormously different; market pastoralism having a profit orientation and subsistence pastoralism a survival orientation. A universal goal in market economics is, in some sense, to maximize net production or profits (see Chapter 3). The corresponding goal in subsistence systems is to maximize the probability of survival. Another goal of subsistence pastoral systems is to maximize the number of people that can be supported on the land through livestock or to maximize the human to livestock ratio. In market-oriented economies the corresponding goal is nearly the opposite, to minimize the number of people required to maintain the livestock operation or to minimize the human to livestock ratio such that the monetary profits per person are maximized.

These survival goals translate into two general and pervasive management strategies in subsistence pastoral systems; namely, to maintain the continuity of production so that there is always food for people, and to avoid or minimize risks to both humans and livestock. Production and the flow of food to the pastoralists must be maintained at some minimum acceptable level both within years (across seasons) and between years. In a subsistence setting, strategies that lead to high but variable food production, where food availability sometimes falls below levels needed for family survival, are not acceptable. Likewise, strategies that increase health risks to people or their livestock can not be tolerated. Arid ecosystems are inherently unpredictable and thus risky. Increasing risk further through strategies aimed at increasing production may be successful in the short term but can frequently court disaster over the long term.

Management Tactics

Given that the goals and strategies of subsistence and economic pastoral systems are so different, it is not surprising that different sets of management tactics have developed to achieve them. If we observe the behavior of subsistence pastoral people through a filter that is conditioned by our understanding of the goals of market pastoral systems, we will misinterpret what we see, and arrive at erroneous conclusions about the suitability of subsistence management tactics.

There is a characteristic set of resource management tactics that reappear time and again in subsistence pastoral societies. The strategies often appear as different variants in different systems, but they almost universally exist in one form or another, and contribute to the success of pastoralism.

Mobility. Mobility of both people and livestock is so common in subsistence pastoralism that it can be considered a defining characteristic of such systems. The two major forms of mobility are nomadism and transhumance. In nomadic systems, herders and their herds wander erratically through their pastures, without any set repeatable pattern. Pastoralist movement seems to be an opportunistic response to the availability of forage and water for the livestock - which normally correlates with the presence of rain. The reasons for movement, however, are more complex and vary from year to year depending upon environmental and social conditions (20). In addition to the need for forage and water, movements may be a response to concerns about diseases in certain areas, or may be politically or socially driven - such as the desire to see friends and relatives or the desire to avoid threatening people.

In transhumant systems, seasonal movements are quite predictable, usually involving movement from a wet-season range to a dry-season range. Pastoral systems in the Sahelian zone of Africa are the classical example of transhumant systems. Among the Turkana of Kenya, the southern Ngisonyoka are nomadic while the Ngikamatak to the west are essentially transhumant, moving from their wet season range on the Turkana Plains to a dry-season range atop the Rift escarpment.

Mobility results in complex patterns of aggregation and dispersal of pastoralists and their herds, and so affects spatial and temporal patterns of grazing intensity. Among the Ngisonyoka Turkana, people and livestock are most strongly aggregated during the wet season, when conditions provide adequate nutritional forage and drinking water for all livestock within a small area. People remain together during this period because environmental conditions permit aggregation. This time is a period of intense social interaction including singing, dancing, courtship and exchange of information about the pastoral environment. As forage resources diminish, with the onset of the dry season, social groups begin to disperse. Typically the families and herds of three to five men will split away from the larger group and move to some other location with better forage. As forage is depleted, these intermediate-sized groups begins to disaggregate further, splitting into individual family units that eventually split into smaller sub-family units. Thus there is a gradual dispersal of pastoralists over the landscape as the dry season progresses. Dry-season use of forage is fairly wide spread and even over the areas not grazed during the growing season. Dry-season dispersal is driven by declining forage as the forage grown during the wet season is eaten. Dispersal is possible because drinking water is fairly wide spread throughout the territory of the Ngisonyoka, though usually not in large sources .

Among the Rendille of Northeast Kenya, just across Lake Turkana from the Ngisonyoka, the patterns of aggregation and dispersal are exactly the reverse. Here, maximum aggregation occurs during the dry season around the few, large, dependable water sources. Livestock can move from these centers to forage only so far before they must return for water. Thus, areas around dependable water holes experience prolonged and heavy use during dry seasons. When the wet season begins, water is widely available in temporary streams and surface accumulations. This permits the Rendille to spread throughout their territory in search of new forage and relieves the pressure on the dry season camps found around permanent water sources. Consequently, the existence of standing water can have profound consequences on pastoral systems.

Mobility is a response, therefore, to maintain the continuous health and production of livestock. When mobility is motivated by concern for disease or other threats to humans and livestock the aim is to minimize risk.

Livestock Diversity. As in most subsistence lifestyles, pastoralists do not specialize in a single variety of livestock, in contrast with market-oriented economic systems. In market systems, concentrating efforts on livestock varieties with the highest average rate of return simplifies management through specialization and may be a good tactic for maximizing profits. For a number of reasons maximizing the diversity of livestock has more advantages in subsistence systems. First, high livestock diversity minimizes competition because the different livestock species utilize different resources. This tends to maximize yields by ensuring that all niches are full and all resources are used. Niche separation occurs not only because different species consume different plants but also because different species can utilize different types of terrain, consume forage from different heights in the plant community, and have different watering requirements, which converts to different traveling distances from water sources. Secondly, unpredictable environmental conditions may favor different species such that a diversity of livestock will ensure that some livestock species are successful as environmental conditions change. Thirdly, livestock diversity also results in a wide range of products being available to the pastoralists and often results in those products being available at different times. Small stock, with their relatively high reproductive rates and small body sizes, are better suited as meat animals than are cattle or camels. Large stock, less suitable for meat production, are excellent sources of milk and blood.

Keeping a range of livestock species also creates a wide range of labor requirements. Small children are often suitable as goat and sheep herders, freeing up young adults to herd larger stock. With sufficient livestock diversity, opportunities exist for almost all family members to contribute productively to the pastoral enterprise.

Grazing Reserves. Most subsistence pastoral systems include areas that are reserved for grazing during dry seasons or more extended droughts. In some cases these reserves are formally designated and recognized by the pastoralists as grazing reserves, as in the case of the hema systems of North Africa and the kallo systems among the Ethiopian Borana. More frequently they are not specifically designated or thought of as reserves but they clearly serve the same ecological function. In South Turkana, for example, the central mountain range which bisects the territory of the Ngisonyoka is rarely grazed. The people cite lack of water, abundance of predators, and a general dislike of mountainous regions as the reasons it is not used. During severe droughts, however, cattle are taken into the mountains where they consume the dead grass that was not used early in the drought. Turkana pastoralists living in the western part of the district often take livestock up the Rift Valley escarpment into and across Uganda during drought periods, opportunistically using forage that may include field crops of Ugandan farmers. These farmers do not think of their fields as being a grazing reserve for the pastoralists, but, in essence, they are.

Frequently there are grazing reserves that have arisen for almost accidental reasons. Certain areas are considered too dangerous because of disease or livestock raiding. Under drought conditions, however, the value of the unused forage makes the risk worth while. Similarly, in some pastoral systems there are often areas that are not currently being used because droughts or other emergencies have reduced pastoral livestock numbers such that they do not fill the land. That is, livestock are currently below the long-term carrying capacity of the land, and, the areas not in use, constitute *de facto* grazing reserves in the event of another drought.

Herd Splitting. Typically, as part of the disaggregation process referred to above, or for other reasons, pastoralists split their herds into smaller units provided there is sufficient shepherding labor. Herd splitting has several advantages. First, species are often best suited to different vegetation, topography and water availability and splitting herds will ensure that each species forages under optimal conditions. Secondly, herd splitting reduces or spreads the risk of livestock loss. If all of the animals are kept together, there is the danger that all will simultaneously fall victim to disease, disaster, or raiding.

Particularly during dry or drought periods, livestock herds are split on the basis of their economic function. Productive females producing milk are kept near the homestead where the majority of the pastoral family, particularly women, children and elders are living. Non-milking females, castrated or surplus males and other non essential animals are sent away from the homestead, typically in the care of young unmarried men. This reduces grazing pressure on forage near the homestead, and improves foraging conditions for animals producing milk for the most vulnerable members of the pastoral family. These two herds are frequently referred to as the "wora" (productive or milking) and "fora" (nonproductive or fallow) herds, from the Borana words for these two herd types.

Restocking Procedures. The drought-prone nature of the arid and semi-arid zones leads to repeated large-scale losses of livestock from malnutrition and thirst. When this occurs, it is in the interest of the affected pastoralists to redevelop their herds as rapidly as possible. Redevelopment of a viable herd is made easier when owners include small, fast-growing stock in their holdings. These animals, with their rapid reproductive rates, can recover their numbers rapidly when good environmental conditions return. Consequently, many pastoralists, who normally keep mixed herds, concentrate their efforts on small stock husbandry. Once an acceptable holding of small stock has been attained the pastoralist will rediversify the herd to include larger stock.

Another traditional method of augmenting livestock numbers following large-scale losses has been inter-tribal raiding. In this way areas less affected by drought or other catastrophes are forcibly subsidizing the recovery of more severely affected pastures. This method of restocking has been sharply curtailed in colonial and post-colonial times. The negative aspects of theft currently make it an unacceptable practice. In the past, however, it probably acted to increase the long-term stability of pastoralism in the region. If this practice is not to be permitted, consideration should be given to finding legitimate restocking procedures.

Emigration. During drought periods, as nutritional stress deepens, those people who are not essential to the herding often emigrate to neighboring areas and attempt to subsist there, returning to their homes when the threat of starvation ends. In Ngisonyoka, mothers, their small children, and the elderly - anyone who can not contribute to the care of livestock - frequently move out of Turkana and into the highlands to the south. Here, they either impose on previously settled relatives or attempt to find work on farms. This tactic improves the probability of survival for both those who leave and those who remain. It is also beneficial to the herds by reducing the demand for food from humans. This reduces the need to slaughter animals for food and leaves a larger proportion of milk for young animals. Temporary emigration is sometimes used by drought-affected pastoralists to find work and money to buy back livestock and other necessities.

Ecosystem Energetics

The most completely studied pastoral system in the arid and semiarid zones of East Africa, in terms of ecological energetics, is that of the Ngisonyoka Turkana of Northwestern Kenya. A large team of ecologists and anthropologists worked there for nearly 10 years, beginning in 1980, characterizing the ecology and culture of these traditional nomadic pastoral people. Data were collected that permitted the investigators to trace the flow of energy from the plant community, through the livestock and their products, and into the human population.

The results indicate a complex network of energy flow through the ecosystem (9). Livestock products dominate the food energy obtained by pastoralists (Fig. 10.4). Milk, meat and blood made up 75% of the energy intake of the average Ngisonyoka on an annual basis. Most of the remaining 25% was obtained by trading pastoral products for food produced outside this pastoral system. Milk, comprising 61% of the diet, was the most important single food, confirming that these pastoral systems are essentially extensive dairy operations. Livestock meat and blood (consumed mostly during the dry season when milk is not available in quantity) each contributed another 7% to the diet. Wild meat and plant foods also contributed 7%. Locally grown sorghum accounted for 3%. The only non-traditional foods of any importance were maize meal (10%) and sugar (3%).

Camels are an important component of this system, and camels contributed more pastoral food energy to human diets (56%) than did any other species (Fig. 10.5). Sheep and goats together accounted for 23%, cattle 19% and donkeys - an emergency food source - 2%. Tracing the human food obtained as livestock

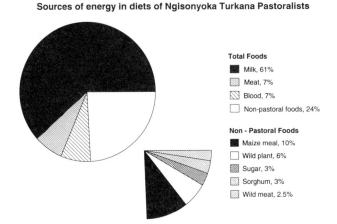

Figure 10.4. Percentage of energy sources in the diets of Turkana pastoralists.

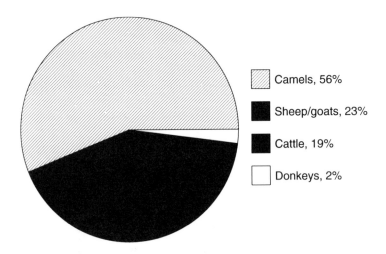

Figure 10.5. Animal sources of pastoral foods found in the diets of Turkana pastoralists.

products back to the plant community responsible for their production reveals the importance of browse in this system (Fig. 10.6). Dwarf shrubs (principally *Indigofera spinosa*) were the most important component of the plant community, being ultimately responsible for 43% of the energy consumed by humans. Herbaceous species provided 36%, leaves and twigs of trees and large shrubs 19% and seed pods of *Acacia tortilis* 2%. The largest single pathway providing food to the humans was from dwarf shrubs to camels to camel milk to humans; accounting for 30% of the energy consumed as food. Obviously other pastoral systems will have different patterns of energy flow. In particular, we would expect the importance of

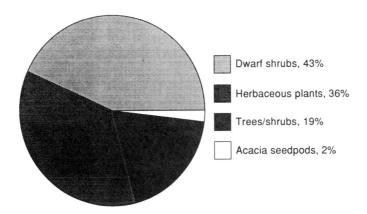

Dwarf shrubs, 43%

Herbaceous plants, 36%

Trees/shrubs, 19%

Acacia seedpods, 2%

Figure 10.6. Contribution of different plant groups to production of livestock products in the diets of Turkana pastoralists.

shrubs versus grasses to be variable among ecosystems and that the role of camels in human nutrition will change from place to place.

Some more general observations can be made from this study. The Ngisonyoka system produced about 2500 megajoules (MJ) of human food per km^2 per year. This is much lower than agricultural systems, which may produce as much as 7,700,000 MJ/km^2/y, or even intensive beef production systems (63,500 MJ/km^2/y). It is comparable to other arid systems in East Africa, however, and requires no subsidy from fossil fuels as do the more intensive agricultural and pastoral systems.

Of course one of the primary reasons for this low level of productivity is the low level of rainfall. Arid systems do not have much rainfall, and generally convert rainfall into secondary production less efficiently than wetter ecosystems (3). Rain-use efficiency (livestock production per unit of rainfall) in Ngisonyoka is 83 MJ/km^2 per centimeter of rain, only slightly higher than the 76 MJ reported for a system dominated by native ungulates in a 300 mm rainfall zone (3). By this measure, the energetics of this pastoral system appear to be very much like those of similarly arid ecosystems where man is absent, despite the high levels of human labor and management used in Ngisonyoka.

Livestock in Ngisonyoka harvest a surprisingly small proportion of above-ground net primary production - only about 7%. Only dwarf shrubs (67% utilization) are utilized heavily here. The subsequent conversion of the consumed plant material into human food is also surprisingly low - around 1%. Much of what is consumed by livestock is of course not digested, and that which is digested is not used with 100% efficiency. Much of the useful (metabolizable) energy obtained by the animals is used for their maintenance, with the activity cost of travel being particularly important. Some potentially human consumable food (milk) is used by young livestock. Sharing of milk between humans and young livestock is one of the most difficult energetic decisions made by pastoralists. If too much milk is consumed by humans, people do well in the short run but herd health suffers and long-term sustainability is endangered.

Livestock diversity in Ngisonyoka allows these pastoralists to use the food web in three ways. First, there is the general and continual dependence upon the most reliable energetic pathway - woody plants to camel milk. Secondly, when weather conditions permit there is opportunistic use of the pathway from herbaceous plants to cattle milk. This pathway, while capable of transforming large quantities of energy when in full operation, is not dependable either within or between years. Finally, there is the contingency conversion of livestock meat and blood to human use when the other pathways are not sufficient to meet the pastoralists' needs.

The foregoing description of energy flow describes the average use on an annual basis. Equally important, however, are the temporal dynamics of energy flow, both within the annual cycle and in multi-year periods involving sequences of drought and recovery.

Typical or normal years in Ngisonyoka are characterized by a unimodal rainfall pattern followed by a single, short period of active plant growth. During this period forage quantity and quality are sufficient for the animals to improve their condition and to produce large quantities of milk. When the rains cease, and soil moisture becomes limiting, plants stop growing and their quality to herbivores declines. Animals dependent on these plants can not maintain high levels of milk production or their body condition, even if the quantity of food is adequate. The production of food to humans begins to decline, and does not reach high levels again until the subsequent rainy season. This pattern is observed for grasses and grazers and for browse and browsers. There are, however, important differences in the timing of events along these two complementary feeding pathways.

Grasses exploit mainly the soil layers near the surface, and so, when rain ceases, their water resources are very quickly exhausted. Growth of grasses stops suddenly, plants senesce and their forage quality plummets. Thus, the diets and patterns of productivity of grazers, such as cattle, reflect the timing of rainfall very closely - a short period of good quality forage and high productivity followed by a long period of sub-maintenance conditions. In normal years, a great deal of energy can flow along this pathway from grasses to cattle to humans, but only for a brief part of the annual cycle. Consequently, a subsistence strategy based solely on this path of energy flow is tenuous indeed.

Where shrubs exist it is because water, for one reason or another, is able to percolate deeper into the soil where it is accessible to woody plants but not grasses. Deep percolation is more likely where soils are coarse and so have low water-holding capacity, and in low-lying areas which receive water in the form of runon from adjacent uplands. Such water-subsidized locations are effectively much wetter than the average of the area. Here, water supplies last longer into the dry period and the growth of shrubs and trees continues for longer periods than grasses. Consequently, the effective period of high forage quality is extended, and browsers such as camels enjoy longer periods of good nutritional conditions and productivity. Typically, energy flow along this pathway (browse to browsers to humans) is never as rapid as it is along the pathway based on grasses because browse quality, at its best, is not as good as the quality of grass during its active growth period. The length of time that the pathway functions close to its maximum production level is longer however. The browse pathway is more dependable and shortens the period of nutritional stress for browsers and their human herders. Both of these ecological pathways are evident in Ngisonyoka. The result is a pulse of energy into the human population during the rainy season, originating from the grass-grazer pathway, and a relatively long,

Box 10.2.   Chaos in Ecosystems

Ecologists exploring mathematical models of plant and animal populations have found that relatively simple deterministic equations or groups of interacting equations created to reflect the process of population growth and decay can produce complex and seemingly unpredictable behavior that has been referred to as chaos (1,2). The three main attributes of these equations are that they are 1) entirely deterministic or have no randomness added to the equations, 2) the modeled variables never settle to an equilibrium or steady state nor do they have repeatable cycles, and 3) the behavior of the equations can be very sensitive to the initial values used to start the simulations (3). Consequently, two simulated populations or ecosystems that start at nearly the same point can, with time, diverge such that they no longer resemble one another. Because one of the main purposes of creating mathematical models was to predict population dynamics, the discovery of chaos in these equations was unsettling to many ecologists and was, therefore, often discarded as a mathematical oddity not relevant to real populations and ecosystems. Many of the population equations required very high reproductive rates before chaos was observed and therefore some ecologists believed that chaos would only be observed in small fast-growing populations. In complex multispecies models or models with feedback's and time delays, however, chaos is more commonly observed because of the complexity of equations. Once ecologists began looking for chaotic behavior in real populations supportive evidence began to accumulate even for larger and slower growing populations (4,5). It has also come to be realized that some chaotic behavior is predictable. Oscillations may never exactly repeat themselves but some studied oscillations have predictable periodic shifts, rises, and falls. It is still too early in the development of ecological science to determine the practical usefulness of studies in chaos and complex dynamics but field observations and models suggest that it cannot be ignored and that we should be prepared to find surprising dynamics in nature.

T.R. McClanahan

References

1.   Gleick, J. 1987. *Chaos: Making a New Science.* New York: Penguin
2.   Pimm, S.L. 1991. *The Balance of Nature: Ecological Issues in the Conservation of Species and Communities.* Chicago: The University of Chicago Press
3.   Schaffer, W.M., Kot, M. 1986. Chaos in ecological systems: The Coals that Newcastle forgot. *Trends in Ecology and Evolution* 1: 58-63
4.   Tilman, D., Wedin, D. 1991. Oscillations and chaos in the dynamics of a perennial grass. *Nature* 353: 653-655
5.   Schaffer, W.M. 1985. Order and chaos in ecological systems. *Ecology* 66: 93-106

attenuated input of energy from the browse-browser pathway, which extends well into the dry season.

Patterns of energy flow over time, associated with sequences of drought and recovery, are not necessarily obvious. If we consider a sequence of two consecutive drought years followed by a return to normal conditions, we find that the third year, the year of recovery, is likely to be the most difficult year for the pastoralists. During the drought period, particularly year two, animals are in very poor condition; and, while milk production is very low, livestock dying from malnutrition provide a substantial source of meat energy to the pastoral people. In the third year, foraging conditions are once again good. Animals are able to regain condition, but very few produce sufficient milk because poor environmental conditions in the preceding year depressed reproductive performance. Once again, there is little milk for the human population, but now animals are no longer dying and food for pastoralists is very low. Herd owners are reluctant to slaughter animals for food because livestock numbers have already been reduced to dangerously low levels by the drought. This recovery year is a year of recovery for the plant and livestock communities, but not for the pastoralists themselves who sit at the top of the food pyramid. Their recovery requires a second year of good rain conditions. The position of the pastoralists in the trophic web, as secondary consumers, introduces a time lag into the drought and recovery process. If famine relief efforts have been initiated to deal with the drought, the first year of recovery is certainly not the time to discontinue those efforts.

Effects of Pastoralists on the Ecosystem

Subsistence pastoralism is not a high technology or energy-subsidized form of resource use. Direct, complete replacement of native plant communities with other domesticated plants, as occurs in agricultural systems, is rare. Grazing systems are expected to stay in natural or near-natural conditions. Nevertheless, pastoralists and their livestock have important effects on the structure and dynamics of the ecosystems they inhabit. Whether the changes that result from human use are beneficial or detrimental is often difficult to say. The answer depends on what we expect the ecosystem to look and behave like, what we feel it should be producing, and for whom.

Plant Communities and Productivity

The structure of plant communities is almost universally modified by pastoral activity; but neither the direction nor the degree of change is the same in all pastoral systems. Pastoral exploitation often affects the balance between grasses and shrubs. In some instances, pastoralism is responsible for the conversion of bushland to grassland and in others instances it results in bush encroachment.

In South Turkana, pastoralism shifts the balance towards grasslands through a variety of processes. The Ngisonyoka Turkana keep a wide variety of livestock, including many goats and camels which are browsers. This reliance on browsers favors the opening of bushland and an increase in grass cover. The pastoralists themselves also favor grasses by harvesting some small *Acacia* species for constructing corrals and homes. In some areas, fires are deliberately set because these fires promote the maintenance of grasslands in areas where brush would otherwise dominate.

On balance, from the perspective of the Ngisonyoka, these changes are beneficial. Browse is generally not in short supply and any increase in grass is useful to cattle and sheep. Heavily bushed areas, particularly those covered by

Acacia reficiens, an unpalatable and well armed species, are of little use to pastoralists, although they often support wildlife populations. Farther to the north in Turkana District, where browse is in short supply, any activity that further reduces browse would be detrimental to these pastoralists.

In the Sidamo region of southern Ethiopia, Borana pastoralists tip the scales in favor of shrubs, and brush encroachment is frequently a problem. The Borana keep no camels and relatively few goats, concentrating on cattle. In addition, the setting of wild fires has been outlawed by the Ethiopian government. These two circumstance combine to favor shrubs over grasses, and dense stands of *Acacia drepanolobium* often develop. This *Acacia* species is of little value as forage and reduces grass where it grows. Here, the changes induced by pastoralism are not favorable to the pastoralist.

In some cases, not only in East Africa, but elsewhere, grazing can convert perennial grasslands to annual grasslands. This is normally taken to be an indicator of degradation, but this is not necessarily the case. The stresses placed on grasses by grazing and by the lack of water are similar and in some sense additive or perhaps multiplicative. Typically, even in the absence of grazing, very dry grasslands are dominated by annual grasses while wetter sites support perennial grasses. The addition of pastoral grazing to sites at the drier end of the environmental range capable of maintaining perennial vegetation may add another stress to the plants, already water stressed. This may tip the ecosystem over into a state dominated by annuals. Presumably, these sites could be maintained as perennial grasslands, but only by eliminating grazing completely. The annual grassland may be seen as a degraded ecological state, but this ecosystem is supporting higher levels of secondary production than could be supported by a perennial grassland - at least during part of the year.

Grazing may affect primary productivity even in the absence of large changes in plant communities. The traditional view has been that grazing reduces primary productivity by removing photosynthesizing surface area (such as leaves), increasing woody and herbivore-defense organs and, thereby, increasing plant respiration. More recently, however, McNaughton and other plant ecologists have demonstrated that at low to moderate levels of grazing, net primary productivity can actually be increased (21). This is referred to as the "grazing optimization" response because it implies that there is some optimal level of grazing at which plant production is maximized.

While there is still considerable controversy surrounding this subject, what does appear to occur with consistency is that grazed plants are able to accommodate grazing losses through "compensatory growth". This compensation may be incomplete, in that the grazed plant does not produce more than an ungrazed plant (a true optimization type response), but the compensation is, at least, partial, meaning that the plant is able to recoup at least part of the losses in production that a perfectly passive plant would suffer in response to grazing. Nitrogen uptake by grasses is often stimulated by grazing and if the optimization issue is reanalyzed in terms of aboveground nitrogen rather than above-ground dry weight production, a positive response to grazing is more likely (33).

The responses of shrubs to browsing has been less studied. Work in South Turkana on *Indigofera spinosa*, however, reports that at even fairly high levels of herbivore removal, no depression in aboveground production or nitrogen yield was found (6,22). This is not to suggest that rangeland can be continuously over-grazed without penalty. No doubt, in areas within the pastoral zone that are chronically grazed at very high intensities, such as around major water sources and permanent settlements, primary production is severely reduced as a result (see Box 10.3). This is true of course whether the chronic over-grazing is being done by domestic or wild animals. Further research is needed to determine the optimal levels of grazing under different environmental conditions.

Nutrient Dynamics

One of the less obvious roles of pastoral livestock is to transport nutrients through the system, strongly modifying spatial patterns of soil fertility. In most East African pastoral systems, livestock graze over large areas during the day and then are corralled at night to avoid predation, theft, and wandering. During the day they harvest nutrients from the plant communities they exploit. Many of these nutrients are then returned to the soils of the corrals with urine and feces deposited during the night. Frequently these corrals are constructed near the edge of ephemeral waterways to take advantage of the shade cast by riverine trees. Such areas typically have better soil water stores, both near the surface and at depth, than adjacent uplands. The effect, then, is to harvest nutrients over a fairly broad area and to concentrate them in "ecological hotspots" where those resources limiting to plant growth can be found in relatively high abundance (37).

In South Turkana at least, this concentration of nutrients is accompanied by a concentration of viable seeds of *Acacia tortilis* which have passed through the digestive tracts of the corralled livestock. These passed seeds find themselves in an ideal seed bed, mulched, fertilized and reasonably moist; and abandoned corrals are often discernible as miniature thickets of young *A. tortilis* (27). In this case, pastoral management increases the populations of a valuable resource which provides a nutritious food for young livestock in the form of seed pods, building and fire wood for pastoralists, and a welcome shade for both pastoralists and livestock.

Pastoralist Effects on Wildlife

Conflicts between pastoral livestock and wild ungulates and their predators is a major, recurring concern in East Africa. Clearly the potential for competition for forage between wildlife and livestock exists. One of the tactics of pastoralism is to fill all of the grazing and browsing niches with domestic stock; and this can not be done without some negative effect on wild herbivores. Still, most subsistence systems contain significant populations of wild grazers, though, we may assume, not in the numbers that would be found if livestock were absent. Our observations based on our work in northwest Kenya and southern Ethiopia suggest that the full expected complement of wild grazers are present but that numbers of most species are not very high and that the animals distribute themselves so as to avoid interactions with pastoralists and their livestock. In both areas, the greatest diversity and numbers of wild ungulates is found in places not currently being used by livestock. Wildlife often "reappear" in areas temporarily abandoned by pastoralists, and their seasonal movements between wet and dry season ranges may be exactly the opposite of the pattern shown by domestic animals.

Pastoralists also affect wildlife directly by hunting them for food. In many pastoral systems, hunting and gathering are still practiced as part of the normal mode of life on a seasonal or opportunistic basis. Large ungulates, small mammals, birds, reptiles and insects may be harvested by pastoralists, particularly young herd boys who are adept with a variety of weapons from bows and arrows to hand-cast stones. In some instances, predatory animals may be killed to protect livestock, but in general, the pastoral solution to predation seems to rely more on corralling livestock and predator avoidance than on direct predator control. Hyenas, cheetahs and to a lesser extent, lions are fairly common in South Turkana, but the Ngisonyoka don't consider them a serious threat and, in general, they are not persecuted.

Livestock and wildlife have some diseases in common, and in some cases one species may not be affected by the disease but will carry it and transmit it to species that are affected. There is the danger that wildlife may act as a reservoir for diseases that can erupt and spread into pastoral livestock or vice versa. In the Ngorongoro crater in Tanzania, wildebeest carry malignant cattarahl fever but are immune to it. Cattle are able to contract this disease from wildebeest when the wildebeest are calving because the microorganisms that cause this disease are released from the afterbirth of the wildebeest. As a consequence, Maasai herders must avoid areas being used by wildebeest for calving. On the other hand, rinderpest was common among wildebeest and buffalo populations in the Serengeti until the early 1960s because the disease was common among pastoral cattle until that time (29). When the disease was eradicated among the livestock in and around the Serengeti, it quickly died out among wildlife populations as well. The low levels of veterinary intervention in most subsistence pastoral systems keeps livestock and wildlife disease as a continuing problem.

Biological Diversity

To our knowledge the effects of pastoralism on the biological diversity of arid and semiarid systems have not been definitively determined. It is possible that the diversity and abundance of the "charismatic megafauna" is reduced by pastoralism. Beyond that, a case can be made that pastoralism, unless it results in serious range degradation, may actually increase biological diversity. Pastoralism, through patterns of differential use, conversion of some bushland to grassland and vice versa, and redistribution and concentration of nutrients tends to create a more patchy environment with more types of habitats within it than would be found in similar landscapes without pastoralists. This should create suitable living conditions for a broader range of small mammals, birds, insects, soil dwelling organisms and others. This pattern, however, may be dependent on the abundance of people and livestock rather than just the simple presence or absence of pastoralism.

Degradation and Desertification

On of the most important questions relating to subsistence pastoralism, and one of the most hotly debated subjects over the last 50 years, is what role, if any, does pastoralism play in rangeland degradation or desertification. This issue was cogently reviewed by Dodd (10), who concluded that pastoralism is not inherently destructive and that scientific evidence for the desertification scenario was weak.

The two process, degradation and desertification, although related, are different. According to Dodd, degradation generally refers to unfavorable change in plant communities or productivity which is reversible over relatively short (ecological)

Figure 10.7. The two photographs show the same location in Lachakall (one of the deep wells in Ngisonyoka, Turkana) after two years of drought (left) and 2 years later after good rainfall (right). These photos show that areas that appear severely degraded if seen in one season often recover quickly when the rains return (Credit D. Swift).

time scales. Desertification is generally taken to be more extreme in degree and essentially permanent in nature. It often refers to the expansion of deserts into adjoining areas, and is frequently suggested to be at least partly anthropogenic in nature, pastoralists normally being the suggested culprits.

One camp holds that, because pastoralists always accumulate livestock beyond their needs (16) and because the grazing lands are held in common (14), pastoralism always and inevitably leads to overstocking, degradation or desertification. Lamprey (17) argued that the boundary of the Sahara had moved south some 100 km in the two decades prior to 1975; and that pastoralism was at least partly responsible for the shift. Others argue that there is, in fact, no compelling evidence that the Sahara is expanding (11,15). For them the question of pastoral responsibility in creating deserts is moot because there is no long-term directional change in the spatial extent of deserts in East Africa. Rather, it is argued, the boundary between deserts and adjacent grasslands shifts back and forth dynamically in response to, primarily, vagaries in rainfall and not livestock numbers.

The issue of degradation, however, remains. Areas that are repeatedly overgrazed become degraded. We can find evidence of rangeland degradation in East African subsistence pastoral systems and in North American market pastoral systems as well, any place that stock concentrate in large numbers for long periods of time. The important question deals with the spatial extent of the degradation and the causes behind it.

A decade of work in South Turkana, Kenya has led us to believe that widespread rangeland degradation is not a universal and unavoidable result of subsistence pastoralism. Areas around permanent, high producing waterpoints are frequently severely degraded as are zones around permanent human settlements in the district. Similar situations can be found in northeastern Kenya, southern Ethiopia, and throughout the pastoral zones of Africa. Degradation around waterpoints is generally minor in extent and could be corrected by a change in the availability of water. Degradation around settlements has more to do with the provision of education, health care and religion than with pastoralism *per se*. Outside of these fairly restricted zones, in the vast majority of the traditionally exploited pastoral area, we find no evidence of significant degradation (13).

Pastoralists may indeed wish to accumulate stock in excess of their immediate needs, but this appears to be a rational hedge against drought rather than an irrational "cattle complex" (16). In many pastoral areas, herd owners are not particularly successful in this accumulation in any event; not successful enough in most cases to build their herds to the point that overgrazing is a general problem. Periodic droughts and other disasters typically reduce herds below the long-term support capacity of the range before serious degradation can occur. This is not to say that pastoralists are not capable of degrading their ranges or that they would not do so if they could. Interventions in pastoral systems that would reduce fluctuations in herd size and permit the development and maintenance of over-large herds might be accepted by pastoral people, but are likely to be prescriptions for disaster.

Certainly there are examples of degraded pastoral systems in East Africa. Consideration of these cases reveals, however, that the degradation has usually followed some sort of disruptive change that has modified the traditional patterns of land use. Rangeland degradation in Baringo District, Kenya has resulted from the fact that the Il Chamus pastoralists have lost much of their traditional dry season ranges to agriculturists (18). The resulting year-long use of what had been wet season ranges has resulted in their degradation. Similarly, when the Mkomazi

Box 10.3. Rangeland Degradation in Arid and Semi-arid East Africa

In the driest ecosystems in East Africa, there is increasing evidence that much of the ecosystem function is driven by rare and unpredictable rainfall events. In these ecosystems, livestock may only rarely exceed the carrying capacity of the land, and degradation due to 'overgrazing' may be rare. There are, however, rangeland conditions under which degradation by livestock is more likely. First, at focal points within arid ecosystems such as boreholes, livestock densities can cause severe localized degradation (1). Second, in less dry ('semi-arid') ecosystems, livestock numbers may be more able to reach levels high enough to cause degradation (4).

In arid regions such as northern Kenya, degraded rangelands are usually closely associated with settlements and sources of permanent water: wells, boreholes and rivers (3). The same pattern occurs in arid central Somalia, where dryland farming also contributes to degradation (2,3). The extent of the degraded area depends on the type of rangeland and on the type, intensity and duration of land use supported by a water source. For example, two relatively low output wells on coastal perennial grassland in central Somalia, where livestock husbandry was the only land use, were surrounded by 0.3 and 0.8 km² of degraded rangeland, whereas a 17-year old high output borehole in deciduous shrubland supporting both livestock and farming was surrounded by 63 km² of degraded rangeland (2,3).

Generally, degraded areas in arid ecosystems are fairly small because water sources are a) few and widely scattered, b) of low productivity, c) too new to have had much effect on the surrounding land, or d) located in a rangeland type, such as annual grassland, that cannot support enough continued use throughout the year to cause widespread degradation. In the Marsabit District of Kenya, where many of these features co-occur, the typical area of degradation around settlements and major water sources is about 10 km² (4).

Where water sources are more numerous, more productive, older, and/or occur in rangeland that can support longer periods of use, the degraded areas become larger. For example, Wajir town in northeastern Kenya, which has numerous shallow wells and one borehole and which is surrounded by eight groups of very old wells, is centered in an area of approximately 250 km² of degraded rangeland (3). The future health and productivity of arid ecosystems in East Africa may therefore be compromised by ongoing efforts to increase the number and output of permanent water sources (4).

In semi-arid rangelands, rainfall is higher and more reliable than in arid rangelands. Natural sources of water are more abundant and rangelands are capable of supporting larger numbers of livestock for longer periods of time. There is, therefore, less need for pastoralists to move long distances to find forage and water, and farming is more likely to be attempted. Both of these lead to increased sedentarization of people. The effects on these rangelands of increasing human and livestock populations are particularly severe because although these ecosystems are mesic enough to encourage growth, development and sedentarization, they are arid enough to have strong limitations in their ability to support these trends.

Subsequently, large areas of communal semi-arid rangelands in northern Kenya have been overgrazed and degraded (4). In these areas degradation is no longer easily identified with particular settlements or water sources. Instead, whole landscapes have been degraded: (in northern Kenya) 1750 km² in southwestern West Pokot

Box photo. Severely degraded area in South Turkana - caused in this case by sedentarization around Lokori town (credit D. Swift).

District, 5200 m² in central Samburu District, and 5600 km² in central Baringo District and the Kerio Valley portion of adjacent Elgeyo Marakwet District (4). Accelerated soil erosion is common and, in some areas, has irreversibly lowered the productive potential of the rangelands (4). The relationship between rainfall, livestock, degradation, and desertification needs to be better understood.

D.J. Herlocker

References

1. Andrew, M.H. 1988. Grazing impact in relation to livestock watering points. *Trends in Ecology and Evolution* 3: 336-339
2. Barker, J.R., Herlocker, D.J., Young, S.A. 1989. Vegetal dynamics along grazing gradients within the coastal grassland of central Somalia. *African Journal of Ecology* 27: 283-289
3. Herlocker, D.J., Ahmed, A.M., Thurow, T.L. 1987. Response of vegetation of the *Acacia recifiens/Dichrostachys* sp. shrubland range site to land use intensity, Ceel Dhere District, Somalia. *Somali Journal of Range Science* 2: 10-24
4. *Range Management Handbook of Kenya*. Nairobi: GTZ/Ministry of Agriculture, Livestock Development and Marketing

Reserve was formed in Tanzania, Maasai lost access to traditional dry-season grazing pastures that became part of the reserve. Here, the conflict was between wildlife and pastoralists rather than agriculturists and pastoralists, but the results were the same: dry season ranges were lost and the remaining land to which pastoralists had access suffered.

Sinclair and Fryxell (30), discussed the role of pastoralists in degradation and desertification in the Sahelian zone of Africa. There, they concluded, desertification was occurring and it was man caused, not due to climatic change. Again, however, the trigger for degradation was changes in land use away from the traditional pastoral pattern. They felt that the causes were the settlement of pastoralists around permanent water points and the expansion of crop agriculture into traditionally pastoral areas.

Any form of land use, applied improperly, has the potential to degrade the resource base, and history is full of examples of this sort of abuse. Degradation is inevitable when we attempt to use the land in ways to which it is not suited; for example if we attempt dry-land farming in areas that receive insufficient rainfall to support crops. If land is being used for purposes to which it is generally suited, the question then becomes one of managing that use properly. The arid and semi-arid rangelands of East Africa are suited to subsistence pastoralism. The history of the last 4000 years demonstrates that pastoralism is sustainable. Degradation of pastoral systems can be avoided by proper management.

Ecosystem Control

The traditional view of ecosystems has been that they possess a strong tendency to approach equilibrium or stable configurations. They vary, to be sure, but the variance is around some equilibrium state or "attractor". Under normal conditions, the variance is produced by fluctuations in external driving forces, usually weather, and the tendency to return to the equilibrium state is conferred by strong interactions among the working components of the system; the producers, consumers and decomposers. These strong interactions result in negative feedbacks within the system that damp perturbation imposed from the outside. In this view, mismanagement of the system is also a perturbation, and will cause the system to move from its equilibrium state. The reinstitution of proper management will, by removing the perturbing force, permit the feedback network to reestablish the equilibrium.

More recently ecologists, particularly those studying arid ecosystems, have begun to question the universal validity of this view (23,35). Noy-Meir described the view described above as the "ecosystem hypothesis" of ecological control and argued that it may not be valid in arid systems. He proposed as an alternative the "autoecological hypothesis", which states that the individual components of the ecosystem respond individually to the stresses imposed by the climatic drivers; and that negative feedback among the components is not important to the dynamics of the system. In this view, the ecology of the system is equal to the "sum of the autoecologies" of its components. Climatic variability is so great that mean values have little significance. The noise is more important than the signal, and individual populations are varying so enormously in response that the opportunity for the development of intricate interspecific interactions never arises.

This is a disturbing idea. It suggests that, if an ecosystem is persistent, it is more by chance than by inherent design. In this view, the system is wandering erratically through some unspecified domain, from which it might suddenly escape. To make matters worse, almost all of what we know about the management of ecosystems relies on the ecosystem hypothesis and the network of negative feedbacks and density-dependent responses that are inherent to it. If the ecosystem hypothesis does not apply, we might not be able to manage the system effectively. Not only might the system escape its bounds, but we may not know how to intervene to prevent its escaping.

Disturbing or not, evidence is accumulating that many systems, particularly arid ones, exhibit these non-equilibrium dynamics (2,13,34). As average annual rainfall increases and variability declines, the dynamics of arid grazing systems grade off into the sort of equilibrium dynamics we are more familiar with (4). In East Africa, it appears that the transition point between the two types of dynamics occurs around 300 and 400 mm of rain per year (12). Thus, while the South Turkana system appears to be dominated by non-equilibrium dynamics, the Borana system of southern Ethiopia appears to have a more or less normal equilibrium point, obtainable by management based on the equilibrium model.

Economic Development in Pastoral Systems

Economic development activities in arid and semi-arid zones of East Africa have been designed to improve the viability or increase the productivity of pastoral economies. High expectations of both donors and pastoralists often precede development plans. Frequently these expectations have not been met. Often, development plans have been ineffective, or worse, have contributed to resource degradation, leaving the pastoralists worse off than before help arrived (31).

In most cases these failures have resulted from misunderstandings about the operation of these ecosystems: a failure to appreciate the limits imposed by low and erratic rainfall, ignorance of the socioeconomic conditions in pastoral societies, and in particular the uncritical development of management prescriptions based on an ecological theory of rangelands developed in temperate regions. Many failed schemes have included attempts to "settle" pastoralists in one form or another, usually in association with a reduction in livestock numbers. Both of these aims, sedentarization and destocking, fly in the face of pastoral management tactics which have proven effective in East Africa for millennia. It is not surprising that pastoralists do not rush to accept them, or that they do not work well if they are imposed.

The attempt to reduce livestock numbers results from the erroneous characterization of pastoralists as irrational accumulators of livestock, and from the belief that the range should be stocked conservatively so that it is not overstocked in years of poor rainfall. The economist Sandford (28) has demonstrated that this conservative approach is itself irrational because it bears large opportunity costs in most years as a result of underutilization of forage resources. He argues that an opportunistic approach to stocking level which attempts to be responsive to changes in the pastoral environment is more appropriate in highly variable systems. This may not always be easy to achieve, however.

While economic development of pastoral systems does not seem to be susceptible to "the quick technological fix", there appear to be three important keys to their successful and sustainable development. First, the space available for exploitation by a pastoral population should not be reduced. Mobility and expansion of ranges during periods of environmental stress is characteristic of pastoral peoples and is important to their survival and to the avoidance of habitat degradation. Second, opportunistic stocking strategies in which the numbers of animals are quickly adjusted as foraging conditions change are preferable to conservative strategies aimed at matching stocking rates to the worst possible environmental conditions. Third, interventions that reduce livestock mortality, from disease or drought, must be accompanied by marketing opportunities and policies that encourage sale of livestock in excess of subsistence needs.

The history of failure or at least poor return on investment that has been the rule for pastoral development schemes has prompted many donors to lose interest in development in the arid and semi-arid zones. Unfortunately, the problems remain. If development in these regions were undertaken with a better understanding of the ecological and social realities that exist there, and if the goals of that development had more to do with improving quality of life for pastoralists than with meat production, success would still be possible. Pastoral development here should be based on an appreciation of the scarcity and uncertainty of precipitation (and the certainty of drought), an appreciation of the non-equilibrium nature of many of these ecosystems and a thorough understanding of the cultural norms and adaptive management strategies of the people.

References

1.   Barkham J.P., Rainy M.E. 1976. The vegetation of the Samburu-Isiolo Game Reserve. *East African Wildlife Journal* 4:297-329
2.   Behnke R.H., Scoones I., Kerven C., eds. 1993. *Range Ecology at Disequilibrium: New Models of Natural Variability and Pastoral Adaptation in African Savannas* London: Overseas Development Institute
3.   Coe M.J., Cummings D.H., Phillipson J. 1976. Biomass and production of large African herbivores in relation to rainfall and primary production. *Oecologia* 22:341-354
4.   Coppock D.L. 1993. Vegetation and pastoral dynamics in the southern Ethiopian rangelands: implications for theory and management. In *Range Ecology at Disequilibrium: New Models of Natural Variability and Pastoral Adaptation in African Savannas,* eds. Behnke Jr., R.H., Scoones I., Kerven C., pp. 42-61. London: Overseas Development Institute
5.   Coughenour M.B., Ellis J.E. 1993. Landscape and climatic control of woody vegetation in a dry tropical ecosystem: Turkana District, Kenya. *Journal of Biogeography* 20:107-122
6.   Coughenour M.B., Coppock D.L., Rowland M., Ellis J.E. 1990a. Dwarf shrub ecology in Kenya's arid zone: *Indigofera spinosa* as a key forage resource. *Journal of Arid Environments* 18:301-312
7.   Coughenour M.B., Coppock D.L., Ellis J.E., Rowland M. 1990b. Herbaceous forage variability in an arid pastoral region of Kenya: Importance of topographic and rainfall gradients. *Journal of Arid Environments* 19:147-159
8.   Coughenour M.B., Detling J.K., Bamberg I.E., Mugambi M.M. 1990c. Production and nitrogen responses of the African dwarf shrub *Indigofera spinosa* to defoliation and water limitation. *Oecologia* 83:546-552

9.   Coughenour M.B., Ellis J.E., Swift D.M., Coppock D.L., Galvin K., McCabe J.T., Hart T.C. 1985. Energy extraction and use in a nomadic pastoral ecosystem. *Science* 230:619-625

10. Dodd J.L. 1994. Desertification and degradation in sub-Saharan Africa. *Bioscience.* 44:28-34

11. Dregne H.E., Tucker C.J. 1988. Desert encroachment. *Desertification Control Bulletin* 16:16-19

12. Ellis J.E, Coughenour M.B., Swift D.M. 1993. Climate variability, ecosystem stability, and the implications for range and livestock development. In *Range Ecology at Disequilibrium: New Models of Natural Variability and Pastoral Adaptation in African Savannas,* eds. Behnke Jr. R.H., Scoones I., Kerven C., pp. 31-41. London: Overseas Development Institute

13. Ellis J.E, Swift D.M. 1988. Stability of African pastoral ecosystems: alternate paradigms and implications for development. *Journal of Range Management* 41:450-59

14. Hardin G. 1968. The tragedy of the commons. *Science* 162:1243-48

15. Hellden U. 1988. Desertification monitoring: is the desert encroaching? *Desertification Control Bulletin* 17:8-12

16. Herskovits M.J. 1926. The cattle complex in East Africa. *American Anthropologist* 28:230-72,361-80,633-64

17. Lamprey H. 1988. Report on the desert encroachment reconnaissance in northern Sudan: 21 October to 10 November 1975. *Desertification Control Bulletin* 7:1-7

18. Little P.D. 1987. Land use conflicts in the agricultural/pastoral borderlands: The case of Kenya. In *Lands at Risk in the Third World: Local-Level Perspectives.* eds. Little P.D., Horowitz M.M., Nyerges A.E., pp. 195-212. Boulder and London: Westview Press

19. Marshall F. 1994. Archaeological perspectives on East African Pastoralism. In *African Pastoralist Systems: An Integrated Approach.* eds. Fratkin, E., Galvin, K.A., Roth, E.A., pp 17-44. Boulder: Lynne Reinner

20. McCabe J.T. 1983. Land use among the pastoral Turkana. *Rural Africana* 15/16:109-126

21. McNaughton S.J. 1976. Grazing as an optimization process: grass ungulate relationships in the Serengeti. *American Naturalist* 113:691-703

22. Mugambi M.M. 1989. *Response of an African Dwarf Shrub, Indigofera spinosa, to Competition, Water Stress and Defoliation.* PhD Dissertation. Colorado State University, Fort Collins

23. Noy-Meir I. 1979/80. Structure and function of desert ecosystems. *Israel Journal of Botany* 28:1-19

24. Patten R. 1991. *Pattern and Process in an Arid Tropical Ecosystem: the Landscape and System Properties of Ngisonyonka, North Kenya.* PhD dissertation. Colorado State Univ., Fort Collins

25. Pratt D.J., Greenway P.J., Gwynne M.D. 1966. A classification of East African rangeland, with an appendix on terminology. *Journal of Applied Ecology* 3:369-382

26. Pratt D.J., Gwynne M.D. 1977. *Rangeland Management and Ecology in East Africa.* Huntington, NY: Robert E. Krieger Publishing Company

27. Reid R.S. 1992. *Livestock-mediated Tree Regeneration: Impacts of Pastoralists on Woodlands in Dry, Tropical Africa.* PhD Dissertation. Fort Collins: Colorado State University

28. Sandford S. 1983. *Management of Pastoral Development in the Third World.* New York: John Wiley and Sons

29. Sinclair A.R.E. 1979. The eruption of the ruminants. In *Serengeti: Dynamics of an Ecosystem,* eds. Sinclair A.R.E., Norton-Griffiths M. pp 82-103. Chicago: University of Chicago Press

30. Sinclair A.R.E., Fryxell J.M. 1985. The Sahel of Africa: ecology of a disaster. *Canadian Journal of Zoology* 63:987-94

31. Talbot L.M. 1972. Ecological consequences of rangeland development in Maasailand, East Africa. In *The Careless Technology: Ecology and International Development*, eds. Farvar M.T., Milton J.P., pp. 694-711. Garden City NY: The National History Press

32. Weltzin J.S., Coughenour M.B. 1990. Savanna tree influence on understory vegetation and soil nutrients in northwestern Kenya. *Journal of Vegetation Science* 1:325-334

33. Whicker A.D., Detling J.K. 1988. Ecological consequences of Prairie dog disturbances. *Bioscience.* 38:778-785

34. Wiens J.A. 1977. On competition and variable environments. *American Scientist* 65:590-97

35. Wiens J.A. 1984. On understanding a non-equilibrium world: myth and reality in community patterns and processes. In *Ecological Communities: Conceptual Issues and the Evidence*, eds. Strong Jr., D.R., Simberloff D., Abele L.G., Thistle A.B. pp. 439-57. Princeton, NJ: Princeton University Press

36. Woodhead T. 1968. *Studies of Potential Evaporation in Kenya.* Nairobi: Water Development Department, Ministry of Natural Resources

37. Young T.P., Patridge N., Macrae A. 1995. Long-term glades and their edge effects in acacia bushlands in Laikipia, Kenya. *Ecological Applications* 5: 97-108

Chapter 11

Savanna Ecosystems

Helen Gichohi, Chris Gakahu & Evans Mwangi

Great herds of wildebeest, big cats, the great tuskers and the all but extinct rhino are the images that epitomize the savannas of Africa. The savannas of Africa are occupied by the earth's' richest and most spectacular large mammal fauna, a sizable proportion of which is found in East Africa. The diverse topography and geology, and the spatial and temporal variability in climate in eastern and southern Africa have created a mosaic of distinct vegetation types that were accompanied by parallel broad ungulate radiations. Wildlife of the African savannas remained relatively intact when most similar large mammals became extinct on other continents (see Chapter 2). Nonetheless, other pressures have recently come to bear on these populations, reducing them to small remnants of more impressive herds that existed several decades ago. Tremendous pressure from a burgeoning human population, a growing demand for agricultural land and heavy poaching have severely reduced the numbers of many wildlife species, particularly elephants, rhinos and major carnivores. The moist savanna ecosystems which constitute areas of higher agricultural potential have been settled and the dry semi-arid zones, where much of the wildlife now occurs, are currently under similar threats. Habitat fragmentation and habitat changes induced by the removal of traditional human agencies (fire, grazing) also continue to affect the savanna's wildlife populations.

Rainfall

Many aspects of savanna ecology can be accounted for by rainfall quantity and variation, as well as by variation in the underlying soils. Rainfall influences the structure and productivity of vegetation, and determines food supply and the availability of water. Rainfall seasonality is also an important factor determining food quality and therefore herbivore distribution.

In East Africa, rainfall is seasonal and is governed by a large scale weather pattern, the Intertropical Convergence Zone (ITCZ). This zone is a belt of low pressure which follows the northward and southward movement of the sun across the Equator. The southward movement is associated with the northeasterly prevailing winds blowing from the dry northern part of East Africa and the Arabian peninsula. These winds are relatively dry. The south easterly winds are associated with the northward movement. These winds blow from the Indian Ocean and usually result in wetter conditions. The ITCZ produces two wet and two dry seasons nearer the equator and one dry and one wet season to the north and south. Rainfall is, however, highly variable among years (see Box 10.1). There are also large rainfall variations due to modifying influences of relief and large water bodies in the region. For example, the Laikipia plateau and Amboseli receive lower rainfall, being on the rainshadows of Mt. Kenya and Kilimanjaro, respectively.

Vegetation of the Savanna

The vegetation of East African savannas has been variously described (7,22). Used in the broadest sense, the term savanna describes a range of vegetation types, from humid woodlands to dry grasslands (9). From the classification of Pratt and Gwynne, savanna ecosystems range from derived grasslands of ecoclimatic-Zone II to the semi-deserts of Zone V. The various forms of savanna vegetation are an expression of the interactions of climate, soils, herbivores, fire and human activities. These have resulted in variations in the vegetation over large geographical areas and variations on a local scale. Savannas were present in many areas before the advent of modern human populations. Their extent increased and decreased with long-term climatic changes, but they became much more widespread as a result of human influences. The change from dry tropical forest to savanna is determined by fire, grazing and other factors that increase with human presence. Savanna vegetation is composed of an open overstory of trees and/or shrubs and a ground layer of grasses. The chief characteristic of the savanna is the dominance of grasses in the understory. The height and spacing of the tree and shrub component determine the categories of the savanna, and are influenced by soil moisture conditions and the incidence of fire and grazing. The composition of the vegetation units within each category varies with nutrient status. Altitude, edaphic factors, slope, water-table level and other influences account for localized vegetation differences. The savanna vegetation is frequently a complex mosaic that provides diverse habitats for animal life.

Savannas range from areas receiving annual rainfall of as much as 1200 mm to those receiving as little as 300 mm. The distribution of moist and arid savanna is also related to the base status of the major soil types. The quality of soil is inversely related to rainfall (37). Higher rainfall results in greater quantities of water passing through the soil, thus removing the mobile fractions by leaching. There is a critical point between 500 and 700 mm rainfall per year below which water does not pass right through the soil profile, and nutrients accumulate within the profile. Above this point, soils become freely draining and continuously lose nutrients. This inverse relationship between rainfall and soil nutrient availability is the basis for the distinction between the moist nutrient-poor ('dystrophic") savannas and the arid nutrient-rich ('eutrophic') savannas. Certain factors do sometimes override this correlation. For example, the western Serengeti is a 'eutrophic' savanna though it is not really arid.

Moist Savannas

The moist savannas can broadly be divided into woodland savanna and 'derived savanna'. The woodlands cover about one third of East Africa although in some protected areas these have more recently been converted into grasslands due to the local increases in elephant numbers, and subsequent tree mortality. In others the interaction between elephants and fire has had the same effect. Most East African woodlands occur in Tanzania. Typical in this category are the Brachystegia or miombo woodlands (see Chapter 12). The dominant genera here are *Brachystegia, Isoberlina,* and *Julbenardia*, all of which are adapted to a long and severe dry season. These deciduous woodlands are a dominant feature of the single rainfall system experienced further away from the equator. The common grasses in this area include *Andropogon* and *Hyperrhenia* spp. (22). Though the moist savannas have higher primary productivity due to a longer growing season and higher rainfall, the herbage is of high bulk and low nutritive value. Consequently, these savannas typically have a low diversity and density of ungulate species, which include gray (bush) duiker, sable antelope, roan antelope and oribi.

'Derived savannas' are formed from forest and woodland communities by the felling of trees, agricultural activity, burning and grazing. Invasion of grasses follows and a savanna develops as a result of continuous suppression of forest trees. Vegetation in these savannas typically consists of a mixture of forest remnants, invading savanna trees and a grass layer which, in East Africa, is often dominated by *Pennisteum* spp. Over much of the moist savanna, a combination of the above factors holds this habitat in a grassland state, preventing regrowth of tree and shrubs. *Cynodon* grasslands may occur at lower elevations, and where forests are frequently burnt and the grazing pressure is not high, *Themeda* grasslands occur. The uniformity of the grassland may be broken by the growth of shrubs such as *Capparis* and *Grewia* and trees such as *Euphorbia candelabrum* and especially several *Acacia* species.

Dry Savannas

An important climatic feature of dry savannas, which have an average rainfall less than 600 mm, is the considerable variability of rain falling from year to year, but unlike the deciduous woodlands, many of these areas receive two rainfall seasons and have both shorter dry seasons and longer growing seasons. The pastures produced are of higher quality and as a result support a more diverse and productive wildlife community. Conditions in these arid savannas are drier and vegetation adapts accordingly. The dry woodlands and wooded grasslands are found in areas with a mean annual rainfall of between 300 to 900 mm with the typical double rainfall maxima (22). These ecosystems fall between ecoclimatic Zones IV and V and are spread over much of eastern Kenya and central Tanzania. They constitute the major wildlife areas in East Africa. Zone IV, the semi-arid zone, is land of marginal agricultural potential. Its natural vegetation are dry forms of woodland, often an *Acacia-Themeda* association. It includes dry *Brachystegia* woodlands and *Themeda-Acacia drepanolobium* wooded grasslands in black cotton soil areas such as in parts of Laikipia, Kajiado and Narok districts.

The drier areas of Zone V cover a large expanse of land in East Africa (see Chapter 10). This covers about half of Kenya and extends well into central Tanzania. In Uganda it is represented in the drier parts of Karamoja district. The rangeland is typically dominated by *Commiphora, Acacia* and allied genera, mostly of shrubby habit. Perennial grasses such as *Cenchrus ciliaris* and *Chloris roxburghiana* can dominate. An example of a wildlife area falling in this zone is Tsavo National Park in Kenya. The vegetation here is mostly thorn-bushland, *Commiphora* woodland and grassland. *Adansonia* is locally abundant. Generally, *Acacia* predominates in the Rift Valley and western areas, and *Commiphora* in the eastern areas. Trees and shrubs tend to drop their leaves in the dry seasons and many have spines. Succulents are also common in this zone.

The high incidence of spines in the arid and semi-arid savannas, mostly absent in moist savannas, is thought to have evolved as a response to heavy browsing from the wide variety of large mammals historically present. Most species of wildlife, including elephant, Burchell's zebra and buffalo, reach their highest population densities in nutrient-rich arid and semi-arid savannas. Most of East Africa's pastoral communities also occupy these areas. Animals found in the more arid parts of these savannas include the lesser and greater kudu, Grant's gazelle, Grevy's zebra, oryx and gerenuk.

Ecological Energetics

The savanna landscape harbors plants ranging from grasses to trees and a complex mix of invertebrate and vertebrate fauna. Many species of insects, birds, and large mammals are readily evident in savannas. The grass, decomposing litter, and soil provide habitats for other hidden and elusive microorganisms, invertebrates, amphibians, reptiles and small mammals. Each species, from the microscopic to elephants and huge trees, has a unique ability to contribute to the savanna ecosystem and also to make a living from it.

All ecological processes require energy. This energy is initially provided by solar radiation and captured by plants through photosynthesis, a process that transforms a small fraction of radiant energy from the sun into organic compounds. It is this energy that fuels ecological processes as it flows from plants through the various organisms of the savanna. This flow is from plants to herbivores to carnivores. Dead plant and animal materials are decomposed by bacteria, fungi, and a wide variety of invertebrates that reduce the material to organic matter and nutrients. This linear process of eating and being eaten is called the food chain. Each stage in the process is termed a trophic (=food) level (see Box 11.1).

Plants are called autotrophs, or primary producers, because of their ability to use radiant energy to convert water and carbon dioxide into energy contained in bonds of organic compounds. Plants use some of the energy they produce for the metabolic costs of respiration, growth and reproduction. What remains is referred to as primary production. Primary production is available for consumption by herbivores, and can either be above-ground, for example stems, branches, leaves, flowers and fruits, or below-ground, such as roots, rhizomes, corms, tubers and bulbs.

Above-ground primary production is eaten by grazers like zebras and

Box 11.1. Trophic Ecology

Virtually all biological energy on the planet has as its source in photosynthesis. Plants (and phytoplankton) have the unique ability to harness the energy of the sun and convert it to biochemical energy, in the form of simple sugars. It is the energy in these sugars that allows plants to make the myriad of chemical compounds and morphological structures necessary for them to survive, grow, and reproduce. All other organisms gain energy from plants either directly (herbivores) or indirectly (carnivores).

It is one of the most basic natural laws that energy tends to dissipate. The use of biochemical energy is inherently inefficient. Only a fraction of the energy value in plants is passed on in the form of energy value in herbivores. The bulk of the energy is lost as metabolic heat. The net result is that there is far more bulk in plants than in plant-eaters in the world. A simple walk through any terrestrial ecosystem will confirm this. Fewer still are the animals that eat herbivores, again because energy is lost at each level.

Warm-blooded animals (birds and mammals) are less efficient than cold-blooded animals (such as reptiles and invertebrates), and small mammals are less efficient than large animals. On average, however, 90% of the energy is lost at each stage and therefore the area needed to support species feeding increases with trophic level.

The overall pattern of energy production and consumption is called a trophic pyramid (see Box 4.1), in which the net production of biologically useful energy at each level is but a fraction of its transfer to the next level. (Note that the value here is the production of energy, not the actual number or amount of organisms at each level. In some aquatic systems, a small amount of very productive phytoplankton can support a relatively large amount of herbivore biomass).

This pyramid has other consequences than explaining the relative abundance of plants, herbivores, and carnivores. It also clearly demonstrates that harvesting an ecosystem at to lowest level (plants) results in more food than harvesting at higher trophic levels. When people fatten cattle on grain crops instead of eating those crops, there is a 90% loss of caloric value.

Another consequence is the concentration of environmental toxins. Although energy is lost at each level, toxins are often not lost, but passed on to the next level very 'efficiently'. The result is that toxins become more and more concentrated as one moves higher and higher on the trophic pyramid. Many marine fishes are on the fourth or even fifth levels and can have dangerously high levels of mercury and other toxins. In terrestrial systems, DDT is similarly concentrated. Levels in plants and herbivores may only be toxic to the insects DDT was designed to control. Concentrated at higher trophic levels, however, DDT becomes deadly. In particular, it results in weakened egg shells in carnivorous birds (the very birds so helpful in reducing crop pests). DDT has been banned in many countries, but not yet in any East African country.

T.P. Young

wildebeests, browsers like giraffes, dik-diks and black rhinoceroses, or mixed feeders like Grant's gazelles and elephants. Insects like bees and butterflies together with sunbirds subsist on pollen and nectar from flowers. Many species of birds, small herbivores and monkeys eat fruits and seeds of many different types of plants (see Box 15.1). Below-ground production is eaten by fossorial animals such as mole rats, and terrestrial animals that excavate, such as warthogs, baboons and porcupines. Many insects are also plant eaters. All these plant eaters are referred to as herbivores. Only a fraction of what they eat is actually digested, and of the food assimilated from the digestive tract, a large proportion is expended on metabolic costs. What remains is turned into animal biomass, termed secondary production.

Carnivores eat other animals, and occur in the third trophic level (and higher). However, the linear food chain over-simplifies feeding relationship because some organisms feed in more than one trophic level. For example, the leopard can feed on gazelles of the second trophic level and also on eggs of the predatory snakes. ackals feed on all trophic levels, from plants to carnivores.

Organic material originating from deaths of organisms and from excreta are also used by a wide range of animals to satisfy their energy requirements. As a result a savanna ecosystem will be characterized by many food chains interlinking the movement of energy and chemicals into a 'food-web'. There are two types of food webs, the grazing food web in which most of the food materials moving from one trophic level to next are in living tissues, and the detritus food web that is based on dead materials. Decomposition is carried out by invertebrates such as dung beetles, termites and earthworms, and microorganisms such as fungi and bacteria. Like all heterotrophs, these decomposers break down carbon-containing compounds to get energy. Another important agent of decomposition is fire, which releases organic materials in the air and minerals into the soil (see section on fire, below).

As energy moves from a trophic level to the next much of it is lost in the form of heat during the breakdown of chemical bonds by digestion and metabolism. Biomass usually decreases gradually outward to form a pyramid of numbers from the green plants to herbivores and from herbivores to carnivores. Often the number of herbivores (or energy fixed by herbivores) determines the number of carnivores. In the savanna the relationships are more complex because carnivores do, to some extent, control the number of herbivores.

Numbers and biomass tend to overemphasize the importance of large and conspicuous organisms. As a result, the role of small organisms such as invertebrates and small mammals is virtually unstudied and apparently underestimated. An example of such organisms are the termites, whose biomass in the African savanna ranges between 10 to 500 kg/ha, often higher than the biomass of mammal species, which range between 5 to 100 kg/ha (36). In the arid and semi-arid savanna of northern Kenya termites consume more than the 80% of net primary production (2).

The unique and diverse large herbivore fauna that ranges in size from dik-diks to elephants is a dominant feature of the East African savanna. The efficiency with which these herbivores exploit the energy fixed by the vegetation depends on their numbers, body size, and the quality and quantity of plant materials (Fig. 11.1). Large single or multi-species aggregations of herbivores harvest a large amount of the net above-ground production of the vegetation. Where plant materials are out of their reach or the number of herbivores is small, exploitation is low and most of the energy produced by plants enters the detritus food chain, which includes termites. To sustain large intake, big-bodied ungulates need areas with large amounts of plant material. This requirement sometimes translates into migrations, as in the

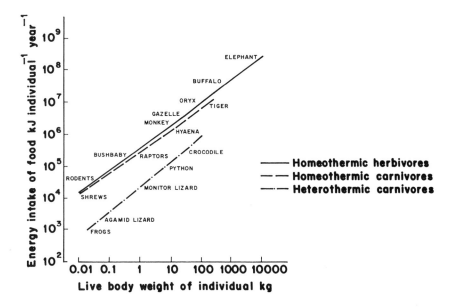

Figure 11. 1. Relationship between the average body weight and energy intake of common herbivores and carnivores of the savanna. Shows the difference in intake between heterothermic and homeothermic animals. Modified from (10).

Serengeti/Mara where hundreds of thousands of wildebeests and zebras move over wide areas to satisfy their food requirements.

Smaller herbivores, such as rodents, hares, warthogs and dik-diks require less bulk. However, they require foods of high nutritive value due to their high metabolic rates. Such foods include young leaves, flowers, fruits and seeds.

Food from animal tissue is more easily digested than food from plant tissue. Carnivores are, therefore, digestively more efficient than herbivores. Hyenas are even more efficient than other carnivores due to unique digestive efficiencies which includes the ability to consume bones and marrow. The dominant savanna carnivores, lions, leopards, cheetahs and spotted hyenas mainly eat newly killed herbivores. However, all except the cheetah scavenge a proportion of their food. In addition to the large carnivores, the savanna has many other meat eaters that include invertebrates, amphibians, reptiles, many species of insectivorous and predatory birds as well as small mammals such as mongooses, genets, civets, jackals and foxes.

In the course of food harvesting, animals play very important roles in the structure and function of the savanna. Elephants take large quantities of coarse grass and the stems and branches of trees. Much of this intake, including many seeds of grasses, shrubs and trees, passes through the digestive tract undigested. As a result, elephants are good seed dispersal agents. Impalas also play a similar role by dispersing seeds of acacia trees. Insects such as butterflies and bees feed on nectar and pollen and pollinate many plant species. Birds act both as pollinators and seed dispersers. Animals that burrow and feed below the ground such as gerbils, mole rats and earthworms aerate the soil and improve nutrient levels and organic matter of the soil, thereby improving plant productivity.

Community Structure

The savanna community contains hundreds of species, each interacting with other species in a variable abiotic and biotic environment. Amidst this complexity ecologists have found what seem to be repeatable patterns in species diversity, dominance, composition, life histories and food webs. These patterns suggest that there are underlying factors or processes that constrain and structure ecological communities. There are many descriptive and experimental data sets that suggest that resources such as food and space are important factors in structuring communities (31). The greater the diversity of resources offered by a habitat, the greater the number of species that it can support. Spatial heterogeneity in a habitat can enhance species richness by simply providing a larger number of resource types among potentially competing species. But in general, there are several mechanisms which may be responsible for structuring communities. These can be broadly divided into physical factors (such as rainfall, soils, and temperature) and biotic factors, (such as predation and competition within and between species).

Non-Human Factors that Control Community Structure

Physical Factors. The distribution of animals in the savannas is often indirectly determined by physical variables such as rainfall, temperature, soil type, and geomorphology, because these variables affect vegetation type, distribution and productivity. In general, the higher the rainfall the greater the primary production. The secondary production of herbivores increases as a function of rainfall across a wide range of habitats, for both large wild mammals and domestic livestock (Fig. 11.2).

Herbivore biomass is, however, also related to geomorphology and soil type. Alkaline geology and fertile soils support higher herbivore biomass than acidic geology and infertile soils. Bell (4) argued that a family of curves relate herbivore biomass to rainfall, one curve for volcanic soils of high fertility, a second for soils of low fertility from basement rocks and a third from basement rocks of the Rift Valley. In higher rainfall areas, the herbivore communities consist of very large animals, particularly elephants and buffaloes, where they can contribute 75% or more of total animal biomass. In more arid grasslands, less massive herbivores such as wildebeest and gazelles dominate.

Biotic Factors. Community structure as defined by environmental factors is often modified by biotic factors. Ungulates may regulate vegetation productivity to partially override the rainfall-driven primary productivity patterns through feedback mechanisms. There is strong evidence to suggest that the large mammalian herbivores, whether wildlife or livestock, are often limited by food and water availability.

In East African savannas, one of the main questions has been how so many species coexist. For two decades, research on the African ungulates has described

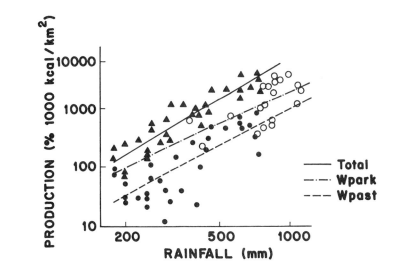

Figure 11.2. The relationship of production to rainfall for mixed livestock and wildlife in pastoral areas (total), wildlife only in pastoral areas (Wpast.), and wildlife in protected areas (Wpark). The regression equations are y= -0.45 + 2.46x, y= -0.85 + 2.56x and y= 0.54 + 1.95x respectively (adapted from 32).

ecological separation of savanna herbivores by habitat (15), by food plant species and by plant parts eaten (27). The underlying premise is that such niche separation was the result of interspecific competition. Early pioneering studies by Lamprey (15) in Tarangire Game Reserve in Tanzania documented a variety of feeding patterns among herbivore species ranging from exclusive grazers such as wildebeest and buffaloes to exclusive browsers such as giraffes and dik-diks. A variety of mixed feeders fell between the two extremes. Feeding at different heights from the ground among these categories of ungulates, the use of different habitats in the wet, dry and transition periods contributed to resource partitioning. The proportion of different plant parts in the diet has also been used as an indicator of trophic distinctions between grazers. Compared with wildebeest, zebras ate relatively more stem than leaf material in the Serengeti, while seeds were more prevalent in the diet of Thomson's gazelles, especially during the dry season when fresh, short vegetation was in short supply.

It has been postulated that the degree of selectivity among herbivore species is related to body size, gut capacity, digestive efficiencies and metabolic requirements (8). Relative metabolic demand increases with body size while gut volume remains a constant proportion of body size. As a result, large herbivores can take a coarser diet, hold the forage longer in the gut and digest it more efficiently than small herbivores (8). On the other hand, smaller herbivores must take higher quality food, which is rapidly digested, in order to meet their high metabolic demand. The allometry of diet tolerance explains the increase in body size with rainfall described above (4). Since vegetation quality is inversely proportional to vegetation biomass, wetter areas produce more biomass of lower quality than drier areas. Large herbivores are therefore at a competitive advantage in moister areas, thus explaining the increase in mean body weight and biomass of a community with rainfall

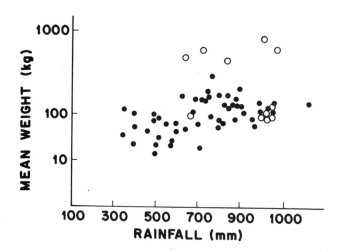

Figure 11.3 The relationship between the mean weight of large mammal communities and rainfall. A best fit regression equation is given by y = 61.32 + 0.18x. The points marked by a circle indicate parks where elephants had recently been compressed by poaching, giving an elevated mean body weight for the community (adapted from 33).

(Fig. 11.3). Large herbivores are less able to meet their metabolic demand on sparse pasture at low rainfall.

Even within ecosystems, body weight is related to forage quality. Thomson's gazelles feed on short highly nutritious vegetation with little fiber, wildebeests on slightly taller vegetation, zebras on the next level of coarseness and the buffalos and elephants on the very coarse and rank vegetation of very high fiber content.

The savanna environment is characterized by continual, stochastic fluctuations of rainfall, grazing, nutrient availability and fire over time and space. Ungulates are therefore faced with a widely distributed forage of varying quality and quantity. Migratory herbivores are able to track rich food sources, which are frequently ephemeral and follow an annual cycle. In the face of this variation, nomadic behavior allows the exploitation of these widely spaced bursts of productivity. The dense herding behavior of wildebeests allows them to crop the grasslands in a way that increases forage yield ensuring the maintenance of highly productive grassland areas. This may account for their dominance in some grassland ecosystems.

Impact of Herbivore Grazing and Browsing. Grazers and browsers, including livestock, can have profound effects on vegetation. Their effects are variable and depend on the numbers of animals involved, the intensity of vegetation offtake and movement patterns. In addition to the actual consumption of biomass, they affect the ecosystem by trampling, urinating and defecating.

Research in many of the East African parks has shown the effects of various intensities of grazing on grasses. Grazers influence the composition and diversity of the vegetation (17). Grasses growing in areas that are continuously and intensely grazed have short growth forms with shorter internodes and smaller leaves. Under continued intense grazing there is a decline and subsequent replacement of perennial grass species by annual grasses and eventually unpalatable herbs. Some of these annuals, such as *Eragrostis tenuifolia*, are eaten by herbivores, but their shallow rooting habit and ephemeral growth makes them more unpredictable food source, resulting in declining total production and plant cover. Areas under heavy grazing by livestock and wildlife bordering Nairobi National Park display many of these ecological changes. In areas of moderate to heavy grazing, the removal of tall grasses by grazers stimulates new shoot production. Old ungrazed grasses have a lower protein content than the new shoots. The grazing-induced increase in nutrient content and density of the grass sward facilitates further grazing resulting in "grazing lawns" or "hot spots" that are grazed and regrazed leaving adjacent areas less utilized. Grasses in these hot spots also have a higher content of N, Ca, Mg, Na, P, Al and Cu (18,19,38) than surrounding areas due to addition of urine and feces from the ungulates. On the other hand tall grass species of poor quality dominate in little grazed or fenced areas. Such areas also have lowered productivity due to intense shading by these tall grasses. Research in Nairobi National Park also showed that fenced areas have low diversity, poor quality vegetation, (Fig. 11.4) and these areas were normally avoided by grazers until better quality vegetation in other areas was used up.

Compared to grazing, the effects of browsing on the tree and shrub vegetation are less well documented. Browse generally has a relatively high mineral content and smaller fluctuations in seasonal protein content than grass. Besides the nutrient content, chemical and structural defenses are important determinants of leaf and twig palatability and can also affect herbivore densities and distribution. Plant species have evolved mechanisms to reduce browsing. These include spines found in many trees and shrubs species in the arid savannas, especially the acacias (7), and a variety of toxic compounds. Spinescent species, such as acacias, respond to browsing by producing longer thorns (21).

The ability of browsers to use browse vegetation depends partly on the height of the plant relative to the height of animal. Tall browsers such as giraffe can feed on the upper part of tall plants and well beyond the reach of other species, creating the commonly seen giraffe browsing lines on trees and bushes. Medium sized browsers such as the kudus use the level below the giraffe feeding line and the short browsers concentrate on the bushes and leaves close to the ground. Severe browsing can causes stunted growth, and in some cases prevents regeneration and recruitment (23).

Despite the spectacular local diversity and abundance of large herbivores, most primary production in savannas is consumed by invertebrates and fire. Fire is more common in wetter savannas, often consuming 50% or more of annual forage production. In semi-arid and arid environments, termites are the most important consumers in environments where they often surpass mammalian biomass (2). Herbivorous insects and small mammals are also likely to be important. Collectively, these organisms are significant in structuring communities, though their impacts are often overlooked and have been rarely studied in East Africa. Large mammals, small mammals and invertebrate herbivores, each have unique niches characteristics but to some extent have overlapping effects on rangeland vegetation (26). Small mammals and invertebrates may have significant grazing effects in terms of competition for green biomass at times of low forage availability, but their defoliation effects may be more significant for browse than grass species (5).

Box 11.2. Elephants: A Keystone Species

Elephants play a critical role in modifying their habitats. Depending on their population densities and whether or not they are free to move, elephants determine the composition of the terrestrial communities in which they live by exerting strong influences on many ecosystem states, processes, and functions.

Elephants diversify savanna and forest ecosystems and create a mosaic of habitats through browsing of woody vegetation and understory plants and in destroying trees, shrubs, and saplings. Savanna woodlands have been changed to wooded grasslands in many parts of Africa by the activities of elephants (2). There is evidence that elephants initiate the destruction of forests by facilitating the entry of fire into the forest edge and elephants may have played a leading role in the retreat of the tropical forest and the advance of savannas in Africa (1). A single animal may knock down an average of 4.2 trees per day. They are also important seed dispersalagents for a number of plant species. Thirty-eight plant species were dispersed through elephant dung in only one season in a small area of the Aberdares National Park and 19 species in Tsavo East National Park. These results imply that many plants may be dependent on elephants for dispersal.

Elephants play a central role in the formation of grassland savanna from African forests. Studies carried out in the Aberdares in Kenya have shown that elephants create and expand gaps in forests and in the process open up more productive and diversified ground layer that is exploited by a wide range of other animal species. The high proportion of gaps and secondary forest species of plant and animal species in African forests suggests that patchiness is a characteristic feature in these ecological biomes and that elephants play a significant role in vegetation patch dynamics.

In many parts of Africa, heavy poaching and habitat fragmentation coupled with high demand for land from increasing human populations have split the distributions of elephants. In some areas, population numbers have increased several-fold while in others, particularly in non-protected areas, numbers have plummeted in response to these factors. This was reflected in the savanna ecosystems where elephant densities within parks averaged five times those outside parks in the late 1980s during the peak of the poaching of the African elephant.

The ecological impacts of elephants on their habitats are chiefly dependent on their numbers, the local eco-climatic factors and whether or not they are free to move and exploit the resources existing within their natural home-ranges. In forests of Aberdares N. Park with rainfall of between 1000 mm and 2000 mm per year, an elephant density of about four to a square kilometer appear optimum in promoting biological diversity while in Tsavo East N. Park with rainfall of about 500 mm, a density of one elephant to a square kilometer is considered to be high (3,4). Where human interference prevents elephants from moving freely within their habitats, their role shifts from that of maintaining essential ecological balance and ecosystem processes to being environmentally catastrophic. Evidence from Aberdares, Tsavo, Laikipia and Amboseli (4,5) shows that elephants at very high or low densities strongly affect vegetation structure, reducing biological diversity, significantly changing the ratio of grazer and browser, significantly alter productivity patterns and reduce ecosystem diversity. Nonetheless, when elephants stem bush invasion and open up forests and woodlands they create more productive communities of grazing and browsing ungulates. Subsistence and commercial livestock economies benefit from the work of elephants in reducing bushland, expanding grasslands and reducing the incidence of tsetse flies. Within the conservation areas, elephants diversify habitats by preventing the savannas from changing into monotonous bushlands and woodlands.

The elephants are a key species in maintaining and promoting the tourism industry. In a survey conducted in Kenya in 1989 to determine the viewing value of wildlife species, the tourist rated the elephant as the single most important species. Many respondents indicated that they would change their travel plans if elephants populations were grossly reduced or exterminated in Kenya.

The ecological role of elephants in savanna, coupled with their economic importance to livestock and tourism, are good reasons to encourage their range expansion beyond protected areas. Nonetheless, elephant increases have created serious human-elephant conflicts that have far-reaching conservation implications. These conflicts include: raiding of crops, killing people, breaking water-pipes, damaging houses, cattle sheds, trees, dams, fences and erosion trenches and disrupting school and other social meetings.

The economic problems of elephants extend beyond farmlands and settlements. In some landscapes where forests are developed for commercial timber production, elephants damage plantations at different stages of maturation, causing large economic losses. This has been observed in many forest plantations bordering the Aberdares and Mt. Kenya National Parks where they have been implicated in damaging as many as 150 mature trees per hectare per year. These conflicts can be reduced by community-based conservation strategies and popularization of wildlife conservation programs. Many countries in eastern, central and southern Africa have taken the challenge of bringing this problem under control after realizing that the absence of active grassroots participation in natural resources management and decision-making is a part of the root-cause of the problems. The elephant, being a flagship species for the conservation of other species, habitats and ecosystems, ought to be given high priority in the conservation agenda. In view of its keystone role and its influence on land-use patterns and economies, African governments must develop practical conservation strategies that balance the ecological needs of the elephants with the aspirations of the local people and the conservation needs of their countries.

This approach is especially critical for the long-term conservation of a species that is now capable of driving itself out of existence by destroying its own restricted habitat or by causing havoc to the people on whose attitudes its future depends.

J. M. Waithaka

References

1. Ford, J. 1966. The role of elephants in controlling the tsetse flies. *Bulletin of the International Union for the Conservation of Nature* 19:6
2. Laws, R. M. 1970. Elephants as agents of habitats and landscape change in East Africa. *Oikos* 21: 1-15
3. Phillipson, J. 1975. Rainfall, primary production and 'carrying capacity' of Tsavo National Park (East), Kenya. *East African Wildlife Journal* 13: 171-201.
4. Waithaka, J. M. 1994. *The Ecological Role of Elephants in Restructuring Habitats.* PhD. Thesis, Kenyatta University, Nairobi
5. Western, D. 1989. The ecological value of elephants in Africa. *Pachyderm* 12: 42-45

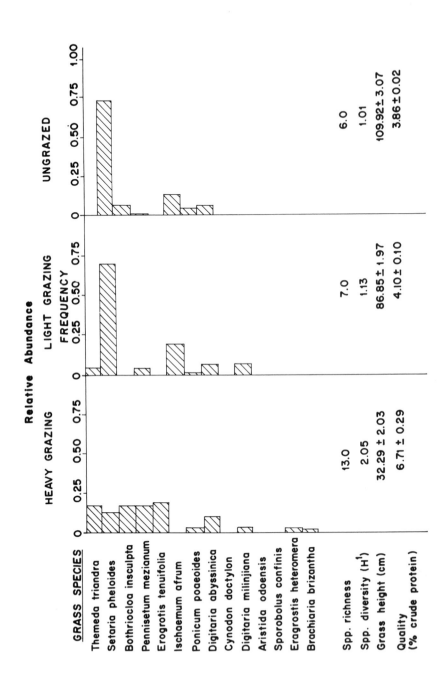

Figure 11.4. Differences in grass characteristics between ungrazed, lightly grazed and heavily grazed plots in and adjacent to Nairobi National Park. Plant species richness, diversity and quality (measured by digestible crude protein) increases with grazing, creating short grazing lawns favored by wild ungulates (adapted from 34).

Migrations. Large-scale movements of wildlife populations are widespread in East African savanna. Examples include seasonal migrations in the Serengeti/Mara in Tanzania and Kenya, Nairobi National Park/Athi-Kapiti plains, Amboseli and Laikipia ecosystems in Kenya, and the Boma ecosystem in southern Sudan. The main factor determining these movements is rainfall, through its effects on food supply and water availability.

Wildlife movements may also be influenced to a substantial degree by mineral heterogeneity. At the landscape level, the differences in mineral nutrients in the vegetation reflect the dichotomy between tall grassland savannas growing on nutrient-poor (dystrophic) soils and the short grass plains growing on nutrient-rich (eutrophic) soils. McNaughton (19) found calcium to be higher in grasses of the Serengeti short grass plains and suggested that this may account for movements of wildebeest to these areas in the wet season particularly for locating females (16).

The best known large scale wildlife migrations are in the Serengeti/Mara where movements of over a million wildebeests occur. zebra and Thomson's gazelles move on a smaller scale, although they number in the hundreds of thousands. In October, the migrants move to the wet season range in the short grass Serengeti plains where they spend November to May. As the dry season progresses the migrants move north and north west along a rainfall gradient to the woodlands and taller grass areas such as the Mara (20). By the following October, they are usually back in the Serengeti plains, having completed a migration of hundreds of kilometers. White-eared kob and tiang migrate in similar massive numbers in the Boma ecosystem of southern Sudan.

Other migrations occur in the East African savannas. Most are smaller and more localized, but are driven by the same basic factors. Resident herbivores also display similar qualities by duplicating scaled-down features of the migrations from short grass hilltops in the wet season to taller grass valley bottoms in the dry months. These are commonly referred to as 'catena' movements. Regardless of scale, migrations result from the need to track a widely distributed and often variable resource base.

Predation. Predation is also an important factor in structuring animal communities. Predators can shift the dynamic relations between competing prey species and therefore influence diversity. Five important large predator species occur in many savanna ecosystems: lion, leopard, cheetah, spotted hyena and wild dog. Wild dogs are highly endangered and mostly confined to the Serengeti/Mara and Selous ecosystems. These different predators make differential use of potential prey, related to habitat, hunting methods, time of hunting, the size and social organization of their prey. In the Serengeti, these factors may cause the relatively low degree of overlap in the food types selected by each predator species. In general, larger predators feed on a wider array of prey. Lions can prey on large animals such as buffaloes and small ones such as warthogs and even rabbits and mice. Group hunting confers certain advantages to medium sized carnivores and makes them as formidable as the largest carnivores. A good example are the hunting dogs, which kill large animals like zebras.

The effect of predators on prey populations depends on the ratio of predator to prey, and the degree of movement of prey population. In systems with a migrating prey population, the impact of predators is limited to the time of year when prey and predator occur together. For the rest of the year predation is minimal. Where predators occur in large numbers, they may have significant effects on prey populations. In the Serengeti/Mara for example, predators may be important in regulating zebra populations, and in the Ngorongoro, the presence of large numbers of hyena may control species diversity (13).

The Role of Humans in Savannas

Humans have long played a role in shaping the savanna landscapes and large mammal populations of eastern Africa through activities such as shifting cultivation, pastoralism and fire. Over the last 3000 years or more, fire and grazing by domestic stock (9) have contributed to the expansion and maintenance of savanna grasslands (see Chapter 2). Pastoralism has been a factor in the East African savannas for at least 3000 to 4000 years, and perhaps a good deal longer, even though the Maasai themselves only arrived in East Africa from the north as little as 500 years ago. The livestock of early pastoralists are thought to have used the savannas in ways ecologically similar to wildlife migrants, and during droughts the herdsmen relied on hunting to sustain them until their herds recovered. Livestock, which comprise cattle, sheep, goats, donkeys and camels, continues to increase rapidly as human population increases, despite the unavailability of large parts of the savanna due to the presence of tsetse flies. The effects of livestock on savanna habitats are profound. Where livestock exists in very large numbers, overgrazing often results in changes in grass species composition and vegetation growth forms. In more arid areas, overgrazing and removal of tree cover have been implicated in desertification (but see Chapter 10).

The ratio of sheep, goats, livestock, donkeys and camels changes along an aridity gradient and is strongly influenced by habitat (22). There is a large complementarity in feeding among these species, between grazers and browsers and between coarse feeders and fine feeders. Cattle commonly convert long grass to short grass which is subsequently used by sheep (and goats) and finally by juvenile stock. A high browse-to-graze ratio favors goats while short grass sward favors small stock rather than cattle. The diversity of habitats governs the range and composition of livestock. Dense bushland provides suitable environments for tsetse fly and conceals predators. Such areas are less preferred than the open grasslands. Management techniques of livestock herds, including their number, have significant effects on community structure. Old boma sites in Laikipia are high in nutrients and favored by both livestock and wildlife. Their intense grazing of these sites has resulted in relatively permanent 'glades' within acacia bushland (38), similar to the grazing lawns described in the Serengeti.

European colonization of Africa had several effects on savanna wildlife populations and the distribution of livestock. Apart from the introduction of weapons used in killing of wildlife through hunting, they introduced a series of previously unknown pests and pathogens. These had drastic effects on people with no prior exposure or resistance (11,14). The famines and epidemics of this time had extreme ecological effects on vegetation, livestock, wildlife and human populations. Around the turn of the century, rinderpest, a bovine virus exotic to the continent spread from the north of Africa to the south. It had a catastrophic effect, killing 90 to 95% of the domestic cattle and decimating many susceptible wild ungulate populations on the way. East Africa was struck by rinderpest in 1890 and in 2 years, an estimated 95% of the buffalo, wildebeest and cattle populations had died. In the Serengeti (28), there were large declines in wildebeest, buffalo and other wild ungulates. The result, over much of East Africa, was widespread human starvation and movement of local populations, due to their dependence on livestock, leaving vast areas of grassland empty. The Maasai and other pastoral people were particularly hard hit. Human diseases, particularly smallpox, struck at the same time, further decimating the human population. This is often referred to as "Emutai", meaning total destruction in Maasai.

The absence of livestock and wildlife grazing, less human activity such as cultivation and burning, and the large reductions in elephants in the late 1800s by hunters was accompanied by the spread of dense woodlands and trypanosomiasis, the extent of which is still debated (32). There was a subsequent decline of tsetse flies due to the absence of suitable ungulate hosts. By 1910, most of the wildlife had developed immunity to rinderpest and begun to increase rapidly. Following this increase in wildlife, there was an increase in tsetse flies in the presence of hosts and dense thickets. Colonial governments used mechanical removal of woodland cover in the 1930s in an attempt to eradicate tsetse fly. This allowed livestock, herbivores, and eventually their predators to expand their range. For example, the elimination of rinderpest in cattle herds resulted in large increases in wildebeest numbers in the Serengeti/Mara ecosystem from 250,000 in 1961 to 500,000 in 1967 to the current more than 1,000,000 .

Fire is another factor known to be important in shaping the savannas. The most ancient fires were natural and depended on lightning for ignition. Due to the strong development of the grass layer and seasonality of the climate, there is often abundant fuel in tropical savannas. Humans have burnt for many thousands of years. The frequency of fires may have increased markedly over the last several hundred years as human population increased. Towards the end of the 19th Century, when settlement became more firmly established, the area of savanna was greatly enlarged by forest clearance, burning and massive increases in the number of cattle. In savanna woodlands, fires are important in maintaining the balance of grass to trees and shrubs where elephants are absent or have changed in their distribution and density.

The Role of Fire in Savanna Ecosystems

Fire has been associated with savanna ecosystems for millions of years and together with grazing, is thought to be among the most important factors determining the structure of these ecosystems. Fire as a powerful factor in natural ecosystems and its use by humans in Africa is ancient. Lightning caused fires as it does today, even before man used fire as a tool. Hunter-gatherers of sub-saharan Africa appeared to have known how to keep and use fire to improve grazing for their livestock and to facilitate hunting of wild game for at least the last 60,000 years, and perhaps much longer (see Chapter 2). Fire was and still is used by pastoral communities and 'modern' range managers to control tall grass and bush and to create grazing lawns, particularly during very wet years when grazing pressure is low. It is also used to remove old growth at the end of the dry season in preparation for the rains. In spite of African savanna ecosystems' long acquaintance with fire, its fire ecology is not yet clearly understood, though its usefulness in the maintenance and improvement of the quality of grasses and in the prevention of bush encroachment is undisputed.

Rainfall is a key factor influencing the timing, frequency and extent of fires in East Africa. Fires normally occur late in the dry season and their incidence and intensity are determined by the amount of combustible material present. Higher rainfall areas have higher plant biomass. Accumulation of combustible materials is therefore rapid and makes the moister savannas more prone to fires than drier, sparsely vegetated savannas. Grass growing in high rainfall areas is coarse and of poor quality (due to its taller and more rank growth forms) and its frequent removal using fire is sometimes necessary to provide high quality grazing. In the moist savannas, fire plays an important role in the prevention of bush encroachment. It favors the development and maintenance of a predominantly grassland community by destroying the juvenile trees and shrubs. Once the trees get beyond a critical height (normally above grass height), fire is far less effective in eliminating them.

In semi-arid savannas, rainfall is lower and more variable. It causes marked fluctuations in the seasonal production of grass material, and often fuel loads are insufficient to support a fire. Fire is further suppressed by grazing, which removes the little grass growth. Such conditions may be conducive for the development of trees and shrubs adapted to the local climatic conditions. Where fires do burn, the tree and shrub species of these savannas are very resistant to fires. They have dormant buds at the base of the stem from which fresh shoots grow after burning. Prevention of bush encroachment may require a combination of methods to be successful.

The effect of fire on the structure and fertility of the soil in tropical savannas is still the subject of much controversy. Research in southern Africa suggests that fire enriches the soil surface layer with nutrients and encourages the growth of *Themeda*, a grass favored by both livestock and ungulates. After a fire, there is an abrupt release of elements such as calcium, phosphorous and potassium held in plant material and normally released gradually as plants decay. Research in Nairobi National Park showed nutrient content was higher in grasses growing on previously burnt areas (12). These soluble forms of nutrients are, however, also subject to removal by water and leaching through the soil profile.

Burning may also reduce soil moisture content. This may not necessarily be serious in high rainfall areas, but in more arid regions, where soil moisture is a serious limiting factor to plant growth, this can adversely affect plant vigor, reduce basal cover and result in increased runoff (6).

The effect of fire on the soil fauna varies with the type of fire. Birds and larger mammals and birds usually flee as fire approaches, and reptiles and small mammals normally stay underground, but many invertebrates perish, with the exception of soil dwelling species. Many jumping and flying insects are eaten by birds attracted to the fire. Fires are hotter at the end of the dry season in higher rainfall areas and affect the soil dwelling fauna more severely.

The Interaction Between Burning and Herbivory

The interaction between burning and grazing is important in savanna ecosystems. Herbivory and burning represent alternate pathways of nutrient cycling and since fire volatilizes (but not herbivory) much of the carbon and nitrogen balance between soil nutrients and their availability will be influenced by the balance of grazing and burning. Fire, like heavy grazing, increases the palatability of grasses by creating a green flush of very high quality and digestibility. In many parks in East Africa, the cessation of human activities including burning and grazing have resulted in major habitat changes. In many parks these changes have been accentuated by the decline in numbers of elephants due to poaching. In Nairobi National Park the main outcome of the removal of human resource-use activities and later, cessation of controlled burning regimes was overmaturation of the grasslands and a decline in herbivore use and diversity. Many East African savanna parks are undergoing similar ecological changes in response to removal of a combination of these factors. Burning experiments in Nairobi National Park reversed these trends by providing grazing arenas for the small grazers, which were most affected (12). Burning conducted early in the dry season provided highly nutritious vegetation for the resident herbivores. Grazers initially concentrated on the burnt plots and later moved onto adjacent areas, thus extending their impact beyond the burn. The concentration of minerals from dung and urine and the effects of burning and grazing result in grazing arenas that last beyond the burn season. Preferences for habitats exposed to heavier grazing and fire regimes are also clearly illustrated by the obvious choice of areas outside the Maasai Mara reserve by returning migrants.

When burning was conducted at the end of the dry season, just prior to the rains, wildebeest and zebra which normally leave the park with the onset of the rains responded by remaining on the burnt plots for a greater part of the wet season. These late season burns demonstrated that the availability of short grass areas in the park, at the beginning of the wet season, could delay outward migration of these species and thereby reduce grazing pressure on privately owned ranches outside the park. Wildlife concentrating on the green flushes included wildebeests, hartebeest, Grant's gazelles, Thomson's gazelles, zebras, warthogs, ostriches and elands. Burnt sites had a higher diversity of wildlife than the unburned sites. When the migrants finally departed, the grass was still short enough to be attractive to the small resident ungulates. The removal of shading, the absence of competition and additional nutrients from ash-fall (30) are thought to contribute to greater grass productivity in the burnt plots. An prime example of these 'grazing lawns' on a large scale is the area just north of Maasai Mara Game Reserve, where livestock grazing by traditional Maasai attracts large numbers of grazing herbivores. The reserve itself with its taller grasses, is shunned by these herbivores until dry conditions force them back.

The interaction of burning and grazing has greater impact on vegetation than each alone. Of particular interest is the rest period between the time of burning and subsequent grazing. Experimental work in South Africa shows that the shorter the interval between the time of burning and grazing, the greater the reduction in the total season's grass yield. In Nairobi National Park grass productivity was lower on small areas subjected to intense grazing after burning than on larger areas where grazing effort was more widely distributed (12).

Another controversy has been on the most suitable season in which to burn. Many researchers advocate burning at the end of the dry season when the grass is dormant. Many farming and pastoral communities normally set fires to grasslands at this time in readiness for the rains. Fires set early in the dry season in Nairobi National Park did not seem to affect the grasses ability to regenerate, while hot fires set in the dry season were better at retaining the migrants at the beginning of the wet season (12). The decision on when to burn should be based on the management goals and controlled experiments in each ecosystem. The experiments in Nairobi National Park show that fire can be used to reverse the impoverishment of many savanna parks if used judiciously.

Box 11.3. Island Biogeography and Species Extinctions

Biogeography is the study of factors that effect the distribution of species on the landscape or regional scale (>1 km²). Island biogeography addresses the questions of how many and which species islands or isolated areas can support. There are two major tenets of island biogeography. First, the number of species is proportional to the area of the island. This is usually well described by the equation, Number of species = C Area^z where C and z are constants. Values for z are usually around 0.25 for large areas so the relationship between area and island size is best described by a saturation curve. On smaller islands, fewer species can maintain populations large enough to be viable - particularly large-bodied and high metabolism species. In addition, smaller islands generally have fewer habitats. Second, more isolated islands have fewer species than islands closer to mainlands or other islands (2,3). The more isolated the island, the fewer the species that are likely to colonize it (or recolonize it after a local extinction).

Island biogeographic theory is not solely focused on small land masses surrounded by water. Its principles also apply to any biotic community of limited size that is surrounded by areas unsuitable for island residents. The lakes and the mountains of East Africa are both sets of natural 'islands'. As with other islands, the numbers and identities of the species they support are related to both their size and their isolation.

In addition, humans can isolate parts of ecosystems. Many of East Africa's savanna and forest parks and reserves are being increasingly isolated from each other by agriculture and other land uses (3,4). This isolation may threaten the biological diversity of these areas by reducing their effective size, and by increasing the distance to other protected areas (1,3). The loss of species as areas become isolated and reduced in size is known as faunal relaxation (the extinction of species) to some expected equilibrium number for the given size island (3). Whether faunal relaxation will occur in East African forests and savanna parks is of considerable interest and debate (1,3,4). In savanna ecosystems some researchers have noted that the species-area relationship of the major herbivores is weak (1,4), and therefore faunal relaxation is less likely to occur, particularly if areas with the most diverse habitats and species are protected (4). The relationship between area and species is much stronger in studied forests (see Chapter 13). Regardless, of the strength of these relationships and the predicted amount of faunal relaxation, most ecologists agree that intensive management of smaller protected areas will be needed to overcome the natural faunal relaxation or species extinction process.

Management should include monitoring endangered species and translocating, breeding or feeding the more threatened species. Additionally, the establishment of corridors between protected areas has been suggested as a way to minimize the effects of isolation. Establishing corridors between areas such as between Tarangire and Lake Manyara, or between Mount Kenya and the Aberdare Mountains have been proposed as ways to reduce species extinction. Such corridors, however, require the displacement of alternative land uses, and will be difficult to establish without moving private land owners.

More successful attempts to protect large conservation areas include sharing the area among two or more countries. Although the protected area of each country may be small, the overall 'island' size is large. Examples include the protected areas of Mount Elgon (Kenya and Uganda), the Serengeti-Mara ecosystem (Tanzania and

Kenya) and the Virunga ecosystem (Rwanda, Uganda, and Zaire). These protected designs can also have problems, however, when there is disagreement between nations about appropriate management of the protected areas.

T.P. Young & T.R. McClanahan

References

(1)   East, R. 1983. Application of species-area curves to African savannah reserves. *African Journal of Ecology* 21: 123-128
(2).   MacArthur, R.H., Wilson, E.O. 1967. *The Theory of Island Biogeography*. Princeton: Princeton University Press.
(3)   Soule, M.E., Wilcox, B.A., Holtby, C. 1979. Benign neglect: a model of faunal collapse in game reserves of East Africa. *Biological Conservation* 15: 259-272
(4)   Western, D., Ssemakula, J. 1981. The future of savannah ecosystems: ecological islands or faunal enclaves? *African Journal of Ecology* 19: 7-19

Threats to Savanna Ecosystems

Insularization and the Impoverishment of Savanna Ecosystems

The integrity of savanna ecosystems is dependent on the maintenance of its key ecological processes. As human populations increase, many savanna systems are becoming smaller in size, fragmented and some totally closed off. The effect of these changes on savanna wildlife ecosystems are tremendous and pose new challenges to conservation. Growing human population, changes in lifestyles among pastoralists and changes in land tenure are among the problems confronting these systems and their magnitude will determine the form in which the ecosystems persist.

People have long assumed that strict protection can best maintain biodiversity, leading to the establishment of national parks as the dominant approach to conservation. Several factors have been recognized as potential threats to the viability of these parks. Among them are limited size, ecological isolation, edge effects, diseases, poaching, pollution and global climatic change and more recently segregation effects (a common suite of threats arising from the creation of a protected area itself; 33). Natural communities have become increasingly reduced and fragmented in many parts of the world. In the African savanna, national parks represent the last large-scale remnants in the world of the tremendous variety of mammals of the Pleistocene epoch (see Chapter 2). Most parks are too small to encompass whole ecosystems and their boundaries were fixed with little regard for the year-round needs of the herbivores. The removal of diversifying factors such as fire and other human agencies can lead to a loss of biodiversity, as can the creation of artificially high concentrations of large mammals, due to increased competition for critical resources. When traditional human influences are removed, large herbivores are displaced, and other key factors governing the systems are affected. Human ecology (livestock grazing, fire, wood cutting and settlement) has played a large role in shaping and maintaining the East African savannas.

'Insularization' is fragmentation of a previously single geographic range into small isolated habitats that are virtual ecological islands. Insularization has been best documented in the temperate regions and tropical woody savannas, mainly resulting from the extensive development of agriculture and forestry. In East Africa, high rates of human population increase, coupled with changes in people's lifestyles over the last 50 years, have resulted in unprecedented conflicts between humans and biodiversity. In an attempt to reduce these conflicts, some areas have been partially or completely isolated from surrounding lands by physical barriers, such as trenches and electric fences, that contain and protect animals within national parks, and to protect people and their property from wildlife damage in areas of conflict. These developments have created new challenges in the management of wildlife.

Concern for the effects of insularization on species richness and the diversity of biological communities is not new. Theory and experience predict that the number of species on an island or a similarly isolated area is an increasing function of the size of the area (see Box 11.2). Fragmentation, by reducing the size of such areas, reduces populations to sizes that make them vulnerable to local extinction (25).

The continued existence of many national parks and reserves as viable conservation units depends on sound management of wildlife resources and concerted efforts to restore damaged habitats. The number of protected areas in Kenya has increased considerably since the establishment of the first national park in 1946. By 1992, the country had gazetted 26 national parks and 29 reserves. The area available for wildlife conservation outside of protected areas decreased considerably between 1950 and 1985. This reduction in habitat areas has resulted in severe pressures being exerted on natural ecosystems.

The problems of ecological segregation and insularization are of relevance in the East African savanna. Habitat disturbances result in the disappearance of species at rates that increase as the islands decrease in area or become more isolated from each other. By virtue of their low population densities, birds and mammals are among the taxa most likely to disappear from isolated fragments. It has been estimated that 11% of the species of large mammals may be lost from East African national parks in the next 50 years (29,35).

With increased isolation, the influence of large mammals on habitats becomes more pronounced. For example, elephants' natural role in opening up woodlands and forests has been recognized for a long time (see Box 11.1). However, not until they are forced to live in unnaturally high densities because of insularization, does their ability to destroy habitats become a real cause for worry. Fragmentation also causes the redistribution of the remaining area into disjunct components, thereby reducing dispersal opportunities for all species. The ultimate result is an increase in the probability of local species extinctions (24). Park establishment has sometimes resulted in severe reductions in species richness and diversity both inside and outside protected areas, without any accompanying loss of total land area. The establishment of a park usually results in redistribution of large herbivores in response to the availability of resources and human activities on both sides of the boundary. Such alterations may only be seasonal but they can have far reaching effects on ecosystem structure and function.

Unsustainable Livestock Practices

Most wildlife in East Africa, however, occurs outside parks, where the land is shared with pastoralists and their livestock. Both are key components in the ecology of these savannas. Conventional wisdom has often equated pastoralist regimes with overgrazing, overstocking and environmental degradation, in contrast to wildlife, which are seen as existing in harmony with their surroundings (1). Although this is true in certain areas, these processes are too often invoked without evidence to back up their existence (see Chapter 10). Large herds of migratory wildebeest in the Athi-Kapiti plains for example usually remove far more vegetation on their migratory path than livestock.

As pressures to adapt to a more modern life have increased, many pastoral societies, including the Maasai, have become sedentary and their populations have grown. Both sedentarization and increasing populations of people and livestock are likely to make 'traditional' use of rangeland less sustainable.

Conversion to Agriculture

Conversion to agriculture occurs at two different scales. Large-scale enterprises have increasingly taken over large tracts on semi-arid lands. Where rainfall is deemed sufficient, groundnuts or wheat have been planted. In drier sites, irrigation schemes have been established. Some of these large-scale conversions have been partial successes, others have failed. All have a devastating effect on local biodiversity, which must be weighed against their possible economic advantages.

Small-scale conversion is also increasing. As land suitable for small-scale farming is fully settled, landless farmers look to drier, less suitable land. Throughout East Africa, this has led to the conversion of rangeland into plots of sub-marginal land even though they are too dry for agriculture, and too small for livestock use. This conversion can be profitable for land 'developers', but it is a disaster for the small-scale buyer, for biodiversity, and for national economies (which lose productive rangeland and potential tourism sites).

Poaching

The illegal killing of animals is the most publicized threat to savanna ecosystems. Elephants and rhinos have already been eliminated from much of their former range. There are, however, a wide variety of plants and animals under similar threats. The illegal orchid and aloe trade may be as dangerous as the illegal trade in birds, reptiles and mammals. Most of this poaching is species-specific, but poaching for game meat locally threatens a broad range of ungulates. When the threatened species is a keystone species, such as elephants (see Box 11.1), even the elimination of a single species can have far-reaching effects on the ecosystem.

One form of semi-arid and arid land poaching that degrades entire ecosystems is illegal charcoal production. Large areas of bushlands have been cleared to convert the trees and shrubs to charcoal, and almost never with information that could make the harvest sustainable (but see 36). Efforts are underway to regulate this industry.

Research and Management Needs

The processes and results of ecological isolation present new challenges to wildlife managers. With total insularization, migration routes previously available for large mammals become sealed off, with attendant build up of animal populations. Such a situation is likely to set off a series of events leading to irreversible changes in the ecosystem. For example, plant-herbivore and predator-prey interactions can be radically disrupted by changes in the composition of plant and animal communities. The authorities charged with the conservation and management of wildlife resources in East Africa are now faced with the task of establishing a network of scaled-down functional units within the existing national park system.

Lake Nakuru National Park provides a unique opportunity to understand the implications of insularization to wildlife management. The Park was surrounded with an electric fence in 1987. Events leading to its complete isolation started earlier this century when man began using the catchment for urban development, commercial farming and livestock rearing. The park is characterized by high herbivore densities and low predation pressure. The numbers of the three most common herbivore species, waterbucks, impalas and warthogs, more than doubled between 1978 and 1990. The oscillations from very high to very low numbers are accompanied by large-scale changes in vegetation structure and pose serious management concerns. The animal and plant communities need to be continuously monitored and managed to avoid ecological catastrophes. The successful management of such an ecosystem depends on availability of ecological and socio-economic information. The isolation of ecosystems may require active management of the entire resource spectrum. Herbivores may need to be controlled when necessary. A similar program would also be required for predators. There is need to understand the effects of habitat fragmentation on territorial and migratory species. Their behavioral responses to confinement and its resultant effects on population dynamics, as well as habitat responses to changing utilization regimes, require constant observation and evaluation.

Research is also needed in the area of population genetics. Insularization may also increase the incidence of inbreeding, causing a reduction in genetic variability and perhaps limit the ability of animal species to adapt to new conditions. Such work should also go hand in hand with veterinary investigations into disease risks. One solution may be to link several isolated ecosystems using corridors to increase space and reduce the incidence of inbreeding.

Information on visitor attitudes, and those of neighboring communities, is important in encouraging a harmonious coexistence between man and wildlife for mutual benefit. Visitor controls may be needed, if backed up by controlled studies of their ecological impacts.

Summary

The savanna ecosystems of East Africa are a unique and invaluable world heritage. They are the birthplace of mankind, have provided a livelihood to countless generations of many traditional cultures, and are today one of the main income earners in the region. The image of East Africa is inextricable tied to these ecosystems. Their ecology is some of the best understood of any in the world. Nonetheless, there is still much to learn if we are to effectively conserve and manage them for future generations. There are very real threats to these ecosystems, and only on-going vigilance and dedicated study can ensure their long-term survival.

References

1. Anderson, G.D., Herlocker, D.J. 1973. Soil factors affecting the distribution of the vegetation types and their utilization by wild animals in Ngorongoro Crater, Tanzania. *Journal of Ecology.* 61: 627-651

2. Bagine, R.N.K. 1982. *The Role of Termites in Litter Decomposition and Soil Translocation with Special Reference to Odontotermes in Arid lands of Northern Kenya.* PhD Thesis, University of Nairobi.

3. Barnosky, A.D. 1979. The late Pleistocene Event as a paradigm for widespread mammal extinction. In *Mass Extinction: Process and Evidence*, ed. Donovan S.K., pp. 235-254. London: Belhaven Press

4. Bell, R.H.V. 1982. The effect of soil nutrient availability on community structure in African Ecosystems. In *Ecology of Tropical Savannas* eds. Huntley G.J., Walker B.H. pp 193-216. New York: Springer Verlag

5. Belsky, A.J. 1984. Role of small browsing mammals in preventing woodland regeneration in the Serengeti National Park. *African Journal of Ecology.* 22: 271-279

6. Booysen, P. de V., Tainton, N.M. eds. 1984. Ecological effects of fire in *Southern African Ecosystems*. Ecological Studies 48. Berling: Springer-Verlag

7. Cole, M.M. 1986. *The Savannas Biogeography and Geobotany.* New York: Academic Press

8. Demment, M.W., Van Soest, P.J. 1985. A nutritional explanation for body size patterns of ruminant and nonruminant herbivores. *American Naturalist* 125: 641-671

9. Deshmukh, I.K. 1986. *Ecology and Tropical Biology.* Palo Alto U.S.A.: Blackwell Scientific Publications

10. Farlow, J.O. 1976. A consideration of a late Cretaceous large dinosaur community (Oldman Formation). *Ecology*: 841-857

11. Ford, J. 1971. T*he Role of Trypanosomiases in African Ecology: a Study of the Tsetse Fly Problem.* Nairobi: Oxford University Press

12. Gichohi, H.W. 1992. The Effects of Fire and Grazing on Grasslands of Nairobi National Park. MSc Thesis, University of Nairobi

13. Homewood, K.M., Rodgers, W.A. 1991. *Maasailand Ecology:Pastoralist Development and Wildlife Conservation in Ngorongoro, Tanzania.* Cambridge: Cambridge University Press

14. Kjekshus, H. 1977. *Ecology, Control and Development in East Africa History.* London: Heinemann

15. Lamprey, H.F. 1963. Ecological separation of the large mammal species in the Tarangire Game Reserve, Tanganyika. *East African Wildlife Journal* 2: 1-46

16. Maddock, L. 1979. The 'migration' and grazing succession. In *Serengeti: Dynamics of an Ecosystem*, eds. Sinclair, A.R.E., Norton-Griffiths, M. pp. 104-129. Chicago: University of Chicago Press

17. McNaughton, S.J. 1970. Mineral nutrition and seasonal movements of African migratory regulates. *African Wildlife Journal* 2: 1-46

18. McNaughton, S.J. 1983. Serengeti grassland ecology: The role of composite environmental factors and contingency in community organization. *Ecology of Monographs* 53: 291-320

19. McNaughton, S.J. 1984. Grazing lawns: Animals in herds, plants forms and coevolution. *American Naturalist* 124 (6): 863-886

20. McNaughton, S.J. 1989. Interactions of plants of the field layer with large herbivores. In *The Biology of Large Mammals in their Environment.* eds. Jewell, P.A., Maloiy, G.M. pp. 15-29. New York: Oxford Scientific Publications

21. Milewski, A.V., Young, T.P., Madden, D. 1991. Thorns as induced defenses: experimental evidence. *Oecologia* 86: 70-75

22. Pratt, D.J., Gwynne, M.D. 1977. *Rangeland Management and Ecology in East Africa.* London: Hodder & Stoughton

23. Pellew, R.A. 1983. The impact of elephants, giraffe and fire upon *Acacia tortilis* woodlands of the Serengeti.. *African Journal of Ecology* 21: 41-74

24 Quinn, J.F., Wolin, C.L., Judge, M.L. 1989. An experimental analysis of patch size, habitat subdivision and extinction in a marine snail. *Conservation Biology.* 3: 242-251

25. Rey, J.R. 1984. Experimental tests of island biogeographic theory. *Ecological Communities: Conceptual Issues and Evidence.* eds. Strong, D.R., Simberloff, D., Abele, L.G., Thistle, A.B., pp. 101-112. Princeton, N.J.: Princeton University Press

26 Sinclair, A.R.E. 1975. The resource limitation of trophic level in tropical grassland ecosystem. *Journal of Animal Ecology.* 44: 497-520

27. Sinclair A.R.E. 1977. *The African Buffalo.* Chicago: University of Chicago Press

28. Sinclair, A.R.E., Norton-Grifths, M. eds. 1979. *Serengeti: Dynamics of an Ecosystem.* Chicago: University of Chicago Press. 389 pp

29. Soule, M.E. 1980. Thresholds for survival maintaining fitness and evolutionary potential. In *Conservation Biology: An Evolutionary Ecological Approach.* eds. Soule, M.E., Wilcox, B.A. pp. 151-170. Sunderland, Massachusetts: Sinauer

30. Tainton, N.M., Groves, R.H., Nash, R.E. 1977. Time of mowing and burning veld: short term effects in production and tiller development. *Proceedings of the Grassland Society of Southern Africa.* 12: 59-64

31. Tilman, D. 1982. *Resource Competition and Community Structure.* Princeton: Princeton University Press

32. Waller, R.D. 1990. Tsetse fly in western Narok, Kenya. *Journal of African History.* 31: 81-101

33. Western, D. 1991. *Climatic Change and Biodiversity. A Change in the Weather.* eds. Ominde, S. H., Juma, C., pp. 87-96. Nairobi: Acts Press

34. Western, D., Gichohi, H. 1994. Segregation effects and the impoverishment of savanna parks: the case for ecosystem viability analysis. *African Journal of Ecology.* 31: 269-281

35. Western, D., Ssemakula, J. 1982. Life history patterns in birds and mammals and their evolutionary interpretation. *Oecologia.* 54: 281-290

36. Wood, T.G., Sands, W.A. 1978. The role of termites in ecosystems. In *Production Ecology of Ants and Termites,* ed. Brain, M.V., pp 55-80. Cambridge: Cambridge University Press

37. Young, A. 1976. *Tropical Soils and Soil Survey.* Cambridge: Cambridge University Press

38. Young, T.P., Francombe, C. 1991. Growth and yield estimates in natural stands of leleshwa *(Tarconanthus camphoratus). Forest Ecology and Management* 41:309-321

39. Young, T.P., Macrae, A., Patridge, N. 1995. Long-term glades and their edge effects in Lakipia, Kenya. *Ecological Applications* 5: 97-108

Chapter 12

The Miombo Woodlands

W.A. Rodgers

This chapter on the miombo woodlands follows from the previous more generalized chapter on East African savannas. The miombo is, however, much more than a mere variant of the typical East African savanna. Covering some 50% of Tanzania it is the largest single vegetation type in East Africa. It links East Africa floristically and ecologically to south-central Africa. The miombo is distinct in its mixture of climatic and soil characteristics. The unimodal rainfall pattern, and prolonged dry season with severe fires, influence miombo ecology differently from that of savannas with two rainfall peaks per year. In species composition the miombo is largely distinct at the generic level for trees and, therefore, in many aspects of its ecology.

The East African miombo woodlands are an impoverished form of the more extensive and more species-rich miombo woodlands of Central Africa. The African miombo as a whole covers an estimated 3.0 million square kilometers (=km²) of Zimbabwe, Zambia, Mozambique, Angola, Malawi, Katanga province of Zaire, and southern Tanzania. The miombo formation is described and mapped as the largest vegetation unit in the Zambezian center of endemism - a distinct plant -geographic unit or 'phytochorion' (52). A phytochorion is a distinctive plant geographical area, recognizable by its species composition rather than by the structure of the vegetation. Figure 12.1 shows the major phytochoria of tropical Africa. The miombo has similarities in the structure of the vegetation and some floristic or evolutionary affinities with the Guinea Savanna formations of West Africa. About 24% of Zambezian tree species also occur in the Soudanian phytochorion.

The word 'Miombo' (Muuyombo) is a vernacular name in use in both Tanzania and Zambia for a species of *Brachystegia*, (*B. boehmii*), the dominant tree genus over the entire area.

In Tanzania, the miombo may once have covered up to 450,000 km², in two distinct blocks, in the south-east and the west of the country. These blocks are separated by a mountainous belt (the Eastern Arc mountains of Ukagurus and Udzungwas, see Box 13.1), and by the adjacent drier communities in the mountain range's immediate rain shadow. This separation barrier is an old one, as evidenced by the sub-speciation of sable antelope (*Hippotragus niger*), a typical miombo large antelope, either side of the mountain range: *H. niger niger* to the west, and *H. niger roosevelti* to the east (Fig. 12.2). The miombo does not extend into Uganda, and

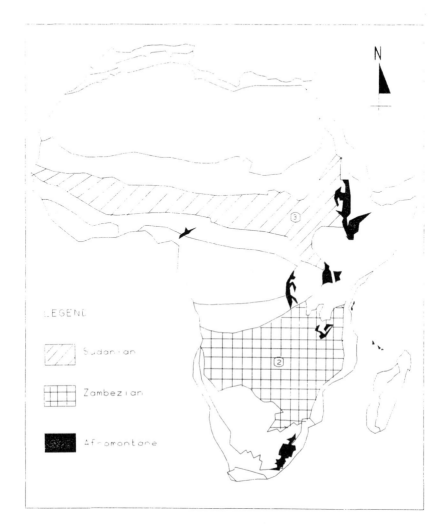

Figure 12.1. The phytochoria of Africa (after 52) showing the extent of miombo woodland.

while the genus *Brachystegia*, (*B. spiciformis*) does enter the south-east of Kenya, it is a member of the coastal savanna-dry forest complex, rather than normal miombo woodland community.

While there are a number of basic floristic and faunistic descriptions of the Tanzanian miombo (8,36,42) there are few detailed ecological studies (see 5,6). Previous published reviews of the African miombo are authored by Malaise (40) Celander (11) and Tuite and Gardiner (50). Most detailed research is from Central and Southern Africa (see the extensive studies by Chidumayo in references 13 through 20). This chapter focuses on the East African miombo; however, by necessity, the literature cited frequently reflects the Central and Southern African situation. The actual patterns of woodland ecology and human use change little across the region.

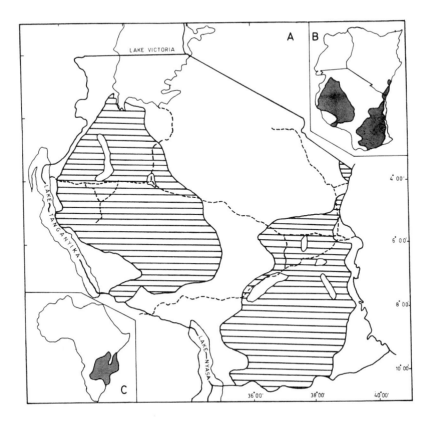

Figure 12.2. The distribution of (a) miombo in Tanzania, (b) sable in East Africa, and (c) Africa.

Vegetation Structure

The miombo is a distinctive vegetation formation, fairly homogenous in structure, function and species composition throughout its range. In East Africa it can be described as:

> *'A deciduous unarmed (having few tree species with spines or thorns) woodland occurring in the unimodal rainfall areas on geologically old, acid, sandy soils. It is characterized by trees in the sub-family Caesalpinoideae, especially species in the genera* Brachystegia *and* Julbenardia. *The shrub layer is variable in density, percent cover and species composition. It is often dominated by* Diplorhynchus *and* Combretum *spp. The ground cover varies from a dense coarse grass growth to a sparse cover of herbs and small grasses. The structure and species composition is largely maintained by periodic dry season fires'* (42).

Structurally, the woodland has a canopy cover between 20 and 60% (East African rangeland classification uses the terms 'wooded grassland' for canopy covers below 20%, and 'forest' for cover over 80%. Thicket is used for woody vegetation of canopy over 80% when canopy height - trees or shrubs - is below 8 m). Tree height ranges between 8 and 25 meters, and two tree height classes can be distinguished - the overwood or canopy dominants such as *Brachystegia*, and a smaller class of underwood tree species such as *Combretum* and *Terminalia*. Higher canopy and tree height values are associated with greater rainfalls in western Tanzania. The shrub layer may be up to 4 meters tall, but is rarely dense enough to reduce the cover of the grass layer. Grasses are tall (1.0 to 1.7 m) coarse tussock species, normally members of the Andropogonae. Termite mounds, rocky outcrops, and shade islands offer refuge for species less tolerant of fires. Epiphytes of all forms (ferns and orchids) are rare, as are mosses and ferns. Species in the Loranthaceae (parasitic mistletoes) family do occur, but are not common. There are woody lianas (such as *Entada* and *Landolphia*) but they are rarely conspicuous.

To give the reader a feel for the miombo vegetation I quote from the writings of a past Tsetse Department botanist, in Tanganyika, B.D. Burtt (8):

'A month before the rains set in, the miombo covered hills burst all at once into reds, salmons, pinks and coppery tinges of all hues as the Brachystegia trees flush into young leaf and, within a week, all this riot of color has blended into a forest of the freshest green carpeted with legions of flowers.

In full leaf the miombo is delightfully cool and shady, and the scanty grass a pleasure to walk through. The miombo woodland greatly resembles the oak woods of the New Forest in England.

In the dry season what a change! The whole miombo forest becomes entirely leafless, while grass fires burn up all the grass and leaf litter. The sun beats down unmercifully and whichever way one turns there is the same view, the gray stems of miombo trees fading into the shimmering distance. The buzz of insect life has vanished, except for the sharp hiss of tsetse flies'.

Distribution, Habitat Factors, and the Limits of the Miombo

In general the miombo formation occupies old, acidic, nutrient leached, shallow soils (see Table 12.1), in areas with a characteristic unimodal rainfall system (Table 12.2). The single rainy season influence is important for the ecology of plant and animal species. The northern extremity of the miombo is where the bimodal rainfall pattern begins or is more frequent. Bimodal rainfall is typical closer to the Equator (around 6° south). The northern equivalent of the miombo (Guinea Savanna) picks up again in unimodal west Africa, where rainfall patterns allow a woodland cover.

Miombo soils are typically shallow; often stony and the rooting zone can be restricted by a 'lateritic' hard-pan. (Laterite is a concretionary layer characterized by high iron-alumina concentrations). Deeper soils have a better quality woodland (taller and denser canopies). The poor nutrient status of the acidic soils is also important to miombo ecology. Where soils are more basic, typical of valleys (for example the Zambezi and Luangwa valleys in Zambia and the Rufiji valley in Tanzania), a different set of plant communities occurs, usually containing more palatable grasses. This more nutrient-rich woodland is called the southern African 'sweet-veld', as distinct from taller coarser 'sour-veld'.

Table 12.1. Soil data for a typical miombo site (Kitulangalo FR, Morogoro, East Tanzania), after (27). Well-drained slopes on old acidic basement rocks (biotite gneiss) dominated by *Brachystegia boehmii* and *B. spiciformis* communities.

Depth, cm	pH	C%	N%	Ex P ppm	CEC meq/100g
0-20	6.2	5.3	0.1	4	9.4
20-55	6.0	1.5	0.1	1	5.8
55-75	6.0	0.7	0.1	4	6.7
75-125	6.2			3	9.1
125-175	6.2			4	6.4

Physical Composition (Fraction as % per weight).

	Soil Depth, cm			
Particle size	20-55	50-75	75-125	125-175
Coarse sand	44	37	30	36
Fine sand	22	15	18	12
Silt	6	8	3	6
Clay	27	40	49	46

Table 12.2. Climatic data for a miombo site in Tabora, West Tanzania (32° 49' E, 05° 02' S. 1265 m asl) with an overall annual precipitation of 892 mm/y. Data from (25).

Month	J	F	M	A	M	J	J	A	S	O	N	D
Rainfall	132	129	166	134	27	2	0	1	7	17	103	174

Temperature, °C	Annual Average	22.9
	Mean maximum temperature	28.9
	Mean minimum temperature	17.0
Sunshine	71%	
Total Radiation	503 cal/cm²/day	
Evapotranspiration	1743 mm/y (Penman method)	

Total rainfall is as important as is the seasonal distribution. As rainfall increases beyond 1200 mm/y and where fire conditions allow, the miombo gives way to closed forest communities. Below 700 mm/y of rainfall, woodland cover reduces to a wooded grassland or thicket - again depending on fire conditions.

Towards the coast, the increasing rainfall with less seasonality, and higher air humidity and temperature lead to more typical coastal woodlands and woodland forest mosaics. *Brachystegia* disappears and there are more plants of the genera *Lannea, Sclerocarya, Tamarindus, Sterculia, Terminalia,* and *Combretum,* as well as a number of palm species. Increasing aridity in the north and in the center of Tanzania reduces tree cover and these areas eventually become bushland or thicket at the lowest rainfall levels. To the south, the Zambezi valley creates a frost hollow that forms a barrier for the dispersal and expansion of several miombo species. With increasing frost at higher altitudes (>2500 m above sea level = m asl), the miombo becomes a dwarf woodland (where miombo trees such as *Brachystegia utilis* fruit at <2 m in height), and then the miombo gives way to plant communities more rich in *Acacia* and *Protea* species.

With the increased intensity of agriculture, which involves clearing and felling of trees, and more frequent fires, the woodland cover decreases to an open wooded grassland or shrubland. Large areas that were historically miombo have been so cleared as to become an 'agricultural steppe' - characterized by annual crops, but with relict natural vegetation on rocky outcrops and along streams, and with scattered

Box 12.1.   Ecological Succession

Communities of plants and animals change through time (1). These changes are called succession, because there is a series of often predictable replacements of one set of species after another. One of the most common kinds of succession is recovery after disturbance. Wave damage, floods, fire, deforestation, the abandonment of farmland, and severe grazing and browsing are all forms of disturbance common in East Africa. On a longer time scale, the retreat of East African glaciers allowed a succession of alpine plants to colonize recently exposed rock.

After any of these kinds of disturbance, the first plants and animals to colonize a site are often short-lived and 'weedy'. These species are capable of long-distance dispersal, and can also tolerate the sometimes severe conditions of the early successional sites. These species are replaced in time with species that are better competitors, longer lived, and often larger. For example, the plant succession after a farmer abandons his field starts with the invasion of annual grasses and herbs. Within a few years, perennial species (including shrubs) begin to crowd out the annuals. In wet forests, these species are often relatively palatable, and a wide variety of terrestrial forest mammals including buffaloes, gorillas, and forest hogs prefer these patches of secondary growth. A decade or so later, trees have become established that will eventually dominate a regenerating forest.

There are, however, other forms of succession in East Africa that are not so directly related to physical disturbance. Bell and Vesey-Fitzgerald first described a grazing succession in the Serengeti-Mara ecosystem. An area of tall grass is first grazed by zebras, which can thrive on coarse plant parts. This opens up the grassland for species that can eat grasses of intermediate height, such as topis and hartebeests. After these species have moved on, the short stubble that remains is prime food for Thomson's gazelles that specialize in eating the delicate and now exposed plant parts. The next rainy season allows the tall grasses to return, and the process begins anew.

Another form of succession are the series of species that scavenge large mammal carcasses or forage in dung piles. In both cases, there are species that arrive early, and others that arrive later, often displacing the early species. In all cases of succession, there are opportunities for a wide variety of species to specialize in different successional stages, adding to the rich diversity of life within ecosystems.

T.P. Young

References

1.   Pickett, S. T. A., Collins, S. L., Armesto, J. J. 1987. Models, mechanisms and pathways of succession. *Botanical Review* 53: 335 - 371

trees such as baobab (*Adansonia digitata*). The clearing of miombo for groundnut farming in Nachingwea (S E Tanzania) and Urambo (West of Tabora in western Tanzania) between 1948 and 1953 are early examples of such large-scale ecosystem transformation. The southern Sukumaland steppe was cleared as a policy for controlling tsetse flies in the 1930s and 1940s, and regrowth of the miombo species was prevented when the land was converted to agriculture. Agricultural clearing is still continuing, and the question of plant regeneration is of critical importance for understanding miombo ecology and conserving this ecosystem.

Vegetation Communities

While the miombo is relatively homogenous over large areas, there are notable differences in plant communities between sites, some due to variations in habitat. Important habitat variables would include slope position (ridges with old, leached sands and lower slopes with more nutrient-rich loams), rockiness and soil depth. But, as will be discussed in greater detail below, many miombo communities are seral, being stages in vegetation succession following disturbances such as intense fires. Such succession often leads to a closed-canopy 'forest' climax - especially on sites with greater nutrient and rainfall abundance. Many plant and animal communities are thus points along a continuum of disturbance frequency, rather than discrete and persistent entities. Some common species groups are recognizable as distinct seral stages - a response to historical features and site-specific physical and chemical factors.

Vegetation scientists in Tanzania and Zambia distinguish two major woodland variants: (1) the *Brachystegia* woodland or miombo proper, and (2) a community typical of greater fire intensities called 'chipya' (or 'chao' in southeast Tanzania). This is in addition to the less common thicket/forest climax community ('msitu' in Tanzania and 'mushitu' in Zambia), which can have several different structural and compositional variants depending on rainfall patterns.

The Communities

Miombo. Miombo has a diverse tree community. *Brachystegia* species are usually the dominant canopy trees, but *Julbenardia* species may become common on nutrient-poor soils. *Isoberlina* is more common in the west. In southeast Tanzania *B. spiciformis* is the most common species, along with *B. boehmii, and B. bussei,* but a greater range of species exists in the west (such as *B. utilis* and *B. taxifolia*). Other trees associated with these dominants include species of the genera *Afzelia, Amblygonocarpus, Burkea, Erythrophleum, Ficus, Monotes, Pterocarpus, Swartzia, Uapaca, Xeroderris.* Miombo has an understory tree layer, rarely exceeding 10 m in height, dominated by several species of *Combretum* and *Terminalia.*

The shrub layer is variable in density, but usually sparse enough to allow a distinct herb layer to grow beneath it. *Diplorhynchus condylocarpon* is a distinct dominant, along with *Xeromphis, Byrsocarpus, Tetracera, Combretum, Ximenia* and *Flacourtia.* Shade islands under larger trees or associated with termite mounds are common, and include climber tangles (*Landolphia*) and thicket precursors (*Leptactina*). The ground layer is shorter and less dense than in the chipya, with a more diverse grass, sedge and herb content. Grasses are less coarse, and include *Panicum* spp as well as *Andropogon, Themeda* and lower-stature species such as *Sporobolus.* Campbell and coworkers (10) discuss the nature of this small-scale variation in the Zimbabwean miombo and attribute it to the effects of shade and fire as well as soil influences, such as termitaria.

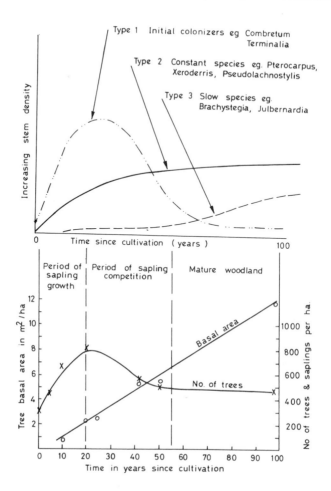

Figure 12.3. Several successional relationships suggested for impacts of clearing and fire on the miombo plant communities.

The suffruticose habitat is characteristic of the miombo and is composed of dwarf woody plants (such as *Cryptosepalum* which is related to *Brachystegia* but only 25 cm tall with a rootstock of 10 cm in diameter). Suffrutex species have evolved within 31 plant families found in southern Africa and have been referred to as 'the underground forests of Africa' because most of their woody content is in the root layer (51). Several factors including frost, waterlogged soils and fire are considered responsible for the development of this habit and the species adaptations.

Chipya. Chipya is more open and less diverse in tree, shrub and ground layers than the miombo. Chipya has less or none of the relatively fire-sensitive species of *Brachystegia*, and more fire-tolerant species in the genera *Pterocarpus*, *Burkea*, *Erythrophleum*, *Terminalia*, *Pseudolachnostylis*, and *Combretum*. In southeast Tanzania two coastal species are common in the chipya -*Millettia stuhlmannii* and *Pteleopsis myrtifolia*. Chipya has less of an understory or shrub layer. Grasses are coarse; taller and denser than in the miombo, with fewer sedges and herbs. *Andropogon* and *Hyperrhenia* are the dominant genera of grasses in this community.

Catena Concept

The recurring vegetation pattern associated with valleys and ridges suggests distinct relationships between geomorphology, soils, and vegetation. These interrelationships led to the 'catena' concept, used initially for mapping and surveying soil resources, The co-occurrence of two or more components is the mapped feature rather than the individual vegetation or soil component.

A typical catena is as follows, from ridge top to valley bottom: sparse chipya woodland on the frequently burnt sandy soil ridge tops, a denser canopy of *Brachystegia* woodlands with a distinct shrub layer on the lower slopes, which are more protected from fire, and a valley community at the base. Steep escarpments often have a closed canopy with a low leaf density dominated by *B. microphylla* above a closed shrub layer.

Valley communities can be of several types. In central Africa they are often referred to as 'dambos' (34). Dambos are broad flat valley bottoms, which flood during the rainy season due to overland run-off of rain as opposed to overspill from a stream or river. There is usually a riverine forest community along a stream-bed at the bottom of the valley. Kokwe (34) gives a detailed account of dambo ecology. In Tanzania, soil drainage is less impeded and a more open woodland develops and is characterized by species within the genera of *Bahia*, *Pericopsis*, *Terminalia* and *Combretum* (42,52). This is referred to as 'upper valley community' (42). As the valleys broaden and merge into the wooded grassland with less acidic soils, the plant community has more *Sclerocarya*, *Sterculia africana*, *Acacia* and *Combretum hereoense* and *Combretum imberbe* on impeded drainage. This is the 'lower valley community'. The drainage line communities, with their distinctive soil, water, geomorphological and vegetation features are of great importance for the resource utilization patterns of man and wild animals.

These plant communities and their degree of relationships can be depicted by association diagrams that are based on calculations of the species shared abundance among communities (35,42). Table 12.3 documents some ecological features of these distinctive communities.

Floristics

Most miombo plant species, especially the woody elements, are very widespread and there is thus little localized endemism. The serpentine soils of southern Africa have, however, allowed considerable speciation on these distinctive soil patches. Brenan (7) discusses miombo endemism in detail. The Zambezian domain is floristically rich. Of the 426 tree species found in the Sudano-Zambezian Region, 171 are found in the Sudanian sub-region compared with 334 in the Zambezian. This richness reaches its highest species levels in Zambia, for example *Brachystegia* has 17 full species in Zambia, 11 in western Tanzania, 6 in eastern Tanzania (Selous Game Reserve), 1 in Kenya (*B. spiciformis*), and none in Uganda (42). Generic endemism is low within the Zambesian region (<15%), but species endemism, to the phytochorion as a whole, is high. White (7) estimated some 8500 plant species in

Table 12.3. Ecological characteristics of miombo and related vegetation types in Upper Shaba Province of Zaire.(40).

Character	Dry forest	Miombo	Savanna
Tree height, meters	20	16	4
Higher plant species	105	480	330
Woody plants/ha	8500	700	55
Basal area m²/ha	40	20	0.5
Total dry weight tons/ha	320	150	10

Table 12.4 Shrub/seedling abundance in the burn plots for the fire experiments begun in southeast Tanzania. Data are presented as woody stems per hectare in 1972, 3 years after the experiment's initiation.

Species	Control 1969	No Burn 1972	Early Burn 1972	Late Burn 1972
Combretun zeyheri	825	4083	1344	311
Terminalia sericea	619	182	392	470
Lonchocarpus bussei	361	728	644	17
Xeromphis obovata	181	182	56	8
Dalbergia melanoxylon	155	182	20	0
Combretum collinum	155	273	56	8
Totals	2580	9100	2800	840

the Zambezian Phytochorion of which 54% are endemic. The majority of species richness is from the herbs, where there has been extensive radiation in some genera such as *Crotalaria* and *Indigofera*.

The thicket patches of south-eastern Tanzania (the climax community resulting from reduced burning of the miombo) can display a very high level of species endemism. Chapter 13 on coastal forests describes this endemism in more detail.

In Tanzania there are over 1300 plant species in the Selous Game Reserve, and 1400 in the Ruaha National Park and surrounds. There are probably some 3000 to 4000 plant species in the Tanzanian miombo as a whole.

Miombo Environment

Fire

Mention has been made of the important role of fire in the ecology of the miombo ecosystem. Most miombo ecologists agree that but for fire and man's clearing for agriculture, much of the present miombo area would be a closed forest or thicket community. Both the *Brachystegia* and chipya communities are thus not climax communities but are prevented from reaching a climax state by fire disturbances, they are termed fire dis-climax communities.

Much of our scientific understanding of the effects of fire on the miombo woodland communities comes from the Ndola fire experimental plots set up in northern Zimbabwe in 1933 (47). Fire experiments had three basic treatments: (1) no burning at all, (2) annual late or hot burning, and (3) annual early or cool burning. In summary, the early or cool burning plot retains a plant community similar to most miombo woodlands. Late burning reduced tree, shrub, and herb species diversity. In this plot, the woody layers are more open than the control plot and all fire-sensitive species such as *Brachystegia* were eliminated. Tree and shrubs have not regenerated and the grass layer became nearly a pure stand of coarse tussock species (degraded chipya). The no burning plot developed into a closed forest with considerable regeneration of forest species and no regeneration of typical miombo species. The grass layer has gone. The relationships between the ecological communities and the influence of fire and clearing are shown in Figure 12.3. Table 12.4 shows the impact of fire on the shrub layer from a similar study in southeast Tanzania.

Fire has been a component in the region for millennia (human-induced fire for at least 55,000 years, fire as evidenced by charcoal in the soil for 300,000 years). Adaptations to fire are common in both animals and plants (43). Miombo trees are characteristically thick barked, and seeds may require 'cracking' by fire to allow germination (such as *Pterocarpus angolensis*). Several species resprout from persistent rootstocks after fires which is an adaptation known as 'die-back' that allows seedlings to escape death due to the frequent dry season fires. Chidumayo (14,17,19) discusses the effects of fire on seedling ecology and subsequent woodland regeneration, and agrees with the overall conclusions of earlier workers; frequent hot fires suppress regeneration and lead to a more open less diverse woodland type. Rodgers (43) discusses the implications of fire for wildlife management. Most species are adapted to an early and cool burn regime which is therefore stressed as an important management tool. Late season hot fires kill more saplings than the early cool fires - resulting in lower species diversity. Maintaining a no-burn or a "natural-burn policy" is presently difficult due to human influences.

Topography, Soil Water, and Nutrients

In East and Central Africa the miombo woodlands are characteristic of old tertiary period planation surfaces, that have soils with a lateritic horizon. Differences in woodland composition, apart from seral stages (due to fire and clearing), are related to soil type and drainage conditions which are a consequence of local geomorphology. Key ecological factors include the extent that erosion has removed top soil and the lateritic material and so increased exposure to the bedrock; soil drainage and the degree of soil formation.

White (52) differentiates two site qualities - drier and wetter miombo woodland types. Dryness is a function of both rainfall, and soil depth with its water-holding capacity. Such site differences are expressed in woodland height, plant density and volume, and the number of species. Bell (4) described the relationship between moisture and nutrients in a tabular format, (Table 12.5). Increasing moisture and decreasing nutrient availability, often due to high soil leaching, leads to coarser grass growth, and a woodland of low productivity and lower wildlife abundance and diversity.

Recent research has shown the importance of ectomycorrhizae and nitrogen-fixing root nodules in the ecology of the miombo (29). Miombo woodlands dominated by the Caesalpinoidae are typically ecto-mycorrhizal in nature, rather than endo-mycorrhizal as in most of the tropics (29,30). There is a significant change in mycorrhizae types from the endo-mycorrhizas of the chipya to the ecto-mycorrhizas

Table 12.5. Ecological relationships of soil and water in African habitats (after 4).

VARIABLES	LOW NUTRIENT	HIGH NUTRIENT
HIGH MOISTURE	Wet miombo woodland. Tall grass, low nutrient content, high fire factor. 'Sourveld'*	Valley communities, often forested
LOW MOISTURE	Low wildlife density. Sahelian types. Short grass, often annual, higher nutrient content. 'Sweetveld'* Water limiting wildlife and stock production.	Serengeti short grasslands Seasonality high wildlife density

*Sourveld is a South African term indicating a tall coarse, low palatability, tussock perennial grass sward. Sweetveld is a term indicating a shorter, often annual, grass sward with a higher palatability, usually found on more alkaline soils.

of the *Uapaca* and *Brachystegia* and *Marquesia* dominated communities in Zambia (30). Such mycorrhizae increase water and nutrient uptake, speed up annual reproduction cycles and make phosphorus available to plants in a low-phosphorus soil environment.

Annual fires, which release much organic nitrogen to the atmosphere, make the importance of nitrogen fixation and other beneficial root symbioses obvious in low-nutrient soils. Miombo dominants do not have root nodules, as is the case with most Caesalpinoideae, but species in the genera *Pterocarpus* and *Cassia,* and *Afzelia quanzensis* are nodulated and nitrogen fixing.

Wildlife

The miombo is relatively impoverished as a large-mammal habitat with a low species diversity and, in general, a low biomass density. Some areas may, however, have a higher biomass due to the large body sizes of miombo herbivorous mammals, largely elephants. This faunal species impoverishment applies to birds and insects as well. Infertile soils are the major cause of the low wildlife diversity but is compounded by the long (up to 7 months) dry season with frequent fires. Surface water limits animal production, as does reduced availability of green fodder during the dry season. Most large herbivore species are facultative grazer/browsers (roan, sable, impala, buffaloes, elephants), or browsers (eland, greater kudu, duikers, rhino). Pure grazers are rare (the non-ruminant coarse grazer zebra, and the selective grazer Lichtenstein's hartebeest). The latter is almost totally restricted to the miombo.

Wildlife densities dramatically increase when the miombo habitat is near to other less harsh or stressful habitats. These habitats include vegetation islands of the more alkaline short grass open woodlands such as the East Selous, Zambezi, Saabi and Luangwa Valley systems, or floodplains which, in Tanzania, include the Kilombero, Rukwa, and Moyowosi drainage systems. Tables 12.6 and 12.7 show the pattern of ungulate population density in pure miombo and mixed habitat wildlife reserves in southern Tanzania (42). Data show that the mixed communities have a greater population density and diversity of ungulates. The greater density of elephants in the tall grass of well wooded miombo results in a greater large-mammal biomass than in the short grass more open savanna.

There are some characteristic miombo animal species. Amongst the ungulates, the sable, Lichtenstein's hartebeest and gray or bush duiker are notable examples. The roan is associated with more open woodlands across much of Africa, but is absent from the southeast Tanzanian miombo block. Several ungulates were originally widespread across several African habitats including the miombo. These species include buffalo, eland, greater kudu and non-bovids such as zebra, black rhino, warthog, and elephant. Presently, the larger species are restricted to protected areas. Some species such as impala and wildebeest use the miombo habitats in the dry season when the tall-grass layer is burnt and grasses begin to sprout. Some typical short-grass species or arid-area species such as the gazelle, oryx and the other hartebeest species are absent from the miombo. Table 12.8 shows the pattern of wildlife habitat preference in Tanzania.

Wildlife habitat relationships are discussed sufficiently in the preceding chapter on East African savannas in general (Chapter 11). The miombo demonstrates an extreme of seasonality in rainfall, with often severe dry season aridity accentuated by frequent burning. These features lead to complex patterns of wildlife ecology (31,42).

The miombo wildlife protected areas of Tanzania (Selous, Rungwa, Ugalla, Katavi and others) did not show the pattern of vegetation over-use by elephant that is associated with many other East African ecosystems - although there was such evidence in miombo in southern Africa (see 3). Several reasons account for this. First, the miombo reserves of Tanzania are relatively large (for example the Selous Game Reserve of 45,000 km^2). Additionally, little compression by human populations around the reserve boundaries was observed. Reserves were new, most formed in the past forty years and perhaps time has been insufficient to observe changes. Finally, high elephant population densities of up to 2 per km^2 (prior to the population crash in 1980s due to poaching) did not appear to cause habitat changes seen in other areas such as Tsavo.

The invertebrate fauna of the Zaire miombo has been described (39) but no such overviews of the Tanzanian woodlands have been published. Observers comment on the relative paucity of insect life and attribute this to the severe dry season.

Standing Crop and Productivity

Phenology

Productivity is extremely seasonal, so a brief summary of the phenology of the miombo is a necessary to understand this discussion on production. Malaise (39,40)

Table 12.6. Wildlife abundance data for miombo and other habitats in the Selous Game Reserve. Population densities are in the units of numbers/km² and biomass in kg/km². Savanna here is a short grass wooded grassland on alkaline soils. (Data from 42).

Species	Miombo	&	Savanna	Miombo		
	Density	Biomass	%	Density	Biomass	%
Elephant	0.5	923	19	2.5	4312	62
Buffalo	1.9	842	17	2.5	1125	16
Wildebeest	6.4	982	20	1.0	153	2
Zebra	4.0	800	16	1.6	320	5
Hartebeest	2.7	361	7	2.0	270	4
Impala	6.7	234	5	3.2	112	2
Sable	0.4	80	2	0.5	80	1
Total herbivore	22.6	4222		13.3	6372	

Table 12.7. Wildlife population density data from Tabora Region in western Tanzania from aerial surveys undertaken between 1978 and 1979 and expressed as numbers/km². Data from Ecosystems Ltd (21,22).

Species	Ugalla GR	Inyonga	Wala/Ugunda
Buffalo	2.18	1.24	1.67
Elephant	0.55	0.15	0.13
Impala	0.09	0.04	0.26
Hartebeest	0.29	0.05	0.23
Roan	0.08	0.10	0
Sable	0.60	0.24	0.18
Topi	0.17	0.41	0.47
Waterbuck	0.52	0.38	0.15
Zebra	0.04	0.20	0.75
Total	5.56	3.43	4.90

provides a detailed review of miombo phenology. The miombo tree layer shows adistinct pre-rain flush of new leaf which may be a defense strategy against insect herbivores. Flowering of trees and shrubs largely occurs in the dry season, allowing fruit-set and dispersal in the rains. Herbs and grasses show a fire induced dry-season flush when soil moisture permits. Otherwise all production is rain produced.

Primary Production

Standing crop data from the East African woodlands are scarce, but more data are available from Zambia and Zimbabwe. Values of 25 m²/ha for tree basal area can be found, but this reduces to close to 12 m²/ha in areas of high elephant damage (3). Total plant biomass of the damaged area was estimated at 23 tons/ha dry-matter (28). Adjacent woodlands of similar structure had standing crops greater than 65 tons/ha. Zambian *Brachystegia* woodlands had a plant biomass of 48 tons/ha, while chipya biomass was 17 tons/ha, (44). Chidumayo (15) showed that tree basal area was related to aridity. Within his study areas, basal areas increased from 11.7 to 16.5

Table 12.8. Large mammal species distribution data, with reference to miombo woodland habitats. Units are the number of quarter degree squares dominated by the particular vegetation type in Tanzania. Data are taken from distribution maps for the species, superimposed on a quarter degree square grid annotated as to the major structural vegetation type. The number of squares of a vegetation type occupied by a species, is expressed as a percentage of the total squares of that type. Data from (42).

Vegetation	Miombo	Tall grass	Savanna/bush	Thicket
Coverage	460	62	211	135
Miombo group				
Sable	59	71	11	13
Roan	32	70	20	23
Lichtenstein Hartebeest	79	76	11	20
Savanna group				
Thomson's Gazelle	0	0	51	73
Common Hartebeest	0	0	39	59
Oryx	0	0	18	53
Widespread group				
Buffalo	89	100	84	93
Zebra	87	100	89	97
Transitional species				
Impala	45	5	43	39
Wildebeest	37	8	86	33

Note that the low representation of roan in miombo is due to the fact that it is totally absent from the south-eastern Tanzanian block (42).

m²/ha, with increasing moisture availability. Basal area and biomass are frequently closely related, so it is likely that tree biomass also increases with water availability.

Forestry studies suggest that the miombo is of low wood productivity (23) even for some of the best Zambian miombo. Studies showed a mean annual basal area increment of 1.2%. Estimated yields of salable woody material were between 18 to 28 m³/ha/y. Consequently, for the purposes of wood production, plantations of fast-growing species and the suppression of fires may be necessary.

The herb layer has a varied standing crop biomass depending on rainfall and the density of the tree canopy. Maximum standing crop biomasses reach 400 grams/m², all of it produced during the three months after rain. Table 12.9 illustrates the monthly patterns of growth. The high proportion of stem is of interest, miombo grasses are coarse and of low palatability, with an overall crude protein content below 6% for much of the year.

Secondary Production

Production of invertebrate animals is reviewed by Malaise (39,40) and data largely originate from his Zaire studies that emphasize soil fauna such as termites. These studies show an overall low species diversity, and, apart from groups such as termites, biomass is low.

Large mammal productivity has been documented in some detail in the miombo of Zimbabwe, but less so in Zambia or Tanzania. These studies are important for determining the comparative benefits of ranching domestic animals and/or wildlife. Tsetse fly infestation affects the ability to ranch cattle. Therefore, wildlife ranching is often preferable in tsetse fly infested areas. Carrying capacities are low, limited by dry season water and adequate forage. Forage values such as protein content are low and biomasses are dominated by low population density mega-herbivores which can use some browse but have slow turn-over rates.

Human Uses of the Miombo Woodlands

Overview of Uses

Thematic maps of the Tanzania produced during the early 1970s clearly show the two miombo regional blocks. The miombo stands out as the area with low human populations, virtually no livestock, ubiquitous tsetse flies, and a great many forest and wildlife reserves (see Figs. 12.3 to 12.6). While that picture is now changing, it does accentuate the fact that the miombo with a long dry season and poor infertile soils, coupled with tsetse fly and the attendant trypanosomiasis, presents an extremely difficult environment for human settlement and existing methods of farming.

Like the wild animal communities, successful colonization and use of the miombo by man was usually near islands of other habitat types - river lines, floodplains or more alkaline and nutrient-rich soil localities. Hunting, honey collection and timber extraction were the principal human resource uses, both by local communities and by the colonial governments.

Tsetse flies, principally *Glossina morsitans* in the miombo woodlands, but also *G. pallidipes* and *G. palpalis* in riverine forests, have exerted a major impact on land-use in the miombo regions by maintaining low numbers of cattle and humans. Landuse patterns and the spread and decline of tsetse are inter-related in many parts of Tanzania (33). The great rinderpest outbreak of the late 1890s caused such serious cattle die-offs that many areas of western Tanzania were deserted, allowing their recolonization by tsetse. The great size of the Selous Game Reserve (45,000 km^2) is due to the colonial government forcefully evacuating settlements because of trypanosomiasis.

Timber. Historically, one species of tree has dominated forestry activities in the miombo. This species is muninga *or Pterocarpus angolensis*, a legume that grows to 18 meters. The timber is prized for furniture throughout east and central Africa, and is also exported. *Pterocarpus* occurs at low population densities with limited regeneration throughout the drier areas of miombo. The species attracted foresters' attention in the 1960s, with attempts to grow it in plantations and to understand its complex relationship with fire (6). Plantation activity has been hampered by the plant's need for an extensive root system as a sapling, and plantation forestry has not been successful.

Table 12.9. Primary production parameters Miombo/Chipya Grass Layer (Selous Game Reserve with rainfall = 760 mm/y, elevation 300m asl). Standing crop biomass of grass layer separated into fractions. Data expressed as air-dry weight in grams/m². Green leaf crude protein expressed as a percentage of biomass. Data from (42).

Month	Total biomass	Green leaf	Dry leaf	Stem	Green leaf protein %
Sept	14	14			12.9
Nov	8	8			9.6
Dec	26	26			7.8
Feb	161	114	8	39	4.3
April	395	105	158	130	5.1
June	297	95	123	79	2.8
Aug	268	61	136	71	3.6
Oct	265	4	176	85	2.9

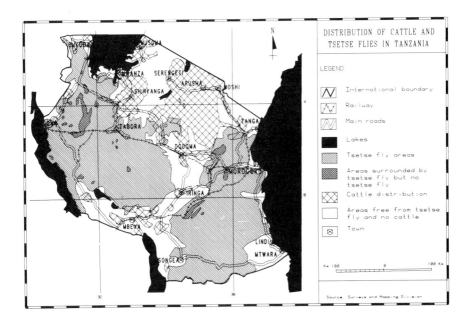

Figure 12.4. Map showing the distribution of wildlife and forest reserves in Tanzania.

The seeds of *Pterocarpus* are resistant to fire, and, in fact, need firing at temperatures of up to 300° C in order to germinate. This provides evidence for the long evolutionary history of the miombo as a fire-adapted environment. The other life stage at risk is the young seedling. *Pterocarpus* demonstrates the die-back adaptation; where every year the above-ground shoots die back, but the root stock continues to grow and store resources. One year there is a burst of shoot activity that attempts to put the sapling's apical meristem above the lethal temperatures of fires.

Pterocarpus has been over-harvested within both Forest Reserves and open areas. Legal minimum girth classes for tree felling were reduced several times, and were set at 45 cm diameter at breast height (=dbh) in the late 1970s. In practice the

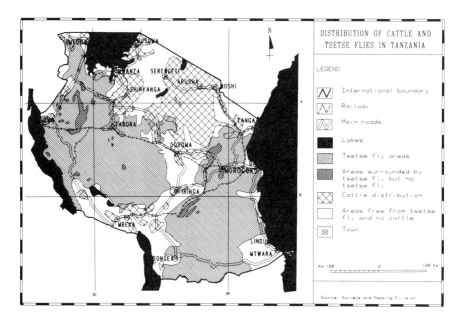

DISTRIBUTION OF CATTLE AND
TSETSE FLIES IN TANZANIA

Figure 12.5. Map showing the distribution of cattle and tsetse flies in Tanzania, (1976).

minimum diameter felled depends more upon the economics of converting small sizes into the needed materials rather than on the observance of regulations, and the cutting of stems of less than the minimum size allowed is common-place (12). Consequently, large trees are now rare.

In Tabora Region of western Tanzania the relative abundance of 'timber' species was:

Julbenardia globifera	31%
Brachystegia spiciformis	21%
Brachystegia boehmii	11%
Pterocarpus angolensis	4%

With a typical miombo timber volume of 43 to 63 m³/ha this suggests a mean extractable timber volume of 2.1 m³/ ha for all timber species - a low figure for economic exploitation (12).

Several other genera of tree such as *Pericopsis*, *Afzelia*, *Millettia* have recently gained recognition as timber. The latter two genera are both in more coastal habitats. For rough timber, several species of *Brachystegia* are used. For example, *B. spiciformis* was extensively used as props in the mine pits of Zambia. Despite the comment in the World Bank's 1993 *Forest Strategy for Africa*, mpingo or African Blackwood (*Dalbergia melanoxylon*), is not a typical miombo species (26) but is found in coastal and drier savanna woodlands.

Fuelwood and Charcoal. Fuelwood is discussed below under the heading Agriculture because tobacco cultivation creates the biggest impact on the woodland for fuel. Charcoal exploitation, however, is seen as a growing problem, especially

Table 12.10. Forest land use in Tabora Region, Tanzania in 1974.

District	Forest Reserve hectares	Public Forest hectares
Igunga	268	231
Nzega	965	1928
Tabora	24,112	10,096
Urambo	9880	5684
Totals	35,225	17,939

Forest Production*	
Item	Amount
Saw logs, m^3	40,631
Charcoal, bags	41,809
Fuelwood, m^3	4979

Domestic and Tobacco Fuel Wood Needs	
Tabora population size	831,642
Per capita fuel use	1.5 m^3
Annual demand for fuelwood	1,247,000 m^3
Fuelwood plantation yield	15 m^3/ha pa
Fuel plantation requirement	83,000 ha
Miombo yield	50 m^3/ha wood every 50 years
Miombo requirement	1,200,000 ha (add 30% to account for amount as timber etc).
Fuelwood need per ha tobacco	64 m^3
Tobacco area	20,720 ha
Total needs	1,300,000 m^3/year
Plantation requirement	86,000 ha (assume 15 m^3/ha/year).

*(Sawn logs are *P. angolensis - mninga* of around 9%; *B. spiciformis* and *J. globifera* reach 30 to 40 %; while *B. boehmii* abundance is <10%).

along access routes to urban centers. Miombo species produce good charcoal, and because they coppice easily, it should be possible to harvest them sustainably. The miombo woodlands close to Dar es Salaam are being converted rapidly into degraded scrub by unregulated and excessive use. The areas are not reserved forests, and while ostensibly village land, the demand for products comes not from within the village community but from urban growth. There is little attempt at sustainable use, temporary cultivation and fire preventing a regrowth of woodland after charcoaling.Recent detailed studies in Zambia provide some basis for resource management (15,18). For example canopy species had greater wood growth rates than understory trees. There is a need therefore to seek enhanced regeneration of canopy trees. Experiments with Joint Forest Management, which involves full participation by local communities, in northern Tanzanian miombo in Babati District suggest that this has potential for sustainable management elsewhere.

Honey. Bees (*Apis mellifera unicolor*) are of great importance to many rural people in the miombo woodlands. 'Tabora honey', from Western Tanzania, is a miombo product that is produced by cooperatives, and marketed and sold all over the world. Honey production is still largely from wild populations although the current practice is to use improved manufactured box hives instead of tree bark cylinders, and

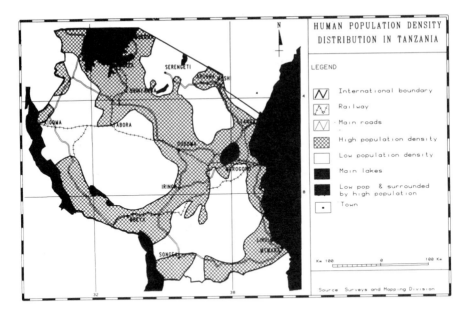

Figure 12.6. Map showing human population density distribution in Tanzania (1976).

to avoid killing the bees. The Tanzanian Bee-keepers Cooperative Society(TBCS) does provide considerable extension inputs. There are two periods of honey production May through July after the rains, and October to November with the tree flower flush.

In 1960 Tabora Honey export was 535 tons of beeswax and 339 tons of honey. Today it is well over twice that. In 1992 wax sold at 600 to 1200 Tanzanian Shilling per kg, and honey at 4500 Tanzanian Shilling per 28 kg bucket. There are some 30,000 hives in the 600 km² of Ugalla Game Reserve, in the south of Tabora, giving an annual revenue of 47 million shillings (Ugalla Game Reserve Management Plan 1994).

Agriculture and Fuelwood Use. The miombo has old, leached and infertile soils, with a single rainy season. Agriculture without adequate fertilizer inputs has low yields on soils with such poor organic matter contents and low NPK status. This was greatly demonstrated by the failure of the British 'Groundnut Scheme' of the post-war 1948 to 1952 period. There were two miombo sites used for planting groundnuts - Urambo in the Tabora region and Nachingwea in southeast Tanzania. Both failed because not only were soil nutrient levels low but because soil structures did not allow the harvesting of the nuts. There had been virtually no soil survey. Today the cleared areas of miombo grow some maize, cassava, pulses, and, as a cash crop, tobacco. Fertilizer inputs could triple tobacco yield!

Table 12.11 show statistics on the growth of the tobacco industry in Tabora Region of western Tanzania from 1975 and 1976 where 18,000 farmers produced

8700 tons. In 1980 40,000 farmers in the Region (mainly the miombo areas of Tabora and Urambo Districts) produced 15,000 tons from 14,500 hectares, which is 65% of Tanzania's total production. In 1992 these two Districts produced 14,000 tons - 60% of national production. Constraints to tobacco production are now ubiquitous soil pests such as nematodes that necessitate expensive control or frequent (every 2 to 3 years) shifts of fields. Tabora tobacco is flue cured. One hectare of tobacco requires 12 hectares of miombo fuelwood for effective and sustained yield. The already degraded woodlands near human settlement need more protection, preferably through joint forest management schemes involving villagers and government forestry. The lack of regulation and tenure means that fuel scarcity resulting from woodland degradation is an increasing problem (46). Coal from south Tanzania and improved kilns are seen as one answer. If successful, however, this could result in the clearing of more miombo woodlands for agriculture as they would not be needed as a source of fuel. Sustainable charcoal, timber and fuelwood use may be one of the best uses for the conservation of miombo woodlands because it does not result in habitat destruction.

The pattern of woodland clearance around Tabora tobacco farms is widespread throughout east and central African miombo, as highlighted by Tuite and Gardiner for example (50). Chidumayo (14) suggests that over half of the 7700 km² of miombo woodland in the Zambian copperbelt was deforested for fuelwood between 1937 and 1983. Studies after 1983 are required to determine the extent of the loss since this survey.

Traditional Farming Systems. In the Zambian miombo woodlands (1,20) two patterns of traditional cultivation are distinguished - locally called *chitemene* and *ntumba*. *Chitemene* is a shifting cultivation technique that is unusual because it involves cutting and burning the branches of trees collected from over a wide area. Crops are then grown in the mineral rich ash left from the burning. Allen (1) remarked on the decreasing time interval presently allowed for the recovery of woodlands. Recently, less than 8 years between crops is typical instead of the 30 years that may be needed for adequate recovery and sustainable use. The second type of cultivation includes manuring systems of several types, *ntumba* a mound system, *chibela* a flat-bed composting system, and *fundikila* a combination of both systems. These are practiced on the coarse grass areas of northern Zambia after miombo has been degraded through other uses. Composting is a more intensive cultivation system than *chitemene* because it involves composting grass in flat beds and/or mounds. Curiously, agricultural historians in Tanzania have not documented similar situations (see 33) although there are links to Wafipa mound cultivation in south-west Tanzania.

The presently reduced fallow periods of *chitemene* and other shifting cultivation practices of the miombo are still a source of concern because of declining nutrient concentrations and crop yields. Shifting cultivation under long fallow periods need not be destructive. For example, Stromgaard's (44,45) analyses of nutrient flows under different agricultural practices found that burning does not destroy organic matter, and that burning does lead to immediate increases in Ca, Mg, K and P. Cropping does, however, lead to rapid decreases in nutrient availability, and levels three to four years after the burn were lower than the preburn concentrations. As human populations increase, simple shifting or even *chitemene* and *ntumba* systems may be insufficient to sustain the population.

Table 12.11. Tobacco production statistics for Tanzania. In 1992 Tabora and Urambo Regions produced 14,000 tons or 60% of the total Tanzanian production.

1975/1976			
District	Farmers	Area, ha	Production, tons
Igunga	110	42	10
Nzega	876	361	154
Tabora	10,742	7380	5911
Urambo	7446	4572	2715
1980/1981			
Igunga	220	80	60
Nzega	1140	720	540
Tabora	22,490	11,470	8600
Urambo	16,000	8000	6000

Tobacco and Fuelwood Production in the 1970s in Tabora Region (after 46)			
Year	Tobacco, ha	Tobacco, tons	Fuelwood, m^3
1969	5538	2903	582,000
1973	10,300	9951	1,081,000
1977	15,902	9200	1,669,000

Clearance and Woodland Regrowth. The available evidence points to increased clearance of miombo woodlands all over Africa. Land-use is rarely sustainable. Of interest therefore is the level at which the miombo plant and animal communities can recover after human and natural perturbations. This has recently attracted some study. MacGregor (37) found that the miombo woodland community exhibited a high degree of stability after cutting - species composition did not change significantly. Most species (of the 94 species analyzed) were widespread and recovered after cutting. Regeneration of coppicing plants is rapid, and cleared sites regenerate from both root suckers and seeds (14,17). Fires and canopy closure do affect growth rates and their manipulation could be criteria for management inputs.

Maghembe (38) reviews the potential for agroforestry in the miombo region of Africa as a whole and concludes that there are benefits to both sustainable agriculture and in reducing the fuelwood crisis through charcoal use. Agroforestry development will need more than the provision of selected species seed. A tree growing culture, involving permanent cropping systems as well as the acceptance of the loss of free goods will be needed.

Cattle from Sukumaland. The miombo until recently has been largely cattle free, due to abundant tsetse flies and cattle trypanosomiasis in the past. The recent spread of cattle in western Tanzania has been at the expense of the miombo woodlands. Extensive tsetse eradication schemes in the 1940s on the southern Sukumaland steppe in Shinyanga region, were achieved through mechanical woodland clearing and followed by cultivation of annual crops.

The late 1970s saw the beginning of a migration of Sukuma agro-pastoralists, southwards from the then densely populated cleared steppe areas into the miombo of Tabora and Urambo Districts, the Rukwa region, and into Zambia. Movement into the miombo was facilitated by the strip clearing along the lines of road and rail, and the increase of clearing for tobacco. The beginning of this move was recorded as both the presence of cattle on cleared patches in the miombo, but also in the empty settlements in Nzega District abandoned due to overstocking (21).

The growth of tobacco cultivation in the 1970s and 1980s has accentuated the move of these agropastoralist peoples. Additionally, the recent breakdown of the collective village movement (ujamaa) and reduction of the land-use controls exerted during the socialist regime of the 1970s and early 1980s have facilitated human settlement into the miombo. Tabora and Urambo Districts now have 336,000 livestock units, a figure growing more rapidly than human population growth rates. Rukwa region to the south of Tabora sees the influx of cattle and their owners as their single biggest environmental problem In 1978 Rukwa had 19,000 livestock units - today there are over 79,000! Protecting the miombo habitat from clearing and excessive grazing by cattle will be a major challenge to conservation and development in the next decade.

Minor or Non-Tree Forest Products. Local people use the miombo for fuel, poles, honey, hunting, fruits and other foods. Fruits important seasonally as local dietary items and as part of the cash economy include *Uapaca kirkiana*, *Parinari curatellifolia* and species of *Syzygium*. None of these species, however, are typical of the continuous blanket of miombo woodland of the western plateau, where the tree layer is almost totally leguminous. In the shrub layer *Flacourtia*, and in south-east Tanzania *Landolphia*, provide valuable fruit resources. There are several mushrooms of importance to the local diet. Malaise and Parent (41) review the variety of miombo vegetable foods and Campbell (9) discusses fruit availability and use.

In Central Africa, the mopane worm (*Gonimbrasea belina*), and a species of *Elaphrodes* (both large caterpillars) are important foods rich in protein for local people in rural areas and are also now sold in urban areas. Large mound termites (species of *Macrotermes* or *unchwa* in Swahili and *inswa* in Zambia), a common feature of the miombo, provide a major food source during the annual swarming of the reproductive alates.

Hunting. The low density of past human settlement and cultivation due to tsetse fly infestation meant that wildlife was a relatively widespread and locally abundant resource. Several trophy or 'game' species inhabiting the miombo were not readily available elsewhere in East Africa (for example sable, roan and greater kudu) and so there was considerable interest from sport hunters. As in most areas of anglophone Africa, the development of tourist hunting led to the exclusion of local hunting and the consequent alienation of local communities towards wildlife conservation activities.

The spread of tobacco and cattle with the opening of woodland canopy has led to a reduction of wildlife abundance and a reduced potential for tourist or local hunting outside the protected areas.

Figure 12.4 shows the extent of Game Reserve and Game Controlled Areas (=GCA) in the miombo. Some GCAs now have little relevance because of greatly reduced wildlife resources due to agricultural spread (such as Igunda). To ensure conservation of wildlife resources the Tanzanian Government has increased the extent of Game Reserves (such as upgrading Mlele Hills and North Rukwa GCAs). The Government has, however, realized that miombo woodlands harbor a multitude of resources beyond tourist trophies. Some exploitation of honey, timber and fishing (such as in the Ugalla river) need not be incompatible with sustainable hunting.

New Game Reserve utilization rules are being devised, and multiple use is being advocated in the management plan of the Ugalla Game Reserve.

Sport hunting is a major foreign currency earner as is documented in the Ugalla Management Plan. The Ugalla Game Reserve alone (600 km²) raised game fees revenues of $127,000 in 1992 - of this $33,000 goes to the Tanzanian Treasury and the balance to the Tanzania Wildlife Protection Fund which supports further management and development programs in Ugalla. Total foreign exchange benefit retained in Tanzania (net foreign exchange earnings) from game hunting in Ugalla is $224,000 per year. These figures contrast with the fees paid by local and resident hunters which for the whole of Tabora region totaled only $670 in 1992.

Tourism. The relative inaccessibility of most Tanzanian miombo areas, coupled with low wildlife densities and tsetse fly infestation have not encouraged tourism, apart from the trophy hunters. Wildlife viewing is confined to areas where the miombo adjoins with other habitats such as the sweet-veld valleys of Luangwa (Zambia), and the north-east Selous and Greater Ruaha in Tanzania.

Conclusions

The contents of this chapter lead to several conclusions about the future of the miombo habitat. These are:

(1) Miombo was the last great wilderness area of Tanzania. The 1976 atlas maps show it as empty of people, agriculture and cattle (see Figs. 12.4, 12.5 and 12.6). Much of the area was wildlife and forest estate. That situation has changed, and continues to change rapidly. Presently, the only wildernesses left are the legal forest and wildlife reserves. The miombo which was open to human settlement and which was left empty, due initially to concerns with tsetse flies and low productivity, and latterly to rigorous containment of people in the planned *ujamaa* villages, is now being settled. The overcrowding of Sukumaland, to the north of Tabora and Shinyanga Regions, has led to a major exodus to the south. Woodlands are being cleared for agropastoralist use, their cattle, mixed cropping and tobacco cultivation. Clearing the woodland removes the tsetse fly threat, in exactly the same manner as the colonial government's clearance schemes in the 1940s and 1950s. This increased rate of settlement of the miombo poses problems for several groups of people :

The smaller indigenous tribes in the southern miombo are pressured by more numerous immigrants and often lose their grazing lands. Better lands (those with water) are settled and cultivated. Wildlife, timber and honey resources are less available because of such settlement. The wave of new settlements has now reached to Zambia, through Mbeya and Rukwa Regions; and impacts many resources such as the Usangu floodplains.

Environmentalists see natural resources being used non-sustainably. These resources include wildlife populations and timber. Planning for integrated natural resource/agricultural management involving, for example, natural woodlots and tobacco curing, requires time for investment. Time is scarce!

Government planners are concerned with the uncontrolled spread of settlement, and District authorities are expected to provide basic services with little or no financial or material resources.

(2) The wildlife and forest resources in the increasingly isolated reserves will be under increasing pressure.

(3) There is inadequate data and information on miombo ecology, including the plant and animal communities and their interactions with man. No monitoring system exists.

(4) Available evidence shows that miombo productivity is low - whether it be forest, wildlife, cattle or crops. Sustainable utilization of the resources will be hard to achieve, given the growing population and the breakdown of traditional resource tenure and management systems with no resource planning and land-use controls to replace them. The miombo species and productivity patterns are, however, well adapted to the harsh environmental conditions - low nutrients, fire and a seven month dry season. Perhaps there are lessons in resource use that humans can learn from the adaptations of wild species.

References

1. Allen, W. 1967. *The African Husbandman*. Oliver and Boyd, London
2. Anderson, B. 1962. *The Soils of Tanganyika*. Bulletin 16. Dar es Salaam: Ministry of Agriculture
3. Anderson, G.D., Walker, B.H. 1974. Vegetation composition and elephant damage in the Sengwa Wildlife Research Area, Rhodesia. *Journal of South African Wildlife Research* 4: 1-14
4. Bell, R. 1982. The effect of soil nutrient availability on community structure in African ecosystems. In *Ecology of Tropical Savannas*, eds. Huntley, B.J., Walker, B.H., pp 193-216. Berlin: Springer-Verlag
5. Boaler, S.B. 1966a. Ecology of a miombo site, Lupa North Forest Reserve, Tanzania. *Journal of Ecology* 54: 465-469
6. Boaler, S.B. 1966b. The ecology of *Pterocarpus angolensis* in Tanzania. London: HMSO
7. Brenan, J.P.N. 1978. Some aspects of the phytogeography of tropical Africa. *Annals of Missouri Botanical Gardens* 65: 436-478
8. Burtt, B.D. 1942. Some East African vegetation communities. *Journal of Ecology* 30: 65-146
9. Campbell, B.M. 1987. The use of wild fruits in Zimbabwe. *Economic Botany* 41: 375-385
10. Campbell, B.M., Swift, M.J., Hatton, J., Frost, P.G.H. 1988. Small scale vegetation pattern and nutrient cycling in miombo woodland. In *Vegetation Structure in Relation to Carbon and Nutrient Economy*, eds. Verboeven, J.T.A., Heil, G.W., Werger, M.J.A., pp 69-85. The Hague: SPB Academic Publishing
11. Celander, N. 1983. *Miombo Woodlands in Africa -Distribution, Ecology and Patterns of Land Use*. IRDC; Swedish University of Agricultural Sciences. Working Paper 16

12. Chaffey, D.R. 1980. *Tabora Rural Integrated Development Programme, Tanzania.* Forestry Consultancy Report. London: LRDC, ODA

13. Chidumayo, E.N. 1987a. Species structure in Zambian miombo woodland. *Journal of Tropical Ecology* 3:109-118

14. Chidumayo, E.N. 1987b. Woodland structure, destruction and conservation in the copperbelt of Zambia. *Biological Conservation* 40: 89-100

15. Chidumayo, E.N. 1987c. A survey of wood stocks for charcoal production in the miombo woodlands of Zambia. *Forest Ecology and Management* 20: 105-115

16. Chidumayo, E.N. 1988a. Regeneration of *Brachystegia* woodland canopy following felling for tsetse-fly control in Zambia. *Tropical Ecology* 29: 24-32

17. Chidumayo, E.N. 1988b. A re-assessment of effects of fire on miombo regeneration in the Zambian copperbelt. *Journal of Tropical Ecology* 4: 361-372

18. Chidumayo, E.N. 1988c. Estimating fuelwood production and yield in regrowth dry miombo woodland in Zambia. *Forest Ecology and Management* 24: 59-66

19. Chidumayo, E.N. 1992. Seedling ecology of two miombo woodland trees. *Vegetatio* 103: 51-88

20. Chidumayo, E.N. 1993. Three traditional farming systems in Miombo woodland. In *African Biodiversity: Foundation for the Future.* Biodiversity Support Programme. Washington DC: WWF pp. 149

21. Ecosystems Ltd 1978. *Aerial Census of Rukwa Region, Tanzania.* Report to the University of Dar es Salaam. Nairobi: Ecosystems Ltd.

22. Ecosystems Ltd 1979. *Tabora Rural Integrated Development Project - Aerial Survey of Natural Resources.* Final Report. Nairobi: Ecosystems Ltd.

23. Endean, F. 1968. *The Productivity of Miombo Woodland in Zambia.* Forest Research Bulletin No 14, Lusaka: Government Printer

24. Ernst, W.H.O. 1988. Seed and seedling ecology of *Brachystegia spiciformis*, a predominant tree component in miombo woodlands in South Central Africa. *Forest Ecology and Management* 25: 195-210

25. FAO 1984. *Agroclimatological Data - Africa*, Vol 2. Rome: FAO

26. FTEA 1971. *Papilionoideae, Vol 1; Flora of Tropical East Africa.* London: Kew Gardens

27. Fleetwood, E. 1981. *Soil Properties Under Five Forest and Woodland Types in the Morogoro Area, Tanzania.* Swedish University of Agricultural Sciences. Uppsala: IRDC

28. Guy, P.R. 1981. Changes in the biomass and productivity of the woodlands in the Sengwa Wildlife Research Area, Zimbabwe. *Journal of Applied Ecology* 18: 507-521

29. Hogberg, P. 1980. *Occurrence and Ecological Importance of Ectomycorrhizas and Nitrogen-Fixing Root Nodules of Trees in the Miombo Woodlands of Tanzania.* Uppsala: Swedish University of Agricultural Sciences

30. Hogberg, P., Piearce, G.D. 1986. Mycorrhizas in Zambian trees in relation to host taxonomy, vegetation type and successional patterns. *Journal of Ecology* 74: 775-785

31. Huntley, B.J., Walker, B.H. eds. 1982. *The Ecology of African Savannas.* Berlin: Springer-Verlag

32. Kikula, I.S. 1986. The influence of fire on the composition of miombo woodland in SW Tanzania. *Oikos* 46: 317-324

33. Kjekshus, H. 1979. *Ecological Control and Development in Eastern Africa.* Nairobi: Longmans

34. Kokwe, M. ed. 1993. *Sustainable Use of Dambos in Southern Africa.* Proceedings of the Regional Policy Workshop, Lusaka, Zambia, January 1993. London: IIED

35. Lawton, R.M. 1978. A study of the dynamic ecology of Zambian vegetation. *Journal of Ecology* 66: 175-198

36. Lind, E.M., Morrison, M.E.S. 1974. *East African Vegetation.* Nairobi: Longmans

37. McGregor, J. 1994. Woodland pattern and structure in a peasant farming area of Zimbabwe: ecological determinants and present and past use. *Forest Ecology and Management* 63: 97-133

38. Maghembe, J.A. 1994. Agroforestry research in the African miombo ecozone. *Forest Ecology and Management* 64: 105-184

39. Malaise, F.P. 1974. Phenology of the Zambesian woodland area with emphasis on the miombo system. In: *Ecological Studies No 8. Phenology and Seasonality Modelling,* eds. Jacobs, J., Lange, O.L., Olson, J.S., Lieth, H., pp 269-286. New York: Springer-Verlag

40. Malaise, F.P. 1978. The Miombo Ecosystem; *In Tropical Forest Ecosystems,* pp 589-606. UNESCO/UNEP/FAO - Paris, UNESCO. 683pp

41. Malaise, F.P., Parent, G. 1985. Edible wild vegetable products in the Zambezian woodland areas: a nutritional and ecological approach. *Ecology of Food and Nutrition* 18:43-82

42. Rodgers, W.A. 1979a. *The Ecology of Large Herbivores in the Miombo Woodland of South East Tanzania.* PhD Thesis, University of Nairobi

43. Rodgers, W.A. 1979b. The implications of woodland burning for wildlife management. In *Wildlife Management in Savanna Woodland,* eds. Ajayi, S.S., Halstead, L.B. pp. 103-112. London: Taylor & Francis

44. Stromgaard, P. 1985. Biomass, growth, and burning of woodland in a shifting cultivation area of south central Africa. *Forest Ecology and Management* 12:163-178

45. Stromgaard, P. 1991. Soil nutrient accumulation under traditional African agriculture in the miombo woodland of Zambia. *Tropical Agriculture* (Trinidad) 68: 74-80

46. Temu, A.B. 1979. *Fuelwood Scarcity and Other Problems Associated with Tobacco Production in Tabora Region, Tanzania.* Forest Division Record 12. Tanzania: Sokoine University of Agriculture

47. Trapnell, C.G. 1959. Ecological results of woodland burning experiments in Northern Rhodesia. *Journal of Ecology* 47: 129-168

48. Tuite, P., Gardiner, J.J. 1990a. *An Inventory of Trees and Shrubs in Miombo Woodland of South Tanzania.* Dublin: University College

49. Tuite, P., Gardiner, J.J. 1990b. *A Survey of Farm Regrowth in Southwest Tanzania.* Dublin: University College

50. Tuite, P., Gardiner, J.J. 1990c. *The Miombo Woodlands of Central, Eastern and Southern Africa.* Irish Forester

51. White, F. 1976. *The Underground Forests of Africa : A Preliminary Review.* Gardner's Bulletin 29: 57-71

52. White, F. 1983. *The Vegetation of Africa.* Natural Resources Research XX, Paris: UNESCO

Section V:

Forest Ecosystems

Chapter 13

Coastal Forests

Neil Burgess, Clare FitzGibbon & Phillip Clarke

On first entering a coastal forest, it is sometimes difficult to appreciate their significance. Unlike tropical rain forests, one is not immediately struck by the size of the trees, the variety of the plants or the sheer amount of vegetation (Fig. 13.1). Instead, the trees are relatively small, and just a few species dominate the canopy in any one place. During the dry season, in particular, the vegetation appears sparse and many of the trees have lost their leaves. The forest can appear empty and devoid of life, particularly in the middle of the day when the high air temperature reduces animal activity. Only the diversity of butterflies and the intensity of insect noise give any indication of the true nature of this unique habitat. As a result, it is not surprising that the conservation significance of the coastal forests has only recently been realized. Now it is known that they support a wealth of species, many of which are found nowhere else in the world; many species found have not even been named, let alone investigated in detail; the ecological processes are poorly understood; and very few studies have been made on the importance of the plants and animals and their value to local people.

What separates the East African coastal forests ecologically from rainforests is the pronounced annual variation in water availability; in the coastal forest most rainfall occurs over a 4 to 5 month period and there is a 3 month long dry season from July to October. This dry season imposes significant water stress on the plants and animals in the coastal forests and many species are adapted to desiccation events. The combination of a long-existing forest habitat and stressful environmental conditions is reflected in large numbers of species unique to the small remaining area of coastal forest. The total area of dry coastal forest remaining in Eastern Africa is probably around 3000 km², and the coastal forests are perhaps the most threatened forest type in Africa. Moreover, as representatives of tropical dry forest they are examples of the most threatened of all tropical forest habitats (17). Despite the scarcity of coastal forest the remaining examples are being degraded by human activities at an ever increasing rate. Conservation of sites with the characteristics of being relatively undisturbed by humans, and having high species diversity and endemism is, therefore, a conservation priority.

Figure 13.1. View of the *Brachystegia* habitat in Arabuko-Sokoke Forest, Kenya (Credit: J. Fanshawe/Birdlife).

History of Coastal Forests

Forest has existed on parts of the Eastern African coast for millions of years (11,12,13). Some of the species now restricted to the coastal forests probably evolved there. Coastal forest has also provided a refuge for forest species which evolved outside the coastal area, especially during the dry periods associated with glacial events of the recent ice age (during the past 1 million years). These factors, and the current isolation of coastal forests from each other and other large forest blocks, have resulted in high levels of endemism in both plants and animals (8,16,31).

Coastal forest was formally extensive along the Eastern African coast but has been destroyed in this heavily populated region to provide wood for fuel, building and carving, and to make way for farmland and tourist developments (11,15). Remaining coastal forests exist as a series of small patches, isolated from each other mainly by agricultural land or degraded scrub habitat, but in some areas by savanna-woodland. The high population density on most of the coast makes the conservation of remaining forests particularly difficult.

Definition of Coastal Forest

Forest is:

- A continuous stand of trees
- With a canopy layer more than 6 meters high
- Crowns overlap or touch

Forests can also have these traits:

- Often festooned with lianas
- Shrub layer normally present
- Ground layer sparse/absent
- Epiphytes in moister areas

The coastal forests in an undisturbed state, or potential climax state, fit this definition but contain a distinct fauna and flora, including many endemics, which tend to occur together within a distinct geographical range (Fig. 13.2).

Eastern African coastal forest is:

- Found on the coastal plain between the mangroves and 'Eastern Arc' Mountains (see Box 13.1) and separated by semi-arid ecosystems.
- Lowland with a maximum altitude of 500 m near coast (at the base of Usambara Mountains), although up to 1000 m further inland.
- Influenced by the Indian Ocean climatic system.
    - Rainfall between 800 and 2000 mm per year,
    - Rainfall strongly seasonal with a single pronounced dry season of 3 to 5 months
    - Temperatures of 25 to 35° C throughout the year.
- Rich in endemic and near-endemic species.

The effects of drainage, topography, altitude, geology, soil moisture content and distance inland all combine to determine sites able to support a coastal forest.

Not included in this review of eastern African coastal forests were:

- Coastal Thicket
  (Derived coastal forest up to 4 m high with no stratification).

- Mid altitude Forest
  (Biologically part of Eastern Arc mountains in Tanzania but at lower altitudes (generally below 1000 m) intergrades with coastal forest, for example areas in the Usambara and Uluguru Mountains)
- Southern African Dune Forest
  (Tongoland - Pondoland forest type mainly but intergrades with coastal forest in Southern Mozambique (around Maputo))

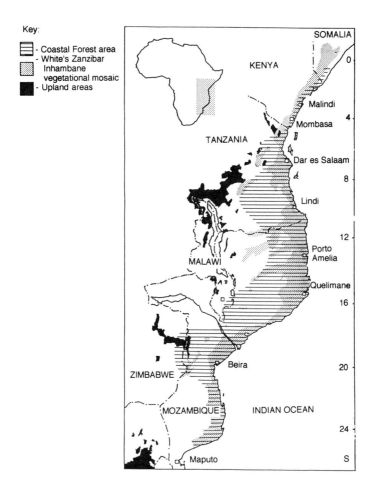

Figure 13.2. Extent of the coastal forest area, White's (34) Zanzibar-Inhambane regional mosaic, and upland areas in Eastern Africa.

Status of Coastal Forest Resources

It is not clear how much coastal forest existed in the past, nor exactly what has been destroyed by human activities to give the present forest distribution (see Fig. 13.2 and 13.3). It is possible that in the past most of the existing evergreen and semi-evergreen coastal forests were surrounded by extensive areas of drier forest or savanna-woodland. This dry forest would have been much easier to clear for agriculture and would have been more susceptible to fires, which are known to have become more frequent as the human population increased and hunting intensified from around 200,000 years ago. The evergreen and semi-evergreen forests are now largely left as islands within areas of savanna-woodland, coastal thicket, or (increasingly) farmland.

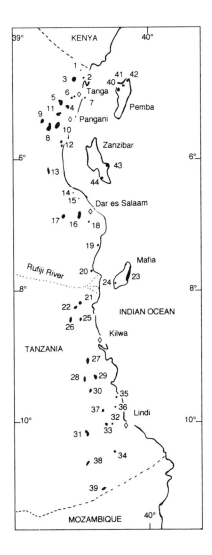

Figure 13.3. Position and rough geographical extent of coastal forest sites in Tanzania (3).

Key to forest localities: 1=Horohoro, 2=Kilulu Hill, 3=Lowland forests of East Usambaras, 4=Tongwe, 5=Kwani/Makinyumbi, 6=Tanga Limestone, 7=Yambe Island, 8=Gendagenda, 9=Mgambo, 10=Msubugwe, 11=Pangani River, 12=Mkwaja, 13=Zaraninge/Kiono, 14=Ruvu North, 15=Pande, 16=Pugu & Kazimzumbwe, 17=Ruvu South, 18=Vikindu, 19=Kisiju, 20=Mchungu, 21=Namakutwa-Nyamuete, 22=Kiwengoma, 23=Mafia - eastern seaboard, 24=Kilindoni, 25=Tong'omba, 26=Mbinga, 27=Mitundembea, 28=Rungo, 29=Ngarama, 30=Pindiro, 31=Rondo, 32=Litipo, 33=Chitoa, 34=Nyangamara, 35=Ndimba, 36=Ruawa, 37=Matapwa, 38=Chilangala, 39=Mahuta, 40=Ngezi, 41=Msitu-Mkuu, 42=Ras Kiuyu, 43=Jozani, 44=Muyuni.

Table 13.1. Number and area (km^2) of coastal forests in eastern and southernAfrican countries based on 194 surveyed forests from a total of 224 known or probable forests (5).

Country	Forests	Area, km^2
Kenya*	97	661
Malawi	2	16
Mozambique*	13	1790
Somalia	2	2
Tanzania	76	721
Zimbabwe	4	3
Total	194	3193

* rough estimates

As recently as 1980, the status of coastal forest in Eastern Africa was poorly known (15), but recent research has vastly improved knowledge of the location and size of remaining forests (2,3,23,28,29). Information from Mozambique remains poor, and most data are from satellite images. Around 3000 km^2 of coastal forest is thought to still exist (Table 13.1). The total area may change, however, with more detailed exploration of potential sites, particularly in southern Tanzania and Mozambique.

Most coastal forests in eastern Africa are located in statutory protected areas, either National Park, National Reserve, Game Reserve, Forest Reserve or National Monument, and consequently have a degree of protection. For example, fifty two of the 76 known forest sites in Tanzania are located in Forest Reserves, and these contain around 82% of the known coastal forest area. The statutory protection afforded by Forest Reserves and other forms of government legislation are believed to be the most important reason for the continued existence of coastal forests, although some small sites may receive protection from the local inhabitants (for example the sacred Kenyan 'Kaya' forests). Most forest outside of protected areas and sacred forest has been cleared.

Factors Controlling the Distribution and Types of Coastal Forest

A variety of factors control the distribution and types of forest in coastal Eastern Africa, among them are climate, topography, geology, and soils.

Climate

The climate of Africa has varied markedly over geological time, with dry periods when forests retreated and wet periods when forests advanced (see Chapters 1 and 2; 11). The coastal forests were, in the distant past, connected to the rain forests of Central and West Africa, as evidenced by the fact that some plant species found in West and Central Africa are found in coastal forests. These species became isolated in coastal forests due to the uplift of the Central Tanganyikan Plateau some 20 million years ago (21), in association with periods of drier climate.

During drier periods the climate at the coast has remained wetter than in many other parts of Africa. This has allowed dry coastal forest to persist and provide a refuge for forest species, when forests were all but eliminated from most other parts of Eastern Africa. The key factor in maintaining a moist climate has been the more constant temperature of the Indian Ocean which has allowed the warm and moist monsoonal winds to transport rain to coastal areas.

The current climate of the Eastern African coast is seasonal and monsoonal, with annual rainfall between 800 and 2000 mm per year (34). The annual rainfall pattern varies with distance from the equator (Fig. 13.4). In northern Mozambique and southern Tanzania, there is a single pronounced dry season from May to early October and a wet season from October to late May. In comparison, in northern Tanzania and Kenya, there are two rainy seasons, the long rains from March to June and the short rains from late October to early December (34). The short rains can, at times, be unpredictable and indistinguishable from background rainfall. When these rains fail the dry season can be extended and these years may be very stressful for the biota.

Temperature variations in these forests are relatively small, with daytime temperatures often over 30° C, rarely falling below 25° C, and night time temperatures rarely falling below 20° C. The temperatures and humidity within the forests are, however, generally different to those of surrounding vegetation. Studies in Tanzania have shown that daytime temperatures can be 10° C lower within the forest than in adjacent areas of savanna-woodland. Humidity in the forest can also remain close to 100% for many days following rain, whereas in adjacent savanna-woodland humidity rapidly declines after rainfall. Coastal forest thus has a generally cooler, more humid and more stable microclimate than adjacent habitats such as savanna-woodland and grassland. Towards the end of the long dry season, however, the forests become quite dry, and there may be little difference between the microclimate of the forest and the nearby woodland. It is this desiccation event which is believed to make the coastal forests so biologically distinct from most other forest types in Eastern Africa.

Topography

Coastal forests are generally located at less than 500 m altitude, and mostly on areas of raised land within the coastal plain. There are at least two reasons for this.

(1)   Higher ground tends to receive increased orographic rainfall (air is uplifted and cooled by the higher ground), allowing the development of forest vegetation (Fig. 13.5). At altitudes below 500 m this forest is all 'coastal forest' in character. At altitudes above around 500 m many forests can be more accurately classified as midaltitude forest (see Chapter 15). Hence in mountain blocks such as the Udzungwas, Ulugurus, Ngurus and especially the Usambaras in Tanzania there is a gradation from coastal forest vegetation in the lowlands to midaltitude and finally montane forest with increasing altitude. In such areas there may be some overlap in plants and animals between the coastal and midaltitude forests (22,34).

(2)   Higher ground is often more precipitous and difficult to farm in comparison with land at lower elevations, reducing the rates of forest clearance.

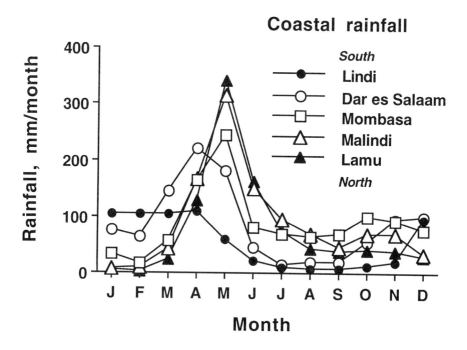

Figure 13.4. Seasonal rainfall patterns at various locations along the Eastern African coast (34).

Forest will also grow in flat lowland areas if there is sufficient moisture. Such forest is often associated with watercourses, such as the Tana River, or is immediately adjacent to the sea where the rainfall is somewhat higher than further inland (for example the coral rag forests at Gedi and Diani Beach, Kenya), or at the base of mountains (for example the lowland East Usambaras in northern Tanzania).

Geology

The varied geology of the Eastern African coastal area is believed to be significant to the coastal forests, as underlying rocks will influence soil formation, which in turn probably influences the vegetation. The solid geology of coastal Eastern Africa comprises freshwater and marine muds, silts, sands and limestones. The reef limestone near the coast was deposited during the last interglacial, 130,000 years ago, when the sea was some 2 to 10 meters higher than at present. Further inland, most rocks date from the Miocene to Pleistocene periods (27 to 2 mya.). However, there are also outcrops of older Jurassic and Cretaceous marine silts and limestones, most notably on the Matumbi Massif (an area of high ground south of the Rufiji river) in Tanzania, inland from Tanga, and around Mombasa. Further inland, ancient rocks of the Pre-Cambrian era rise to form a chain of mountains termed the Eastern Arc (21; Fig. 13.5; Box 13.1).

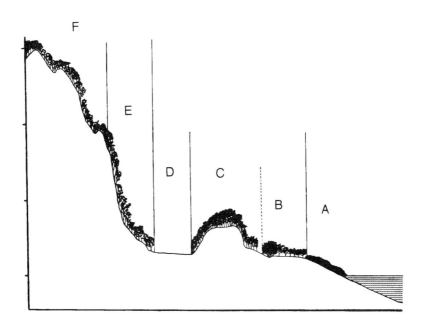

Figure 13.5. Profile of coastal plain and location of coastal forest (29).

A) mangrove forests on the coast in locations with a strong saline influence.

B) coastal forests on flat ground close to the sea, and hence with increased rainfall, but without much saline influence.

C) coastal forests on low coastal hills comprised of rocks of generally less than 30 million years age. Taller forest is generally found in the seaward (eastern) side of the hill, reflecting a higher rainfall.

D) savanna-woodland vegetation (not forest) in the drier areas inland of the coast of eastern Africa, especially low and flat land somewhat inland of the coast.

E) coastal forest at the foothills of the Eastern Arc mountains which are generally comprised of ancient rocks of more than 600 million years age (Pre-Cambrian). Forest vegetation is present here because of the locally higher rainfall associated with the mountains. Vegetation closely similar to the coastal forests extends up to around 500 m altitude in such mountains, where it grades into midaltitude forest types.

F) midaltitude and montane forests.

Soils

The soils of Eastern Africa are highly variable, related to the underlying geology, precipitation, position along a slope (catenary succession), past disturbance, and fire (24). Soil type seems to influence forest vegetation composition. This is illustrated in the Arabuko-Sokoke Forest, Kenya. For example, on white well-drained sandy soils, the vegetation is a deciduous *Brachystegia spiciformis* forest, whereas on adjacent red lateritic soil, the vegetation is a semi-evergreen canopy dominated by *Cynometra webberi* and *Manilkara sulcata*.

Box 13.1. The Eastern Arc Mountains

The Eastern Arc is defined as the ancient crystalline mountains in eastern Tanzania and south-east Kenya under the direct climatic influence of the Indian Ocean. From north to south they are the: Taita Hills, North and South Pare, East and West Usambara, Northern and Southern Nguru, Northern and Southern Uluguru, Ukaguru, Malundwe Hill, Udzungwa and Mahenge Mountains.

To understand how this geological and climatological definition is related to the remarkable biota living on these mountains, it is important to understand the long-term environmental history of eastern Africa (see Chapter 1). The rifts that formed the Eastern Arc data back to the break-up of the super continent, Gondwanaland, 100 million years ago. The mountains have undergone cycles of erosion and uplift since then, with the most recent period of uplift occurring about 7 million years ago. The age of the Eastern Arc contrasts with the relative youth of Africa's volcanic mountains. The oldest lavas of towering Kilimanjaro are a mere 1 million years old, with the last great eruptions occurring only 11,000 years ago.

As well as being geologically relatively stable, the Eastern Arc have also been under a stable climate. During the last ice age, when the African climate was thought to have been cooler and drier than at present, Indian Ocean sea surface temperatures off the coast of eastern African did not change. This suggests that, while the climate may have changed in other parts of Africa, it did not change on the Eastern Arc. These mountains can be more than 2000 m tall and are the first high ground encountered by the rain-bearing winds coming off the Indian Ocean; annual precipitation can exceed 3000 mm. This rain supports the growth of luxuriant lowland to montane tropical forests, and has apparently done so for many millions of years.

Biological evidence of the great age of the forests comes from the plants and animals living in them. For example, the giant Msambo tree (*Allanblackia stuhlmannii*), which is endemic to the Eastern Arc, has its nearest relatives in western Africa. This tree can be more than 40 meters tall and bears fruits weighing up to 7 kilograms. Today the forests of eastern and western Africa are separated by a corridor of arid land running from the Horn of Africa to the Namib desert. Long distance dispersal of Msambo fruits seems unlikely, suggesting that formerly there was a forest connection between eastern and western Africa. This connection would date back to the time when rain forest cover was continuous across Africa, before uplift of the Central African plateau and formation of the arid corridor. Potentially even older biogeographic links exist with Madagascar and Asia, which either must be explained by unusual long-distance dispersal events, continuous forest cover across the Arabian Peninsular, or direct connections pre-dating the break-up of Gondwanaland!

The Eastern Arc mountains' unusually long period of environmental stability and isolation both from other forest areas and from each other has meant that they have accumulated a rich biota. An astonishingly high percentage of the species are only found on the Arc: as much as 30% of the 2000 or so plant species and more than 80% of some groups of spiders and millipedes are endemic to these mountains (1). Perhaps one of best-known endemics are the African violets in the genus *Saintpaulia,* which as a whole, only occur naturally in the Eastern Arc and adjacent Coastal forests. Many of the 20 or so African violet species are only known from one or two mountains. In addition to their obvious values for biodiversity, the

Eastern Arc mountains are important areas of water catchment, and many forests have cultural values as traditional sacred sites and as sources of locally used non-timber products.

Administratively, most of the forests are in catchment forest reserves under the Tanzanian Forest Division. In all, there are more than 130 of these reserves, with a total area exceeding half a million hectares. Not all of the reserved area is forest, and many of the reserves also include woodland and grassland. Small areas are on private land, such as on tea estates, or are under public or district control. Areas strictly protected are the recently created 200,000 ha Udzungwa National Park, 450 ha on Malundwe Hill in Mikumi National Park, the 450 ha University Forest Reserve at Mazumbai in the West Usambara mountains and the 30 ha nature reserve at Luisenga Stream on the Brooke Bond Mufindi Tea Estate. Although increasing population in the well-watered fertile mountains is placing a pressure on the forest, most recent large-scale deforestation has been due to commercial logging, cash crop plantation and excision of forests from government protection. Forest values are now, however, well recognized by policy makers, and an Eastern Arc conservation program is a keystone project in the recent Tanzania Forest Action Plan.

J.C. Lovett

References

1.   Lovett, J.C., Wasser, S.K. eds. 1993. *Biogeography and Ecology of the Rain Forest of Eastern Africa.* Cambridge: Cambridge University Press

Vegetation changes down a forested slope are also ascribed to variations in soil. For example, the Pugu Hills coastal forests of Tanzania have three distinct forest types dependent on position down steep valleys within the forest. In valleys the vegetation is 35 m tall and dominated by species such as *Malancantha alnifolia* and *Antiaris toxicaria*. On the sides of such valleys the canopy is often dominated by *Scorodophloeus fischeri*, or with *Dialium holtzii* and *Baphia kirkii*. Finally, on the ridge-tops, the forest often has the stature of thicket and is dominated by *Manilkara sulcata* and *Brachylaena huillensis* (Fig. 13.6).

Ecology of Coastal Forests

Our understanding of the ecology of coastal forests is still rudimentary although some progress has been made towards describing the various biological communities, such as plants (12,13,26,27), and a number of detailed studies on particular faunal groups have been carried out, such as elephant-shrews. Researchers are only just beginning to investigate the main ecological processes within coastal forests, such as energy production and energy flow, species interactions and competition. Most research on the tropical forests of Eastern Africa has focused on rain forests. These are known to be more diverse in species than the dry coastal forests. For example, Kibale forest in Uganda supports nine primate species, but Arabuko-Sokoke, the largest coastal forest, only supports five species. Similarly while some tropical rainforests may support 200 different plant species in a 0.1 ha plot (14), coastal forests may have less than 50. Many of the species that occur in

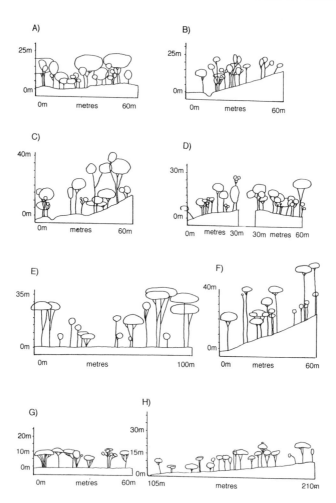

Figure 13.6. Forest profile diagrams for example coastal forests in Tanzania (12).

A) Gendagenda Forest Reserve, transect F, dry evergreen; B) Gendagenda Forest Reserve, transect B, moist valley; C) Gendagenda Forest Reserve, transect D, moist riverine; D) Gendagenda Forest Reserve, transect G, dry evergreen; E) Ruvu South Forest Reserve, dry semi-evergreen; F) Tongwe, transect A, moist hill-side; G) Mkwaja forest, dry ridge-top; H) Kazimzumbwe Forest Reserve, dry ridge-top.

coastal forests, however, are unique to that environment, and do not occur in other forest types, including vertebrates as Clarke's weaver (*Ploceus golandi*), Ader's duiker (*Cephalophus adersi*) and the golden-rumped elephant-shrew (*Rhynchocyon chrysopygus*), and up to 500 species of plant. Hence, there is considerable interest in the ecological interactions of coastal forests and how these might have influenced this high endemism, and what makes dry coastal forests different or similar to tropical rainforests.

The leaf litter of a coastal forest supports many species of bacteria, fungi, earthworms and arthropods, particularly termites and large millipedes. They are primarily involved in breaking down the leaf litter. Although poorly studied it seems that many of these species are unique to coastal forests. There are also a few specialized vertebrates which inhabit the leaf litter including unique species of amphibians (toads and frogs) and reptiles (snakes and lizards). The rich invertebrate assemblage of the litter also provide food for ground-dwelling birds, such as spotted ground-thrush (*Turdus fischeri*) and red-tailed ant-thrush (*Neocossyphus rufus*), elephant-shrews and aardvarks (*Orycteropus afer*).

The forest floor is also home to bushpigs (*Potamochoerus porcus*) which grub about for tubers and fallen fruits, duikers (*Cephalophus* spp.) which feed on low vegetation and fallen fruits, bushbucks (*Tragelaphus scriptus*) which grazes and browses vegetation, and yellow baboons (*Papio cynocephalus*) which search for fruits, seeds, tubers, invertebrates, eggs and nestlings. Elephants (*Loxodonta africana*) are also found in some forests. Where they occur they are probably the most important herbivores due to their large size, and they also have a significant effect on the forests ecology through their habit of uprooting or ringing trees (see Box 11.2). Larger ground predators include caracals (*Felis caracal*) and mongooses which hunt for elephant-shrews and other small mammals and birds, and occasionally leopards (*Panthera pardus*) and lions (*Panthera leo*).

The dense understory of shrubs provides suitable habitat for a wide range of small birds, such as gray-backed camaroptera (*Camaroptera brachyura*) and Fischer's greenbul (*Phyllastrephus fischeri*), which glean for insects from the vegetation. Slightly higher up in the low-middle canopy, fruit-eating bird species such as tinkerbirds (*Pogoniulus* spp.) are more common, along with tree-climbing rodents, such as the lesser pouched rat (*Beamys hindei*) and the red-bellied coast squirrel (*Funisciurus palliatus*).

The mid-top canopy is frequented by the red-legged sun squirrel (*Heliosciurus rufobrachium*), Syke's monkey (*Cercopithecus albogularis*) and black and white colobus monkey (*Colobus angolensis*), two or more species of bush-baby (*Galago*, and fruit-eating birds such as orioles (*Oriolus* spp.) and hornbills. Several specialized insect-eating birds inhabit the canopy, such as plain-backed sunbird (*Anthreptes reichenowi*), helmet-shrikes (*Prionops* spp.) and flycatchers such as ashy flycatcher (*Musicapra caerulescens*). Larger predators in the trees include genets and wild cats, and birds such as African goshawk (*Accipiter tachiro*) and great sparrowhawk (*Accipiter melanoleucus*) perch on trees and wait for prey to pass underneath, and crowned eagles (*Stephanoaetus coronatus*) actively search for moneys and other prey.

Seasonality

Although there is relatively little temperature variation through the year, coastal forests experience highly seasonal rainfall, which has a major effect on forest ecology. Specifically, the production of leaves, flowers and fruits is highly seasonal which in turn has an important effect on the abundance and breeding behavior of forest animals. Most birds and mammals breed in response to the rains. Numbers of flowers, young shoots, fruits and insects gradually decline as the rains cease and the dry season progresses. Trees such as baobabs and figs, which are either able to store water or have long tap roots, provide a critical source of food during the dry times. As pools of water dry up and leaf-litter dries out, many of the frogs aestivate, digging themselves into the ground to reduce dehydration. Intra-african migrant birds such as spotted ground thrush, red-capped robin-chat (*Cossypha natalensis*) and African pitta (*Pitta angolensis*) are also thought to migrate in response to the rains; they breed during rains in coastal forests of north Mozambique and southern Tanzania and then move north to catch the rainy season in north Tanzania and Kenya, arriving between May and June.

Nutrient Cycles

Although little is known about nutrient cycling in coastal forests, it is assumed to be similar to other tropical forests (Fig. 13.7). Some nutrients enter the system from rainfall. Many nutrients are derived from the smoke of fires associated with farm and scrub clearance, but also from air-born dust. Rain falling through the vegetation leaches out nutrients so that the water reaching the ground has higher concentrations of nutrients than the rainfall. The forest floor, however, receives the majority of its nutrients from the leaves that are naturally shed from the canopy trees. As the litter decomposes, the nutrients pass into the soil and may be taken up by the roots or leached out into streams or ground water. Nutrients are also derived from the weathering of rocks. Moreover, some trees, such as *Brachystegia spiciformis,* enhance their nutrient-uptake by having fungi associated with their roots (mycorrhizae), and there are other trees and shrubs which have formed associations with bacteria in their roots to enhance nitrogen uptake.

Vegetation Structure

The canopy of coastal forests is in a constant state of change. A number of factors such as lightning storms, tree felling and old age cause trees to fall, creating gaps. These gaps are colonized by seedlings which grow up and eventually form mature trees. As a result, the forest is a mosaic of patches at different stages in the growth cycle.

Even without this mosaic of differently aged patches, the vegetation structure of the coastal forests would be highly variable, with the stature of the trees being controlled by factors such as position on slope, soils, rainfall, ground-water and drainage. Below the canopy the forests have a distinct shrub layer and a few specialist ground herbs which only partially cover the forest floor (see Fig. 13.6). Lianas are common, especially in areas which have been disturbed by man. Epiphytes (plants living directly on the trees) are generally scarce, but may be abundant on trees over water-courses.

In reality, most forests have been extensively disturbed by humans, which has influenced their structure. There are often large canopy gaps where timber trees have been removed, and extensive areas of impenetrable thicket and liana tangle, indicating highly disturbed areas such as old farms.

Species Richness

Species that breed in coastal forests, or which rely on them for part of their life cycle, include around 50 species of mammals, 200 species of birds, 1000 to 1500

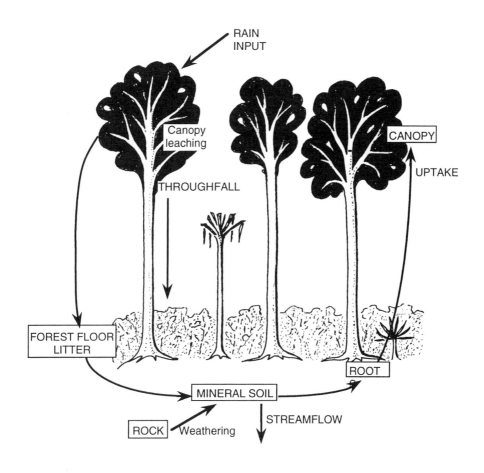

Figure 13.7. The nutrient cycle in coastal forests (35).

species of higher plants, and an unknown number of invertebrate species, but probably several tens of thousands. In comparison with other terrestrial habitats in Eastern Africa, the overall species-richness of the coastal forests is probably only exceeded by rainforests.

Four factors are thought to influence patterns of species richness (and indeed rarity) in the coastal forests. These are:

(1)   Existence of forest cover over many thousands or perhaps millions of years.

(2)   Position of the forests over a large longitude north to south in eastern Africa.

(3)   The small and fragmented extent of remaining forest.

(4)   The long history of extensive disturbance of the forest by man.

Duration of Forest Cover. Different coastal forests have persisted at particular sites for different lengths of time. Forests on higher ground would have escaped flooding during sea-level rises associated with the warm periods (Interglacials) and the desiccation events associated with cold periods (Glacials; see Chapters 1 and 2; 9,11). Hills also have a higher diversity of potential microclimates due to varying topography, rainfall and drainage patterns, which will also help explain patterns of species-richness. Coastal forests with the highest species diversity are probably those in the lowlands of the East Usambara Mountains - due to their large size, large number of fragments, variable rainfall, complex topographic patterns, and low levels of disturbance. The next most important forests in terms of species-diversity are probably the Arabuko-Sokoke forest in Kenya and the Rondo forest in southern Tanzania.

Location Along The Coast. There are distinct differences in the species composition of the coastal forests from north to south in Eastern Africa. This is best known for the flora, where several broad vegetation groups have been defined from multivariate statistical analysis of plant data collected from standard plots within various forests (12,13). These vegetation groups include a northern type stretching from Malindi in Kenya, down to the Tanga area in Tanzania, an intermediate group of forests around Pangani in northern Tanzania, and a further group in the area around Dar es Salaam. The vegetation of Somalia, northern Kenya, southern Tanzania, and Mozambique were not sampled and, therefore, additional broad vegetation types probably exist. For example, forests around the Rondo Plateau inland of Lindi in southern Tanzania are known to be highly distinct from other areas and contain at least 50 endemic plant species (33).

There are also differences in the fauna between northern and southern coastal forests. For example, the golden-rumped elephant-shrew is only found in the coastal forests of Kenya north of Mombasa, whereas in similar forests of southern Kenya and Tanzania this species is replaced by the black and rufous elephant-shrew (*Rhynchocyon petersi*). Similar patterns exist for other animals. A major break in the species-composition of the fauna exists north and south of the Rufiji River in Tanzania, with both species and sub-species differences between these regions (18).

Forest Area. The area of forest remaining is a significant factor determining both the species richness of a particular forest and its value in terms of rare species (Fig. 13.8). Bird species which are unlikely to be found in the smaller forests include those with large home ranges, such as crowned eagles, and those which live at low population densities. While it might be expected that plants could survive in much smaller areas of habitat, the available data suggest that small forests of around 2 km² also support only a few rare forest plants. This effect is most likely caused by a reduction in the population sizes of rare species below viable levels, and the degradation of the forest microclimate or the increasing edge effect in increasingly small sites. Both factors will cause the extinction of species requiring continuous forest and an increase in the dominance of forest-edge, fugitive, and woodland species.

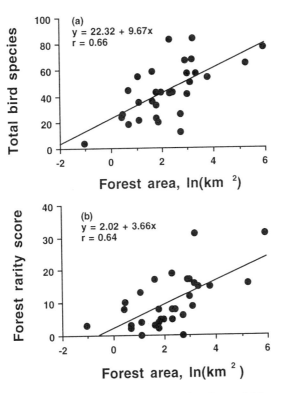

Figure 13.8. Scatter-plot graphs with fitted regression lines of (a) total number of bird species against the natural logarithm (=ln) forest area (km²) and (b)bird rarity scores against ln of forest area (km²). Rarity scores are calculated from the standard scores (8), with the score of 0.5 added for candidate Red Data Book species.

Forest Disturbance And The Edge Effect.
Disturbance of the coastal forest habitat by man has a number of deleterious effects on the forest. Some of the major effects of disturbance are outlined below, with examples of what they mean for coastal forests.

The edge of a forest is different to the interior in that the microclimate is not as stable on the edge, levels of insolation are often higher, the ground may become parched and desiccated and a drying wind may reach into the forest causing a higher level of exposure for many species. Such edge effects may extend up to 300 m into the forest (36), hence small forest sites (such as most of the Kayas in Kenya) may be regarded as entirely 'edge'. Even in many larger forests the 'edge effect' may be very important, as many sites have been subjected to heavy disturbance which removes the canopy and creates 'forest edge' conditions more widely throughout the forest. Thus even some of the relatively large coastal forests may possess only small areas with the microclimate and structure of 'normal' coastal forest.

Forest-dependent species often avoid forest edge, as a result of habitat change and human disturbance (19,20). For example, little yellow flycatchers (*Erythrocercus holochlorus*) and olive sunbirds (*Nectarinia olivacea*) are more common in the center of Diani Forest, Kenya than near the edge. As a result, forest species may become concentrated in a central core, and the overall number of individuals that the forest can support will be reduced.

Following disturbance, the understory at the edge of the forest is exposed to increased levels of sunlight and wind which alter the physical environment, usually making it much brighter, hotter and drier. This may result in a rapid increase in the density of the undergrowth as well as increased levels of tree mortality (20). Certain adaptable non-forest species may actually benefit from such changes, but many others do not, and these are frequently the species of greatest conservation concern. For example, the spotted ground thrush prefers dense shade and is therefore expected to avoid disturbed areas, but the grass *Olyra latifolia* invades recently disturbed areas. The greater the quantity of disturbance the greater the change in the microclimate and the more likely it becomes that sensitive species will be lost (27).

Plant and animal species adapted to hotter, high light conditions (mostly nonforest species) tend to invade damaged forest. As an example, vervet monkeys only occur on the edge of Sokoke forest in Kenya. Often common and successful, such invasive species are able to cope with modified secondary habitats, while the forest-adapted species they replace occur at low population densities. Exotic species may also invade from the surrounding agricultural land and *Lantana camara,* an invasive shrub from India, is presently invading many coastal forest patches in Kenya. Moreover, certain predators exploit edge habitats and the breeding success of forest-dependent birds occupying edge forest can be reduced (6,10). Birds that parasitize other species nests are also more common in fragmented forests (1).

Species Endemism

Eastern African coastal forests contain hundreds of endemic and near-endemic species of plants and animals (Table 13.2); many are threatened with extinction as their habitat disappears or is degraded. At least 4 species of mammals, 5 species of birds, 4 species of amphibians, 27 species of reptiles, 40 species of butterflies, 20 species of millipedes, 400 species of plants, and an unknown number of beetles, flies and other invertebrates are dependent on the survival of coastal forest if they are not to become globally extinct.

For the birds, a relatively well studied group in the coastal forests, the main center of endemism appears to be between the Tana River in Kenya and the lowland forests of the East Usambaras in Tanzania (16). Forests between Tanga Region and Kilwa District in Tanzania also possess some of the coastal forest endemic birds, but not all. The forests in the extreme south of Tanzania lack many of the species typical of the coastal forests to the north, but there is one species here that is not present further north - hence this area is also important for bird conservation, in line with the pattern found for plants (4,16,33).

Distribution of Species Between Individual Forests.   Many of
the species endemic to coastal forests are not widespread within this habitat, and are found in only a small percentage of the remaining forests. Even forests of fairly close geographic location may have few endemics in common. For example, up to

Table 13.2. Species thought to be endemic or near-endemic to the coastal forests.

Taxon	Endemic	Near-endemic
1. Mammals	4	10
Cephalophus adersi	x	
Kerivoula africana	x	
Pteropus voeltzkowi	x	
Rhynchocyon chrysopygus	x	
Up to 4 new species of shrew	?	
Galago sp. nov. A	?	
Beamys hindei		x
*Cercocebus galeritus**		x
Dendrohyrax validus		x
Galagoides zanzibaricus		x
Myonycteris relicta		x
Otolemur garnettii		x
*Procolobus rufomitratus**		x
Pteropus seychellensis		x
Rhinolophus deckenii		x
Rhynchocyon petersi		x
2. Birds	5	18
Anthus sokokensis	x	
Batis reichenowi	x	
Erythrocercus holochlorus	x	
Otus sokokensis	x	
Ploceus golandi	x	
Alethe fuelleborni		x
Anthreptes neglectus		x
Anthreptes pallidigaster		x
Anthreptes reichenowi		x
Apalis chariessa		x
Bubo vosseleri		x
Campethera mombassica		x
Circaetus fasciolatus		x
Macrosphenus kretschmeri		x
Oriolus chlorocephalus		x
Phyllastrephus debilis		x
Phyllastrephus fischeri		x
Pogoniulus simplex		x
Prionops scopifrons		x
Sheppardia gunningi		x

continued

Table 13.2 - Continued

Taxon	Endemic	Near-endemic
Swynnertonia swynnertoni		x
Tauraco fischeri		x
Turdus fischeri		x
3. Amphibians	4	4
Afrixalus sylvaticus	x	
Hyperolius rubrovermiculatus	x	
Mertensophryne micranotis	x	
Stephopaedes sp.nov.	x	
Hyperolius puncticulatus		x
Leptopelis flavomaculatus		x
Spelaeophrgne methneri		x
Stephopaedes loveridgei		x
4. Reptiles	27	17
Aparallactus turneri	x	
Bradipodion mlanjense	x	
Dasypeltis medici	x	
Dendroaspis angusticeps	x	
Dipsadoboa werneri	x	
Gastropholis prasina	x	
Gastropholis vittata	x	
Leptotyphlops sp. nov	x	
Lygodactylus broadleyi	x	
Lygodactylus conradti	x	
Lygodactylus kimhowelli	x	
Lygodactylus rex	x	
Lygodactylus ulugurensis	x	
Lygodactylus viscatus	x	
Lygodactylus williamsi	x	
Melanoseps rondoensis	x	
Philothamnus macrops	x	
Prosymna janii	x	
Prosymna sp. nov.	x	
Rhinotyphlops lumriciformis	x	
Rhampholeon chapmani	x	
Rhampholeon k. kersteni	x	
Scelotes duttoni	x	
Scelotes insularis	x	
Scolecoseps sp. nov.	x	
Sepsina tetradactyla	x	

continued

Table 13.2 - Continued

Taxon	Endemic	Near-endemic
Typhlops rondoensis	X	
Aparallactus guentheri		X
Aparallactus werneri		X
Bradipodion tenue		X
Cnemaspis barbouri		X
Cnemaspis uzungwae		X
Crotaphopeltis tornieri		X
Holaspis guentheri laevis		X
Lygosoma lanceolatum		X
Lygosoma mafianum		X
Lygosoma pembanum		X
Melanoseps loveridgei		X
Natriciteres variegata sylvatica		X
Philothamnus natalensis		X
Rhampholeon brachyurus		X
Rhampholeon brevicaudatus		X
Rhampholeon platyceps		X
Urocotyledon wolterstorffi		X

5. Invertebrates

at least 20 millepedes	X	
40+ butterflies	X	

6. Plants

~400 plant species	X	
50 plant genera	X	

* Species known only from the Tana River forests, which are closely related to coastal forests

13 species of plant are endemic to the Pugu Hills coastal forest (Pugu and Kazimzumbwe) in Tanzania, but none of these species are found in Pande coastal forest which is only 20 km to the north. Conversely, there are at least three endemic plant species in Pande which are not found in Pugu. Such examples of site-specific endemism are also found in other groups of plants and animals. For example, the Sokoke scops owl (*Otus sokokensis*) and Clarke's weaver are quite common in Sokoke but do not appear to inhabit neighboring forests in southern Kenya, such as the Shimba Hills.

Many species found in more than one site also have highly disjunct distributions, for reasons that are not understood. The east coast akalat (*Sheppardia gunningi*), a bird which is relatively common in some coastal forests sites, is absent from many others (4), and the Ader's duiker is only recorded from Jozani Forest on Zanzibar and Arabuko-Sokoke Forest in Kenya. These rather unpredictable distribution patterns emphasize the extent to which the coastal forests vary. As a result, protecting a small selection of forests may not protect a representative selection of coastal forest fauna and flora.

There is currently great interest in assessing whether centers of endemism for various taxa overlap, so that conservation priorities can be selected efficiently to conserve maximum biological diversity (16). Preliminary information from coastal forests, such as for Pugu Hills in Tanzania, suggests that at least some sites are indeed important for many groups of organisms, and hence endemism does overlap. There are, however, other sites important for one group, but are not important for others. The Sokoke forest in Kenya is extremely important for birds and mammals, but is not particularly important for plants. The Matumbi Hills forests in Tanzania are very important for invertebrates, plants and mammals, but are not particularly important for birds. Further work is urgently required to assess whether a few coastal forest sites can be selected for intensive conservation efforts which will save the majority of the coastal forest species.

Conservation of Coastal Forests

Because the coastal forests are areas of high biological diversity, with large numbers of endemic and rare species, the conservation of these forests has become a priority. The conservation problems that will have to be tackled if coastal forests are to be saved are outlined below, with some of the possible solutions.

Problems

Increasing Human Population.
Over the last century the whole coastal belt has experienced an increase in population, vastly increasing the demand for natural resources, and land for farming. In the Coastal Province of Kenya, population growth was estimated at ~3.8% per year after independence, and a large fraction of the growth was due to migration into coastal areas. Moreover, the populations of Malindi and Mombasa have increased by up to 20% per year during the 1980s, as people come in search of work in the tourist industry. The most recent statistics from the late 1980s censuses suggest continued high population growth but a slight decline in the annual rate (see Chapter 3). Similar movements of people are occurring in Tanzania, especially into the East Usambara mountains, Tanga, Dar es Salaam and the Matumbi Hills. Such population increases put more and more pressure on the natural resources of these coastal areas. Coastal forests and other habitats with woody vegetation suffer heavily as they provide building materials, commercial timber, and a stock of 'unclaimed' farmland.

Land Clearance.
Forest clearance for agriculture is occurring in most unprotected areas of coastal forest and also in a few Forest Reserves. Clearance of unreserved land is currently taking place in the East Usambara Mountains, Tanzania and at Mangea Hill, Kenya. Forests are normally cleared by ring-barking the large trees, cutting the shrubs and small trees, piling these around the large trees and setting fire to them. This process kills the large trees and they are left to fall naturally. Crop yields in the first year following clearance are high because of the nutrients contained in the leaf litter, ash, and roots, but decline in subsequent years.

Removal of Timber and Fuelwood. Removal of building poles from coastal forests has a significant effect on the structure, regeneration potential and species composition of forests. Local communities around the Arabuko-Sokoke forest in Kenya remove over 30,000 poles each year for building purposes, mainly from the *Afzelia* and *Brachystegia* habitats. This equates to approximately one in 20 stems of the 6 to 10 cm diameter range being cut per year (25). Pole cutting is presumed to affect the regeneration of trees and forest. Although it may be possible to harvest poles sustainably from coastal forests, in practice pole removal from the small patches of coastal forest remaining is an unsustainable activity (Table 13.3).

Removal of fuel wood also has a significant effect in many sites. For example, the communities around Arabuko-Sokoke currently remove over 610 tons of fuel wood, mainly dead *Brachylaena*, from the forest each year, while seven licensed commercial fuel wood collectors remove an additional 2800 m^3 for sale in Mombasa, Malindi, and Kilifi (25). Removal of dead wood may affect invertebrate populations and birds such as woodpeckers that feed on the insects in dead timber, and also hole-nesting birds, such as barbets.

The Forest Reserves around Dar es Salaam in Tanzania (Pande, Pugu, Kazimzumbwe and Vikindu) have also suffered extensive damage from people cutting all woody species over large areas and burning the wood to produce charcoal to supply the domestic market. Surveys in 1990 indicated that approximately 200 bags (20 to 30 kg each) of charcoal were being removed from the Pugu hills Forest Reserves (total 22 km^2) per day. Such levels of exploitation were leading to the removal of virtually all canopy trees and all other larger shrubs from extensive areas within the Forest Reserves. Increased protection of the forest initiated in 1991 has reduced this problem to some extent, but it still continues.

Commercial Forestry. During the colonial and post-colonial era in both Kenya and Tanzania, many Forest Reserves were set up with the aim of extracting their valuable timber. Many of the larger Forest Reserves were subjected to heavy logging pressure during the 1950s and early 1960s and logging continued in most areas until the early 1980s, and in some areas beyond this date. The felling of mature trees in the *Brachystegia* habitat of the Sokoke forest has reduced the canopy density and consequently altered the animal community. Populations of golden-rumped elephant-shrews have been reduced, and forest bird species such as little yellow flycatcher and forest batis (*Batis mixta*) are replaced by non-forest species.

Logging may be particularly harmful to coastal forests, because they are believed to be less adapted to disturbance than many other tropical forest types, particularly those where typhoons and other natural disasters are common (35). In coastal forests there are very few native pioneer species that will invade large disturbed areas. Hence regeneration following intense natural or logging disturbance may be slow - although no studies on rates of coastal forest regeneration exist. An open canopy and slow regeneration will place the forest at risk from fire. This can destroy regenerating forest species and lead to their replacement by more fire-adapted woodland species. Logging is also known to have an effect on the forest microclimate (7) since subcanopy species are exposed to increased insolation and air temperature, and decreased humidity. Sensitive understory species are likely to be

Table 13.3. Social, cultural and natural resource values of coastal forests and the sustainability of their use (5).

USE	User	Gender	Production	Economy	Season	Sustainability
Timber	Sp	M	H	C	S	NS
Charcoal	Sp	M	H	C	Y	NS
Fuel Wood	G	F	H	C	Y	NS
Building Poles	G	M	H	C/D	S	NS
Wood Carving	Sp	M	H/L	C/D	Y	NS/S
Medicinal Plants	Sp	F/M	L	C/D	Y	S/NS
Edible Wild Plants	G/Sp	F	L	D	S	S
Hunting	Sp	M	H/L	D	S/Y	S?
Bee Keeping	Sp	M	L	D	S	S
Water Collection	G	F	-	D	Y	S
Cultural Sites	Sp	M/F	-	D	S/Y	S
Tourism	Sp	M/F	L/H	C	S	S
Agricultural Clearance	G/Sp	M/F	H	C/D	S	NS
Building Sites	Sp	M	H	C	Y	NS
Mining	Sp	M	H	C	Y	NS
Grazing *	Sp	M/F	L	D	S	S
Conservation/Research	Sp	M/F	L	C/D	S/Y	S
Education	Sp	M/F	L	D	S	S
Plantations	Sp	M	H	C	Y	NS

KEY:

1. Who does activity (Sp= specialist user, G = generalist user/whole community)
2. Gender of those doing activity (M = male, F = female)
3. Volume of material generated by activity (H = high, L = low, - = not applicable)
4. Commercial basis of activity (C = commercial, D = domestic)
5. Seasonality of activity (S = seasonal, Y = year-round)
6. Sustainability of activity (S = sustainable, NS = non-sustainable)
* Not generally encountered in coastal forests

particularly affected and regeneration of the forest species may be impeded (27). Some species of animals are also known to avoid crossing sunlit patches (32).

Domestication of Natural Forest. The colonial and post-independence governments of Kenya and Tanzania have pursued policies to replace some areas of natural forest with plantations of domestic species. These species, it was hoped, would provide a larger volume of commercial timber than natural species. For example, a large area of Mvule *(Milica excelsa)* forest on the Rondo Plateau in southern Tanzania was logged in the 1950s, and replaced with exotic pine trees. Other plantations of species such as *Eucalyptus* and teak *(Tectonia grandis)* were also planted in cleared areas of several coastal forests. These plantations were only partially successful, due to disease and poor growth rates. Consequently, the wholesale replacement of natural vegetation with plantations has largely been abandoned.

On the edge of coastal forest reserves in Tanzania and Kenya plantations of species such as *Cassia siamea* were set up in the 1960s and 1970s as a source of poles and timber. These programs, however, have rarely met the demands of local people and many planned programs have not been implemented.

Carving Timber. As the number of tourists visiting Kenya has increased over the last 20 years, the demand for carvings has rocketed. Much of the wood for these carvings derives from the coastal belt. In 1988, the carving industry was estimated to have earned 30 million Kenya shillings in foreign exchange, and in 1976, it was estimated that over 1000 trees were being cut each month for carving. The main species now used are muhuhu (*Brachylaena huillensis*) and mkongolo (*Combretum schumannii*), both from coastal forests. Over-exploitation in Kenya means that forests (and woodlands) supporting species suitable for carving, even as far south as the Rufiji River in Tanzania, are being exploited to supply the Kenyan market.

Subsistence Hunting. A wide range of mammal species are hunted for food in coastal forests, including duikers, primates, elephant-shrews, bushpigs, giant pouched rats, and other small rodents. Information from Arabuko-Sokoke Forest in Kenya suggest that harvesting of yellow baboons, Sykes monkeys, and the larger ungulates has reduced their populations and that the current hunting pressure is not sustainable. A similar situation exists in Pugu Forest Reserve just outside Dar es Salaam, Tanzania; here intense exploitation has eliminated black and white colobus and hippopotamus (*Hippopotamus amphibius*) and has reduced populations of other edible species (duiker, suni and bushpig).

Land Ownership. The problems of conserving forest on the coast have been compounded by changes in traditional land ownership, and the fact that few people own the land they farm. In Kenya, a ten mile strip extending inland from the coast was historically held by Swahili Sheikhs from Mombasa. In the early 1800s this land came under the Sultanate of Zanzibar, and large areas were then registered by Arab settlers, many from Oman. After independence these title deeds have been honored despite the fact that many of the owners reside outside Kenya. Therefore, many of the people currently living on the Kenya coast are squatters, and, as a consequence, have little incentive to conserve forest resources as they have no long-term right of ownership.

Similar problems arose in Tanzania during the Ujamaa collectivization and villagization programs in the 1960s and early 1970s. In the Matumbi hills south of the Rufiji river in Tanzania, people were moved by these resettlement programs and the traditional land practice of farming on ridges and leaving forest on steep slopes and valleys was abandoned. Recently, large numbers of people are moving back to these areas because the soils are fertile and the rainfall reliable and these hills are more productive than adjacent areas. Those moving back to the area, however, no longer follow traditional land use practices. The remaining forest land, outside Forest Reserves, is rapidly being cleared for agriculture with little concern for the conservation of natural resources. Valuable trees within the Forest Reserves have also been logged heavily.

The best known traditional forest conservation practice is that of the Mijikenda peoples of southern Kenya and northernmost Tanzania (30). These people originally occupied the hinterland around Lamu, but moved south across Tana and Sabaki rivers partly to flee from the Galla people. They took shelter in the forests of the coastal hinterland of Kilifi and Kwale districts, building fortified villages. Gradually the Mijikenda spread out from these villages. The forest refuges, however, were still used for ceremonial purposes and were considered sacred because of the shelter they had provided to the village. The Kaya elders have enforced the protection of these 'Kaya' forests for many years, but mounting pressure for land and forest products, the breakdown of traditional beliefs, the increasing influence of Islam and Christianity, and the immigration of people from other tribes and nations now threatens their existence.

Attitudes Towards Forest Conservation. Local communities often suffer the direct costs of forest conservation, such as the loss of potential agricultural land, and crop damage by forest animals, while current forest management policies often prevent local people from deriving short-term benefits. For example, elephants annually cause damage to the crops of 50% of the households living adjacent to Arabuko-Sokoke forest (estimated value of lost crops Kshs 663,000 per year in 1992), while yellow baboons, Sykes monkeys and bush-pigs are also a major problem. Moreover, the substantial benefits that local populations around Arabuko-Sokoke derive from the collection of forest products (over 8 million Kshs in 1992), is mainly from the illegal collection of resources the people feel that they have a right to anyway (Table 13.4). It is therefore not surprising that many local communities have negative attitudes towards forest conservation, and, probably because of the immense pressure for land, many consider that it would be more beneficial to them if the most of the remaining forest was cut down and converted to agriculture (25).

Tourism. Tourism is an important source of income for East Africa, particularly Kenya and Tanzania, and has assisted the protection of natural resources in protected areas. Tourism, however, has also contributed to the destruction of coastal forests in Kenya as a result of clearance for hotel development and increased demand for wood products. For example, Diani forest on the south coast of Kenya is now almost entirely owned by hotel companies, and most forest has been cleared. Less than 100 ha of forest remains, under a quarter of the area a few decades ago. Mass tourism is less developed outside Kenya, and, as yet, there are few examples of conflict between tourism and coastal forest conservation.

Poor Information. Up until the last few years the coastal forests of Eastern Africa were poorly known biologically. Although recent research has started to reveal the importance and complexities of the coastal forest ecosystem, further analysis of research results is needed to identify priority sites for conservation, based on the distribution of endemic, rare and threatened species. In addition, sustainable yields for log removal, pole cutting, charcoal collection and trapping of fauna are not available because the regenerative capacities of the coastal forests are not known. If these forests are to be managed sustainably in the future, this information is vital.

Table 13.4. Estimates of the financial benefits derived by local communities from Arabuko-Sokoke Forest in 1992 (25), both from cash income and from savings on expenditure.

Resource	Number of people benefiting	Value in Kenya Shillings
Income		
Employment in Forest Department	89	1,326,420
Timber	160	519,000
Medicinal plants	216	1,036,000
Fuelwood licensees	7	420,840
Handicraft	25	37,940
Primary savings		
Water	400	34,515
Poles	730	122,518
Game meat	730	2,851,911
Fuelwood	730	1,752,351
Honey	730	40,963
Medicines	730	54,728
Total Financial Benefits		8,197,186

Due to the above factors coastal forest is today largely confined to areas with traditional sanctity or controlled by the national government. This largely equates to sacred groves and Forest Reserves. Few private individuals or companies have made an attempt to conserve natural forest. Large-scale clearance of forest for agricultural land is now rare, although it still occurs in some areas such as in the East Usambara Mountains, Tanzania and Mangea Hill, Kenya. The area of forest remaining is in most cases determined by the edge of the sacred grove or official boundary of the protected area.

Solutions

Reducing Dependence on Indigenous Timber. Providing alternative sources of timber and fuelwood is a high priority along the entire East African coast. While plantations and on-farm planting of fast-growing exotics, such as *Eucalyptus*, *Cassuarina* and *Cassia* species, can provide fuelwood and timber, there are, at present, no good alternatives for trees favored by the carving industry. Research is underway to find alternative timbers and to investigate the feasibility of growing indigenous tree species in plantations. Unfortunately, these species are generally slow growing and therefore costly in terms of land and resources.

Protection. There is so little coastal forest remaining that we believe that all remaining habitat outside a protected area should be gazetted, probably as Forest Reserves. This does not guarantee their future but past experience, that nearly all remaining forest is in Forest Reserves, suggests that gazettement vastly enhances the chance of survival. Also, in theory, it allows the controlled harvesting of forest

Box 13.2.   The Kipepeo (Butterfly) Project and Forest Conservation

'The race is on to develop methods to draw more income from the wildlands without killing them and so to give the invisible hand of free-market economics a green thumb". (E.O. Wilson)

Arabuko-Sokoke Forest makes up over two-thirds of the total area of coastal forest in Kenya and has been ranked as the second most important forest in Africa for bird conservation (8). But this means little to the rapidly growing human population on its borders. The forest-adjacent community is hard-pressed for land, jobs and sources of income, frustrated and impoverished by wildlife crop raiding, threatened with death and injury by forest elephants, and denied legal access to forest resources. In a 1992 attitudinal survey, 93% of local residents were unhappy with the forest's presence and 53% wanted it totally cleared for agriculture (25). The forest was twice invaded by hundreds of squatters in 1993 and 1994, and in the following year the Kilifi District Development Committee recommended the clearing of 1200 ha for settlement. If the long-term future of Arabuko-Sokoke Forest depends on the support of local people and their leaders and politicians, then the current outlook is bleak.

The Kipepeo Project is one of several efforts to improve this situation. Set up in 1993 with funding through UNDP from the Global Environmental Facility Small Grants Program, and administered by the East African Natural History Society with the support of the National Museums of Kenya, Kenya Wildlife Service, and the Forest Department, this project has introduced butterfly farming to the local community. Villagers living immediately adjacent to the forest have been trained to rear forest butterflies which they sell to the project for export to live butterfly exhibits overseas. By enabling them to benefit economically from the presence of the forest, the project hopes to encourage their support for its conservation.

Participants in the project were identified by the villagers themselves using Participatory Rural Appraisal maps drawn on the ground at community meetings. One hundred and forty-two households were subsequently interviewed along a 45 km stretch of the eastern margin of the forest, and 134 of them took part in the project in 1994. By mid-1995 the number had grown to 144, and the farmers had started to form self-help groups to consolidate their butterfly farming efforts and to establish savings and credit accounts using proceeds from butterfly sales.

In the first year of operations, the farmers were supplied with week-old caterpillars from a central farm facility at the National Museums Gede Ruins site near Watamu. They reared a total of 7 species (four *Papilios* and three *Charaxes*), feeding the caterpillars on leaves from bags tied on trees and plastic containers kept at home. Each farmer was visited on a weekly basis by a project extension officer who brought the caterpillars, gave technical advice, and purchased the pupae at a dollar each on a cash-on-delivery basis. A little over 10,000 pupae of 14 species were exported in 1994, earning around US $16,000 in export revenues for Kenya, of which over US $5000 was paid directly to the community farmers. By mid-1995, exports had accumulated to some 17,000 pupae of 20 species earning about US $26,000, and community earnings reached US $10,000.

Future plans for the project include the construction of a live exhibit of butterflies and other invertebrates from Arabuko-Sokoke Forest. This will help to increase the number of tourists visiting the forest and will make its biodiversity more accessible, in addition to providing opportunities for conservation education and local markets for butterflies produced by the villagers.

I. Gordon

products. Gazettement programs are being pursued in Tanzania to ensure that remaining lowland forests areas in the East Usambaras mountains are protected and in Kenya to protect remaining Kaya Forests. Gazettement, however, is not useful unless regulations are enforced. In many cases increased resources are required to implement protection. These resources are often not available, so a more flexible and cooperative approach to protection will be required.

In many cases community-based conservation will be needed. In this case forest managers need to cooperate with local villages, providing them with benefits and incentives to not cut the remaining forests. If the benefits of protection are clear than local villagers are more likely to play an active role in protecting the forests from exploitation by outsiders, unscrupulous companies and corrupt government officials. Under this management regime, the requirements and objectives of national governments and the reasonable needs and requirements of the local people might be combined. This approach, however, will take considerable time and resources to implement in all coastal forests, and, therefore, sites might have to be prioritized according to their biological importance, degree of threat, or other attributes.

Restoring Local Control of Forests. It is possible that restoring the rights to use forest products back to local communities will allow these communities to better manage these resources. This approach may be best for the smallest forests such as the Kaya Forests in Kenya that have a sacred history. It is also possible that this lack of control could lead to the tragedy of the commons' if selfish behavior prevents the achievement of community goals. Conservation programs using community-based management are in their infancy and it is still too early to know whether these methods will work. The possibility remains that the pressure for land may override the perceived benefits of harvesting resources (such as poles, timber and medicinal plants) at sustainable levels, and the forests will be first degraded and finally cleared, by local people, commercial operators or even the government itself.

The Value of Tourism. There are some instances where the significance of a forest is enhanced by giving it additional values. An obvious example is tourism. The number of tourists visiting the Kenya coast is considerable (the number of tourist bednights on Kenya coast was over 2.5 million in 1989), and is a huge potential source of income. In the Arabuku-Sokoke forest, Kenya, tourists from beach hotels are encouraged to visit the forest, providing employment to local people as guides. Already tourism is an important source of income for Gedi National Monument (36,000 visitors in 1990), and Shimba Hills National Park (30,000 visitors in 1990), both of which contain coastal forest habitat. However, one problem with tourism is the fluctuation in numbers (for example the 25% decrease in visitors to Gedi between 1988 and 1989), and its reliance on political stability. Tourism is, however, currently an insignificant factor in other countries possessing coastal forests. In Tanzania, where tens of thousands of tourists visit Dar es Salaam and the coast, almost none visit coastal forests.

Conclusion

The forests along the coast of eastern Africa support a wealth of species, many of which are found nowhere else in the world and some of which have not yet been named - let alone investigated in detail. The ecology of these forests and that of individual species within them is poorly understood, which hinders the development of conservation programs. Due to pressure for agricultural land and wood for timber, fuelwood, carving, charcoal burning, and house building, the coastal forests have been cleared and degraded by human activities. The coastal forests of eastern Africa are perhaps the most threatened forest type in Africa; total area of forest is probably around 3000 km². Gazettement of remaining forest as Reserves, increased protection of Reserves, increased production of forest products outside reserves, and sustainable use of those forests already gazetted represents the best approach to the conservation of the remaining forest patches.

References

1. Brittingham, M.C., Temple, S.A. 1983. Have cowbirds caused forest songbirds to decline? *BioScience* 33: 31-35
2. Burgess, N.D., Mwasumbi, L.B., Hawthorne, W.D., Dickinson, A., Doggett, R.A. 1992. Preliminary assessment of the status, distribution and biological importance of the coastal forests of Tanzania. *Biological Conservation* 62: 205-218
3. Burgess, N.D., Dickinson, A., Payne, N.P. 1993. Tanzanian coastal forests - new information on status and biological importance. *Oryx* 27: 169-173
4. Burgess, N.D, Mlingwa, C.O.F. 1993. Forest birds of coastal forests in East Africa. *Proceedings VIII Pan-African Ornithological Congress* 295-301
5. Burgess, N.D., Muir, C., eds. 1994. *The Coastal Forests of Eastern Africa: Biodiversity and Conservation needs.* London: Society for Environmental Exploration and University of Dar es Salaam (Unpublished)
6. Chasko, G.G., Gates, J.E. 1982. Avian habitat suitability along a transmission line corridor in an oak-hickory forest region. *Wildlife Monographs* 82 (1)
7. Chew, W.L. 1968. Conservation of habitats. *IUCN Publications New Series* 10: 337-339
8. Collar, N.J., Stuart, S.N. 1988. *Key Forests For Threatened Birds in Africa. ICBP, Monograph No.3.* Cambridge: International Council for Bird Preservation
9. Cooke, H.J. 1970. The cave systems of the Tanga Limestone in north-east Tanzania and the influence of former sea-levels on their formation. *Studies in Speleology* 2: 2
10. Gates, J.E., Gysel, L.W. 1978. Avian nest dispersion and fledging success in field-forest ecotones. *Ecology* 59: 871-883
11. Hamilton, A.C. 1981. The Quaternary history of African forests: its relevance to conservation. *African Journal of Ecology* 19: 1-6
12. Hawthorne, W.D. 1984. *Ecological and Biogeographical Patterns in the Coastal Forests of East Africa.* PhD Thesis, University of Oxford

13. Hawthorne, W.D. 1993. East African coastal forest botany. In: *Biogeography and Ecology of the Rain Forests of Eastern Africa.* Lovett, J.C., Wasser, S.K., eds. Cambridge: Cambridge University Press

14. Holm-Neilsen, L.B., Nielsen, I.C., Balslev, H. eds. 1989. *Tropical Forests: Botanical Dynamics, Speciation and Diversity.* San Diego: Academic Press

15. Howell, K.M. 1981. Pugu Forest Reserve: biological values and development. *African Journal of Ecology* 19: 73-81

16. ICBP 1992. *Putting Biodiversity on the Map: Priority Areas for Global Conservation.* Cambridge: International Council for Bird Preservation

17. Janzen, D.H. 1988. Management of habitat fragments in a tropical dry forest: growth. *Annals Missouri Botanical Garden* 75: 105-116

18. Kingdon, J., Howell, K.M. 1993. Mammals of the forests of Eastern Africa. In: *Biogeography and Ecology of the Rain Forests of Eastern Africa.* Lovett, J.C., Wasser, S.K., eds. Cambridge: Cambridge University Press

19. Lovejoy, T.E., Bierregaard, R.O., Rylands, A.B., Malcolm, J.R., Quintela, C.E., Harper, L.H., Brown, K.S., Powell, A.S., Powell, G.V.N., Schubart, H.O.R., Hays, M.B. 1986. Edge and other effects on Amazon forest fragments. In: *Conservation Biology.* ed. Soulé, M.E. Massachusetts: Sinauer Press

20. Lovejoy, T.E., Rankin, J.M., Bierregaard, R.O., Brown, K.S., Emmons, L.H., Van der Voort, M.E. 1984. Ecosystem Decay of Amazon Forest Fragments. In: *Extinctions.* ed. Nitecki, M.H.. Chicago: University of Chicago Press

21. Lovett, J.C. 1986. The Eastern Arc Forests of Tanzania. *Kew Magazine* 2: 83-87

22. Lovett, J.C., Wasser, S.K., eds. 1993. *Biogeography and Ecology of the Rain Forests of Eastern Africa.* Cambridge: Cambridge University Press

23. Madgwick, J. 1989. Somalia's threatened forests. *Oryx* 23: 84-101

24. Milne, G. 1947. A soil reconnaissance through parts of the Tanganyika Territory, Dec. 1935-Feb. 1936. *Journal of Ecology* 35: 192-220

25. Mogaka, H. 1992. *A Socio-economic Survey of Arabuko-Sokoke Forest, Kenya.* Nairobi: Kenya Indigenous Forest Conservation Project, National Museums of Kenya (Unpublished)

26. Moomaw, J.C. 1960. *A Study of the Plant Ecology of the Coast Region of Kenya.* Nairobi: Government Printer

27. Mwasumbi, L.B., Burgess, N.D., Clarke, G.P. 1994. The vegetation of Pande and Kiono coastal monsoon forests, Tanzania. *Vegetatio* 113: 71-81

28. Robertson, S.A., Luke, W.R.Q. 1993. *Kenya Coastal Forests: Report of the NMK/WWF Coast Forest Survey.* Nairobi: National Museums of Kenya/World Wide Fund for Nature (Unpublished)

29. Sheil, D. 1992. Tanzanian coastal forests - unique, threatened and overlooked. *Oryx* 26: 107-114

30. Spear, T. 1978. *The Kaya Complex a History of the Mijikenda Peoples of Kenya to 1900.* Nairobi: Longmans

31. Stuart, S.N., Adams, R.J., Jenkins, M.D. 1990. *Biodiversity in Sub-saharan Africa and its Islands: Conservation, Management and Sustainable Use.* Occasional papers of the IUCN Species Survival Commission No.6. Gland & Cambridge: IUCN - The World Conservation Union

32. Terborgh, J.W., Weske, J.S. 1969. Colonisation of secondary habitats by Peruvian birds. *Ecology* 50: 765-782

33. Vollesen, K. 1992. *Trichaulax* (Acanthaceae: Justicieae), a new genus from East Africa. *Kew Bulletin* 47: 613-618

34. White, F. 1983. *The Vegetation of Africa.* Paris: UNESCO

35. Whitmore, T.C. 1990. *An Introduction to Tropical Forests.* Oxford: Oxford University Press

36. Williams-Pinera, G. 1990. Vegetation structure and environmental conditions of forest edges in Panama. *Journal of Ecology* 78: 356-373

Chapter 14

Riverine Forests

Kimberly E. Medley & Francine M.R. Hughes

'Riverine forest' is an ecosystem that exhibits a composition, structure, and function dependent on the river processes of erosion, sediment transport, flooding inundation, and alluvial deposition (6). Other terms such as 'floodplain', 'riparian', 'gallery,' 'fringing,' and 'alluvial' are synonymous with the term 'riverine', but may be used in reference to a particular section along a river course as it flows from the highlands to the delta. Some terms refer to specific ecosystems such as the 'igapo' and 'varzea' forests along the Amazon (40). In all situations, the physical dynamics of the river strongly influence the biotic dynamics of the ecosystem. 'Forests' within the riverine environment are dominated by trees, which may range in height from 3 to 35 meters. If one views the system from above, at the landscape scale, riverine forests most often exist as linear patches in a mosaic of other natural communities and human land uses. Understanding the relationships among all these communities is very much a part of studying the riverine forest ecosystem and its conservation.

All riverine forests are wetland ecosystems. They are characterized by having saturated soils at frequent (more than once per year) or infrequent (>1 to 100+ years) intervals. This water can come from surface flow, such as during a flood, or from a rise in the saturated zone beneath the soil surface (groundwater table). The river not only provides moisture in a region of low precipitation, but also may create waterlogged and often anaerobic conditions. Its hydrologic regime therefore determines the environmental setting for plant establishment and community development. Riverine forest extent depends on the factors that most strongly affect flooding patterns and the position of the water table. A high water table, extensive flooding, or great curves in the river channel favor the development of a broad riparian zone.

Riverine forests occupy a transitional zone, or ecotone between aquatic and terrestrial ecosystems. Not only do species adapted to upland-terrestrial and aquatic environments meet within the zone, but the ecotonal environment allows for species not suited for either of the adjacent ecosystems (8). These regions are often rich in species. The riverine ecosystem is linear in form as it parallels the stream channel, has either diffuse or sharp edges attributable to the nature of species interactions across the ecotone, and is often broken or fragmented in response to the dynamic nature of the aquatic stream system.

By definition, therefore, riverine forests are confined to moist stream corridors, show strong linkages between local environmental factors and biotic response, and have dynamic disturbance regimes. These factors not only make this ecosystem unique, but also jeopardize its long-term conservation. Natural isolation and fragmentation, coupled with the resource value of riverine corridors for human activities, reduce forest area and the biological diversity of the ecosystem. The objectives of this chapter are to better understand the ecology of riverine forest ecosystems, their patterns of diversity over regional and local spatial scales in East Africa, and their value as a resource for conservation. We rely heavily on completed research, and focus, as a case example, on results from studies conducted in the Tana River forests of Eastern Kenya and especially the Tana River Primate National Reserve. A primary goal is to identify distinctive characteristics, processes, and conservation concerns for comparison with other East African ecosystems.

A Geographic Perspective

Riverine forests will exist in all geographic areas where sufficient precipitation and elevational decline occur to permit surface water to concentrate and flow in a single channel and provide a streamside environment suitable for tree growth. Rainfall patterns and topographic relief are both important in understanding the distribution and ecology of riverine environments. In East Africa, stream flow is primarily determined in the highlands bordering the Great Rift (Fig. 14.1). The region is divided into two main drainage basins. To the west, relatively short rivers flow off the highlands into rift lakes such as Lakes Nyasa, Rukwa, Tanganyika, Victoria, and Turkana. Water may remain in these lakes, forming large evaporative basins (such as in Lakes Rukwa and Turkana), flow north as the headwaters of the Nile River (Lakes Victoria and Tanganyika), or south into the Zambezi River basin (Lake Nyasa). Riverine environments along these streams are usually restricted to the banks of deeply eroding streams. To the east, rivers flow much longer distances from the highlands to the Indian Ocean. These rivers account for much of the diversity in riverine ecosystems, creating environments that are both broad and complex.

Rainfall in the highlands is the principal and sometimes only source of water for stream flow. Equatorial East Africa is unusually dry and has seasonal precipitation (see Chapter 3). During the rainy southeast monsoon times unstable winds bring much moisture to the highland regions (>1400 mm), but a stable easterly flow off the Indian Ocean results in high rainfall along the coast (~1000 to 1400 mm) with a sharp decline inland (1000 to 400 mm; 38). This bimodal precipitation pattern is clearly shown along the Tana River in Eastern Kenya (Fig. 14.2). Northeastern Kenya and southern Somalia have less rainfall due to their relatively greater distance from the Indian Ocean coast and absence of significant highlands in the west.

Nearly all river systems in East Africa, therefore, rise in humid regions and flow through arid regions. As rivers flow through these hot arid areas, tributaries decrease or become nonexistent, water is lost in evaporation and soil infiltration, and in essence, the river loses volume and force. Depending on the amount of water received at the headwaters they may continue as perennial streams (such as the Tana River) or flow intermittently in response to seasonal rainfall patterns (such as the Uaso Ng'iro which dries up downstream in the Lorian Swamp). These rivers provide moisture to the surrounding land through floods and groundwater flow, and thereby provide an environment, linear in form, suitable for ecosystem development atypical of the local climate. Local precipitation may only be important in

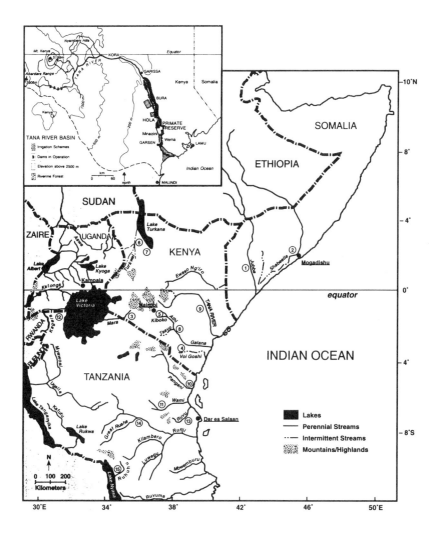

Figure 14.1. Major riverine ecosystems in East Africa. Circled numbers show the approximate locations of the study areas described in Table 14.1. Top left inset is a more detailed map of the Tana River basin in eastern Kenya (modified from 29).

influencing early tree establishment or herbaceous growth. By nature of their semiarid environment, the riverine ecosystems of East Africa are more similar to those along the Nile and Niger rivers of Africa and the Colorado River of the United States. They differ from those along rivers of humid regions such as the Congo, Amazon, and Mississippi rivers.

Variations in riverine corridor width and vegetation composition and stature are dependent on the moisture regime established by the river (Table 14.1). Rivers that descend quickly from the highlands into lake basins are characterized by narrow 'fringing' forests (such as the Mara River and tributaries of the Ruaha and Ruhuhu rivers). In contrast, broad floodplain forests may develop along rivers flowing across

Table 14.1. Descriptive characteristics from some major riverine forest ecosystems in East Africa. The approximate locations of the descriptive accounts are provided in Figure 14.1 by their number.

River System	Ecosystem Characteristics
Somalia	
1. Jubba River at Shoonto and Barako	Northern-most outlier of moist evergreen forest in lowlands of E. Africa; estimated area at 505 ha; several meters to 1 km m wide corridor with an open canopy at 20 m (24,25).
2. Shabeelle River (seasonal, not reaching the delta)	Fragmented into small isolated patches, 10 to 50 m in width; open canopy at 15 to 25 m and dense understory with abundant creepers (10).
Kenya	
3. Mara River System- Grumeti, Orangi, and Mbalageti	Narrow, often discontinuous evergreen forest (41).
4. Voi River (seasonal) in Tsavo East NP	Narrow forest corridor; semievergreen vegetation, 18 m in stature (41).
5. Kiboko River mile 102 on Nairobi-Mombasa Road	Greater tributary (perennial)- deep, narrow erosional channel, with fringing vegetation a few meters in width; Lesser tributary (seasonal)- emerges from spring and flows to an open plains swamp, 1.6 km in width; semievergreen forest and swamp palm (*Phoenix reclinata*) vegetation; floristically poor (5,41).
6. Turkwel River	Floodplain forest > 1 km in width; semideciduous with palms (*Hyphaene compressa*); diversity of forest types associated with floodplain landforms (35).
7. Streams of Gregory Rift Valley- Turkana	Riparian zone about 10 m in width; semievergreen; species richness increases with precipitation (42).
8. Athi-Galana River System	Evergreen-semievergreen vegetation; presence of riverine endemic *Populus ilicifolia* (16,41).
9. Tana River (Bura to Garsen)	Closed floodplain forest about 1 km in width; evergreen-semievergreen trees 14-30 m in height; high diversity; presence of riverine endemic *Populus ilicifolia* (14,29).
Tanzania	
10. Pagani River	Papyrus in bordering permanent swamps; fringing forest with *Acacia* spp. in less flooded areas (16).
11. Wami River	Forms swamp of the Mkata plain; rivers with fringing forest or patches of groundwater forest dominated by *Chlorophora excelsa* and *Barringtonia racemosa*; floodplain dominated by grasses (16).
12. Kagera River	Bordered by papyrus in permanent swamps with only patches of forest (16).
13. Ruvu River	Closed gallery or fringing forest about 200 m in width; evergreen trees at 15-30 m ht; thick wall of climbers along the edge (7).

Table 14.1. Continued

14. Ruaha/Rufigi System	Semi-evergreen forest at 15 to 30 m height, often with scattered emergents and a thick understory--best developed along upper reaches of Ruaha; areas of inundation are dominated by a single species (*Combretum constrictum*) (7,39).
15. Ruhuhu River	Dense fringing forest confined to narrow valleys (7).

the eastern plateau and coastal plain (such as the Jubba, Tana, Athi-Galana rivers). Rivers of low or intermittent flow maintain fewer species (for example Kiboko, streams of the Gregory Rift) and may be unsuitable for establishment by evergreen trees (for example certain streams in the Selous Game Reserve). Papyrus-dominated swamps occur where a near absence of topographic decline limits channelized flow (for example Pangani and Kagera rivers). Note that variations in the factors that establish the river system (topography, precipitation, and river flow) also determine the great diversity.

Biogeographic Patterns of Biological Diversity

Riverine forests in East Africa are isolated remnants of a continuous rain forest belt that extended between the Congo (Zaire) basin and the Indian Ocean coast. Forest expansion occurred during warm wet periods between times of glacial ice advance (around 31,000 to 26,000 BP) and also during a warm post-glacial period (around 8000 BP). Severe climatic drying later (around 4000 BP) isolated evergreen forest in montane, coastal, and riverine localities where they were further fragmented by human land use and natural disturbances (such as fire and river meanders). These past and ongoing processes are critical factors in the current distribution of plants and animals (23). Today, riverine corridors in East Africa maintain isolated islands of tropical forest vegetation in a semiarid environment. Biogeographic patterns of diversity among the rivers and even along a river corridor are a result of species dispersal, establishment, and persistence. Species distributions are disjunct. Depending on the time since isolation and rates of speciation, some species may only occur in or be endemic to a portion of a single river system. Each isolated patch is a refuge for the species that are present (23); if the habitat disappears, species will become locally extinct because the distances required for recolonization are too great and the intervening habitat too inhospitable.

An examination of species distributions, therefore, provides insight into the history of a particular site (for example how long since isolation) and the dispersal routes or continuity among riverine forest patches. Most species distributions in the woody flora of East African riverine forests show one of the following patterns:

(1)   Pan- African- no affinity with a particular location;

(2)   Guinea-Congolian- species showing an affinity with west-central Africa;

(3)   Zanzibar-Inhambane- species showing an affinity with coastal East Africa; or

(4)   Somalia-Masai- species showing an affinity with the more arid lands of the northeast (Table 14.2).

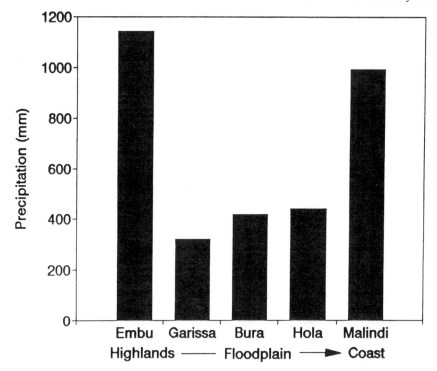

Figure 14.2. Mean annual precipitation for four locations along the Tana River and one location at the coast.

'Affinity' suggests that species originated and radiated from that location. Evergreen trees from the tropical moist forests of west-central Africa (*Pachystela msolo, Garcinia livingstonei*) or coastal Africa (*Sorindeia madagascariensis, Mimusops obtusifolia*) occur on the floodplains of large, perennial streams, while woody plants of the Somalia-Masai center (*Acacia elatior, Capparis* spp., *Maerua* spp.) predominate in the more arid stream environments. Species composition in a particular riverine setting is dependent on the interplay among factors influencing dispersal and persistence. Of the woody plants in the Tana River Primate National Reserve, approximately 30% are Pan-African, 31% Zanzibar-Inhambane, 12% Guinea-Congolian, 16% Somalia-Masai, and <6% endemic (29). These data suggest a favorable environment for evergreen forest species, dispersal from three biogeographic centers into the Tana River floodplain, and a relatively short period of isolation and speciation.

Bird and mammal species show similar distribution patterns and variation among the riverine environments. The Tana River forests support a rich array of animals that are now biogeographically isolated from the tropical moist forests of the coast and west-central Africa (2). Examples include the red duiker and several forest primates. Most of the large mammals found in riverine forest reside on the adjacent savanna and show varying relationships with the forests from extensive use (for example waterbuck, buffalo, elephant) to little use (oryx, zebra). Endemism, especially in birds and primates, is more common than for plants. Of special importance in the Tana River forests are the endemic and endangered subspecies of

Table 14.2. Biogeography of representative woody plant species in riverine forest ecosystems of East Africa. All plants are trees >10 m height unless otherwise indicated.

Guinea-Congolian Center (West-Central Africa)
Acacia robusta subsp. *usambarensis (*Mimosaceae*)*
*Albizia gummifera (*Mimosaceae)
*Albizia glaberrima (*Mimosaceae*)*
*Garcinia livingstonei (*Guttiferae*)*
*Pachystela msolo (*Sapotaceae)
*Oxystigma msoo (*Caesalpiniaceae*)*

Zanzibar-Inhambane Center (Coastal East Africa)
*Alangium salviifolium (*Alangiaceae)
*Lecaniodiscus fraxinifolius (*Sapindaceae*)*
*Majidea zanguebarica (*Sapindaceae)
*Mimusops obtusifolia (*Sapotaceae*)*
*Newtonia erlangeri (*Mimosaceae*)*
*Polysphaeria multiflora (*Rubiaceae)
*Rinorea elliptica (*Violaceae*)*
*Sorindeia madagascariensis (*Anacardiaceae*)*
*Spirostachys venenifera (*Euphorbiaceae*)*
*Sterculia appendiculata (*Sterculiaceae*)*

Somalia-Masai Center (Northeast Africa)
Cadaba spp. (woody vines) (Capparaceae)
Capparis spp. (woody vines) (Capparaceae)
Maerua spp. (woody vines or shrubs) (Capparaceae)
Indigofera schimperi (shrub) (Papilionaceae)
Phyllanthus somalensis (shrub) (Euphorbiaceae)
Polysphaeria parvifolia (shrub) (Rubiaceae)
Terminalia brevipes (small tree) (Combretaceae)

Pan-African
*Afzelia quanzensis (*Caesalpiniaceae)
*Antidesma venosum (*Euphorbiaceae*)*
*Cassia singueana (*Caesalpiniaeae*)*
*Diospyros mespiliformis (*Ebenaceae)
Ficus spp. (*F. capraefolia, natalensis, scassellatii, sycomorus, bubu*) (Moraceae)
*Lannea schweinfurthii (*Anacardiaceae*)*
*Oncoba spinosa (*Flacourtiaceae*)*
Phoenix reclinata (palm) (Palmae)
Trema orientalis (Ulmaceae)

the red colobus monkey (*Colobus badius rufomitratus*) and crested mangabey (*Cercocebus galeritus galeritus*).

Species diversity in riverine ecosystems may be described by the number of species (species richness), proportions of species from different biogeographic centers, and by the number of rare occurrences. Compared with tropical rain forests, the isolated, more environmentally stressed riverine environments support fewer species. In the Tana River forests, field studies have documented about 175 woody plant species in 49 families, 250+ species of birds, and at least 57 species of mammals (22 ungulates). This species complex is primarily derived from three biogeographic centers of diversity, and consists of four endemic primates (2 are

Box 14.1. Floodplain Hydrology and Geomorphology

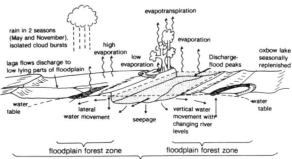

FLOODPLAIN HYDROLOGY AND GEOMORPHOLOGY

Hydrological inputs and outputs in a semiarid floodplain are shown in the diagram above. Hydrologic inputs to riverine forest ecosystems are principally from the stream channel through lateral seepage to the underlying groundwater table. Local precipitation during the rainy seasons is scattered and unpredictable, possibly resulting in some channelized laga flows but little to no water input to the stream. During flood periods, which can last several weeks, the water level rises above the river bank and flows laterally across the floodplain. Local water tables are recharged and floodplain sediments are replenished. Evaporation from the soils and river and evapotranspiration from the vegetation are important hydrological outputs in semiarid areas. These outputs will reduce downstream flows; the river will lose volume and force.

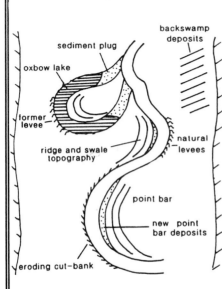

To the left is a plan diagram of meandering floodplain features. As a river channel meanders through the floodplain, it erodes, transports, and deposits alluvial sediments. Natural levees adjacent to the channel are composed of coarse sands deposited during overbank floods. They are the highest features in a floodplain, approximately 3 to 5 m meters above the river channel. In contrast, fine clays are deposited in backwater swamps, where lateral flow velocity is very low. Ridge and swale topography, also known as meander scrolls, is formed from progressive deposition on a point bar as the river migrates laterally across the floodplain. Periodically, meanders are cut-off during the floods. They contain oxbow lakes, which are gradually filled in as sediments are deposited during successive floods.

> Following meander cutoff, there is a temporary decrease in river sinuosity and increase in local channel gradients until a new meander forms.
>
> K. Medley

endangered), several rare endemic birds (for example white-winged apalis, *Apalis chariessa*), and some unusual plant occurrences. These include *Populus ilicifolia* found only along three river systems in Eastern Kenya, *Pachystela msolo* and *Uncaria africana*, found elsewhere only in the Usambara mountains in Tanzania and west-central Africa, and *Cynometra lukei*, a newly described tree endemic to riverine environments (29). From a biogeographic perspective, riverine forest ecosystems are an important refuge for a diverse range of species.

Physical-Environmental Determinants of the Ecosystem

A unique feature of riverine ecosystems is their linear form along rivers. Accordingly, riverine forests process large amounts of energy and materials that flow via the river channel from upstream highlands to a downstream basin, lake, or ocean. Stream energy depends on water volume and the topographic gradient, and material transport depends on the erosion and deposition of sediments that range in texture from very coarse sands to very fine clays. Functionally, therefore, upstream land areas are closely linked with areas downstream through their common interactions with stream processes.

Seasonal climatic conditions determine hydrologic inputs to the stream system through precipitation, and outputs through evaporation (Box 14.1). River discharge is a measure of the volume of water passing through a given section of the stream in a given unit of time. In response to rainfall in the upper catchment, water level will rise and fall, causing vertical changes in river levels. In the Tana River basin, the average annual discharge at Garissa is 6105 cubic meters for the period 1944 to 1978 (12; see Fig. 14.1). Below Garissa, accounting for approximately 74% of the basin, the river gains no additional water from tributaries. Some ephemeral streams (lagas) bring water to the floodplain following localized storms, but they do not reach the main river. In this lower basin, therefore, the river loses water continuously through seepage and evaporation. Annual rates for evaporation are between 2200 mm and 2400 mm with little seasonal variation in the warm temperatures. These rates are similar to values recorded in the Turkwel floodplain area of northern Kenya (35) and are representative of semiarid East Africa.

Along a downstream gradient, there is a decrease in the ratio of erosion-to-deposition in response to the decreasing topographic gradient and increasing sediment load. In upstream highland areas, erosion is the dominant force. Stream influences are restricted to a narrow bordering fringe and limited to rises and falls in the river level; little lateral seepage occurs in these erosional gullies and overbank flooding is rare. As the rivers descend from the highlands, erosion becomes less dominant. A more gradual slope and greater sediment load reduce the velocity and consequent energy force of the river. Creation of the downstream floodplain zone begins once a threshold sediment load is reached and deposition becomes a dominant process. At the mouth of the river, stream velocity is significantly reduced and a delta is formed through deposition of the sediment load.

Whereas in the upstream areas, stream energy is directed principally in a vertical direction as the river erodes its channel, in the floodplain zone considerable energy is directed laterally in response to river meanders and overbank flooding. The formation and downstream migration of meander belts depend on the magnitude and frequency of floods (as defined by 6). High magnitude, infrequent floods result in channel repositioning and meander movement. Low-to-medium magnitude, frequent floods build the floodplain vertically through the deposition of sediments from overbank lateral flows.

The combination of different magnitude/frequency flooding patterns and associated sedimentation means that floodplain areas are not only physically active but are characterized by a complex mosaic of landforms (see Box 14.1). Successive floods deposit sediments of different texture on the floodplain causing local elevational gains and changes in drainage characteristics. Coarse sediments are deposited near the stream channel (called levees) in response to the decreasing stream velocity of overbank lateral flows. Fine clays are deposited in the more distant backwater swamps, where lateral flow velocity is very low. Sediment load and its textural composition depends primarily on the geology and the erosive abilities of the river in the upstream basin and a continual reworking of downstream alluvial deposits. Data from the Tana River show a mean suspended sediment load of 8.5 million tons/year for the period between 1948 and 1965 at Garissa (data from the Ministry of Water Development, Republic of Kenya), which is high relative to the catchment size. This suspended load consists of about 30% fine sand (0.06 to 0.2 mm), 40% coarse silt (0.02 to 0.06 mm) and 20% fine or medium silt with little clay. Data collected from a flood in the early 1980s, however, show that a high percentage of the suspended sediment was finer than 0.06 mm, with little sand. Dam construction in the upper basin of a river may directly affect sediment load and composition by effectively capturing the coarser sediments (12).

Alluvial soils, by nature of their continual replenishment by river sediments, remain mostly undeveloped. A core through the floodplain shows layers of sediment rather than the development of well-defined soil horizons. While mineral concentrations in the sediment are usually high, frequent flooding and high decomposition rates reduce organic carbon percentages throughout the floodplain area. In the Tana River floodplain, organic content of surface soils averages between 1 and 4%, and less than 1% at depth (12).

The downward loss (leaching) of dissolved nutrients and the evaporative accumulation of calcium (calcification) or salts (salinization) at the surface are the principal soil-forming processes in semiarid floodplains. If the substrate is well-drained, as in the sandy levees, easily dissolved nutrients such as potassium, calcium, and magnesium are leached from the surface horizons following flooding. Neither salts or moisture are found near the surface. This is in contrast to clay soils, which are common in backwater swamps and oxbows. Poor drainage not only inhibits the leaching of nutrients, but also, through evaporation, may result in the accumulation of nutrients and salts at the surface. In the Tana River Primate National Reserve, there is little variation among freshly-deposited sediments from different locations but significant differences in the subsequent nutrient leaching and retention patterns of pre-flooding surface soils (Fig. 14.3). Surface soils 1, 4, and 5 are from levees and point bars that are sandy and well-drained, whereas surface soils 2, 3, and 6 are from clay-backwater depressions or clay-covered levees on the edges of cut-off meanders. Much higher levels of calcium at the surface of the clay soils were documented, suggesting an accumulation of this nutrient. Alkaline pans, that are also observed in the Tana River Primate National Reserve, may form in extreme cases of evaporation and nutrient accumulation, limiting nearly all plant growth and clearly sharpening the ecotone between forest and plains vegetation.

Just as the arrangement of floodplain landforms shift in response to stream processes, soil conditions change rapidly from year to year and through the year in association with the flooding cycle. In the case of the Tana and other East African rivers receiving a bi-annual flooding regime, these spatial and temporal changes are complex and difficult to monitor.

Ecosystem Structure, Function, and Dynamics

Strong associations between the physical environment and biota result in gradients of change in the ecological characteristics of riverine forests over regional and local spatial scales. Water availability is of principal importance in determining community structure. Along gradients of decreasing moisture, the growth form of trees may be classed as:

(1) Broadleaved-Evergreen- trees do not experience synchronous leaf fall;
(2) Broadleaved-Semievergreen- trees that lose leaves synchronously but replace them immediately;
(3) Broadleaved-Deciduous- trees that lose leaves synchronously and remain dormant for an extended time period.

The relative proportions of trees in a forest exhibiting a particular growth form determine overall forest physiognomy (Table 14.1). Similarly, the relative height and closure of the forest canopy vary in response to available moisture, ranging from emergent trees above 30 m in height, to closed canopies 18 to 20 m tall, to open woodlands with trees <10 m tall. Trees will often produce multiple basal stems (coppice) rather than a single large stem. Most tree species acquire water by extending their roots to the groundwater table. They vary in their tolerance of flooding conditions (14,15).

Ecotonal edges between open stream and savanna environments are characterized by a dense understory of small trees, shrubs, vines and possibly palms. Understory density within forest patches decreases with closure of the upper-canopy, distance from the edge, increased disturbance by large mammals (such as from elephant and buffalo), and long periods of inundation by flood waters. Ferns, mosses, and epiphytes are usually absent because of shortages in surface moisture and local rainfall. When compared to tropical rain forests, even the best developed riverine forests in East Africa have a more open canopy of lower height, smaller trees that are often coppiced, great variability in the density of the understory, and an absence of plants dependent on a fairly continuous supply of surface moisture.

Differences in forest composition and structure are best explained by the variable conditions of riverine landforms (14,26,29,33). In the Tana River Primate National Reserve, forests vary in community structure from point-bar pioneer vegetation with *Populus ilicifolia*, to low-levee positions dominated by stands of *Pachystela msolo* and *Ficus sycomorus* along the river, to high-levee positions with a more open and diverse canopy, to clay-backwater swamps dominated by trees adapted to

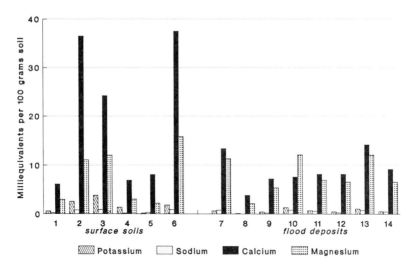

Figure 14.3. Soil nutrients measured in surface soils (12) and in 1988 flood deposits (28) collected in the Tana River Primate National Reserve, Kenya. Columns 1, 4, and 5 are from levees and point bars; columns 2, 3, and 6 are from clay backwater basins or from abandoned clay-covered levees adjacent to cut-off meanders.

long periods of inundation (*Garcinia livingstonei, Mimusops obtusifolia, Cynometra lukei*), to the plains edge with an open canopy of more drought-resistant species (*Acacia rovumae, Salvadora persica*). Variations in forest structure are also apparent in areas of recent disturbances (such as oxbows, clearings). The arrangement of these community types may be observed along a transect crossing the respective floodplain landforms (Fig. 14.4).

Cover from intense heat, protection from predators, access to water, and a variety of food resources make riverine forest areas important for animals. Permanent residents may include several primate species, other medium-sized mammals (such as red duiker, bush pigs, bush squirrels, bushbuck, genets) and many bird species (see 2 for Tana River, Kenya and 25 for the Jubba River, Somalia). Marsh (26) calculated that the estimated biomass of primates in one forest area of the Tana River Primate National Reserve was nearly equivalent, on a per area basis, to the large-mammal biomass of the Serengeti ecosystem. Most of the tree species (such as *Ficus* spp., *Sorindeia madagascariensis, Garcinia livingstonei, Mimusops obtusifolia, Acacia* spp.) are dispersed by mammals or birds, illustrating close interrelationships between the fauna and flora (see Box 15.1). Seasonally, riverine ecosystems become a refuge for animals that migrate in from the surrounding savanna-bushland. These animals have a profound effect on community structure. Elephants, for example, often completely destroy the understory vegetation and also serve as a principal dispersal agent for some tree species (such as doum palm, *Hyphaene compressa*). The recent population crash in elephants along the Tana River, and local extinction of the black rhinos, have allowed for much recent growth in the forest understory (27). Community structure may be as greatly modified by the biota as by the physical environment.

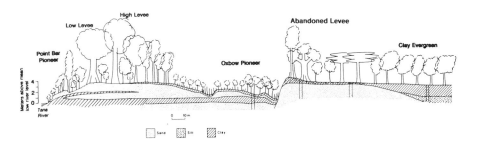

Figure 14.4. Representative profile across the Tana River floodplain, Kenya (compiled from 12).

Little empirical data exist on production, decomposition, and nutrient cycling for riverine forests in East Africa. Under a similar temperature regime, regional differences in these processes are determined mostly by the soil moisture regime, duration of floods, and the nutrient content of the sediment load. Compared with surrounding semiarid uplands, riverine localities (fringing or floodplain) are productive. Factors that lower production within riverine forests are long periods of drought (such as limited access to groundwater), long periods of inundation by flood water, and poor nutrient concentrations in river water and deposits. Production is higher along perennial streams in contrast to intermittent streams, on well-drained levees rather than on poorly drained backwater swamps, and on floodplain soils of high nutrient content (dependent on upstream geology) and low concentrations of salts. On optimal sites tree growth is impressive; annual diameter increment averaged >1 cm/year for open-grown trees (such as *Acacia robusta* subsp. *usambarensis* and *Ficus sycomorus*) in the Tana River Primate National Reserve (12,31).

Mineral cycles are driven by the erosional and depositional processes of the river. In the Tana River forests, infection of tree roots by symbiotic mycorrhizae appear high, further facilitating access to nutrients (33). Soils are basic, but not saline under mature forest. When forests are cleared, or along the upland edge, high evaporation promotes the capillary (upward) transport of salts to the surface and sharply limits growth by riverine-forest species.

Responses by the vegetation to seasonal changes in the climate are complex. Most tree species are dependent solely on lateral seepage from the stream. The resulting water table fluctuates in response to changes in the river level, which is responding to precipitation received in the highlands. There are obvious time lags between rainfall events, local environmental conditions, and vegetation response. Furthermore, seasonal changes in moisture may be less pronounced for trees that have deep roots. Examples of seasonal patterns that were observed in the Tana River forests include: negative tree growth (shrinkage) during periods of low river flow in *Sorindeia madagascariensis* (28), synchrony in the phenology of many tree species with flowering most abundant during times of low river flow (19), and production of fruits coincident with the long and short rainy seasons in *Sorindeia madagascariensis* and *Garcinia livingstonei* (19). Different seasonal responses in different species complicate any generalizations at the community level.

Changes in the composition and structure of a riverine forest through time depend on changes in the site conditions and the habitat preferences or tolerances of the tree species. Community development, or succession, is an expression of establishment, growth, and persistence by different plant and animal species. In riverine environments, a successional pathway is both determined and disrupted by hydrologic and geomorphologic factors that greatly influence the competitive abilities of the biota. There is a complex interplay between environmental (allogenic) and biotic (autogenic) factors that influence successional dynamics in floodplain forests.

Successional changes on the floodplains of meandering streams result in an array of community types in response to hydrologic-geomorphic changes associated with alluvial deposition (floodplain building) and meander movement. Based on observations and studies in the Tana River forests, relationships between hydrologic and geomorphic influences and the vegetation response were modeled for a floodplain-forest ecosystem (Fig. 14.5). Low-to-medium magnitude floods contribute to floodplain building through the creation of point bars and deposition on low-levee and backwater swamp areas. Forest development follows a primary succession sequence: a) pioneer, shade-intolerant species (such as *Populus ilicifolia, Ficus sycomorus*) establish on point bars; b) progressive invasion, establishment, and recruitment by a greater diversity of shade-tolerant species, which are adapted to well-drained sands that form low-high levees; c) possible development of species adapted to the clay and long periods of inundation associated with low backwater basins and behind areas of more recent levee establishment; and d) an open forest along the plains edge dominated by species adapted to the low water table levels (very deep root systems) or the low precipitation regime. High magnitude, infrequent floods disrupt progressive trends in succession by drastically displacing a forest community's position with the stream channel. One example is through a meander cut-off, which not only opens an old river channel to primary succession, but also will either improve or degrade conditions for forest development at all adjacent locations. Local precipitation, while not directing successional pathways, may be important for all communities in early plant establishment. Local rainfall differences may partially explain the floristic changes along river courses (for example the Tana River from Garsen to Bura; 29) or among intermittent tributaries within a region (streams of Gregory Rift valley; 42).

The riverine ecosystem along floodplain corridors is very dynamic. Environmental changes, through floodplain building and meander movements, constantly alter the relative competitive abilities of particular tree species. No site is characterized by a stable community type, and indeed it has been speculated that the persistence time of a particular community is often under 100 years. Furthermore, conditions suitable for regeneration by species, or even a community-type, relies on infrequent geomorphic-hydrologic conditions. Establishment may be episodic and related to particular flooding conditions that have no predictable time period for recurrence. While over a regional scale, the relative proportions of community types may remain similar over long time periods, the mosaic is constantly shifting. It is very difficult to predict forest conditions at a location through time.

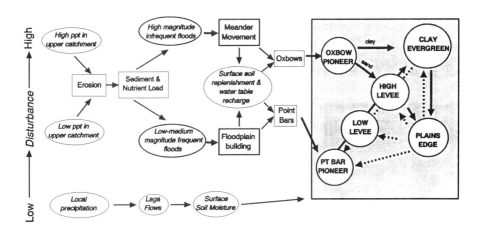

Figure 14.5. Descriptive model showing hydrologic (*ellipses*) and geomorphic (*rectangles*) influences on successional relationships among forest-community types (*circles*) on the Tana River floodplain. Within the shaded box, solid lines show primary succession pathways and dotted lines indicate community changes in response to major riverine disturbances (such as meander movements).

Human Influences on the Ecosystem

An examination of human activities and their effects on the distribution and ecology of riverine forest ecosystems must consider both local and regional activities. Local human populations have direct effects on the distribution of forests through their management of land resources for agriculture and grazing, and on forest composition and structure through their use of extractive resources. At a regional scale, alterations of the river hydrologic regime and consequent effects on forest ecology are imposed through the construction of dams. Changes are also imposed through the development of irrigation schemes in the vicinity of riverine forests. The upstream-to-downstream linkages along a river corridor and the importance of rivers as productive zones in otherwise arid environments complicate relationships between humans and riverine forest ecosystems; often there is a coupling of local and regional influences.

Floodplain agriculture relies mostly on the moisture and nutrient replenishment provided by the biannual rises and falls in the river level. Crops often correspond with particular landforms. For example, a typical planting regime by the Pokomo along the lower Tana River may include the cultivation of rice near the river or a recently-formed oxbow, a mixture of corn and beans on the point bars, and bananas and mangoes on the lower levees. Flood recession agriculture necessarily competes with pioneer forest on the prime sites for establishment (see Fig. 14.5). The probability of flooding declines with distance from the river and crop production depends more on the less predictable rainfall regime. Typically a family will have rights to several different farm plots as a strategy for dealing with changes in the flooding regime over the long term or even during a given year (33). When population and/or environmental pressures increase, mature forests are cleared in the effort to reestablish land rights and potentially obtain crops during the long or short rainy seasons. While agricultural activities shift in response to changing hydrologic conditions, resting land for nutrient and moisture replenishment is not typical. Cultivation can continue at a site that floods for many years (33). Any given floodplain is characterized by a mosaic of land cover types in response to river meanders and corresponding shifts in agricultural activities and forest community types (Fig. 14.6).

Along river sections in more arid areas (for example Turkwel River, upstream along the Tana River; ~<300 mm of rain per year) pastoralism becomes the more dominant land use. Furthermore, all river systems in East Africa are vital to pastoral groups during dry seasons for pasture, fodder, and water. Over 30% of all livestock in the Turkana district of northern Kenya rely on riverine woody vegetation (3). Family group ownership of individual *Acacia tortilis* trees, 'Ekwar,' is a unique management strategy employed by the Turkana that better ensures a sustainable grazing resource (3,34). Along the Tana River, the Orma and Wardei are two pastoral groups that reside and graze their cattle primarily off the floodplain, but rely on the river during the dry seasons. Burning of riparian zones, in an effort to improve grassland development, has been documented for pastoral groups along the Turkwel (see 34) and observed along the Tana River. While burning may kill young trees, regeneration from seed banks in the soil may be improved and some forest species (such as the palm *Hyphaene compressa*) benefit from the fires. It is interesting to note that grazing by livestock near the Turkwel controls a serious pest of *Acacia tortilis*, the bruchid beetle, enhancing successful establishment by the tree in this region (22). Unlike indigenous herbivores, which are normally regulated in their number by limiting factors in the environment, the livestock of pastoral groups are less vulnerable and can impose significant impacts when their number exceed the carrying capacity of the rangeland (21; but see Chapter 10 in this volume).

Pastoral and agricultural groups rely on the riverine forest for an array of botanical resources. Based on their knowledge of uses, local populations assign names to plants that often correspond directly with their respective taxonomic classification at the species or even subspecies level (see 3 and 34 for the Turkwel River; 25 for the Jubba River; 10 for the Shabeelle River; 11, 17, and 30 for the Tana River). Plant uses are diverse and may include: food for humans and livestock, firewood, construction material (for example poles, fencing, furniture), technology (for example rope, glue, tools), medicinal remedies, and many miscellaneous uses (for example beehives and perfumes). Only a few products (mats, baskets, and honey) are marketed for extra income (25,30). A study of Pokomo plant uses in the Tana River Primate National Reserve identified 98 plant species with one or more uses, including 52% of all woody plants recorded in the forest flora (30), and approximately 77 uses were identified for plants along the Jubba River (25). Resources are acquired through the removal of a whole plant, such as the felling of a large tree for a canoe, or a plant part, such as a single tree stem for a pole or a few leaves for a medicinal remedy. These activities are extractive, resulting in more subtle influences on the composition and structure of riverine forest rather than a change in its land area. Forest degradation from extraction is clearly related to the intensity of use, as determined by individual rates of selection and population pressures. For instance, in more arid zones, human pressures on riverine woody

Figure 14.6. Aerial view of a riverine landscape mosaic along the Tana River, Kenya between the Tana River Primate National Reserve and Hola (see Fig. 14.1). Note especially on the photo the small patch of riverine forest near the recently formed oxbow, the small village settlement in the photo center, and the cultivated point bars left of the village and along the river in the lower left.

vegetation for fuelwood and/or charcoal may not only degrade forest structure significantly but also reduce its potential for regeneration (13).

Regional activities of particular importance in riverine ecosystems include upstream dam construction and large-scale irrigation schemes. Dams are constructed in steep upstream catchments in order to capture high velocity flows for the generation of hydroelectric energy. The five dams constructed between 1968 and 1987 on the Tana River and the Turkwel Gorge Dam completed in 1990 account for 80% and 20% of the hydroelectric energy in Kenya, respectively (17). One primary effect of these dams is a reduction in peak flows. For example, hydrographic flows simulated for the Tana River after operation of the Masinga dam show high flows that are only 30% higher than average, in contrast to 58% higher without the dams (1). A second effect is on the sediment load. Sediments, especially coarse-textured sediments, eroded in the headwaters are now mostly trapped behind the dams. Reservoirs are rapidly filling and the composition of sediments downstream is greatly altered. A third effect is greater evaporation from upstream reservoirs and therefore lower mean flows downstream. The Turkwel floodplain and river bed are now completely dry for most of the year. These consequent changes in hydrologic flows and sediment load have serious implications for meander movement, floodplain development, and water table recharge. Some predicted concerns for forest regeneration are: a) pioneer species (for example riverine endemic *Populus ilicifolia*) that only establish on fresh point-bar deposits or oxbows may be reduced; b) a decrease in river meanders reduces the overall length of the channel and significantly reduces the land area suitable for forest development; c) the absence of major river movements as imposed by high-magnitude floods will reduce the complexity of the floodplain environment; and d) the width of the corridor suitable for forest (or agriculture) will be reduced by lower water tables. The whole riverine forest belt could be reduced in both extent and diversity.

Irrigation schemes developed near river floodplains modify large land areas, reduce traditional wet-season grazing areas, and increase pressure on riverine woodlands by the grazing of domestic stock and for fuelwood by the new occupants. The Bura Irrigation Settlement Project, established in 1977 by the Government of Kenya, World Bank, and other donors is an example of these influences. Approximately 6700 ha of land were developed for the scheme, on grazing lands formerly used by the Orma, and it is now occupied by about 2300 tenant families (~15,000 people; 17). While Bura is the only scheme with a program for irrigated forestry to provide fuelwood, plantations were established late. The pressure on local resources has been intense with an estimated loss or degradation of 8000 ha of riverine forest along the west bank and considerable impacts on the regeneration potential of mature forest (12,17). More recently, development of the Tana River Delta Irrigation Scheme near Garsen isolated riverine forest patches and now jeopardizes the southern range limits of the endangered Tana River red colobus and crested mangabey and local forest resources (32). Even though irrigation projects may indeed improve the sustainability of crop production along semiarid rivers, their potential consequences on surrounding resources and the carrying capacity of the landscape need to be carefully considered. In the long term they may reduce the sustainability of natural ecosystems.

Conservation Issues

Conservation of East African riverine forests must include careful management of species populations, community diversity, and the landscape mosaic. The biota, coupled with a dynamic physical environment, form the special qualities of this ecosystem. Riverine forest ecosystems, by virtue of their dependency on the moisture regime provided by the river, serve as a refuge for many plant and animal species adapted to more humid environments. While overall species richness is not as high as some tropical forest locations, the characteristic groupings of species with affinities to different geographic localities provides insight on the biogeography of East Africa. The protection of rare riverine endemics, such as *Populus ilicifolia* and the endangered primates in the Tana River Primate National Reserve is clearly important because their extinction will reduce global species diversity. Equally important, however, is the protection of species that are 'keystone' or 'critical' to the maintenance of riverine communities. For instance, a significant decline in *Acacia tortilis* or *Ficus* spp., common along many rivers in East Africa, would have serious implications because of their value as a food resource for many animals, including livestock. A diversity of forest-community types occurs among river systems, upstream-to-downstream along a river, and locally according to floodplain geomorphology. Ecological surveys of community structure and composition show distinct differences between intermittent and perennial streams, erosional and depositional channels, and point bar, levee, oxbow, and backwater swamp settings. The disturbance regime, imposed by meander movements and human activities, shifts these forest communities and fragments forest areas within the floodplain corridor. The size, shape, and number of individual patches, and their degree of isolation are important parameters describing the occurrence of forest vegetation in the landscape mosaic. Forest area and fragmentation are important determinants of community diversity, which in turn effect population numbers. Because of these interdependencies, species populations, community diversity, and the position of forest patches in the landscape mosaic deserve equal attention in resource conservation.

Riverine environments, especially in semiarid East Africa, represent productive zones that are of critical importance to human populations. While human activities may be viewed as a negative impact on diversity, their dependency on the land and forest resources of riverine environments make them an integral part of the ecosystem. The construction of dams and implementation of irrigation schemes are designed to promote national development. In Kenya, dams were constructed on the Tana and Turkwel rivers in order to reduce electricity imports from Uganda, and the goals of the Bura Irrigation Settlement Project were to provide a source of foreign exchange and subsistence support for a growing rural population (17). Along rivers, human activities are not only concentrated in a narrow corridor, but are also linked longitudinally. Upstream activities influence downstream processes. Increasing pressures on riverine forest ecosystems are driven principally by the resource needs of a rapidly increasing human population. Unless alternative approaches are found for meeting these needs, human-environment conflicts will continue to jeopardize sustainability of this ecosystem.

Box 14.2.   Tana River Primate Conservation

The Tana River Primate National Reserve was established in 1976 to preserve the best remaining riverine forest along the Tana River, Kenya and the primary populations of the endangered Tana River red colobus (*Colobus badius rufomitratus*) and crested mangabey (*Cercocebus galeritus galeritus*; 26). Tana River Primate National Reserve is one of the smallest reserves in East Africa (171 km^2), and the forest area, which occurs as patches near the river, is much smaller (9.5 km^2 in about 26 patches). A resurvey of the primates, conducted ten years after the establishment of the Tana River Primate National Reserve, showed an 80% decline in the red colobus and a 25% decline in the crested mangabey populations (27), a crash that may have been due to declines in forest cover before the reserve was established and severe drought in the early 1980s (9). Between 1960 and 1975 forest cover declined 56% in the south-central sector of the reserve, fragmenting five forest patches into fifteen. Forest loss and fragmentation, which have direct and indirect effects on species survival, were the result of major meander movements and a security-driven move to the west bank by the Pokomo. These factors continue to be a threat on resource protection because the reserve is probably too small relative to the dynamics of the river system, and also because of shifting and expanding agricultural activities by the resident Pokomo families. Long-term preservation relies on managing the dynamics of landscape change in order to sustain species populations.

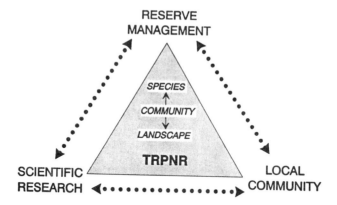

One approach, now proposed for support by the World Bank Global Environmental Facility and the Kenya Wildlife Service, seeks to integrate reserve management with scientific research and the local community (20). Scientific research on the mechanisms of change at the species, community, and landscape scales and applied research on forest restoration or local propagation and marketing of ethnobotanical resources contributes to reserve management, employs local people in the monitoring of ecosystem processes, and promotes sustainable resource use. Reserve management provides infrastructure and security to promote the study of reserve resources and contributes toward local socio-economic development. The local community contributes significantly to our scientific understanding of riverine forest resources and is a principal participant in resource stewardship. Presently, the primate populations appear stable to increasing (18). Through the collaborative efforts of scientific research, reserve management, and the local community, the conservation of natural resources in this special and very dynamic riverine ecosystem may be achieved.

K.E. Medley & F.M.R. Hughes

Conservation strategies, however complicated, should aim to couple protection of ecosystem quality with compensation for the resource needs of local and regional populations. Virtually all rivers in East Africa pass through national parks and reserves: Kora, Arawale, & Tana River Primate National Reserve/Tana River, Samburu/Buffalo Springs, Tsavo East and West/Galana River and Voi Goshi, Masai Mara and Serengeti/Mara River, and the Selous Game Reserve/Ruaha River. These reserves are important in the protection of 'core' riverine forests, as defined by their overall community diversity and the occurrence of rare populations. The linear form of rivers, however, limits the length of the riverine corridor and actual forest area under protection. Upstream-to-downstream patterns of diversity and connectivity among forest patches are lost due to greater agricultural and grazing pressures on adjacent land areas. Furthermore, meander movements may shift prime floodplain areas outside of the reserve, and upstream dam construction alters the flooding regime and land area for forest development. Community diversity and species populations are necessarily in jeopardy because of regional and local disturbances acting on the forest mosaic.

In view of the disturbance regime, the establishment of a reserve around a diverse but small portion of a riverine corridor does not ensure ecosystem preservation. Management strategies for conservation need to be directed at the whole river basin or watershed, including resources both inside and outside of reserve boundaries. Forest restoration promotes species populations, through the direct establishment of their respective habitat (see 31 for the Tana River red colobus and crested mangabey). Regionally, dams may be managed to ensure periodic floods downstream, and thereby promote forest regeneration processes and better support productive grazing and agriculture along the riverine corridor (see 37). Including local people as active participants in the stewardship of riverine forests couples their sustainable development with protection of the landscape mosaic (see 4, 36). An integrative approach, which combines reserve management with a scientific understanding of ecological processes and consideration for human-resource needs, clearly broadens the scale and potential for conservation (see Box 14.2). Unique in their linear form, dependency on a dynamic physical environment, and interrelationships with human activities, the riverine forest ecosystems in East Africa represent a special and most worthy challenge in resource conservation.

References

1. Adams, W.M., Hughes, F.M.R. 1986. The environmental effects of dam construction in tropical Africa: Impacts and planning procedure. *Geoforum* 17:403-410
2. Andrews, P., Groves, C.P., Horne, J.F.M. 1975. Ecology of the lower Tana River floodplain (Kenya). *Journal of East African Natural History Society and National Museums* 15:1-31

3.   Barrow, E.G.C. 1990. Usufruct rights to trees: the role of Ekwar in dryland central Turkana, Kenya. *Human Ecology* 18(2):163-176

4.   Bell, R.H.V. 1987. Conservation with a human face: conflict and reconciliation in African land use planning. In *Conservation in Africa: People, Policies, and Practice*, ed. Anderson, D., Grove, R. pp. 79-102. Cambridge: Cambridge University Press

5.   Bogdan, A.V. 1958. Some edaphic vegetational types of Kiboko, Kenya. *Journal of Ecology* 46:115-126

6.   Brinson, M.M. 1990. Riverine forests. In *Forested Wetlands*, eds. Lugo, A.E., Brinson, M., Brown, S. pp. 87-114. Ecosystems of the World. Volume 15. Amsterdam: Elsevier

7.   Burtt, B.D. 1942. Some East African vegetation communities. *Journal of Ecology* 30:65-146

8.   Décamps, H, Naiman, R.J. 1990. Toward an ecotone perspective. In *The Ecology and Management of Aquatic-Terrestrial Ecotones*, ed. Naiman, R.J., Décamps, H. pp. 1-5. Man and the Biosphere series. Volume 4. Paris and Carnforth, UK: Unesco and Parthenon Publishing Group

9.   Decker, B.S., Kinnaird, M.F. 1992. Tana River red colobus and crested mangabey: results of recent censuses. *American Journal of Primatology* 26:47-52

10. Douthwaite, R.J. 1987. Lowland forest resources and their conservation in southern Somalia. *Environmental Conservation* 14:29-35

11. Gichathi, F.N. 1987. The state of indigenous forest in Bura, Tana River District. Bura Forestry Research Project Working Paper. No. 27. Kenya Forestry Research Institute, Nairobi

12. Hughes, F.M.R. 1985. *The Tana River Floodplain Forest, Kenya: Ecology and the Impact of Development*. Ph.D. Dissertation, University of Cambridge

13. Hughes, F.M.R. 1987. Conflicting uses for forest resources in the lower Tana River basin of Kenya. In *Conservation in Africa: People, Policies, and Practice*, ed. Anderson, D., Grove, R. pp. 211-228. Cambridge: Cambridge University Press

14. Hughes, F.M.R. 1988. The ecology of African floodplain forests in semi-arid zones: a review. *Journal of Biogeography* 15:127-140

15. Hughes, F.M.R. 1990. The influence of flooding regimes on forest distribution and composition in the Tana River floodplain, Kenya. *Journal of Applied Ecology* 27:475-491

16. Hughes, R.H., Hughes, J.S. 1992. *A Directory of African Wetlands*. International Union for the Conservation of Nature, United Nations Environment Program, World Conservation Monitoring Center

17. Johansson, S. 1992. Irrigation and development in the Tana River Basin. In *African River Basins and Development Crises*, ed. Darkoh, M.B.K. pp. 97-112. Helsinki: OSSREA, Research Programme Environmental and International Security, Departments of Human and Physical Geography, Uppsala University

18. Kahumbu, P., Davies, G. 1993. Tana River Primate National Reserve: Primate Census, March 1993. *East African Natural Society Bulletin* 22(3):35-44

19. Kinnaird, M.F. 1992. Phenology of flowering and fruiting of an East African riverine forest ecosystem. *Biotropica* 24(2a): 187-194

20. Kiss, A. 1993. The Global Environmental Facility and the project for conservation of the Tana River Primate National Reserve. *East African Natural History Society Bulletin* 22(3):34

21. Lamprey, H.F. 1983. Pastoralism yesterday and today: the over-grazing problem. In *Tropical Savannas*, ed. Bourlière, F. pp. 643-666. Ecosystems of the World 13. Amsterdam: Elsevier Scientific Publishing Company

22. Lamprey, H.F., Halevy, G, Makacha, S. 1974. Interactions between *Acacia* bruchid seed beetles and large herbivores. *East African Wildlife Journal* 12:81-85

23. Livingstone, D.A. 1982. Quaternary geography of Africa and the refuge theory. In *Biological Diversification*, ed. Prance, G.T. pp. 523-536. New York: Columbia University Press
24. Madgwick, J. 1989. Somalia's threatened forests. *Oryx* 23(2): 94-101
25. Madgwick, J., Wood, B., Varty, M., Maunder, M., eds. 1988. *Somalia Research Project* (Final Report). University College, London
26. Marsh, C.W. 1976. *A Management Plan for the Tana River Game Reserve.* Report to the Kenya Department of Wildlife Conservation and Management, Nairobi. New York: The Wildlife Conservation Society
27. Marsh, C.W. 1985. A resurvey of Tana River primates and their forest habitat. *Primate Conservation* 7:72-81
28. Medley, K.E. 1990. *Forest Ecology and Conservation in the Tana River National Primate Reserve, Kenya.* PhD. Dissertation. Michigan State University
29. Medley, K.E. 1992. Patterns of forest diversity along the Tana River, Kenya. *Journal of Tropical Ecology* 8:353-371
30. Medley, K.E. 1993. Extractive forest resources of the Tana River National Primate Reserve, Kenya. *Economic Botany* 47(2):171-183
31. Medley, K. E. 1994. Identifying a strategy for forest restoration in the Tana River National Primate Reserve, Kenya. In *Beyond Preservation: Restoring and Inventing Landscapes,* eds. Baldwin, A.D., de Luce, J., Pletsch, C. pp. 154-167. Minneapolis: University of Minnesota Press
32. Medley, K.E., Kinnaird, M.F., Decker, B.S. 1989. A survey of the riverine forests in the Wema/Hewani vicinity, with reference to development and the preservation of endemic primates and human resources. *Utafiti* 2:1-6
33. Njue, A. 1992. *The Tana River Floodplain Forest, Kenya: Hydrologic and Edaphic Factors as Determinants of Vegetation Structure and Function.* PhD. Dissertation, University of California, Davis
34. Oba, G. 1990. Effects of wildfire on a semidesert riparian woodland along the Turkwel River, Kenya, and management implications for Turkana pastoralists. *Land Degradation and Rehabilitation* 2:247-259
35. Oba, G. 1991. The ecology of the floodplain woodlands of the Turkwel River, Turkana, Kenya. TREMU Technical Report D - 2, Nairobi: UNESCO
36. Pearl, M.C. 1989. The human side of conservation. In *Conservation for the Twenty-first Century,* eds. Western, D.C., Pearl, M.C. pp. 221-225. New York: Oxford University Press
37. Scudder, T. 1980. River-basin development and local initiative in African savanna environments. In *Human Ecology in Savanna Environments*, ed. Harris, D.R. pp. 383-405. New York: Academic Press
38. Trewartha, G.T. 1981. *The Earth's Problem Climates.* Madison: University of Wisconsin Press
39. Vollesen, K. 1980. Annotated check-list of the vascular plants of the Selous Game Reserve, Tanzania. *Opera Botanica* 59:1-117
40. Walter, H. 1972. *Ecology of Tropical and Subtropical Vegetation.* Edinburgh: Oliver and Boyd
41. White, F. 1983. *The Vegetation of Africa.* Paris: UNESCO
42. Wyant, J.G., Ellis, J.E. 1990. Compositional patterns of riparian woodlands in the Rift Valley of northern Kenya. *Vegetatio* 89:23-37

Chapter 15

Mid-elevation Forests: A History of Disturbance and Regeneration

Colin A. Chapman & Lauren J. Chapman

There is an increasing awareness of the value of tropical forest resources. Trees play significant roles in forest ecosystems by performing functions such as maintaining watersheds, preventing flooding and erosion, and aiding in the stability of long-term climatic patterns. More recently, however, the particular importance of tropical forests as centers of biological diversity has become apparent. Although tropical forests cover only 10 to 15% of the earth's' surface, they are estimated to harbor over 50% of all plant and animal species (30).

East African Forests and Their History

East Africa does not have large expanses of forest. Considering all types of moist forest, there is presently approximately 70,000 km² of forest in Uganda, Kenya, and Tanzania, which account for only 0.7% of the world's moist forest (11). As a result, the resources in such forests are very small. Unfortunately, their management has not reflected their scarcity, and deforestation in East Africa has averaged 1.2 % per year (Uganda = 1.3%, Kenya = 1.7%, Tanzania 0.7%; 21). The conversion of forests to agricultural land is not only decreasing the extent of forest, but is increasing their insular nature by dividing forest blocks. Thus, what forests these countries do possess exist as isolated islands, surrounded by savanna or agricultural land.

Regional-Scale Patterns

The forests in East Africa, although small and insular, are particularly important for a number of reasons. First, these forests contain a large number of endemic species. For example, based on the Flora of East Africa, Lovett (20) estimated that there were 2085 plant species in 801 genera in the montane and lowland forests of Tanzania (excluding the coastal forests). Twenty-four percent of these species were endemic to Tanzania. On a broader scale, the montane forests of Tanzania contain 7% of the endemic plant species of Africa in only 0.05% of the area. Translating these figures for comparison, Tanzania has 0.54 endemic plant species/1000 km², while Kenya has 0.45 endemic plant species/1000 km² and Uganda has 0.13 endemic plant species/1000 km². Second, the countries of East Africa have generally dry and seasonal climates; thus, the role of forests in watershed dynamics becomes particularly vital for harvesting clean water. Third, such forests may maintain the stability of long-term climatic patterns (30). In dry habitats, such as those found in Northern Kenya and Uganda, a slight decrease in annual rainfall patterns could have devastating effects. Finally, these forests contain vital resources, such as building materials, firewood, and medicinal plants, which, if they are managed carefully, could be valuable resources for present and future generations.

There are a large number of forest islands in East Africa, ranging in size from a few hectares to over 1000 km² (Fig. 15.1; Table 15.1). The majority of the forests occur where a rise in elevation results in a local increase in rainfall. Such areas include the Western Escarpment at the base of the Ruwenzori Mountains (such as the Kibale National Park), and the mountains that stretch in an arc around eastern Tanzania (such as the Usambaras, Fig. 15.1). Because of the general elevational rise of the Western Escarpment, Uganda has a greater extent of medium altitude forest than Kenya or Tanzania (Table 15.1; Fig. 15.1). The forests running along this escarpment are similar in species composition to forests that stretch from southern Nigeria to Angola. Kakamega Forest in western Kenya also has its closest affinities with these forests, and appears remarkably similar to such forests as Kibale Forest in Uganda, while being strikingly different from the forests further east and south. Because of its rarity in Kenya, Kakamega forest represents an important national conservation area.

In Kenya and Tanzania, east of Kakamega Forest, all of the major forest blocks that we are concerned with in this chapter (not riparian, high montane, or coastal forests) have very different plant assemblages when compared to the forests to the west. The forests of Kenya and Tanzania have been the focus of a number of floristic studies (20,24). One particularly significant finding of these studies has been the documentation of high levels of endemism. As a result, such areas should be considered of particular importance for global conservation efforts. The level of endemism found in the forests of Kenya and Tanzania (excluding Kakamega) appears to be largely a result of the degree of isolation they have experienced in their past. Many of these eastern forests, have largely been islands of forest surrounded by savanna (see Chapters 1 and 2). Such a situation isolates plant and animal populations and increases the likelihood that species will diverge from their original state - thus the high endemism in eastern forests.

Ugandan Forests. This is not the case for the forests of Uganda. These forests have a rich and dynamic history of being isolated during one period, but being continuous in the next period. Pollen diagrams suggest that prior to 12,000 years ago, the whole region was much drier and cooler than today. For example, in the area of the Ruwenzori Mountains mean annual temperatures were estimated to have been about 6° C cooler than now (13). These conditions resulted in the elimination of forests in many areas along the foothills of the Ruwenzori Mountains and reduced major forest blocks to small isolated forest islands. However, around 12,000 years ago, with the end of the last glacial period, the climate rapidly became warmer and wetter and resulted in conditions very similar to

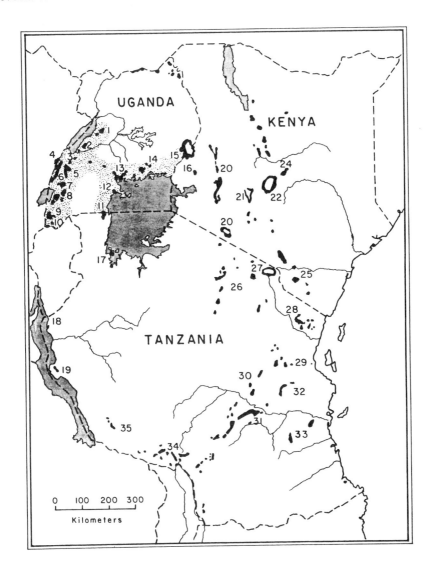

Figure 15.1. Mid-elevation forests of East Africa (this is not a map of all forests: coastal forests riverine forests, and forest blocks that are very small in size have not been included). An estimate of the original extent of forest in Uganda is indicated by the shading (13). Based on the map of Russell (27) and Polhill (24). (1=Budongo, 2= Bugoma, 3=Itwara, 4=Semliki, 5=Kibale, 6=Ruwenzori, 7=Kayoya-Kitomi, 8=Marimagambo/Kalinzu, 9=Bwindi, 10=Mgahinga, 11=Sango Bay/Minziro, 12=Lake Side and Sese Islands, 13=Mengo, 14=Mbira, 15=Mt Elgon, 16=Kakamega, 17=Rubondo Island, 18=Gombe, 19=Mahale Mountains, 20=Cherangani/Mau/Nguruman, 21=Aberdares, 22=Mt. Kenya, 23=East Turkana Mts., 24=Nyambeni Hills, 25=Chyulu and Teita Hills, 26=Masi/Mbulu/Meru Mts., 27=Mt. Kilamanjaro, 28=Usambara Mts., 29=Nguru Mts., 30=Ukaguru/Rubenho/Image Mts., 31=Uzungwa Mts./Maludwe Hills, 32=Uluguru Mts., 33=Kichi and Libangani Hills, 34=Rungwe/Ndumbi Poroto Mts., 35=Mbisi).

Table 15.1. Statistics of the major forest blocks found in Uganda, Kenya and Tanzania. Based on findings presented in (15,23,24,27,32). Rainfall values may very greatly with elevation and mean values are given. Some high-altitude forest also contains mid-altitude forest.

Forest	Area	%Forest	Rainfall mm	Disturbed area, %	Conservation status
Uganda					
Mid-Altitude					
Semiliki	219		1250	30	National Park
Kibale	766	60-75	1400		National Park
Itwara	87	78	1350	50	Forest Reserve
Bugoma	250	82	1200	31	Forest Reserve
Budongo	793	53	1325	90	Forest Reserve
Kashoya-Kitomi	390	82	1325	20	Forest Reserve
Marimagambo/					National Park
Kalinzo	580	85	1275	20	(51%) & Forest Reserve
Sango Bay	151		1275	100	Forest Reserve (Swamp forest)
Mbira	306	91	1325	~100	Forest Reserve
Mengo	728	63	1325	~100	Forest Reserve
High-Altitude					
Ruwenzori	996		1500	5	National Park
Bwindi/					
Impenetrable	371		1650	90	National Park
Mount Elgon	1145		2000	34	National Park
Mgahinga	30		>1500		National Park
Kenya					
Mid-Altitude					
Kakamega	100		1325		National Park
High-Altitude					
Mt. Elgon	170		2000		National Park
Cherangani/Mau/					
Nguruman			885		Forest Reserves
Aberdares			1350		National Park
Mt. Kenya	1400		>1500		Forest Reserve/ National Park
E. Turkana			600		Forest Reserves
Nyambeni Hills	53		>1500		Forest Reserve
Chyulu/ Taita Hills					National Park/
			885		Forest Reserve/ Private
Tanzania					
Mid-Altitude					
Gombe	52		1600		National Park
Mahale			1600		National Park
Rubondo Island	220	81	>1500		National Park
Minziro	250		1275		Forest Reserve (swamp forest)
High-Altitude					
Usambara			1950	70	Forest Reserve
Nguru	250		>1500		Forest Reserve
Ukaguru/Rubeho/					
Image Mts			>1500		Forest Reserve

continued

Table 15.1. continued

Uluguru	100	>1500	Forest Reserve
Uzungwa/ Maludwe	450	>1500	Forest Reserve
Masai/ Mbulu/ Meru	500		Mostly unprotected
Kilamanjaro		>1500	National Park/ Forest Reserve
Kichi and Libangani Hills	760		Unprotected/ Selous Game Reserve
Rungwe/ Ndumbi/ Poroto	275	>1500	Forest Reserve
Mbisi	30	1000	Forest Reserve

that which we experience today. At this time, these isolated forest islands expanded and joined together and lowland forest replaced the grassland communities that had existed for thousands of years (13). In Uganda evidence suggests that forests spread across much of the southern half of the country and all around the shores of Lake Victoria on the Ugandan side (Fig. 15.1). The only exception to this is in the region near the present city of Mbarara, in the south-central region which because of low regional rainfall patterns would have remained grassland.

The situation was not to remain the same for long, however. This time, human clearance and not climate change would be the factor causing forest reduction. East African forests have been affected by agriculturists for many centuries. Evidence from pollen analyses, archeological digs, and linguistic studies suggest that widespread forest destruction occurred at least as far back as 2000 to 5000 years ago (13,14, Hamilton pers. comm). The early cultivators are believed to have been Bantu-speakers that entered the area from North Angola-Katanga. Therefore, the forests that we see today are only remnants of much more extensive forest blocks that existed prior to clearance by man.

It is difficult to determine the long-term history of a forest because the pattern of human forest clearance and environmental history is complex. Some forested areas may have been cleared for agriculture hundreds of years ago and then abandoned. If we just consider recent recorded history, there are a number of well-documented cases that demonstrate the complexity of forest change. For example, between 1902 and 1906, the Sesse Islands in the Ugandan waters of Lake Victoria were evacuated because of a sleeping sickness epidemic, and the forests on the islands were allowed to regenerate. Today the islands support closed canopy forest dominated by *Uapaca guineensis*, and represent a major area for the extraction of timber (16).

Changes in certain animal populations can also have similarly dramatic effects on forest regeneration. Human populations left the area of Murchison Falls National Park because of increased levels of sleeping sickness in 1912. As a result, the elephant population of the area increased dramatically and subsequently de-barked and girdled many trees. Areas that were once forests were transformed into treeless grasslands. The forested area of the National Park decreased by 55 to 60% between 1932 and 1956 (2,18,29).

The example from Murchison Falls National Park also highlights a very important nonhuman factor in forest disturbance and regeneration: elephants (2,17, Box 11.2). Elephants can influence the structure of a forest in a variety of ways: 1) they can easily knock over trees with a diameter less than 20 cm; 2) they can girdle even relatively large trees, causing the tree to die, and 3) they selectively feed on specific types of seedlings and saplings, influencing forest composition. As a result, fluctuations in elephant populations can lead to remarkable changes in forest structure. Elephant populations have exhibited radical fluctuations at the hand of man and a major population decline during the past two decades. Elephant numbers have declined in most areas, but, as a result of confining some populations to small islands of forest, the effective population in some areas are higher than in recent history. Even prior to human induced changes in elephant populations, elephants were probably a significant structuring force in many forest communities. For example, Caughley (4) has suggested that in the Luangwa Valley in Zambia there is a cyclical relationship between elephants and trees. He proposes that when tree density is high, elephants have large amounts of available food, and their populations increase. When the elephants reach high population densities, they destroy the trees, removing an important food source, which inevitably leads to a decline in their numbers. This decline permits tree density to recover, and the cycle repeats itself.

As a consequence of creating intermediate scales of disturbance, elephants, in combination with natural events such as landslides, fires, and lightning strikes, are suggested to maintain the species richness of East African plant communities (5). By disturbing trees, elephants create a mosaic of forests of different ages. When elephants browse in very old and mature forest, they can often prevent plant species that are competitively superior from dominating all other plants but, by these same means, a forest might also be largely composed of the few species of plant that are most resistant to elephant damage. The influence of elephant browsing was quantified in a series of experimental plots established in grassland and woodland areas of Murchison Falls National Park, Uganda where elephants and other large herbivores were excluded from study sites for over 14 years (29). Comparing these experiments to overgrazed areas with high elephant numbers demonstrated that both the long-term removal of grazing and browsing animals and overgrazing resulted in floristic impoverishment. This suggests that an intermediate level of grazing maintains the highest plant diversity. In areas that are overgrazed, forest was degraded and replaced by grassland. At the other extreme, complete protection from grazing and browsing also led to a floristically poor environment.

Forest Classification

The term "rain forest" was first used by Schimper in 1903 (28) who defined it as a forest that is "evergreen hygrophilous (moisture loving) in character, at least 30 m high, rich in thick-stemmed lianas and in woody as well as herbaceous epiphytes". This is a rather broad definition, but one that depicts the character of a rain forest. In this chapter, we will consider major forest blocks within East Africa, excluding the coastal forests, riverine forest strips, and high montane forests (see the appropriate chapters). This includes a variety of forest types, and it is often difficult to determine where one forest type ends and another begins. For example, as one ascends any of the mountains in East Africa, be it the Ruwenzori Mountains of Uganda (Mt. Margherita 5108 m), or any of the mountains in the chain running down the middle of Tanzania, like the Usambaras or Mt. Kilimanjaro (5895 m), there are transitions between a number of forest types, culminating on the low mountains in montane forest, or on higher mountains, in afroalpine grasslands or even areas of rock and glaciers.

Patterns Among Forests

The forests of East Africa are diverse in nature. In general, the higher (non-coastal) closed forests of East Africa can be divided into medium altitude forests and montane forest. The distinction between these forest types is, however, unclear, and both can be considered transitional between lowland rain forests and non-forest montane habitats (13). In addition, there is little agreement on the altitude where medium-altitude forest stops and montane forest starts. Greenway (10) sets the boundary at 1350 m, while Trapnell and Langdale-Brown (in 27) view it to be at 2000 m; and Hedberg (1951, cited in 18) put the boundary between 1700 and 2300 m. Elevational differences may be due, in part, to the aspect of the mountain and the direction of monsoon wind and rain. Regardless of the altitude, however, all medium-altitude forests are closed canopy tall evergreen or semi-deciduous rain forests, with canopies generally over 30 m (12,13). Because of the suitability of medium-altitude forests for conversion to agricultural land, these areas have been heavily degraded and insularized.

Forest classification is a difficult task and a number of different regimes have been proposed. Following Langdale-Brown and coworkers (16) and Hamilton (12), four types of medium-altitude forests can be defined in Uganda which are named after major canopy (often timber) trees in the areas: 1) *Parinari* Zone found predominantly in Western Uganda (for example Kibale National Park), but also in such areas as the lower slopes of Mt. Elgon, the East Usambaras, 2) the *Celtis-Chrysophyllum* Zone found in both western Uganda and in the Lake Victoria region, 3) *Cynometra-Celtis* Zone confined to Western Uganda, and the 4) *Piptadeniastrum* Zone found along the shores of Lake Victoria. The emphasis that has been placed on categorizing areas into "zones" or "types" has in many ways disguised the gradual change in community species composition that exists.

The factors underlying the transition between zones appear to be closely linked to altitude, temperature, and rainfall. The exact processes governing the transition, however, are poorly understood. Even within a single forest block, as one descends an altitude gradient, one gradually travels from one forest zone to another. For example, within Kibale National Park in Western Uganda there is a gradient of decreasing altitude from north to south. Above approximately 1400 m is the *Parinari* Zone, the *Celtis-Chrysophyllum* Zone is between 1000 and 1400 m, and the *Cynometra* Zone is at approximately 700 to 1200 m. Nowhere, however, is there a clear-cut boundary, and at transitional elevations it is often very difficult to unambiguously determine the zone.

Although there do exist some general tendencies for specific forest types to be found within a range of altitudes, there is little information and considerable debate about the factors that create changes in forest composition with altitude. For example, Eggling (6) argued that the four types of forest he recognized were simply different stages of plant succession. He suggested that two types of colonizing forests (*Maesopsis* on richer soils and woodland forest, with *Olea welwitschii*, on poorer soils), both developed into a mixed-species forest, which in turn develops into a forest dominated by *Cynometra alexandri*. Augmenting Eggling's suggestions, Laws and coworkers (17) argued that since *Cynometra* is less susceptible to damage by elephants than many other forest trees, *Cynometra* forest is a climax induced by elephants. In contrast, Synnott (1971, cited in 18) argued that mixed forest is a climax on good soils, while either *Cynometra* or *Parinari* are climax on poorer soils. One piece of evidence in favor of Eggling's successional argument is that there appears to be poor recruitment or no recruitment of seedlings into the populations of a number of canopy-forming species (for example *Parinari* seeds in Kibale had extremely low germination success, Chapman unpublished data, 12,16), suggesting that these species represent a transition stage between early-successional species and those climax species that can recruit under a forest canopy. It is clear that little quantitative information is available to distinguish between the alternative hypotheses, making species-specific management of such areas difficult. This research, however, illustrates the potential importance of disturbance in creating forest species composition and structure, rather than the more static environmental approach of naming and describing forests according to average rainfall or elevation characteristics.

Within Forest Patterns: An Example From Kibale Forest

Since there is little information on the factors that govern the transition between major forest types, it is not surprising that there is so little information on factors contributing to differences between areas within a forest type. To illustrate the potential magnitude of floristic differences between two neighboring forests of similar altitude and rainfall, we describe the composition of two contiguous areas in the Kibale National Park, located in western Uganda (0 13' to 0 41' N and 30 19' to 30 32' E) near the base of the Ruwenzori Mountains. The Kibale Forest is one of the most intensively studied mid-elevation forests in East Africa. It was gazetted a crown forest in 1932 and since 1971 there has been continuous research activities in the area.

Two areas in the park have been the primary focus of research activities. These areas, known locally as Kanyawara and Ngogo, are 12 km apart, sit at approximately the same altitude, and receive similar rainfall (3). Despite their similarities in rainfall and elevation, Butynski (3) noted that species composition of the canopy trees differed between Ngogo and Kanyawara. To illustrate the magnitude of differences that can be found within a forest block, we expand on Butynski's data and document the size, population density, and distribution of tree species at both localities.

Vegetation transects were established at each study site providing a sample of 2111 trees at Kanyawara and 2622 trees at Ngogo in an area of 5.2 ha and 4.8 ha, respectively. Density, distribution (Coefficient of Dispersion between transects; CD), and size (mean Diameter at Breast Height = DBH) were calculated for each tree species at each locality.

In general, the density of trees is lower at Kanyawara (403 tree/ha), than Ngogo (546 tree/ha). Interestingly, the densities of certain species are very different at the two sites (Fig. 15.2). For example, two of the ten most common trees at Ngogo, are not found at Kanyawara. The density of the most common tree species, *Uvariopsis congensis*, is more that twice as common at Ngogo than Kanyawara. *Chrysophyllum* spp. is very abundant at Ngogo, but is found at low density at Kanyawara. The average size of each of the common tree species is similar between Ngogo and Kanyawara (Fig. 15.2). However, there are marked differences in the

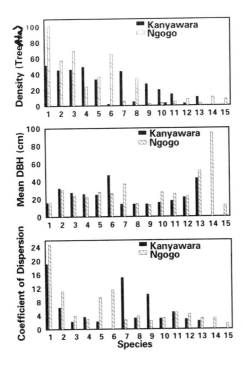

Figure 15.2. The density, mean diameter (DBH), and distribution of the 10 most common trees (> 10 cm DBH) at two neighboring unlogged forest sites in the Kibale National Park, Uganda. The tree species are 1) *Uvariopsis congensis*, 2) *Diospyros abyssinica*, 3) *Celtis durandii*, 4) *Markhamia platycalyx*, 5) *Funtumia latifolia*, 6) *Chrysophyllum* spp., 7) *Bosqueia phoberos*, 8) *Conopharyngia holstii*, 9) *Leptonychia mildbraedii*, 10) *Teclea nobilis*, 11) *Chaetacme aristata*, 12) *Neoboutonia macrocalyx*, 13) *Strombosia scheffleri*, 14) *Pterygota mildbraedii*, and 15) *Dasylepis eggelingii*.

distribution pattern of some species between the two sites (Fig. 15.2). For example, both *Bosqueia phoberos* and *Leptonychia mildbraedii*, are clumped at Kanyawara, but are much more uniformly distributed at Ngogo (Fig. 15.2).

There are a number of functional hypotheses that could be constructed to explain the differences in these two forests although data from more study sites is sorely needed. It is possible that the differences in altitude, temperature, and rainfall, although slight, were sufficient to induce changes in the vegetation. Alternatively, G. Isabirye-Basuta has suggested that the forest structure found at Ngogo may represent later stages of plant succession. From a botanical perspective, this hypothesis is supported by the large numbers of late succession trees at Ngogo (such as *Warburgia stuhlmanni* and *Funtumia latifolia*) and the higher density of small trees. In addition, grinding stone and kitchen implements have been found in exposed areas of soil in the Ngogo Forest (Isabirye-Basuta, personal communication). If this hypothesis proves correct, then the Ngogo forest would, once again, illustrate that the forests of East Africa have a rich and diverse history of disturbance and regeneration. The difficulty of determining what is climax forest and what is late successional forest has long been recognized. In 1964,

Box 15.1. Mutualism

Interactions between any two species are often negative for at least one of the species. Most species compete with others, and most are eaten or parasitized by others (1). It was the recognition of these antagonistic relationships that caused Darwin to suggest that nature was 'an entangled bank', and 'dyed red in tooth and claw'. There do, however, exist species relationships that appear to benefit more than one species. These relationships are called mutualisms, because they are thought to be mutually beneficial. Examples include pollinators and the flowers they visit, tick birds and large ungulates, and the algae and fungi that make up the mutualisms we call lichens.

Even though each species in a mutualism benefits from its interaction with the other, that does not mean that each evolved in order to benefit the other. In fact, each species has evolved adaptations that favor the success of its own individuals. In mutualisms, these 'selfish' adaptations do benefit another species. For example, flowers that produce more nectar (and attract more bees) will tend to be more successful than others. Nonetheless, it is in the interest of both the bees and the flowers to maximize their benefits and minimize their costs in any relationship.

The relationships between fruiting trees and fruit-eating primates are examples of these 'uneasy partnerships' (1). Many tree species surround their seeds with fleshy fruits, which are nutritious and eaten by a variety of animals, including many species of primates. In the process of eating these fruits, animals often disperse them, either by carrying them away and later dropping the seeds, or by consuming the fruits and defecating the seeds elsewhere. The latter also gives the seeds a rich nutrient source during early growth. Both the trees and the monkeys benefit from their relationship.

There is tension in mutualism. Fruits that are eaten too soon contain seeds that are not ready to be dispersed. Therefore, unripe fruits are initially protected by toxic compounds, and only become readily edible ('ripe') later. Unfortunately for the trees, there are many monkeys in a forest competing with each other and with other species, such as hornbills, for these ripe fruits (Kibale Forest has twelve species of primates, and Nyungwe has thirteen!). There is strong selection to eat the fruits before others do, and therefore some animals that can, will eat unripe fruits, even though they may taste bad (at least to us). The red colobus monkey, for example, is a folivorous primate that has adaptations of the gut for digesting leaves that are more difficult to digest by other, more frugivorous primates. These adaptations also allow it to eat unripe fruit, while frugivores must wait until the remaining fruit becomes riper (2). The more unripe the eaten fruit, the less both the tree and the monkey benefit from their relationship. Even apparently harmonious mutualisms are often underlain with this kind of evolutionary and ecological tension.

T.P. Young

References

1.   Howe, H.F., Westley, L.C. 1988. *Ecological Relationships between Plants and Animals.* New York: Oxford University Press
2.   Struthsaker, T.T. 1975. *The Red Colobus Monkey.* Chicago: University of Chicago Press

Langdale-Brown and coworkers (16) stated that the situation was so complex that in many cases it was not yet possible to determine or describe a climax forest.

Forest Fauna

An attribute of the forests that cannot be overlooked is the diversity of animal life that they support. Where hunting has not greatly effected large animal populations, mid-elevation forest harbors a diversity of large animal species including forest elephants, sitatunga, giant forest hog, golden cat, leopard, chimpanzees, a variety of primates and birds. A number of the species are endangered and their only remaining populations are sustained by these East African forests. For example, Kibale Forest is the only place in Uganda to harbor sizable populations of the rare and endangered red colobus monkey (*Colobus badius*).

The abundance and diversity of animal life in these forest is best illustrated by data from Kibale Forest, one of the most thoroughly studied areas in East Africa. Kibale is the home to over 325 species of birds (this equals half the number found within the entire United States). These birds come from 46 different families, and 58% of them can be considered "true" forest birds. The birds from Kibale include such spectacular species as the black-and-white casqued hornbill (*Bycanistes subcylindricus*), the great blue turaco (*Corythaeola cristata*), Ross's Turaco (*Musophaga rosae*), black-billed turaco (*Turaco schuetti*), Cassin's Hawk Eagle (*Hieraaetus africanus*), crowned hawk eagle (*Stephanoaetus coronatus*), gray parrot (*Psittacus erithacus*), and Narina's trogan (*Apaloderma narina*). Kibale has one species, the "Kibale ground thrush" that has been described only from this area, and a number of other species that are considered rare throughout their distribution (Cassin's hawk eagle-*Hieraaetus africanus*; white-naped pigeon-*Columba albinucha*; black bee-eater-*Melittophague fularis*; red-crested owlet-*Glaucidium tephronotum*; superb sunbird-*Cinnyris superbus*; Congo flowerpecker-*Parmoptila jamesoni*).

The mammal fauna in Kibale is particularly diverse and abundant. Kibale harbors 12 species of primates, making it one of the richest primate faunas in Africa (dwarf bushbaby - *Galago demidovi*; inustus bushbaby - *Galago inustus*; potto - *Perodictus potto*; red colobus - *Colobus badius*; black and white colobus - *Colobus guereza*; redtail monkey - *Cercopithecus ascanius*; blue monkey - *C. mitis*, vervet monkey - *C. aethiops*; l'hoesti monkey - *C. l'hoesti*; mangabey - *Cercocebus albigena*; olive baboon - *Papio anubis*; chimpanzee - *Pan troglodytes*). Furthermore, these animals are found at one of the highest population densities ever recorded. It is estimated that there are over 550 primates per square kilometer in the Kibale Forest. Kibale is also the home to one of the largest populations of forest chimpanzees (*Pan troglodytes*).

Kibale is one of the better protected and most thoroughly studied forests in East Africa. A second example would be Nyungwe Forest in Rwanda. Nyungwe is, in many ways, similar to Kibale, but it located at a slightly higher elevation (1600 to 3000 m) and the slopes are typically steeper than those in Kibale. Nyungwe Forest is one of the most important forests in the region; although more than 33% of Rwanda was formerly covered by mid-altitude and montane forest, less than 6% remains. Thus, the 970 km^2 reserve represents an important element to protect the biodiversity of the region (its effective size is 1140 km^2 since Nyungwe connects with Kibira National Park in Burundi). The biological richness of Nyungwe is evident by the fact that it contains one-fifth of all African primate species, more than 260 bird species (17 of which are regionally endemic), and an extremely rich flora.

The rich biodiversity and the abundance of animals should be considered particularly valuable resources, not just in terms of the biodiversity they represent, but also as a particularly important feature for drawing ecotourists.

Conservation of East Africa's Forests

The East African countries of Uganda, Kenya, and Tanzania are experiencing rapid ecological and economic change. Each of these countries is striving for sustainable economic growth, while at the same time advocating both the conservation of existing habitats and maintenance of the processes and species that maintain the stability of their ecosystems. At the present time these are incompatible goals; demands for food and materials are growing at a rate that is greater than growth in the economic sectors that provide these products. The forests of East Africa are particularly hard hit, because they provide vital resources such as building materials and firewood.

Deforestation in East Africa has averaged 1.2% per year (Uganda = 1.3%, Kenya = 1.7%, Tanzania 0.7%; (21)). The major source of this deforestation has come from the demand by the local people for building material, firewood, and agricultural land. This is ultimately the consequence of population increase, which has averaged 3.7% per year in the three East African countries (Uganda 3.4% (doubling time 20 years), Kenya 4.1% (doubling time 17 years), and Tanzania 3.6% (doubling time 19 years; 31), Table 15.2). In addition, a large proportion of this population is rural (for example 91% of Uganda's population is rural), making contact with forests more probable and more difficult to regulate.

A simple solution to increasing population and decreasing forest area is difficult to envision. However, it seems clear that preservation of the biodiversity of East African forests, while still encouraging economic growth requires a strategy with two distinct approaches. First, there must be effective protection of forest blocks deemed to be important for whatever reason, whether it is high levels of endemism, or because they maintain a watershed. Secondly, programs must be developed to take pressure off of existing forest blocks, providing goods normally extracted from the forests through alternative means.

Protected areas may be viewed as preserves of biological diversity to be maintained as intact as is possible. These reserves will house future options for the region, when the animals and plants they protect are no longer found in the neighboring exploited lands. In addition, they have the immediate benefit of serving as a source pool for restored or reconstructed areas. Since the extent of the remaining forests in East Africa is so small, no extractive exploitation should be permitted in such areas. Often multiple use reserves are advocated. If such endeavors are to be encouraged they should not be considered as protected areas or included in these forest classification statistics. Once exploitation is permitted in a protected area, it becomes progressively more difficult to argue that increased exploitation is not reasonable if an economic argument for continued exploitation can be substantiated. Non-extractive use of these areas should be investigated, with the provision that such uses will not alter the ecosystem. One such non-extractive use of protected areas, that is often more profitable than extraction, is ecotourism.

Table 15.2. Vital statistics of countries in East Africa including their size, population, forests, and deforestation rates.

Statistic	Uganda	Kenya	Tanzania
Population			
Country size (km^2) [1]	236,578	582,645	939,762
Population (1989), million [1]	17	24.1	26.3
Annual growth rate [1]	3.4%	4.1%	3.6%
Doubling time [1]	20 years	17 years	19 years
Forest			
Original extent closed canopy Tropical forest (km^2)[2]	103,400	81,200	176,200
FAO (1980) estimate of Remaining forest	7500	6900	14,400
Protected areas (km^2) [3]	6084	13,148	77,008
Deforestation rate (% per year) [4]	1.3	1.7	0.7

[1] Stuart and coworkers (31), [2] Groombridge (11), [3] Tanzania and Kenya are from Groombridge (11) and likely include many parks with little area of forest, Uganda is from the data in Howard (15), and includes only forested area, [4] McNeely and coworkers (21).

International tourism currently generates more than $40 billion dollars per year worldwide (excluding airfares; 8), and estimates of nature tourism's share of this figure range from $2 billion to $12 billion (19). In 1988, there were approximately 400 million international tourists, tourism accounted for nearly 6% of total world exports, and represented approximately 25% of international trade in services (1, see Box 16.2). In terms of East Africa, Kenya has already developed an extensive tourist industry. In 1988, Kenya received more than 600,000 visitors and earned an estimated $400 million from tourism, which represented 30% of Kenya's foreign exchange (1). Elephant viewing alone is worth $25 million annually to Kenya (25). Amboseli National Park is estimated to earn $247/ha/year (25). While these values are certainly open to debate, the point is clear; tourism has the potential of generating large amounts of foreign currency revenue. Uganda is currently taking an active approach to encourage ecotourism, with the establishment of new National Parks, and with the development of a number of tourist attractions, such as chimpanzees (*Pan troglodytes*) and mountain gorillas (*Gorilla gorilla*). If properly managed, ecotourism has the potential of generating income that far exceeds that possible through traditional exploitative means. Thus, it seems advisable to encourage different forms of ecotourism, as a means of decreasing destructive exploitation of the forests. It should also be appreciated that it is very difficult to have both forest exploitation and ecotourism in the same forest.

While protection is one avenue that must be followed, alternative means of producing goods typically acquired from the forest should be investigated and developed. Current and planned programs for woodlots and reforestation fall far short of projected needs (30). Hamilton (12) estimated that 90 to 95% of the wood consumed in Uganda is used for fuel, 7% for building poles and 1% for sawn timber. These demands are increasing at a rate of 3 to 7% annually. These estimates suggest that sources outside of the existing natural forests should be developed and geared to provide local fuelwood. Such woodlot programs could emphasize the use of a variety of tree species, both exotic and indigenous. The exotic softwoods like *Pinus*, and the fast growing medium-weight wood of *Eucalyptus grandis* represent valuable resources because of their fast growth and multiple uses. However, indigenous trees should be considered as well. For example, *Sesbania sesban* is a small fast growing nitrogen-fixing tree species that can provide high quality firewood (30). It is often available for women as a cooking fuel, whereas the larger *Eucalyptus* is often exploited by men as a cash crop (Kasenene, personnel communication).

As well as increasing the production of timber and firewood resources outside natural forests, efforts should be placed on reducing per capita use of these resources. Presently, the brick industry is taking a heavy toll from the forest and from existing woodlots, since bricks are fired with fuelwood. Emphasis should be placed on producing non-kiln fired bricks and in finding alternative construction materials. The use of improved stove design can reduce fuelwood requirements 5 to 10 fold (30). The use of charcoal should be discouraged since 50 to 70% of the energy is lost in the process of producing the charcoal (30).

For non-protected forests, efforts should be placed on identifying and determining the economic value of non-timber forest products. Tropical forests have often been assumed to have little or no economic value other than as a source of timber. However, tropical forests do produce many other products including wild foods, materials for construction, medicinal plants, and much more (26). The value of these non-timber tropical forest products is just now being investigated. Based on rattan cultivation, Godoy and Feaw (7) estimated that forests in Borneo can yield $220 to $530/ha/y. Myers (22) estimated that the value of plant-derived drugs and pharmaceuticals during the 1980s in the USA was $16 billion. In East Africa, it has been estimated that there are more than 1300 species of traditional medicinal plants found in the forests of East Africa (30). Examples from common trees in East Africa include *Spathodea nilotica* used for treating ulcers and kidney ailments and *Rauvolfia vomitoria* containing compounds used to treat cancer and hypertension (30). It is estimated that 100,000 kg of wild coffee (*Coffee canephora*) could be harvested annually from a forested area of just over 300 km², with an estimated market value between $100,000 and $200,000 dollars (based on 1985 prices; 30). It seems likely that as there is more awareness of such products, new markets will open. Thus, future management plans for non-protected forests should investigate the feasibility and marketability of non-timber forest products.

Summary

The forests of East Africa, although small and insular and only remnants of much more extensive forest blocks, harbor a large number of endemic plants and animals, contribute to watershed maintenance, aid in the stability of long-term climatic patterns, and contain vital resources such as firewood, building materials, and medicinal plants. The diverse history of disturbance and regeneration that typifies East African forests has led to considerable debate as to the factors accounting for change in forest composition. What is undebatable is the incompatibility between present exploitation of these forests and their preservation. A combination of protected areas and managed exploitation areas may preserve and strengthen the value of forest ecosystems to the human economy while maintaining options for future economic growth.

References

1. Boo, E. 1990. *Ecotourism: The Potentials and Pitfalls.* Washington: World Wildlife Fund

2. Buechner, H.K., Dawkins, H.C. 1961. Vegetation change induced by elephants and fire. *Journal of Ecology* 50:528-58

3. Butynski, T.M. 1990. Comparative ecology of blue monkeys (*Cercopithecus mitis*) in high- and low-density subpopulations. *Ecological Monographs* 60:1-26

4. Caughley, G. 1976. The elephant problem - an alternative hypothesis. *East African Wildlife Journal* 14:265-83

5. Connell, J.H. 1978. Diversity in tropical rain forests and coral reefs. *Science* 199:1302-10

6. Eggling, W.J. 1947. Observations on the ecology of the Budongo rain forest, Uganda. *Journal of Ecology* 34:20-87

7. Godoy, R., Feaw, T.C. 1989. The profitability of smallholder rattan cultivation in Central Borneo. *Human Ecology* 16:397-420

8. Groom, M.J., Podolsky, R.D., Munn, C. 1992. Tourism as a sustained use of wildlife: A case study of Madre de Dios, Southeastern Peru. In *Neotropical Wildlife Use and Conservation*, eds. Robinson, J.G., Redford, K.H., pp. 393-412. Chicago: University of Chicago Press

9. Greenway, P.J. 1955. Ecological observations on an extinct East African volcanic mountain. *Journal of Ecology* 43:544-566

10. Groombridge, B. 1992. *Global Biodiversity.* London: Chapman and Hall

11. Hamilton, A.C. 1984. *Deforestation in Uganda.* Nairobi: Oxford University Press

12. Hamilton, A. 1974. Distribution patterns of forest trees in Uganda and their historical significance. *Vegetatio* 29:21-35

13. Hamilton, A., Taylor, D., Vogel, J. 1986. Early forest clearance and environmental degradation in south-west Uganda. *Nature* 320:164-67

14. Howard, P.C. 1991. *Nature Conservation in Uganda's Tropical Forest Reserves.* Switzerland, Gland: IUCN

15. Langdale-Brown, I., Osmaston, H.A., Wilson, J.G. 1964. *The Vegetation of Uganda and its Bearing on Land-Use.* Kampala, Uganda: Published by the Government of Uganda

16. Laws, R.M., Parker, I.S.C., Johnstone, R.C.B. 1970. Elephants and habitats in North Bunyoro Uganda. *East African Wildlife Journal* 8:163-180

17. Lind, E.M., Morrison, M.E.S. 1974. *East African Vegetation.* London: Longman

18. Lindberg, K. 1991. *Policies for Maximizing Nature Tourism's Ecological and Economic Benefits.* Washington: World Resources Institute

19. Lovett, J.C. 1989. Tanzania. In *Floristic Inventory of Tropical Countries.* eds. D.G. Campbell, Hammon, H.D., pp. 232-235. New York: New York Botanical Garden

20. McNeely, J., Miller, K., Reid, W., Mittermeier, R., Werner, T. 1990. *Conserving the World's Biological Diversity.* Switzerland, Gland: IUCN

21. Myers, N. 1983. *A Wealth of Wild Species.* Boulder: Westview Press

22. Myers, N. 1980. *Conversion of Tropical Moist Forests.* Washington: National Academy of Sciences.

23. Polhill, R.M. 1989. East Africa (Kenya, Tanzania, Uganda). In *Floristic Inventory of Tropical Countries*, eds. Campbell, D.G., Hammon, H.D., pp. 217-231. New York: New York Botanical Garden

24. Rasker, R., Martin, M.V., Johnson, R.L. 1992. Economics: Theory versus practice in wildlife management. *Conservation Biology* 6:38-349

25. Robinson, J.G., Redford, K.H. eds. 1991. *Neotropical Wildlife Use and Conservation.* Chicago: Chicago University Press

26. Russell, E.W. 1962. *The Natural Resources of East Africa*. Nairobi: East African
     Literature Bureau
27. Schimper, A.F.W. 1903. *Plant Geography Upon a Physiological Basis*. Oxford:
     Oxford University Press
28. Smart, N.O.E., Hatton, J.C., Spence, D.H.N. 1985. The effect of long-term
     exclusion of large herbivores on vegetation in Murchison Falls National Park,
     Uganda. *Biological Conservation* 33:229-245
29. Struhsaker, T.T. 1987. Forestry issues and conservation in Uganda. *Biological
     Conservation* 39:209-234
30. Stuart, S.N., Adams, R.J., Jenkins, M.D. 1990. *Biodiversity in Sub-saharan
     African and its Islands: Conservation, Management, and Sustainable Use*.
     Occasional Papers of the IUCN Species Survival Commission
31. Tabor, G.M., Johns, A.D., Kasenene, J.M. 1990. Deciding the future of Uganda's
     tropical forests. *Oryx* 24:208-214

Chapter 16

High Montane Forest and Afroalpine Ecosystems

Truman P. Young

"Saving Africa's mountain communities is one of the most pressing priorities for world conservation. It is a real test of values and resolve" (13)

East Africa's high mountains are central to its climate, its cultures, and its economy. They produce both essential rain catchments (most strongly on their windward sides) and arid 'rain shadows' (on their leeward sides). Their rivers supply electricity and scarce water for agricultural and industrial development. Their forests have provided the timber and fuel on which rural development has long depended. Their heights have inspired religious reverence and national identity (the Republic of Kenya takes its name from its highest mountain). They have a critical combination of traits with respect to conservation: they are environmentally invaluable, they house numerous endemics, they cover very little land area, and they are endangered by attractive alternative uses, such as agricultural conversion and timber exploitation. There are few ecosystems in the world that are both as valuable in their natural state and as tempting for exploitative development. Their conservation remains a high priority in East Africa.

Geologic History and Biogeography

Different East African mountains were created by opposing forces of tectonic movement. The plates underlying the continent of Africa are often in movement, sometimes crushing together, and sometimes pulling apart. The buckling of the African land mass caused by the collision of its plates pushed up the East African and Ethiopian highlands and the Ruwenzori Mountains (5109 m) over 20 million years ago. The remainder of its high mountains are volcanic in origin. This volcanism is related to the proximity of the great Rift Valley and is associated weaknesses in the earth's crust. As plates pull apart, they create thin areas in the earth's crust, often associated with subsidence and rift valleys. These weaknesses also allow the escape of molten magma to the earth's surface. These escapes of what is then called lava produce volcanoes in the vicinity of the Rift Valley, which can grow to immense height. The oldest of these mountains, Mount Elgon (4320 m), is

Table 16.1. The primary high mountain land masses of East Africa, and their vertically projected land areas (km²). Not included are numerous shorter and less extensive mountains. In particular, the total amount of land 2440 to 3050 m in East African is greater than the amount listed here.

Country	Mountain Name	>2440 m	>3050 m	>3660 m	>4270 m	Altitude km²
Kenya	Mt. Kenya	1228	665	259	45	5200
	Aberdares	2546	560	69	-	4000
	Mau	4367	27	-	-	3100
	Cherangani	1406	310	-	-	3370
Kenya/Uganda	Mt. Elgon	1555	612	194	1	4310
Uganda/Zaire	Ruwenzoris	1070	660	350	210	5109
Zaire/Rwanda	Virungas	480	125	25	2	4507
Tanzania	Kilimanjaro	1140	740	420	105	5895
	Mt. Meru	105	50	30	2	4565
	Kitulo Plateau	~300	-	-	-	2960
Total		>13,900	3750	1340	365	

25 million years old. Kilimanjaro (5895 m), Mount Kenya (5199 m), and Mount Meru (4565 m) are much younger (2 to 5 million years old). The Virungas (4507 m) are still active and may even be growing. Mount Kenya was once much taller (7000 m), but glaciers over the last 250,000 years have worn it down, as they have also done on the Ruwenzori Mountains.

There are eight separate land masses in East Africa that attain heights of over 2440 m (8000 feet), and cover at least 400 km² at this altitude (Table 16.1). In Kenya, the amount of land area above a given altitude decreases by a half for each 350 m of increasing elevation (34; Table 16.1). This means that high forests cover far more land area than alpine ecosystems. Although high mountain ecosystems (>2440 m) account for less than 1% of East Africa's land area, they have far-reaching climatological and hydrological effects, outlined below.

These mountains form an archipelago of isolated montane islands throughout the region. Especially at higher altitudes, this isolation has produced considerable endemism among the plants (80% of afroalpine plant species are endemic; 7) and presumably also the invertebrates. Another consequence of their isolation is that local extinctions are less likely to be reversed by recolonization from other sites (see Box 11.3).

The biogeographic relationships of these high mountains are a mixture of elements from African biotas (with which they share proximity) and temperate biotas (with which they share aspects of their climate; 7). With increasing elevation, the plant families and genera become increasingly similar taxonomically to temperate floras, even if some of the growth forms become increasingly bizarre.

Afroalpine ecosystems are renowned for their unusual plant growth forms. The most striking of these are the giant caulescent (having tall stems) and acaulescent (stemless) rosettes, particularly in the genera *Lobelia, Senecio* and *Carduus*. Other characteristic plant growth forms include massive tussock grasses, gigantism in the form of tree-sized heather *(Erica, Phillipia)* and St. John's wort *(Hypericum)*, and 'moss balls'. These growth forms are evolutionary adaptations to the unique climate in which they live (8,26).

Mountains have profound effects on weather, both locally and regionally. As air rises up mountains, it cools, and this cool air drops its moisture as rain. This is

Box 16.1. Ethiopian Highlands and Mountains

'The Ethiopian highlands rise up from the wastes of equatorial Africa like an island from the sea. There they stand in lonely isolation, for the surrounding wilderness has cut them off from the outer world more effectively than the sea itself' (1)

If you could look across the continent of Africa, as across a flat board, the north-east corner would present a dark mass rising above all it surrounds, a mass more extensive than any other highland area in Africa. Over 50% of the land in Africa that lies above 2000 meters in altitude is in Ethiopia, together with a staggering 80% of the land over 3000 m altitude (2). These highland areas do not occur as isolated peaks, as in other parts of Africa, but as two major blocks separated by the Rift Valley. The Bale Mountains dominate the eastern block, and the Simien Mountains the western block.

The effects of varying topography, altitude and rainfall have produced considerable variation in the soils, and therefore the vegetation. The highlands have been occupied by dense human populations for several millennia, such that it is now difficult to assess 'natural' vegetation. From the vestiges that remain, it is evident that highland and lowland vegetation were very different. While lowland Ethiopia may appear undifferentiated from neighboring parts of Africa, the great extent of highlands and their isolation 'may be a major reason why so many endemic animals are found in Ethiopia' (2).

Description	National Parks		Total Ethiopia
	Simien	Bale	
Area (km^2)	179	2,471	1,200,000
Total mammal species	33	66	277
Mammals endemic to Ethiopia	10	17	31
Mammals endemic to Ethiopia, %	32.3	54.8	11.2
Total bird species	125	262	861
Birds endemic to Ethiopia	12	16	28
Birds endemic to Ethiopia, %	42.9	57.1	3.3

The Ethiopian highlands provide ecosystems ideal for grain cultivation; sixteen percent of the country's total area is composed of highlands suited for intensive cultivation. Over the past 2000 to 4000 thousand years, a number of cultures based on agriculture rose and fell, during which several plant species were domesticated, including the grain teff *(Eragrostis teff)*. Most important of these was probably coffee *(Coffea arabica)*, and Ethiopia remains the most important natural repository of wild genetic resources for this crop. The highlands are important to the survival and support of every Ethiopian. They are by far the most productive areas for cultivation, and also provide considerable fuelwood. As watersheds, they are crucial for Ethiopia's water and hydroelectric needs.

Much of the highland area, however, is now suffering from over-exploitation of its natural resources. An informal robber economy exists, consuming renewable and non-renewable resources unsustainably. Major modification of the remaining natural ecosystems is occurring through the cultivation of previously unused areas, and the supply of fuelwood to urban and rural areas. Overgrazing is also a problem.

Only two Wildlife Conservation Areas exist in the highlands - the Bale and Simien Mountains National Parks, both of which have attracted considerable attention and development due to their spectacular scenery and the presence of several large mammals found nowhere else in the world. The Simien Mountains are home to the endangered and endemic Walia Ibex *(Capra ibex walie)* and the Ethiopian Wolf *(Canis simiensis),* and the Bale Mountains are home to the Ethiopian Wolf and the Mountain Nyala *(Tragelaphus buxtoni).*

The Simien Mountains range in altitude from just under 2000 m to Ras Dashen at 4620 m - the highest point in Ethiopia. The altitude of the main plateau averages 3500 m. Most of the plateau is covered by afro-alpine moorlands, with fringing giant heather *(Erica arborea)* on steeper slopes, leading down to isolated patches of alpine grassland on ledges and relict wooded patches on the lower and less sheer slopes. Intensively cultivated slopes dominate the lower levels. It is a measure of land hunger that much of even the high plateau has been occupied by cultivators for a long time, both inside and outside of the Park. Soil erosion from cultivated land has been estimated at 80 tons per hectare per year, one of the highest in the world (4).

The Bale Mountains National Park contains a complete cross-section of natural habitats from 1500 to 4300 m altitude. At 3800 to 4100 m altitude, the main plateau lies above the frost line and considerably higher than the main part of Simien, and is unsuitable for cultivation. The area of these high altitude moorlands in the Park amounts to almost 1000 km^2, the largest such area in Africa. The rest of the 2400 km^2 National Park is comprised of an extensive fringing belt of giant heather, 1000 km^2 of high montane forest in the south (the Harenna Forest), and 100 km^2 of high altitude grassland and woodland in the north.

The endemic mountain nyala had been reduced to very low numbers by 1968, when an intensive conservation effort began. Their numbers rose to over 1200 in the northern Park and its vicinity by the mid-1980s, but have dropped again due to human disturbance and incursion at the time of the change in government in 1991. Until recently, the Ethiopian wolf occurred in high numbers, with at least 440 animals, estimated to be ~70% of the world's total (3). Again, their numbers have recently declined greatly.

As land hunger in Ethiopia has increased, there has been increasing frustration with the restrictions put on human use of these highland areas. The mountain nyala and Ethiopian wolf have recently been considered by the local people as the *raison d'être* for the Parks, and many have concluded that if they simply eliminated these species, a Park would no longer be required.

Ethiopia's endemic plant and animal species are not only found nowhere else, they represent unique adaptations to existing conditions over an extensive area that is also highly suitable for human use. These species thus represent a stockpile of genetic material ideally suited to the country's conditions, and play a vital part in maintaining the natural ecological processes upon which humans depend for their survival, now and in the future. Therefore we must conserve and use the environment, but not abuse and lose it through a robber economy. Protected areas have an important role to play, but are only a part of the overall conservation ethos. In the past, in Ethiopia, protected areas have acted and been treated as separate from everyday life, with disastrous results for their conservation in recent years.

Jesse C. Hillman

References

1. Buxton, D. 1949. Travels in Ethiopia. London: E. Benn
2. Yalden, D.W. 1983. The extent of high ground in Ethiopia compared to the rest of Africa. *Sinet* 6:35-40
3. Gottelli, D., Sillero-Zubiri, C. 1990. *The Simien Jackal: Ecology and Conservation.* Bronx: Wildlife Conservation International
4. Humi, H. 1986. *Management Plan - Simien Mountains National Park and Surrounding Rural Area.* Addis Ababa: UNESCO World Heritage Commission, and Wildlife Conservation Organization

why the slopes of mountains even in dry regions are often clothed in moist forests, particularly on their windward sides. Air temperature continues to decrease with altitude, at a rate of about 6° C per 1000 vertical meters. Rainfall, however, reaches a maximum at intermediate altitudes. This maximum occurs at 2700 to 3200 m, depending on mountain and aspect (31), with lower values occurring at drier sites. Above this, rainfall steadily decreases because the air has lost most of its moisture at lower altitudes. Precipitation can be extremely low at the highest altitudes, where it falls mainly as snow.

As this cool and dry air drops down the leeward sides of mountains, it not only does not produce rain, it can even extract moisture from the environment. The upper leeward slopes are somewhat protected from this dryness by spill-over from the wet air on the windward side (31). However, at elevations below 2200 m leeward sides of mountains can be extremely dry, a phenomenon known a 'rain shadow'. In East Africa, these rain shadows generally lie to the west and north of mountains, because wet weather generally comes from the east and south, off the Indian Ocean. Rain shadows can be very extensive, reaching up to a hundred kilometers and more from the mountain. Examples include the Amboseli basin north of Kilimanjaro and the dry Laikipia plateau northwest of Mount Kenya. In wetter parts of East Africa, such as the Ruwenzoris and the Virungas, these rain shadows are less pronounced because of overall higher rainfall and more rain originating from the east.

In summary, East African mountains are wetter than the surrounding land, especially at intermediate altitudes and eastern and southern aspects. At higher altitudes, low temperatures and low rainfall produce severe climatic conditions. The thinness of the air at higher elevations also means that solar radiation is greater during the day, and radiant loss is greater at night, compared to lower altitudes. Above timberline (the upper limit of forest trees), nights can be exceedingly cold-well below freezing. At dawn, surface temperatures can climb 30° C in a matter of minutes, although air temperatures remain lower (10 to 20° C maximum).

Hedberg (8) has referred to this extreme climate as 'summer every day, and winter every night'. It is generally believed that the bizarre plant growth forms characteristic of afroalpine ecosystems (and other tropical alpine areas) are evolutionary responses to the unique combination of low mean temperatures, large diurnal fluctuations in temperature, severe drought stress, and a lack of bitter cold and windy winters (such as those in temperate areas; 8,26).

Because these mountains are volcanic and of recent origin, their soils are relatively rich in nutrients. Soils tend to be more leached on the wetter slopes, but even here the soil fertility is more than adequate for agriculture (31). Most of the land on these mountains is sloping, often steeply. Valley bottoms tend to be more fertile than slopes (31,33), and slopes are prone to erosion if their forest cover is lost.

Community and Population Ecology

The plant and animal communities of East African mountains reflect the environmental gradients associated with altitude, aspect, and slope. At the simplest level these ecosystems are comprised of belts of vegetation wrapping around each mountain (6; see also Fig. 16.1). Broadly, these are the forest, ericaceous, and alpine belts. There is of course considerable variation within these broad categories. Detailed community descriptions are available only for the Aberdare Mountains (25), Mount Elgon (10), the Jibat Forest on the Central Plateau of Ethiopia (2) and alpine Mount Kenya (20,39).

Community Descriptions

Forest Communities. Within a given mountain, there are numerous forest types whose distribution depends on local temperature and moisture. At lower elevations and wetter aspects there are mixed forests that are rich in biodiversity. Among the many species of plants they support are such economically valuable trees as camphor *(Ocotea usambarensis)*, olive *(Olea capensis* and Meru oak *(Vitex keniensis)*, which is endemic to Mount Kenya (this means that it is found nowhere else in the world). Many of these forests have already been cleared for agriculture and tree plantations, and the remaining patches are heavily exploited for their timber (Meru oak is now in danger of extinction).

At higher elevations on drier mountains (Mount Kenya, Aberdares, Mount Elgon), increased rainfall and decreased temperatures are associated with forests dominated by two commercially important species: podo *(Podocarpus latifolius,* a timber tree) and bamboo *(Arundinaria alpina).* These two species interact in complex and interesting ways, as yet not fully understood. Bamboo often occurs as monospecific stands up to 10 m and more in height, especially in wetter sites. In drier sites, bamboo becomes progressively less dominant, forming a mosaic with podo, and at its driest sites is represented only by an understory scrub form (35). On wetter mountains (Jibat Forest, Ruwenzoris, Virungas), *Podocarpus* is far less common in bamboo forest, often replaced with *Rapanea* spp. (2).

Associated with bamboo and podo are often herbaceous shrubs in the genus *Mimulopsis*, which, like bamboo can produce nearly pure stands that suppress recruitment of other plant species. At intermediate altitudes on Mount Elgon, species richness is apparently greatly reduced by these two species (10).

An intriguing aspect of the biologies of both bamboo and *Mimulopsis* is that entire populations flower synchronously and die back, allowing other species to establish themselves. It appears that individual *Mimulopsis* plants completely die (28), but the underground rhizomes of bamboo may survive (35). In any case, these rare events are associated with profound changes in the community ecology of these ecosystems (1,35,36). Estimates put the age of flowering for *Mimulopsis* at 7 to 13 years (28) and for bamboo at perhaps 40 years (30) in East Africa.

This remarkable life history of synchronous reproduction coupled with population die-back occurs only rarely, in a variety of species in tropical and

Figure 16.1. Timberline on the Aberdare Mountains, Kenya. The lower slopes are clothed with a mosaic of bamboo (lighter area on upper right) and podocarpus forest. There is a band of *Hagenia* and *Hypericum* trees at timberline. Just above, there is a narrow belt of dark ericaceous scrub. In the foreground is alpine grassland with a few giant rosette plants (*Senecio battiscombei*) two to three meters tall.

subtropical forests throughout the world (36). Combined with the ability to form single species stands, it represents a form of endogenous disturbance that can result in multiple quasi-stable community states in a given locality (Fig. 16.2). Throughout the lifetime of a population, bamboo and to a lesser extent *Mimulopsis* may be able to suppress and even displace other community types, in particular *Podocarpus* forests. However, after flowering and die-back, opportunities arise for

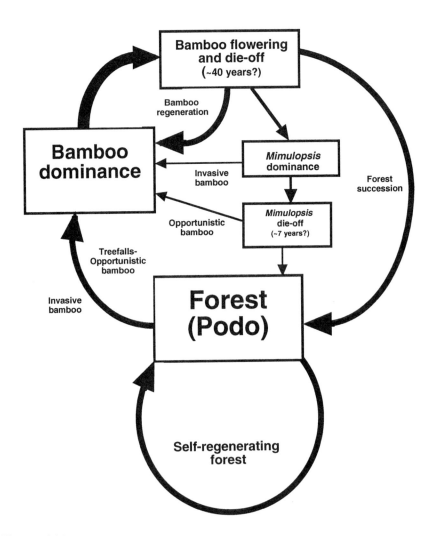

Figure 16.2. Putative patterns of endogenous vegetation change on East African mountains involving the interactions of bamboo and *Mimulopsis* (each which flowers and dies synchronously) and Podocarpus forest.

other community types to invade and establish themselves before the re-establishment of the die-off species. The result may be a dynamic temporal and spatial mosaic of community types, such as those seen on Mount Kenya and Mount Elgon. This mosaic may also be affected by other forms of forest disturbance such as herbivory, fire, and treefall gaps. In other words, the community type found at a given site may be more related to natural historical factors than to climatic and soil factors. Details of this complex community dynamic and the mechanisms that determine shifts between community states await future research.

The driest montane forests are dominated by pencil cedar *(Juniperus procera)*. This species is also of great economic importance, being used for timber, poles, and fuelwood. Some cedar forests are characterized by open, park-like understories. Cedar and olive *(Olea africana)* forests once were common on a variety of mid-elevation sites, at the drier limit of forest cover. Many of these have now been lost to repeated burning and to agricultural conversion.

Timberline Communities. Above the podo/bamboo forests, there is often a zone of parkland dominated by *Hagenia* and *Hypericum* trees 5 to 10 m tall. These trees may form complete canopy cover, or be represented by only scattered individuals. The understory is characterized by shrubs in the genus *Alchemilla*, and by a wide variety of grasses and herbs. Above this parkland is a more or less distinct timberline. This timberline may be less obvious if one is looking for a sudden decrease in average plant height, because the ericaceous shrubs immediately above the Hagenia/Hypericum parklands are examples of plant gigantism, characteristic of East Africa's high mountains. Nonetheless, many alpine herbaceous elements are immediately present just above the last *Hypericum* stands, among these tall ericaceous shrubs.

Timberline varies from 2800 m on drier Kilimanjaro to 3300 m on the wetter aspects of Mount Kenya and the Aberdares, to as high as 3500 m in the Virungas and the Ruwenzoris. Timberline corresponds with soil temperatures of 8 to 9° C (31). On the dry northern slopes of Mt. Kenya the forest band is discontinuous, and so there is no true timberline. This 'treeless forest gap' has provided access routes for plains animals, of which there are many on alpine Mount Kenya. This includes the rock hyrax *(Procavia capensis johnstoni)*, which are absent from all other afroalpine areas.

Immediately above timberline is usually a band of ericaceous vegetation dominated by shrubs in the genera *Erica, Phillipia* and *Alchemilla* on wetter aspects, and by *Artemisia, Protea* and *Euryops* on drier aspects. These are plants with small thick leaves (thought to be adaptations to a cool, dry environment; 8) and the ability to coppice after fire. Indeed, this vegetation is highly reminiscent of fire-adapted Mediterranean chaparral, especially on drier sites. Fires are not uncommon here, and are often attributed to humans (honey hunters, herdsman, poachers, campers). However, fires caused by lightning may have occurred occasionally even before humans became frequent visitors to the alpine zone. Most of Mount Kenya's northern slopes at this elevation burned in the late 1970s, and again in the early 1990s. We do not know whether this represents a natural cycle or what its normal frequency might be. Fires may favor grasses over shrubs, at least on Mount Elgon.

Alpine Communities. There have been numerous attempts to describe the various afroalpine plant communities (5,6,8), but only recently have vegetation maps been drawn (20,25). Only Mount Kenya has had a quantitative analysis of alpine vegetation (39). This quantitative analysis reveals a continuum of vegetation change on Mount Kenya, correlated with altitude and soil moisture. The latter is associated also with a topographic gradient. Moister sites are located on valley bottoms and on ridge tops (except at the highest altitudes). This has implications for management. Considerable trail proliferation has occurred on Mount Kenya in wetter sites, and plans to develop a new trail in the 1980s were initially directed towards putting trails on the supposedly drier ridges. Because vegetation analysis showed that these ridges were also wet and therefore prone to degradation, the trail was instead put on the drier valley slopes (39).

This vegetation analysis also demonstrated what appears to be a natural cycle of vegetation change associated with the establishment, dominance and senescence of single-aged stands of the giant rosette tree *Senecio keniodendron*. Like several tree species in the semi-arid lowlands of East Africa (38), *S. keniodendron* occurs mainly as stands of a single size, and presumably a single age. These individuals all reach maximum age at approximately the same time, and so entire stands disappear through senescence over relatively short periods. Grasses (particularly *Festuca pilgeri*) are favored after the opening of the canopy, but are later replaced by *Alchemilla* shrubs as the next generation of senecios begins to dominate (39). A similar cycle occurs near timberline on the Aberdares, but there in sites without giant senecios (1).

At altitudes above 4200 m, vegetation becomes increasingly sparse, and eventually only isolated sheltered plants are found. Here severe erosion is prevented by frozen soil, infrequent precipitation, and the fact that it falls as snow. Daily frost thaw-cycles continually churned the upper soils in a process called frost heaving or solifluction. This daily churning uproots young seedlings, greatly restricting their ability to survive (8,34).

Population Biology of Key Species. Population biologies of only a few montane and afroalpine species have been studied. Despite their economic importance, we know only the rudiments of growth and mortality rates for key commercial forest species, such as bamboo (30), podo(12), camphor (13) and cedar (11), and even for these only at single sites.

Several (noncommercial) alpine species are better known. The giant rosette species *Senecio keniodendron, S. brassica, Lobelia keniensis* and *L. telekii* have been the subject of considerable long-term research (33,23). Growth rates are slow, and large *S. keniodendron* plants may be hundreds of years old. Since these plants are the only source of wood in the high alpine, their exploitation by visitors for fires has been unsustainable, and current bans on cutting should be extended and enforced. Seasonal and spatial variation in soil moisture have strong effects on growth and survivorship. Experiments show that competition between and within species can be strong (23).

Herbivores. Herbivores are important components of both high montane forest and afroalpine ecosystems, but, with the exception of mountain gorillas, have been largely unstudied (5,23,24,37). We do not yet have reliable estimates of the population sizes of any other animals on these mountains, although there have been recent attempts to census forest elephants in East Africa. There are a wide variety of vertebrate and invertebrate herbivores in the forest and alpine ecosystems, but we are still far short of comprehensive surveys. Some large mammals regularly move between the forest and alpine, or between mountain ranges, including elephants, buffaloes, and several bird species (37).

The dominant alpine herbivore is probably the rodent *Otomys otomys*, which feeds on a grasses and small herbs. Elephants coming up from the forest on Mount Kenya have converted *Senecio keniodendron* woodlands into open grasslands in the past (23). Elephants and hyrax have been accused of threatening the very existence of giant rosette species of *Lobelia* and *Senecio*, but this is probably not the case. Although these herbivores can devastate populations locally, their herbivory occurs only at occasional sites and at rare intervals (23).

Roads through forests encourage secondary vegetation that is particularly attractive to large mammalian herbivores and frugivorous birds, and may increase the overall density of these animals, with possible effects on the remaining forest. Bamboo shoots provide food to several species, including the endangered mountain gorilla. *Mimulopsis* may be a key food for bongo (*Boocercus eurycerus*), which is endangered in East Africa. The large scale, long-term dynamics associated with these plant species (Fig. 16.2) can be expected to have profound effects on the animals that feed on them. Are alternate foods sufficient? If not, do these animals populations crash, or migrate, after the die-off of their major food plants? There are reports of both, but no one knows for sure how these animals respond to changes in vegetation.

There are several large herbivore species that once moved between mountains, or between mountains and lowlands, including elephants and buffaloes. There are numerous bird species that also move up and down the altitudinal gradient, as local conditions change, both seasonally and over multiple years (18). For large mammals, these age-old patterns of movement are no longer possible. Populations of elephants and buffaloes are increasingly 'trapped' on particular mountains. Those that drop down below the forest and raid farms risk being shot. Elephant damage to plantation forests is also a problem, and has been severe in some sites. The existence of intensive agriculture in close proximity to indigenous forests and their wildlife is an unsolved problem for virtually all East African montane forests. Ambitious fencing schemes are currently underway (at least in Kenya). While the fencing of semi-arid parks is of limited conservation value, fencing around mountain reserves may be inevitable, though not without ecological risk. In particular, such fencing will completely block the traditional migration corridors between mountains, and between mountains and lowlands. Such corridors provide buffers against locally unfavorable conditions.

The Aberdare mountains of Kenya provide an excellent example of a blocked corridor. A traditional migration route between Mount Kenya and the Aberdares was cut off by agricultural development in the middle of this century. The Aberdare segment of this route, called the 'Salient', was fenced off in the 1960s. Adding to the problem was the fact that much of the salient was secondary forest, and hence supported unusually high numbers of large herbivores. The game fence acts as a game dam, trapping hundreds of animals in the Salient. Their numbers have retarded forest regeneration and even converting some of the lower salient to heavily grazed grassland (25). Nowhere is the effect of herbivores on montane forest in East Africa more evident. These large herbivore populations also support unusually large carnivore populations, such as hyenas and lions. The latter were only recently introduced into the Aberdares, with devastating consequences for such species as the rare giant forest hog.

Plains animals once moved between alpine Mount Kenya and the grasslands and bush lands to the north, through the 'treeless gap', perhaps as part of a seasonal migration. Now they too are trapped on the mountain. The survivors seem to be doing well, reproducing and showing no obvious signs of over-grazing their habitat.

The establishment of corridors between nearby mountains, and between mountains and protected lowlands, would be an invaluable (although perhaps expensive) conservation strategy. These corridors need not be especially wide;

elephants passing over alpine sections of the Aberdare Mountains (moving from the forest on one side to the other) use the Park road for many kilometers, rarely leaving it.

Ruwenzori Mountains National Park in Uganda is a rare example of a mountain ecosystem whose animals have protected access to lowland areas (and vice versa). As such it would be an ideal system in which to examine the importance of lowland sites for mountain species, and the importance of mountain refuges (during droughts?) for lowland species.

Ecosystem Function

There have been no detailed studies of the ecosystem function of East Africa mountains (26, but see 21), and such studies are desperately needed. I will give a brief summary of what we do know about the ecosystem function of these forest and alpine ecosystems.

Temperature and moisture are the most limiting abiotic factors in these ecosystems (26,33). Soil fertility may also limit productivity. Soil weathering is slowed somewhat by cold temperatures, though leaching does reduce the fertility of soil on wetter aspects. Similarly, low temperatures slow nutrient cycling, especially at the highest altitudes (26). Rates of decomposition of detritus or animal dung are not yet available. There are no published measurements of ecosystem productivity or energy flow for any East African mountain ecosystem. Predictions based on known regressions from moisture and temperature data are possible, but risky. Net primary productivities of natural mountain ecosystems at the very highest altitudes are near zero, while at lower elevations they may be 1000 g(dry) m²/y or more.

In intact forests, much primary productivity can be expected to pass through a number of higher trophic levels before reaching a detritus stage. However, in several forests some large herbivores or carnivores have been greatly reduced. In undisturbed afroalpine ecosystems, it appears that much of the primary production is not eaten either by large or small herbivores, and is only turned over at a slow rate by decomposers.

An exception to this slow turnover may be the periodic fires that sweep the ericaceous belt just above timberline. How often do they occur? How do fires affect the plants and animals, and rain catchment? How fast does the vegetation recover? What happens if fires are not allowed to burn? We do not have any idea at present. Such fires should rapidly cycle many nutrients back to the soil, although nitrogen losses through volatilization may be great. This may be a particular problem in these habitats where cold and dry soils may limit the bacteria responsible for nitrogen fixation, nitrification, and denitrification, as well as the bacteria and fungi responsible for decomposition.

Although seasonal variation is not as obvious in wet forests as it is in the drier lowlands, it is clear that dry seasons affect forest and alpine ecosystems in myriad ways. During periods of drought, there is increased plant and animal mortality, changes in herbivore diet and habitat use, and movement up the altitudinal gradient, particularly by birds. There is also climatic variation across years. A multi-year drought in the 1980s resulted in major reductions in alpine *Lobelia* population sizes on Mount Kenya (23), and in die-offs of buffaloes and warthogs in the Aberdare Salient. At an even larger temporal scale, pollen records show large variations in the altitudes of vegetation belts on East African mountains, with all belts having been much lower in the past (9,17). Such large-scale changes in population and community ecology have probably always put populations at risk. The dangers are even greater now that recolonization and altitudinal migration are now largely prevented by increased development below the forests (blocking historic corridors), and with the recently recognized threat of global warming.

Past and Future Use by Humans

Humans evolved from ape-like ancestors in the shadow of these great East African mountains, and even saw the birth of some of them. It is likely that people have used these ecosystems for a very long time. Until as recently as a few hundred years ago, this use was almost entirely limited to hunting and gathering. These activities probably altered these ecosystems, but in ways that we will never know. It is likely that such use was less at higher altitudes than at lower ones, if for no other reason than the lower productivity of higher ecosystems. Most of these hunter-gatherer cultures are long gone, including the Gumba of Mount Kenya, but some still remain, vestiges of a dying economy- the BaTwa of Rwanda and Uganda, and the Okiek of the Mau Forest.

Pastoralists have occasionally made use of forest and, on Mount Elgon, alpine ecosystems. As is common in the lowlands, this has included burning to improve range. However, the biggest change in montane land use came with the expansion of agriculturists to East African mountain forests in this millennium (and in some forests, as recently as this century): the Kikuyu, Meru and Embu in Kenya, the Chagga in Tanzania, the Bakiga and Bakonjo in Uganda, and the Hutu and Banyarwanda in Rwanda and Burundi. These people stumbled on virgin ecosystems that ideally suited their economy, with rich soils, abundant rainfall, and moderate temperatures. They cleared the forests, planted crops, and greatly increased in numbers and extent. By the time the Europeans arrived a hundred years ago, much of the original montane forests were gone at lower altitudes.

European administrators began to set aside the remaining forests, some as wildlife parks and reserves, but most as forests reserves, where harvesting of indigenous tree species and establishment of plantations of exotic species were planned and carried out. To reduce soil erosion in the agricultural areas, they banned the clearing of vegetation in the vicinity of rivers and streams, and encouraged such practices as terracing (a technique already in use in some parts of East Africa before the arrival of Europeans). These restrictions were considered irksome by many, and have not survived well the transition to independence, leading to increased soil loss and siltation (Fig. 16.3).

Many thousands of hectares of indigenous forest were converted to plantations, mostly exotic softwoods. Many thousands more hectares were 'selectively' harvested for their native timber, both legally and illegally. In recent years, it has become clear to scientists, administrators and political leaders that the remaining forest cover is too valuable to be misused. New restrictions, many of which are more severe than during colonial times, have come into place. New parks and reserves are being created for forest preservation, not exploitation. In Kenya, there is an official ban on the cutting of indigenous forest species, and on conversion of forest to plantation. Unfortunately, these practices continue, often with governmental knowledge.

Figure 16.3. Soil erosion on steep slopes cleared of indigenous forest in the Aberdare Mountains.

Non-Extractive Values of Montane Forest and Afroalpine Ecosystems

Many East African nations are beginning to restrict economically rewarding activities, such as timber extraction and conversion to agriculture, in favor of conservation (4). What alternative values of these ecosystems are behind these promising policies? There are several answers, all of which combine to provide a compelling rationale for continued conservation of East Africa's montane and alpine ecosystems (22,27).

Perhaps the most crucial value of these ecosystems is as watersheds. Most of East Africa is arid and semi-arid, and watered only by unpredictable rains and a few permanent rivers. All of these rivers have their origins in montane ecosystems. If water is the life blood of East Africa, mountains are its heart.

It has often been said that forests actually create rain. Although this may be true, it is also true that forests drink up vast amounts of water. The true value of forest cover is in the quality, not the quantity of runoff. There are two aspects of this quality. First, siltation loads are far less from indigenous forests than from plantation forests or agriculture, the latter being particularly bad (3; Fig. 16.3). In East Africa, where expensive hydroelectric dams provide most of the electricity and much of the irrigation, siltation has been devastating. It greatly reduces the life of reservoirs by filling them, and greatly reduces the life of hydroelectric plants by fouling their turbines. This silt also represents the invaluable lost fertility of agricultural systems upstream. The simple expedient of restricting agriculture to some small distance from rivers would costs little in terms of production, but save considerably in the long run.

Second, forests regulate the flow of water in ways that minimize high and low water levels. An example is provided by rain and river records from Ontilili on Mount Kenya (31). This watershed underwent transition from indigenous forest to plantation forest to agriculture over a period of several decades. At each transition, the regulatory ability of the system was reduced. There was an increasing correlation between rainfall and river flow, both seasonally (Fig. 16.4) and across years (Fig. 16.5). Both floods and low river levels become more common when forests are cleared, and both come at the worst times: floods when rain are plentiful, and dry rivers during droughts.

A second value of these ecosystems is biodiversity. Many species found here are endemic. As stated earlier, 80% of the afroalpine flora is endemic to East Africa and Ethiopia. This includes four endemic genera. Approximately 10% of the afroalpine plant species are found on only one mountain each (7). Many forest plants are also endemic, although no tally has yet been made.

One notable forest endemic is the valuable and endangered tree, Meru oak, known only from the eastern slopes of Mount Kenya. Many species may represent as yet undiscovered medicines, foods or gene pools. Miraa (*Catha edulis*), which occurs on lower mountain slopes is used as a stimulant, and it has been a major export crop in Kenya. East Africa's forests continue to be major sources of traditional medicines.

Undoubtedly, many of the mountain invertebrates also occur nowhere else. There is at least one endemic genus of moth, *Saltia*. There are a number of endemic vertebrates, the most celebrated being the mountain gorilla, an endemic subspecies. Other endemic mammals include two species of mole shrew in endemic genus *Surdisorex*, and the Mount Kenya mole rat. In East Africa, only these mountains (and a few forests at slightly lower elevations) are home to the golden monkey, bongo, yellow-backed duiker, Abbott's duiker, giant forest hog, and the golden cat. There are a number of endemic bird species, including the scarlet-tufted malachite sunbird and the Abyssinian ground thrush (14,18).

This rich biodiversity is largely due the island nature of these ecosystems, making them a natural laboratory for the study of evolution and biogeography. The unique afroalpine climate and plant life forms have attracted scientists from all over the world to study the nature of biological adaptation. The complex dynamics of the forests ecosystems are another mystery awaiting research. In short, these ecosystems are invaluable scientific sites that have helped us to better understand life on earth, and can continue to provide important answers to basic questions (14,29).

This biodiversity also has a more immediate and valuable use. Tourism in East Africa continues to grow, and has already become the largest earner of foreign exchange in Kenya. Forest lodges are an important part of the wildlife industry there. The mountain gorillas of Rwanda have been a major economic and public relations boon, and visitation to gorilla sites in Uganda and Zaire is rapidly growing (4, see Box 16.2). Mountaineers and hikers continue to flock to Kilimanjaro, Mount Kenya, and the Ruwenzori Mountains.

These many values are increasingly convincing local governments of the value of conservation, but there is always pressure to increase the exploitation of these rich mountain ecosystems. I will now briefly examine the potential of some of these exploitative uses.

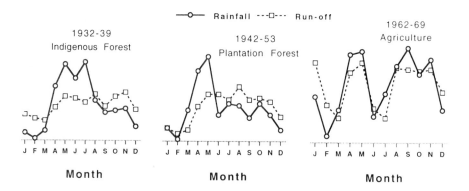

Figure 16.4. The relationship between mean monthly rainfall and mean monthly river runoff with increasingly exploitative human use of forest at Ontilili, Mount Kenya (adapted from 31).

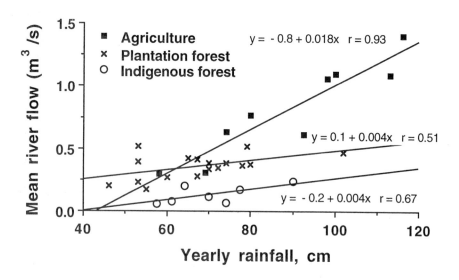

Figure 16.5. The relationship between yearly rainfall and mean river runoff with increasingly exploitative human use of forest at Ontilili, Mount Kenya (calculated from 31). Each point represents one year.

Extractive Uses

Sustainable Use? Legal and illegal timber extraction continue to be major uses of East Africa's montane ecosystems. It may seem reasonable that such productive forests could, if managed properly, produced timber sustainably with minimal ecological effects. There are however at least three serious difficulties. First, we still do not have the basic biological information available on key species

Box 16.2   Ecotourism and the Conservation of Montane Forest Ecosystems

'Use it or lose it' is a common contemporary theme in conservation. The same message is central to the related concepts of sustainable use, and integrated conservation and development. Tourism has traditionally been cited as one form of such use that permits wildlife to 'pay its way' in a generally non-destructive way. Using biodiversity to attract tourists is not a new concept in Africa. On the contrary, Africa has been a world leader, with a long history of wildlife tourism that has been widely developed in eastern and southern African savanna ecosystems. Complementary trekking circuits have existed for decades in the afro-alpine environments of Kilimanjaro, Mount Kenya, and the Ruwenzori Mountains. While these tourism activities have sometimes been criticized for their negative local environmental and socio-economic effects, they have generated considerable revenue and undoubtedly have been a major force in ensuring the conservation of these areas. Ecotourism is a new concept that has emerged over the past decade. Combining aspects of nature and adventure tourism, it seeks to assure appreciation of natural ecosystems, while limiting degradation of the environment. In its most comprehensive form, it also actively promotes the well-being of local populations through revenue-sharing and job creation.

Ecotourism has seen its greatest development in the montane forest environments of east-central Africa. One of the oldest examples is the Mountain Gorilla Project in Rwanda's Parc National des Volcans. This project was conceived in 1979 to counter a dramatic drop in mountain gorilla numbers from roughly 500 individuals in 1960 to barely 260 twenty years later. Primary reasons for the decline were direct poaching and conversion of forest to farmland. Further research suggested an underlying cause in perceptions of the citizenry that gorilla and park conservation were of little value (3,4). In response, the Mountain Gorilla Project placed a primary focus on generating revenue and employment through controlled tourism, along with parallel programs in education and improved security. The resulting program attracted an average of more than 6000 visitors per year throughout the 1980s, who spent several million dollars each year and generated hundreds of local jobs. Most importantly, poaching and encroachment were brought to an end, and gorilla numbers climbed to 320 individuals by 1989 (2). In recent years, a civil war suspended gorilla-based tourism in Rwanda. All warring factions, however, respected the gorilla sanctuary of the Parc National des Volcans, largely because of its economic value.

Building on the Rwandan experience, ecotourism has been successfully implemented in neighboring reserves in Zaire and Uganda, starting in the 1980s. In each instance, the primary emphasis was on tourism that has minimal effects on gorillas and their habitats by limiting the group size, frequency and duration of tourist visits. In Zaire, more direct benefits to the people were also achieved through the use of tourist revenues to build clinics and schools in villages bordering the Virunga park. The most ambitious effort in this regard is now underway in Uganda, where national legislation calls for direct revenue-sharing with the people around M'Gahinga and Bwindi/Impenetrable reserves (1). Gorilla-based ecotourism has demonstrated the ability to generate considerable local income, employment, and political support, with little effect on the gorillas. Chimpanzee-based tourism has also been an effective argument for conservation sites in Zaire and Tanzania.

Chimpanzees, however, have been more difficult to habituate for tourism, and they do not yet generate revenues comparable to those earned by gorillas (3).

Where charismatic species such as gorillas and chimps do not exist, there are nonetheless alternative models of forest-based ecotourism. The most promising have again sprung up in the Kivu-Ruwenzori highlands. The earliest of these was developed in Rwanda's Nyungwe Forest. Nyungwe's 1000 km^2 of rugged montane forest habitat supports 13 species of primates and more than 240 species of birds. With access provided by a network of walking trails, thousands of tourists have come to see these monkeys and birds, or simply to experience the forest

An equally successful ecotourism program has now been developed in Uganda's Kibale Forest. Highly visible monkeys and birds have been the prime attractions, although roughly 20% of tourist groups also see chimpanzees in the course of their forest walks. In addition, this more recent initiative includes support for two community-based cooperatives that control key tourism concessions and allocate the resulting income to local development activities. Gorilla-based tourism, however, still generates far more direct revenue than sites that emphasize other primates, birds, and general forest walks. Revenues from Kibale entry fees were approximately $45,000 in 1994. This is similar to well-known chimpanzee sites such as Gombe Stream, but is only one tenth the revenue generated from the two gorilla ecotourism sites in the region, which charge higher entry fees per visitor. Indirect tourist revenues (from lodging, vehicle rentals, purchases) are, however, probably less different.

Despite their lower income, ecotourism programs along the lines of the Nyungwe and Kibale models have the greatest potential for application elsewhere. Even without gorillas, most tropical forest sites in Africa have diverse primate and bird faunas. East Africa does not yet have any forest canopy walkways of the type that have been so successful in other tropical areas, but the same topographic features (ridges and overlooks) that favor wildlife observations also provide excellent views of plant communities and the three-dimensional structure of the forest. Individual attractions are multiplied by the concentration and relative accessibility of a network of tourist sites in the east-central African highlands. Finally, the attraction of hiking in a cool, scenic environment, with fewer insects and disease risks, is great.

Afromontane forest reserves also confront the common conservation problem of being relict islands of natural forests surrounded and separated by highly modified agricultural landscapes with the highest rural human population densities on the continent. The revenues and employment from tourism can provide tangible and sustainable benefits to local and national economies. These benefits, combined with the high value of virtually all montane forests for water catchment, represent the strongest arguments for their conservation.

On a more cautionary note, there are many limits to successful ecotourism development. First, the natural attraction itself must compete with other sites for visitors and, therefore, attention to several qualitative factors may be essential to retain a competitive edge. Too many visitors, poorly trained guides, badly maintained trails and facilities, and harm to the focal species or ecosystem all diminish the visitors' experiences and may lower their willingness to pay. Failure to ensure that local people benefit directly from tourism may reinforce unsustainable forest and wildlife use. Finally, central government ministries such as forestry or mining, and land hunger in general, are often stronger than ministries concerned with biodiversity and tourism, and threaten the conservation of these ecosystems. These constraints must be considered by anyone promoting tourism development in a forest environment. With careful planning, however, and a long-term commitment to monitor the implementation process, forest ecotourism can be a powerful tool in the cause of conservation.

W. Weber & A. Vedder

References

1. Butynski, T., Kalina, J. 1993. Three new national parks for Uganda. *Oryx* 27(4): 214-224
2. Vedder, A., Weber, W. 1990. The Mountain Gorilla Project. In *Living with Wildlife: Wildlife Resoure Management with Local Participation in Africa,* ed. A. Kiss, pp. 83-90. Washington. D.C.: World Bank
3. Weber, W. 1993. Ecotourism and primate conservation in African rain forests. In *The Conservation of Genetic Resources,* ed. C. Potter, pp. 129-151, Washington, D.C.: AAAS
4. Weber, W. 1987. Socioecological factors in the conservation of afromontane forest reserves. In *Primate Conservation in the Tropics,* eds. Marsh, C.W., Mittermeier, R.A. pp. 205-229. New York: Alan R. Liss

and communities that we need to formulate harvesting quotas (27). How fast do plants grow in indigenous forest? What affects their regeneration? What effects do logging operations have on community structure and soil erosion (Fig. 16.7)? Second, even in systems elsewhere in the world where we think we have many of these answers, we have still not been able to demonstrate sustainability (15). Forest management is often associated with irreplaceable losses in biodiversity, soils, and catchment values (Fig. 16.7). Third, without proper enforcement, even the best management plan is doomed to failure (22,27). Until proper management is based on good information and is properly enforced, bans on indigenous timber exploitation may be the best policy (and even these need increased enforcement).

Grazing in these ecosystems has been a traditional use for centuries, both within all forests (especially during drought), and in the alpine grasslands of Mount Elgon and the Kitulo Plateau. As in other ecosystems, the sustainability of grazing is related to its intensity, and current pressures to increase grazing in mountain ecosystems is leading to habitat degradation (16).

Water extraction is a use of forests that occurs outside the ecosystems themselves. Increasing irrigation is an environmental time bomb in East Africa. Numerous irrigation schemes have cropped up over the past three decades, many of them locally successful. Unfortunately, many of them also have resulted in overuse of this limited and invaluable resource, causing hardship and even death downstream. For example, the Uaso Nyiro river dried up for the first time in living memory in the late 1980s, because of irresponsible irrigation upstream. Livestock and pastoralists paid for this mismanagement with their lives. Such disasters have been largely ignored, and are likely to increase in the future. Any increase in irrigation permits must be accompanied by education about water laws, and strict enforcement of those laws. Enforcement of legal restrictions on current water permits is also essential.

Effects of rural activities on the environment

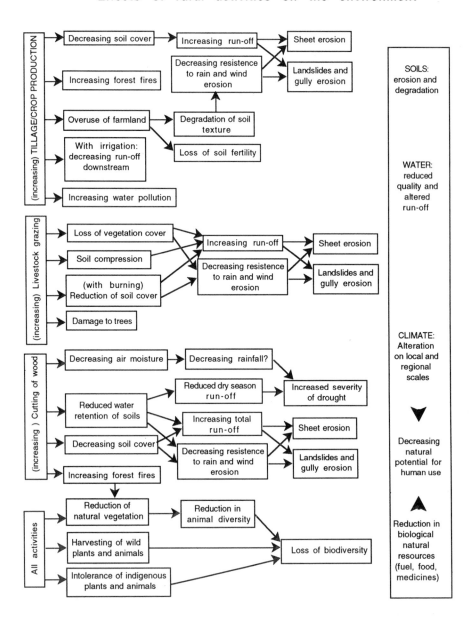

Figure 16.6. Effects of agricultural development on environmental quality in East African high elevation ecosystems (adapted from 31).

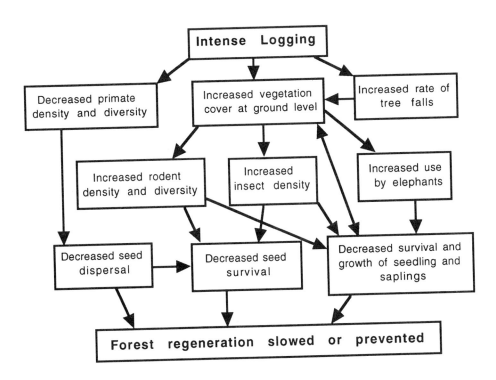

Figure 16.7. Effects of logging on forest ecology and regeneration (adapted from 27).

There is always pressure in a primarily agricultural society for more land to be made available for farming, and it is often politically expedient to do so. Unfortunately, most of the available land in East Africa is either too dry for agriculture, or is currently in indigenous or plantation forest. Unfortunately, these restrictions do not stop unscrupulous land developers/politicians from offering up this land for profit or political advantage. The remaining forest ecosystems cover so little land area in East Africa that the small gains to agriculture are far offset by devastating losses in water catchments and biodiversity (27; Fig. 16.6). There has been an ongoing conversion to farmland at the edges of the remaining forests (19), and this should, of course, be stopped.

There has been increasing interest in encouraging 'traditional' use of forest ecosystems worldwide. While such an approach may be aesthetically pleasing to outsiders, there is no evidence that it can make more than a trivial contribution to national development, and the costs of control may even exceed the value of the harvested forest products. This strategy also runs the risk that enterprising individuals will always find more profitable and less sustainable ways to 'traditionally' exploit the forest. There is also no reason to believe that even the most traditional forest uses are actually sustainable (15). Does anyone really believe that traditional fuel wood collection by thousands of forest neighbors has no long-term effects on ecosystem function?

Even tourism is not without its risks. Claims of park degradation by overuse are usually more aesthetic than ecological, but tourism clearly has the potential for

Box 16.3.   East African Glaciers and Global Warming

Evidence continues to accumulate that the earth's mean air temperature is rising and that this rise is associated with increased 'green-house gases' such as carbon dioxide and methane that are being emitted from the burning of fossil fuels and forests (4,5). These gases absorb long-wave radiation (heat) and can change the mean temperature of the earth, climate variation, and the timing of the seasons. A recent analysis of long-term temperature records suggests that in the 1940s the earth's temperature control switched from control by small changes in the earth's orbit around the sun to control by the accumulation of carbon dioxide (5). The rise in temperature since this time, although small, appears to be increasing exponentially.

Collaborative studies are available from the glaciers on Mount Kenya that have been mapped and studied for over 100 years (1,2,3). These studies show a continual decrease in the size of glaciers since they were studied and a rapid rate of decrease in the last 3 decades. During the interval 1986 to 1990 glacial thickness has decreased by 1 m in upper Lewis Glacier to 7 m in lower glacier (3). Modelling studies suggest that the Lewis Glacier may completely or nearly disappear by the end of this century. These studies are just one well-documented indication of the effect of global warming on the physical and biological structure of the earth's surface. Most of the ecological effects of global warming in East Africa will largely go undocumented because of the paucity of long-term field studies.

T.R. McClanahan

References

1.   Hastenrath, S. 1989. Ice flow and mass changes of Lewis glacier, Mount Kenya, East Africa: Observations 1974-86, modelling, and predictions to the year 2000 A.D. *Journal of Glaciology* 35: 325-332
2.   Hastenrath, S. 1992. The dramatic retreat of Mount Kenya's glaciers between 1963 and 1987: greenhouse forcing. *Annals of Glaciology* 16: 127-133
3.   Hastenrath, S. 1992. Ice-flow and mass change of Lewis glacier, Mount Kenya, East Africa, 1986-90: observations and modelling. *Journal of Glaciology* 38: 36-42
4.   Graham, N.E. 1995. Simulation of recent global temperature trends. *Science* 267: 666-671
5.   Thomson, D.J. 1995. The seasons, global temperature, and precession. *Science* 268: 59-68

damage. Disease transmission from tourists to endangered gorillas appears to occur, but has not been shown to be a serious threat. Attracting wildlife with artificial salt licks undoubtedly shifts local forest ecology in a number of ways, and these may reverberate deeply into the surrounding forest. Nonetheless, I suspect forest tourism at fixed lodges is less ecologically intrusive than savanna tourism, with its hundreds of vehicles, extensive collection of fuel wood, and potential wildlife harassment. And, whereas savanna parks may soon run out of suitable sites for expansion, forest wildlife resources have been barely touched by tourism.

Prospects and Conclusions

In the past decade there has been both an increase in understanding of the conservation value of East Africa's high mountains, and an increase in efforts to actually conserve them. The desire to act seems to be there, but the ability to act is still hampered by a lack of money, a lack of administrative enforcement, and a lack of political will, especially at middle levels of government. It is no exaggeration to say that future economic development in East Africa will depend in no small part on the degree to which these critical mountain ecosystems are protected and understood. Research in these ecosystems, especially the forests, is sorely lacking (35). One bright light has been the hydrological and soils research in Mount Kenya's forests carried out by the Laikipia Research Programme (31,32). If such quality research were expanded to broader ecological questions and to other mountains, it would go a long way toward filling our embarrassing lack of knowledge about the some of the most important ecosystems in East Africa.

References

1. Agnew, A.D.Q. 1984. Cyclic changes of vegetation in the plant communities of the Aberdare Mountains, Kenya. *Journal of the East African Natural History Society* 75(183): 1-12
2. Bekele, T. 1994. Phytosociology and ecology of a humid Afromontane forest on the Central Plateau of Ethiopia. *Journal of Vegetation Science* 5: 87-98
3. Blackie, J.R., Edwards, K.A. 1979. General conclusions from the land use experiments in East Africa. *East African Agriculture and Forestry Journal* 43: 273-277
4. Butynski, T.M., Kalina, J. 1993. Three new mountain national parks for Uganda. *Oryx*, 27: 214-224
5. Coe, M.J. 1967. *The Ecology of the Alpine Zone of Mount Kenya*. The Hague: W. Junk
6. Hedberg, O. 1951. Vegetation belts of East African mountains. *Svensk Botanishe Tidskrift*. 45: 140-202
7. Hedberg, O. 1961. The phytogeographic position of the afroalpine flora. *Recent Advances in Botany* 1: 914-919
8. Hedberg, O. 1964. Features of afro-alpine plant ecology. *Acta Phytogeographica Suecica* 49: 1-147
9. Hamilton, A.C. 1982. *The Environmental History of East Africa: a Study of the Quaternary*. London: Academic Press
10. Hamilton, A.C., Perrott, R.A. 1981. A study of the altitudinal zonation in the montane forest belt of Mount Elgon, Kenya/Uganda. *Vegetatio* 45: 107-125
11. Kigomo, B.N. 1980. Crown-bore diameter relationship of *Juniperus procera* (cedar) and its application to stand density control and production survey in natural stands. *East African Agriculture and Forestry Journal* 46: 27-37
12. Kigomo, B.N. 1985. Diameter increment and growth of *Podocarpus* trees in natural forests. *Kenya Journal of Science and Technology Series B* 6: 113-121
13. Kigomo, B.N. 1987. The growth of camphor (*Ocotea usambarensis* Engl.) in plantation in the eastern Aberdare range, Kenya. *East African Agriculture and Forestry Journal* 52: 141-147
14. Kingdon, J. 1990. *Island Africa*. London: William Collins Sons
15. Levin, S. 1994. Forum: perspectives on sustainability. *Ecological Applications* 4: 545-589
16. Lovett, J.C., Prins, E. 1995. Estimation of land-use changes on Kitulo Plateau, Tanzania using satellite imagery. *Oryx* 29

17. Mahaney, W.C. 1987. *Quaternary and Environmental Research on East African Mountains.* Rotterdam: Balkema

18. Moreau, R.E. 1966. *The Bird Faunas of Africa and its Islands.* New York: Academic Press

19. Ochanda, N., Doute, R., Epp, H. 1981. Monitoring forest cover changes of selected forests in Kenya using remote sensing techniques. Nairobi: *KREMU Technical Reports Series* No. 46

20. Rehder, H., Beck, E., Kokwaro, J.O. 1989. The afroalpine plant communities of Mt. Kenya (Kenya). *Phytocoenologia* 16: 433-463

21. Rejmankova, I., Rejmanek, M. 1995. A comparison of *Carex runsoroensis* fens on Rwenzori Mountains and Mount Elgon, Uganda. Biotropica 27: 37-46

22. Rodgers, W.A. 1993. The conservation of the forest resources of eastern Africa: past influences, present practices and future needs. In *Biogeography and Ecology of the Rain Forests of Eastern Africa*, eds. Lovett, J.C., Wasser, S.K. pp. 283-327. Cambridge: Cambridge University Press

23. Rundel, P.W., Meinzer, F., Smith, A.P. eds. 1994. *Tropical Alpine Systems: Plant Form and Function.* Cambridge: Cambridge University Press

24. Salt, G. 1945. A contribution to the ecology of upper Kilimanjaro. *Journal of Ecology* 42: 375-423

25. Schmidt, K. 1992. *Vegetation of Aberdare National Park, Kenya.* Innsbruck: University of Innsbruck Press

26. Smith, A.P., Young, T.P. 1987. Tropical alpine plant ecology. *Annual Review of Ecology and Systematics* 18: 137-158

27. Struhsaker, T.T. 1987. Forestry issues and conservation in Uganda. *Biological Conservation* 39: 209-234

28. Tweedy, E.M. 1965. Periodic flowering of some Acanthaceae on Mt. Elgon. *Journal of the East African Natural History Society* 25: 92-94

29. White, F. 1981. The history of the afromontane archipelago and the scientific need for its conservation. *African Journal of Ecology* 19: 33-54

30. Wimbush, S.H. 1945. The African bamboo. *Empire Forestry Journal* .22: 33-39

31. Winiger, M., ed. 1986. *Mount Kenya Area: Contributions to Ecology and Socioeconomy.* Geographica Bernensia, African Studies Series A1, University of Bern, Switzerland

32. Winiger, M., ed. 1991. *Mount Kenya Area: Differentiation and Dynamics of a Tropical Mountain Ecosystem.* Geographica Bernensia, African Studies Series A8, University of Bern, Switzerland

33. Young, T.P. 1984. Comparative demography of semelparous *Lobelia telekii* and iteroparous *Lobelia keniensis* on Mount Kenya. *Journal of Ecology* 72: 637-650

34. Young, T.P. 1984. Kenya's alpine and high forest ecosystems. In *Endangered Ecosystems for Development,* ed. Gilbert, V.C. pp. 117-138. Nairobi, National Human Settlements Secretariat

35. Young, T.P. 1991. Mount Kenya forests: an ecological frontier. In *Mount Kenya area: Differentiation and Dynamics of a Tropical Mountain Ecosystem,* ed. M. Winiger, pp. 197-201. Geographica Bernensia, African Studies Series A8, University of Bern, Switzerland

36. Young, T.P., Augspurger, C.K. 1991. Ecology and evolution of long-lived semelparous plants. *Trends in Ecology and Evolution* 6: 285-289

37. Young, T.P., Evans, M.E. 1993. Notes on the alpine vertebrates of Mount Kenya, with particular notes on the rock hyrax. *Journal of the East African Natural History Society* 202: 55-79

38. Young, T.P. Lindsay, W.K. 1988. Role of even-aged population structure in the disappearance of *Acacia xanthophloea* woodlands. *African Journal of Ecology* 26: 69-72

39. Young, T.P. Peacock, M.M. 1992. Giant senecios and the alpine vegetation of Mount Kenya. *Journal of Ecology* 80: 141-148

Concluding Remarks

Several themes resound through these chapters, and are unifying concepts for the ecology and conservation of East African ecosystems. We are impressed both by the level of knowledge of each of these ecosystems, which until recently were essentially unstudied, and by the sophistication of current research. We are also humbled by how many question remain unanswered and the task of conserving the many threatened species and their habitats.

Interactions Among Ecosystems

Even though we have divided the biota of East Africa into a series of biomes, it is clear from these chapters that the boundaries of these biomes are often gradual and that many of these distinctions are arbitrary. In addition, the different biomes affect each other, often strongly. Because animals, nutrients and water move across the landscape, changes in one ecosystem can have profound effects that are felt far away. Aquatic systems are strongly influenced by the ecology and management of lands upstream, and many problems in aquatic ecosystems are directly related to problems in related terrestrial ecosystems.

Elephants

One pervasive common element of terrestrial ecosystems is the importance of elephants, discussed in virtually every terrestrial chapter. Elephants have both a broad ecological tolerance and the ability to change ecosystems dramatically. They can be either a keystone species that help to maintain ecosystem health and diversity or a problem species that can damage ecosystem function and threaten human life and livelihood. Elephants also appear to be critical species economically, being the most important component of one of East Africa's most important industries, tourism. The necessity for the large-scale movement of elephants poses special problems for the conservation and management of this singular species.

The Human Element

Humans have interacted with East African ecosystems for literally millions of years, and today even the most inaccessible and inhospitable areas are under human influence. It is a measure of our ingenuity and fecundity as a species that no ecosystem is still free from our use and misuse. Conservation problems and their solutions are essentially human: political, economic, and social. Many of the region's successes and failures can be attributed to the quality of resource management and to the conservation attitudes of the populace. As East Africa enters an exciting era of political freedom and pluralism, these forces will undergo tremendous flux. It will require both tremendous will and innovation to solve these problems. But solve them we must if we are to leave the next generation even a substantial fraction of the tremendous biological heritage that we have inherited.

Human Populations

The most pervasive features of human presence in East Africa are its numbers and rates of increase. Gone are the days when larger populations were thought to be the key to economic growth. In fact, the opposite appears to be the case in this region where dependence on arable land and natural and renewable resources is strong. For example, over the last thirty years, Kenya has maintained an almost incredible record of continual economic growth, yet the larger portion of this growth has been used only to maintain the equally massive increase in the human population - increasing from 7 to 25 million people in only one generation since Independence. Consequently, most per capita measures of economic wealth indicate declines in the last decade.

The recent acknowledgment by regional governments of the value of reducing this population growth is heartening, but even if such attitudes and associated policies are rapidly implemented, East Africa can expect to experience large increases in population for some years to come. And, a potentially greater difficulty arises from the reasonable expectations of residents for a standard of living that will require far higher rates of per capita resource use (especially nonrenewable energy), when even current rates are often unsustainable.

The Decisive Generation

World-wide, and especially in East Africa, humans are at a unique and critical moment of decision. Never have we experienced such threats to the future of biodiversity, never has there been such widespread recognition of its value, never had we had more information at hand, and never have we had such broad-based support from implementing this knowledge. The majority of the world's species and intact ecosystems will be either lost or saved in the next thirty years. We truly believe that the present generation will determine the fate, for all generations to come, of the rich and valuable diversity to which we owe not only our growing appreciation of nature, but our very lives. We hope this volume, and the dedication of its readers, will be a part of this difficult but essential struggle.

Glossary

Abbreviation and Symbols

B.P. = before present
^{o}C = temperature in degrees centigrade
cm = centimeter
g = gram
ha = hectare
kg = kilogram
km = kilometer
l = liter
m = meter
m asl = meters above sea level
mg = milligram
mm = millimeter
mya = million years ago
s = second
y = years

Definition of Terms

Abiotic - referring to nonliving components of the environment or ecosystem.

Acaulescent - without a stem. When describing rosette plants, the growth form of having the base of the rosette at ground level.

Afroalpine - referring to areas in Africa above the limit of tree growth (see timberline).

Alkaline - water or a solution having an excess of the negative hydroxide anions or with a pH greater than 6.

Allochthonous - material produced outside a system and transported into that system.

Alluvial - water-transported.

Anaerobic - respiration or metabolism with oxygen.

Anoxic - water having no dissolved oxygen.

Anthropogenic - having a cause or origin of a change attributable to humans.

Aquaculture - growing and culturing living organisms in water for human use.

Archipelago - islands or a group of islands.

Autochthonous - material produced within a system.

Basal area - the total area of woody tree stem per unit area (usually expressed as m^2/hectares) based on a measure of the sum of tree's circumference at breast height. Used as an estimate of tree and shrub abundance .

Biomass - the weight (usually wet but also dry) of an organism or group of related organisms in some known area (for example grams/m^2)

Biotic - referring to living organisms and their influence.

Autotrophic - organisms able to produce their own energy from physical energy

Biotope - an environmental region characterized by certain conditions and populated by characteristic organisms (=biota).

BOD - biochemical oxygen demand. The rate of oxygen consumption in a parcel or sample of water.

Catchment - area drained by a stream or river.

Cattarahl fever - also known as malignant Cattarh - a fatal viral disease of cattle and sometimes sheep. May be carried by wildebeest.

Caulescent - with a stem. When describing rosette plants, the growth form of having the base of the rosettes on top of a woody stem up to several meters high.

Collector - invertebrate which eats fine detritus (FPOM).

Conductivity - the reciprocal of the specific resistance of a solution to electrical current. Expressed in mhos (reciprocal of ohms; mmhos/cm) or, equivalently, Siemens (mS/cm).

Coppice - to produce new shoots as the base of old stems.

CPOM - coarse particulate organic matter; detrital fragments with diameters > 1 mm; leaves, twigs, branches, logs.

Deciduous - plants that drop their leaves annually or during a particular season.

Depauperate -having few species.

Deposit-feeder - type of collector which ingests FPOM deposited on the bottom of a body of water.

Detritus -dead organic matter.

Diatoms - microscopic algae most frequently living in the water column that deposit small skeletons on the lake or ocean bottom.

Disequilibrium -an unstable state or status of an ecosystem or biological community.

Disturbance - any relatively discrete event which removes organisms and opens up space that can be colonized by individuals of the same or different species.

Dystrophic - nutrient poor.

DOM - dissolved organic matter.

Ecotone - transition zone between ecological community types.

Edaphic - referring to conditions in the soil such as texture or nutrient content.

Emergent - aquatic macrophytes (large plants) which are rooted or anchored to the bottom of a body of water, but which have some or most of their vegetative structures exposed to the air.

Endemism - the state of occurring in only one locality or region.

Endogenous - inherent to a system; not caused by external forces, such as weather.

Endorheic - water bodies without a river outlet.

Ephemeral stream - stream containing water for only a few weeks or months each year.

Epiphytes - organisms that grow on the surface of a plant.

Ericaceous - comprised of shrubs in the genus Erica (heath), and of related species.

Euphotic -the depth of water to which light penetrates. The upper surface of a water body.

Eutrophic - refers to an environment or ecosystem with high nutrient or productivity levels.

Exotic - a species whose place of origin is outside of a given environment.

Filter-feeder - type of collector which filters FPOM out of the water column.

Floating-leafed vegetation - aquatic macrophytes which are rooted in bottom sediments, but have leaves floating at the surface.

Floodplain - area adjacent to a river which is inundated during floods.

Fossorial -burrowing.

Fouling -colonization of animals onto a substrate such as rocks or vegetation.

FPOM - fine particulate organic matter, diameter < 1 mm.

Free-floating vegetation -aquatic macrophytes which float on the water's surface and are not rooted to the bottom.

Grazer - invertebrate which feeds on periphyton or a terrestrial animal that eats grass.

Heterotherm - organisms requiring or using heat from an external source to control their metabolism.

Heterotrophic - organisms that feed on other organisms for their energy source.

Homeotherm -organisms that can produce their own heat to control their metabolism.

Hominid -refers to the family Hominidae which includes humans and their extinct ancestors. The three genera included in this family are *Homo*, *Australopithecus* and *Paranthropus*. Evidence is accumulating, however, to support the contention that African apes and humans are more closely related than previously thought and may be contained in the same family or superfamily.

Hominoid -refers to the great apes and their ancestors (see above).

Hydrologic -pertaining to the cycling of water at a global or local scale, including the balance between precipitation and evaporation, and surface and groundwater flow.

Hypoxia - levels of dissolved oxygen in water that are below the levels required for most complex organisms.

Ichthyology - the study of fish.

Insolation -sunlight or sunlight intensity.

Interannual - between years.

Interfluvial - between streams or rivers.

Intermittent stream -stream reduced to a series of isolated pools with no flow between them, or entirely dry, in the dry season.

Isotopes -any of two or more forms of a chemical element having the same number of protons in the nucleus but having a different number of neutrons and therefore a different atomic weight. Isotopes have similar chemical properties.

Lacustrine - referring to a lake environment.

Laga - ephemeral stream that forms in response to local rainfall.

Lentic - standing water: lakes, ponds.

Limnology - the study of lakes.

Littoral - referring to the shore of lakes or the intertidal zone of the ocean.

Lotic - running water: streams, rivers.

Macrophyte - higher plant.

Macrofauna - animals that can be easily seen with the naked eye.

Malaria -a human disease caused by protozoans (*Plasmodium*) and transmitted by female mosquitoes (*Anopheles*). Symptoms include a cyclic fever and chills.

Microbes -bacteria, fungi.

Microfossils - microscopic fossils such as pollen, grass cuticles, and the skeletons of diatoms.

Monospecific - consisting of only one species.

Neap tide -a period of about one week when tidal ranges are least variable corresponding to the 1st and 3rd lunar quarters of each lunar month.

Nomad - species that move in search of resources in an unpredictable pattern.

Nutrients -chemicals required for organismic growth such as carbon and phosphorus.

Onchocerciasis -river blindness. A human disease caused by a filarial worm carried by female blackflies (*Simulium*). Adults in nodules in human skin produce millions of microfilariae which spread through the skin. Dead microfilariae cause lesions in the skin and eyes resulting in itching, in the former case, and blindness, in the latter.

Paedophages - carnivores that specialize in eating recent hatchlings or juveniles.

Periphyton -bacteria, fungi, algae attached to submerged surfaces.

Phenology -the study of the timing of important life-history events such as growth, reproduction and dispersal.

Physiognomy or **Physiognomic** - outer appearance, gross external appearance or features.

Phreatophytic - characteristic of trees and shrubs with deep roots which are able to reach, and may depend on ground water.

Polynology -the study of pollen and other microscopic biological components of sediments.

Pool -deep, slow zone in streams.

P/R - gross photosynthesis divided by total respiration for a particular system (such as a section of stream).

Potamodromous -fish which live in lakes most of the time, but which ascend inflowing streams to spawn.

Precipitation -chemicals changing state from a dissolved to solid form. Also to form rain or snow.

Pteridophyte - plant in the group of ferns, horsetail and club mosses.

Quasi-periodic - having periods of change that are not always predictable

Quasi-stable - apparently stable or stable for limited periods of time.

Rain shadow - The dryness associated with being on the leeward side of a mountain. In East Africa, generally on the western and northern aspects.

Regime - ecologists use this term to mean the series of events or processes that effect an ecosystem

Riffle - shallow, fast zone in streams.

Rinderpest - an acute or fatal viral disease that effects cattle, sheep, and other hoofed animals that is characterized by high fever, diarrhea and lesions of the skin and mucous membranes.

Riparian - alongside a stream or river.

River Continuum Concept (RCC) - A construct describing changes in community and ecosystem characteristics of rivers proceeding from headwaters to mouth.

Run - fast, deep zone in streams.

Schistosomiasis - bilharzia. A debilitating human disease caused by a fluke carried by particular species of snails (*Biomphalaria, Bulinus*).

Secondary vegetation - The vegetation that arises after disturbance of the primary, or natural vegetation.

Shredder - invertebrate which eats CPOM.

Spring tide - periods of about 1 week when tides experience their most extreme fluctuations during the full and new moon periods of each month.

Standing crop - the weight of an organism in a known area (see Biomass).

Stochastic - randomness or a random process.

Stream order - an index of stream size. Headwater streams are first order streams. When two streams of equal order join they become a stream of the next highest order; however, when a stream of low order joins a stream of higher order the order of the latter stream does not change. For example, when two first order streams join, the resultant stream becomes a second order stream, and the junction of two fourth order streams produces a fifth order stream; however, a fourth order stream with second order tributaries remains a fourth order stream.

Submergent - aquatic macrophyte completely covered by water.

Succession - the series of replacements of species following disturbance.

Tectonic - related to the movement of the plates on the earth's crust.

Timberline -the upper altitudinal limit of tree growth; the dividing line between forest and alpine.

Transhumanance -humans that move between a number of fixed locations usually on an annual cycle.

Trypanosomiasis -sleeping sickness. Disease of bovids and humans caused by the parasitic Trypanosome protozoan carried by the tsetse fly.

Turbidity - cloudiness; opposite of water clarity.

Volatilization -conversion to a volatile, usually gaseous, state.

INDEX